T0402083

Agricultural Bioinformatics

Kavi Kishor P.B. • Rajib Bandopadhyay
Prashanth Suravajhala
Editors

Agricultural Bioinformatics

Springer

Editors
Kavi Kishor P.B.
Department of Genetics
Osmania University
Hyderabad, Andhra Pradesh, India

Rajib Bandopadhyay
Department of Biotechnology
Birla Institute of Technology
Ranchi, Jharkhand, India

Prashanth Suravajhala
Bioclues Organization
IKP Knowledge Park, Picket
Secunderabad, Andhra Pradesh, India

ISBN 978-81-322-1879-1 ISBN 978-81-322-1880-7 (eBook)
DOI 10.1007/978-81-322-1880-7
Springer New Delhi Heidelberg New York Dordrecht London

Library of Congress Control Number: 2014942334

© Springer India 2014
This work is subject to copyright. All rights are reserved by the Publisher, whether the whole or part of the material is concerned, specifically the rights of translation, reprinting, reuse of illustrations, recitation, broadcasting, reproduction on microfilms or in any other physical way, and transmission or information storage and retrieval, electronic adaptation, computer software, or by similar or dissimilar methodology now known or hereafter developed. Exempted from this legal reservation are brief excerpts in connection with reviews or scholarly analysis or material supplied specifically for the purpose of being entered and executed on a computer system, for exclusive use by the purchaser of the work. Duplication of this publication or parts thereof is permitted only under the provisions of the Copyright Law of the Publisher's location, in its current version, and permission for use must always be obtained from Springer. Permissions for use may be obtained through RightsLink at the Copyright Clearance Center. Violations are liable to prosecution under the respective Copyright Law.
The use of general descriptive names, registered names, trademarks, service marks, etc. in this publication does not imply, even in the absence of a specific statement, that such names are exempt from the relevant protective laws and regulations and therefore free for general use.
While the advice and information in this book are believed to be true and accurate at the date of publication, neither the authors nor the editors nor the publisher can accept any legal responsibility for any errors or omissions that may be made. The publisher makes no warranty, express or implied, with respect to the material contained herein.

Printed on acid-free paper

Springer is part of Springer Science+Business Media (www.springer.com)

Foreword

Life on Earth would be unsustainable without plants. It has been repeatedly emphasized that plant biotechnology has the potential to provide our farmers with another green revolution. For graduate students who are studying agriculture and life sciences, it is imperative to recognize the knowledge and use of plant genomics. This information will help in the development of "new agriculture", and for the research community to drive technological advancement in agriculture. There is very high potential in plant research to translate genomics technology into agronomic advancement. However, this requires an understanding of computational biology, to comprehend biological data and phenomena through innovative applications of bioinformatics and biostatistics.

Bioinformatics focuses on developing and applying computationally intensive techniques such as data mining, pattern recognition, 3-dimensional visualization and machine learning, which helps to quickly and efficiently study large amounts of genomic information, chemical structure and other biological data, as well as to extract biological meaning from large data sets. During the past decade, there has been a tremendous increase in genomic tools available in key crop plants, including expressed sequence tags (ESTs), bacterial artificial chromosome (BAC) libraries and physical maps, genetic sequence polymorphisms, mutant collections and expression profiling resources. Whole genome sequences, well supported with genome annotations and information browsers to enable cross-genome comparisons, are now available for several reference plants, representing diverse families in dicots, monocots as well as lower plants. At the same time, genomic tools like next generation sequencing platforms are becoming increasingly more accessible and reasonably priced. This wealth and accessibility of new genomic resources has caused a paradigm shift wherein researchers whose motivation is crop improvement are no longer constrained by model systems.

Agricultural bioinformatics addresses agricultural problems directly. Genomic resources have created the means to speed the improvement of plants – agronomic, horticultural, and forest tree species. For example, this knowledge can contribute towards development of crops with desirable characteristics such as drought, disease and insect resistance, or crops that require less fertilizer and have higher nutritional content. Apart from plant and animal genomes, *in silico* biology encompasses the use of parasitic plant genomes and hundreds of available 'pathogen' genomes, including that of viruses, bacteria, phytoplasmas, protozoans, fungi and nematodes, providing an opportunity to study plant-pathogen interaction in order to enable disease diagnosis, management and cultivation of disease resistance transgenic crops.

It is truly an exciting time in agricultural research, and I think this book captures the scintillation of the evolving role of bioinformatics in agronomic advancement. With the advent of technological innovations through genomics and bioinformatics, many important questions of economic importance will now become increasingly addressable, bringing forth newer funding opportunities, and laying fruitful ground for many scientific careers.

I appreciate the efforts put in by Prof. Kavi Kishor P.B. and other contributors in bringing out this book.

National Institute of Plant Genome Research
Jawaharlal Nehru University, New Delhi
Asis Datta

National Institute of Plant Genome Research (NIPGR)
Aruna Asaf Ali Marg, JNU Campus
New Delhi-110067, India

Preface

This book is a compact, cohesive review highlighting some important topics that streamline bioinformatics in a nutshell for the applications in agriculture. It is intended both to informed researchers and upper graduates who wish to take up Agricultural Bioinformatics. Wide and interesting topics have been invited from researchers practicing agriculture, primarily plant bioinformatics. The value of bioinformatics in general has caught interest for researchers worldwide for various reasons. One primary reason is the added value in predicting the outcome using *in silico* based approaches, further defusing the wet lab experiments. However, there is much more consensus to be reached using the tools of bioinformatics. The last decade has seen innumerable tools and databases developed from which applied bioinformatics research has been exploited in the agricultural sector. The contributors in this book we believe have brought some delightful reviews to read and showcased which can be majorly applied for agri computing research besides common applications to all practices of bioinformatics research.

Sameera Panchangam et al. in their review "EST derived proteins in plant genomes: Where are we heading?" focus upon the need of understanding the ESTs better especially for those genomes that are not completely sequenced. Comprehensive commentary has been dealt on these sequences that are used to build the plant expressed transcripts and on how *de novo* the functional repertoire can be established. Saikumar and Dinesh Kumar in their chapter brings out an ideal review on the plant microRNAs describing several bioinformatics platforms and approaches on rise that has led to discovery of plant miRNAs, validation and need for understanding miRNAs from their progenitor messenger RNAs (mRNAs) have arisen. With the introduction of sequence-specific miRNA signatures recently found, he discusses myriad of dimensions where miRNAs from oilseed crops are being associated with several putative functional and evolutionary events. Priyanka James et al. review Medbase, a compendium of medicinal plants. Bioinformatics approaches leveraging plant-based knowledge discovery has offered plethora of new tools for the identification of genes and pathways involved in the production of secondary metabolites. Their database is aimed at bioinformatics strategies associated with important ethnic medicinal plants and it helps to identify therapeutically important active compounds. They further suggest the use of this database in helping farmers involved in cultivating medicinal plants.

Khyati and Sivaramaiah in their book chapter, titled "Plant-Microbial Interaction: A Dialogue between Two Dynamic Bioentities" discuss the role of ecology that plant-microbe association has influenced plant's diversity, metabolism, morphology, productivity, physiology, defense system and tolerance against adversities. Further, they discuss the omics approach being used to unveil the role of complex cryptic signaling process in the plant and microbes interaction. Sohini Gupta et al. discuss a brief overview of the viral silencing suppressors that have been identified in their chapter, titled "The Silent Assassins" and further delve on possibilities of the bioinformatics analyses of the viral silencing suppressors with specific case studies. They also conceptualize a future framework for application oriented use of these silent assassins. Subarna Thakur et al. in their chapter titled "Exploring the genomes of symbiotic diazotrophs with relevance to biological nitrogen fixation" discuss accessibility of new computational tools for genomic and proteomic analysis which have accelerated nitrogen fixation research predominantly in the areas of comparative genomics, protein chemistry and phylogenetic analysis of nitrogen fixation genes. Alternative phylogenetic approaches and protein structure based studies according to them have been quite prolific to divulge the unfamiliar aspects of symbiotic nitrogen fixation. Raj Pasam and Rajiv Singh in their chapter, titled "Association mapping – A new paradigm for dissection of complex traits in crops" discuss the genetic basis of important agronomic traits and the scientific challenges and numerous efforts that are underway to understand and decipher the complexity of these traits in crops.

Katsumi Sakata et al. in their chapter, titled "Mining knowledge from omics data" give a gist of extracting knowledge from complex data in multiply layered biological information, a multiple omics-based approach would be a powerful method. They introduce a representative approach to integrate multiple omics data, and discuss topics that relate to mining knowledge from the omics data and further explain the meaning of the *p*-value and introduce applications of statistical testing to expression analyses for proteins specific to early seedling stage of soybean. Tiratha Raj Singh in his chapter, titled "Machine learning with special emphasis on support vector machines (SVMs) in systems biology: A plant perspective" discuss the progress of systems biology with integrative genomics and tools such as bioinformatics. He further narrates the role of recent developments in high-throughput techniques that led to the accumulation of a deluge of biological data. Logical applications from machine learning and how to deal with state-of-the-art techniques for managing data are discussed. Uma Devi et al. in their chapter on "Bioinformatic tools in the analysis of determinants of pathogenicity and ecology of entomopathogenic fungi used as microbial insecticides in crop protection" highlight the proteins involved in deciphering the effect of the inundative application of an entomopathogenic fungus on the native soil fungal diversity as extensively described in their works. The chapter highlights the bioinformatics bolstered investigation of the factors that influence the affectivity of insect pathogenic fungi as microbial biopesticides. Hima et al. describe in brief how bioinformatics can help in the task of understanding the genes associated with abiotic stress tolerance. Sujay

Rakshit and Ganapathy in their review describe comparative genomics of cereal crops with special reference to *Sorghum*. Chavali et al. in their article address the need for cloud platforms with support for database design related virtualization enhancements. While discussing the designing databases specific for cloud in large enterprises, the authors describe database applications which need to be delivered using Cloud Platform while addressing SaaS (software-as-a-service) a dominant service model in cloud computing. The latter can be designed for end users and delivered over web.

We expect that all readers will find these articles interesting in this book; many might feel the lack of a livestock informatics article but we promise to bring that during our next edition.

Happy reading!

Hyderabad, India — Kavi Kishor P.B., Ph.D.
Ranchi, India — Rajib Bandopadhyay, Ph.D.
Secunderabad, India — Prashanth Suravajhala, Ph.D.

About the Editors

Dr. **Kavi Kishor P.B.** is currently working as Emeritus Professor at the Department of Genetics, Osmania University, Hyderabad, India. He has 38 years of teaching, research and administrative experience. Published 160 papers including book chapters, produced 25 Ph.D.s and authored 5 books. He is a fellow of the National Academy of Sciences and also National Academy of Agricultural Sciences. Recipient of several gold medals, winner of Prof. Hiralal Chakravarthy Award from the ISCA, and the best teacher award from the Government of Andhra Pradesh. Received Rockefeller Foundation Fellowship from USA, Visiting Scientist Fellowship from Emory University, Atlanta, USA, Visiting Scientist Fellowship from Linkoping University, Sweden, and from Leibniz-Institute of Plant Genetics and Crop Plant Research, Gatersleben, Germany.

Dr. **Rajib Bandopadhyay** is working as Assistant Professor in the Department of Biotechnology, Birla Institute of Technology, Mesra, Ranchi, Jharkhand, India, for the last 8 years. He did his Ph.D. in Botany (Plant Molecular Biology) in 2004 from the University of Calcutta. He did his post-doctoral research work in Crop Biotechnology in India as well as in South Korea and USA. He has teaching experiences at the undergraduate and post graduate level in the field of Agriculture Biotechnology and also in Bioinformatics. Dr. Bandopadhyay is BOYSCAST fellowship recipient from DST, Government of India in 2007. He also participated in Southern Ocean Expedition in 2011. He has published more than 47 research papers in journals of international repute and authored a number of reviews, book chapters and manuals.

Dr. **Prashanth Suravajhala** is a Post-Doctoral Scientist and a virtual entrepreneur who founded Bioclues.org in 2005. He is also serving as an Associate Director of Bioinformatics.org, and has wide interests in lieu of functional genomics and systems biology of hypothetical proteins in human, specifically targeted to mitochondria. He loves mentoring undergraduates who want to pursue Bioinformatics. Dr. Suravajhala has completed his Ph.D. from Aalborg Universitet, Denmark. He has published 20 papers in international peer-reviewed journals.

Contents

Association Mapping: A New Paradigm for Dissection of Complex Traits in Crops

Raj K. Pasam and Rajiv Sharma

Abstract

This book chapter provides basic information about association mapping (AM), also called as linkage disequilibrium (LD) mapping. Association mapping has emerged as one of the successful, faster and high-resolution method for investigating the genetic architecture of complex traits in plants. Apart from unlocking the genetic information underlying the traits, association mapping also helps in exploiting the existing genetic diversity for crop improvement. The potential of the genetic diversity is not yet fully explored in plant breeding, and association mapping can aid in detecting and including more useful alleles into breeding germ plasm. The recent advent of cheap genotyping and sequencing technologies has eased the marker availability in the majority of the crops contributing to the increased interest in association mapping. Population structure resulting in false associations is one of the major constrains in reliable interpretation and usage of association mapping results. However, the advancement in the appropriate statistical methods for whole-genome association mapping has resulted in mitigating the population structure problems to an extent. Here we discuss basic concepts of AM and LD decay measurements and in depth the principles and steps involved in association mapping. The chapter discusses the advantages of LD mapping over linkage mapping, constrains involved in LD mapping, statistical approaches used for LD mapping and determination and interpretation of LD mapping results.

R.K. Pasam (✉)
Sainsbury Laboratory, University of Cambridge, Cambridge, UK
e-mail: raj.pasam@slcu.cam.ac.uk

R. Sharma
Leibniz-Institute of Plant Genetics and Crop Plant Research (IPK), Gatersleben, Germany
e-mail: sharmar@ipk-gatersleben.de

K.K. P.B. et al. (eds.), *Agricultural Bioinformatics*,
DOI 10.1007/978-81-322-1880-7_1, © Springer India 2014

Keywords
Association mapping • Crop genetics • Genome-wide association studies • Linkage disequilibrium • Heritability • Mapping populations

1 Introduction

Determining the genetic basis of important agronomic traits has been one of the major scientific challenges, and numerous efforts are under way to understand and decipher the complexity of these traits in crops. In the past couple of decades, numerous genetic mapping studies have been reported in crops and other plant species (Bernardo 2008). The aim of the majority of these studies was to understand the genetics underlying the traits and to narrow down the genomic regions responsible for the phenotype variation and in an ideal case identifying the causal mutations. The perplexity in determining the genetics of these traits of agronomic importance is partly due to the quantitative nature of the majority of the important traits in plants and their environmental interactions. However, the emergence of molecular markers in conjugation with the availability of powerful biometric methods has resulted in the surge of quantitative trait locus (QTL) studies in various crops. QTL mapping is a key tool for assessing the genetic architecture of the underlying complex traits and facilitating the estimation of the number of genomic regions affecting the trait. The detection of genes or QTL is mainly possible due to genetic linkage analysis which is based on recombination during meiosis (Tanksley 1993). Till recently, the majority of QTL mapping studies were based on linkage analysis using biparental mapping populations. An alternative approach, association mapping (AM) known as LD (linkage disequilibrium) mapping, relies on existing natural populations or designed populations to overcome the constraints inherent to linkage mapping. Both linkage mapping and LD mapping (or association mapping) strategies exploit the fact that recombination breaks up the genome into small fragments that can be correlated to the phenotype (Myles et al. 2009). Today, association mapping is increasingly used in several crops to detect QTL/genes affecting the trait (Pasam et al. 2012; Breseghello and Sorrells 2006; Yan et al. 2011; Robbins et al. 2011; Fusari et al. 2012; Abdurakhmonov et al. 2009). Association mapping approaches circumvent the limitations in linkage mapping and also benefits from the LD arising from historical recombination in the broader germ plasm.

A majority of the plant breeding efforts have so far neglected vast plant genetic resources and focused extensively on limited germ plasm, resulting in progressive decline and narrowing of genetic diversity in cultivated crops (Tanksley and McCouch 1997). The AM, on the other hand, has the potential to exploit the existing plant genetic resources for detecting QTL and beneficial alleles and to incorporate them into the crop breeding schemes. One of the major factors for determining the resolution of mapping in AM is the extent of LD decay in the population (Flint-Garcia et al. 2003). Genome-wide LD decay is very dynamic and varies among species and within species between different gene pools and populations (Hamblin et al. 2010; Caldwell et al. 2006; Hyten et al. 2007; Ranc et al. 2012). The varying trend of LD in different gene pools provides an opportunity for finer-resolution association mapping using landraces and wild germ plasm. Several studies hinted and proved that the wild and exotic germ plasm with extensive marker coverage can result in fine mapping of QTL or dissecting the QTL to the resolution of gene (Ranc et al. 2012; Waugh et al. 2009; Huang et al. 2012b; Hufford et al. 2012). In lieu of the recent developments in genetics and genomics and changing paradigm of plant breeding, association mapping plays a crucial role in comprehending genetics of traits directed towards further crop improvement and conservation of genetic variation. We present a comprehensive overview of association mapping techniques and statistics involved, along with the emerging new mapping tools and the future prospects for crop improvement.

2 Mapping Approaches

Traditional and modern plant breeding schemes emphasise the importance of location of the genes/QTL affecting the trait of interest for crop improvement. Mapping quantitative traits was first described in common bean by Karl Sax (1923) while proposing a method to locate and enumerate QTL on linkage map (Sax 1923). Molecular genetics including both forward and reverse genetic approaches were used for identifying QTL and genes (Takeda and Matsuoka 2008). During the last decade, biparental linkage mapping has been extensively used for mapping quantitative traits. The AM was initially used in human genetics and was later adopted in many plant studies as a successful tool. QTL mapping populations can be broadly classified into two types: experimental populations, also called family-based linkage populations, and natural populations that use linkage disequilibrium mapping approaches (Semagn et al. 2010; Mackay and Powell 2007). In contrast to the biparental mapping approaches, association mapping populations (panels) are carefully sampled diverse lines representing the diversity of natural or breeding populations of the crops (Zhao et al. 2007; Zhu et al. 2008). Recently, more advanced mapping populations termed as next-generation populations (NGPs) are being developed for various crops to overcome the limitations of both linkage-based and LD-based mapping approaches. These next-generation population designs involve crossing of multiple parents and/or advanced generation intercrosses with further advancement for generations to improve genetic resolution of mapping (Morrell et al. 2011). Multiparent advanced generation intercross (MAGIC) population, nested association mapping (NAM) populations and several designs of advanced intercross recombinant inbred lines (AI-RILs) are the potential NGP designs that can be used in crop improvement. The detailed description of NGP mapping approaches is beyond the scope of this review. Here, we provide an overview of AM populations and illustrate in detail the methods and approaches involved in AM and its implications for crop breeding. We also highlight the pitfalls in AM and discuss the measures to mitigate the disadvantages.

3 Association Mapping Populations

The AM populations ideally should be representative collections of diverse accessions (Fig. 1a) while it relies on available natural genetic variation present in germ plasm and hence doesn't suffer the lack of variation that is characteristic

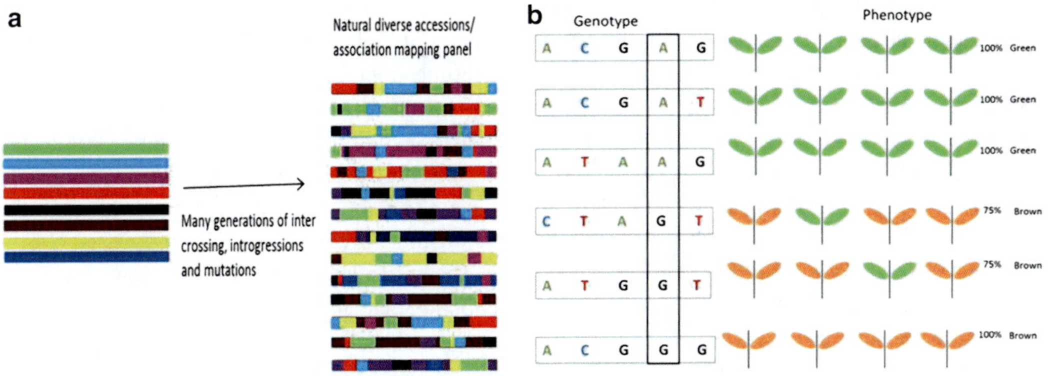

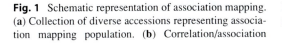

Fig. 1 Schematic representation of association mapping. (**a**) Collection of diverse accessions representing association mapping population. (**b**) Correlation/association between the genotype (nucleotide polymorphism) and phenotype results in significant marker-trait association

of several biparental populations (Hall et al. 2010). The AM involves detecting the significant marker-trait correlations among the diverse collections using different statistical approaches (Fig. 1b). The AM was first introduced in genetic mapping studies in humans (Hastbacka et al. 1992; Lander and Schork 1994) and later on has been considered for plant research (Flint-Garcia et al. 2003). The AM exploits ancestral recombination events that occurred in the population and takes into account all the alleles present in the population to identify significant marker-phenotype associations. By exploiting non-random associations of alleles at nearby loci (LD), it is possible to detect significantly associated genomic regions with a set of mapped markers. The success of mapping depends on the quality of phenotypic data, the population size and the degree of LD present in a population (Flint-Garcia et al. 2005; Mackay and Powell 2007; Pasam et al. 2012).

Association mapping can be broadly classified into two categories: (1) candidate gene-based association mapping, where selected candidate genes are sequenced and the sequence polymorphism is correlated to the phenotype variation, and (2) genome-wide association studies (GWAS), which associate marker polymorphisms across the genome to the phenotype. Genome-wide association studies (GWAS) have become increasingly popular and powerful over the last few years in human and animal genetics. In the last few years, an increasing number of association studies based on the analysis of candidate genes have been published (reviewed in Gupta et al. 2005). Some of the candidate gene association studies in various crops include flowering time genes in barley (Stracke et al. 2009), *Psy1-A1* locus in wheat (Singh et al. 2009), frost tolerance genes in rye (Li et al. 2011), *Dwarf8* (Thornsberry et al. 2001) and the phytoene synthase locus in maize (Palaisa et al. 2003), photoperiodic flowering time genes in sorghum (Bhosale et al. 2012) and the *rhg-1* gene in soybean (Li et al. 2009).

Due to the advent of various cost-effective high-throughput genotyping platforms, GWAS has become an increasingly attractive approach for dissecting genetic basis of complex traits in various crops. Several successful GWAS results were reported in various crops like in rice (Huang et al. 2010; Agrama et al. 2007; Zhao et al. 2011), maize (Remington et al. 2001; Beló et al. 2008), barley (Comadran et al. 2009; Pasam et al. 2012; Cockram et al. 2010), sorghum (Brown et al. 2008; Murray et al. 2009), soybean (Jun et al. 2008), lettuce (Simko et al. 2009), potato (Gebhardt et al. 2004) and sugar beet (Würschum et al. 2011).

4 Linkage Mapping vs. Association Mapping

Linkage mapping, also referred to as family mapping, and association mapping, also referred to as linkage disequilibrium mapping or population mapping or GWAS (Myles et al. 2009), rely on the LD between markers and the functional loci. In linkage mapping, a linkage map is generated by establishing a population from a cross between two diverse parental lines. And then in usual cases, it is followed by powerful interval mapping approaches to identify the co-segregation of alleles of mapped marker loci and phenotypic traits. This allows the identification of QTL and the linked markers. Due to the restricted number of meiotic events that are captured in a biparental mapping population, the genetic resolution of QTL maps often remains confined, to a range of 10–30 cM (Zhu et al. 2008). AM relies on existing natural populations or designed diverse populations to overcome the constraints inherent to linkage mapping. LD in mapping population is a reflection of the germ plasm collection under study (Semagn et al. 2010). Several alleles in a population are tested at the same time in AM, while in linkage-based mapping, only alleles derived from the contrasting parents are tested.

Theoretically, AM should allow for finer mapping than the linkage mapping approaches, wherein the resolution depends on the several number of generations of historical recombination. In AM, the recombination is the result of several generations of historical recombination events and need not require any new mating designs, while in biparental linkage-based mapping, large numbers of segregants are required for accumulating sufficient recombination and it is time consuming to generate these populations. Often quantitative trait measurements are only possible once homozygosity is reached and needs several generations of selfing that involves careful handling. The time required for generating linkage-based mapping populations is large, while AM populations don't require much time. Moreover, the results from a biparental population may provide information that is specific to the same population or related populations, while the results from AM are applicable to much wider germ plasm. However, AM requires availability of large number of markers compared to normal linkage mapping approaches and also is more complex due to the high number of false positives associated with the analysis resulting from the effects of population structure. The issues of associated false positives and population admixture in AM are discussed below in detail due to their major impact on the results. Nevertheless, the potential of AM to map QTLs in collections of breeding lines, cultivated lines, landraces and wild germ plasm provides the scope for trait improvement in plant breeding.

The major advantages of AM can be summed up as follows: (1) no need of population generation as in biparental populations, (2) use of same population for various traits, (3) use of diverse and broader germ plasm for detecting new favourable alleles and (4) possibility of achieving high-resolution mapping. Similarly, the major limitations of AM can be summed up as follows: (1) false positives due to population structure, (2) requires high-density marker coverage which is not available yet in all crops, (3) power limitations to detect low-frequency functional alleles in the AM populations, (4) difficult to estimate the effect of a detected QTL on the genetic architecture of phenotype variation and (5) low explained heritability of detected QTL (Morrell et al. 2011; Myles et al. 2009).

5 Linkage Disequilibrium Implications for Association Mapping

Linkage equilibrium (LE) and linkage disequilibrium (LD) terms are used to define the linkage relationships in population genetics. LE is the random association of alleles at different loci. LD is the non-random association of alleles at separate loci or can also be referred to as the historically reduced level of the recombination of specific alleles at different loci (Flint-Garcia et al. 2003; Hill and Robertson 1968; Lewontin and Kojima 1960). In a genome, tightly linked loci are generally in linkage disequilibrium due to limited recombination between these loci. AM is dependent on LD because even high-density marker genotyping does not genotype all the polymorphisms. It is likely that the functional polymorphism is not among the genotyped markers. In such cases, it is expected that the genotyped marker is in high LD with the functional polymorphism and this genotyped marker is detected in the analysis. In general, the power of association studies also depends on the degree of LD between the genotyped markers and the functional polymorphisms. The decay of LD varies greatly between species, among different populations within one species and also among different loci within a given genome (Caldwell et al. 2006; Gupta et al. 2005). In regions with extensive LD, it is relatively easy to detect a QTL/associated marker because there is a high probability that the genotyped markers are in high LD with causal variant. Nevertheless, the downside of extensive LD regions is that several markers from this region are associated to the phenotype with similar significance and it will be difficult to detect the original causal variant. In cases where LD decays at very short distance and all the variants are independent of each other, complete sequencing of the region is necessary to pinpoint the genetic variant. Several contributing factors including genetic drift, mutations, regional variability in recombination patterns, diversity and population admixture, chromosomal composition and the pattern of mating within a population can significantly affect the patterns of LD within and different populations (Cardon and Bell 2001; Flint-Garcia et al. 2003).

6 How to Quantify LD?

Many measures of LD were proposed (Gupta et al. 2005), and most commonly used measures are *D, D'* and r^2 values. Two loci on a chromosome with alleles 'A/a' and 'B/b', respectively, are said to be in LE, when each haplotype frequency is equal to the product of the corresponding allelic frequencies. LD quantification was first described by Jennings in 1917 (Jennings 1917). The basic component of many measures of LD is the difference between the observed and expected allelic frequencies ($D = \pi_{AB} - \pi_A \pi_B$, where π_{AB} is the frequency of gametes carrying alleles A and B at two loci and π_A and π_B are the product of the frequencies of the alleles A and B, respectively). LD measure r^2, also sometimes referred to as Δ^2, introduced by Hill and Robertson (Hill and Robertson 1968) is the square of the correlation coefficient between the two loci ($r^2 = D/\pi_A \pi_a \pi_B \pi_b$). The LD measure r^2 value can range between 0 and 1. The value $r^2 = 0$ indicates the loci are in complete LE and value $r^2 = 1$ indicates the loci are in complete LD. The choice of appropriate LD parameter depends on the purpose of the study. Most of the plant studies use r^2 value for quantifying and visualising LD in the mapping populations (Gupta et al. 2005). Several LD visualisation methods and software are used in plant and animal studies. The two prominent methods used are LD triangular heat map generated by Haploview (Barrett et al. 2005) and LD scatter plots. LD heat maps are simple and provide information on patterns of LD distribution across the chromosome. LD scatter plots determine the rate of LD decay with genetic (cM) or physical distance (bp). LD scatter plots can be drafted for each chromosome or for the whole genome to determine the overall average distance at which LD decays beyond a critical threshold. The LD r^2 values are plotted against genetic distance and a nonlinear smoothed loess curve through these data points determines the effective genetic distance beyond which LD is likely due to linkage. The background threshold LD can be calculated as the 95th or 75th percentile of the unlinked r^2 values (Breseghello and Sorrells 2006; Mather et al. 2007). The intersection of the loess curve and background LD is considered as the extent of LD in the population. As an example, LD scatter plot (Fig. 2a) and heat plot (Fig. 2b) generated from 3,000 SNP markers across 280 *Arabidopsis* lines for chromosome 1 are presented here. Figure 2a shows the LD decay in the population, while Fig. 2b shows the extent of LD and LD distribution pattern across the chromosome.

7 Population Structure and Consequences for Association Mapping

Population structure leads to spurious trait associations and is the major confounding factor for association studies in both plants and animals. In association mapping, the complex genetic relatedness and the population structure affect the mapping of the phenotype. Population structure is inevitable in all natural populations and results in unequal distribution of alleles among different subpopulations. This systematic difference between allelic distributions among the subpopulations and the correlation of the phenotype to the subsequent population stratification might directly lead to several false positives in the analysis (Flint-Garcia et al. 2005; Myles et al. 2009). The spurious associations are high in adaptation-related genes because of their high correlation with the environmental variables under which they have evolved. For example, traits like flowering time are highly related to their latitudinal differences, and the population substructuring in natural populations is also correlated to their latitudinal differences and hence results in several spurious associations (Zhao et al. 2007). The problem of population

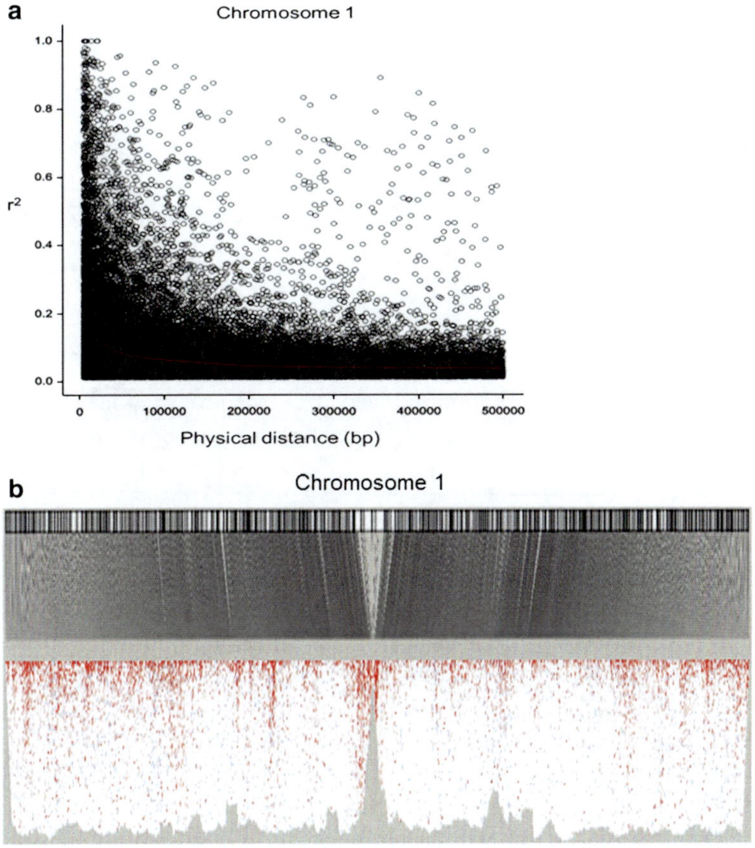

Fig. 2 Methods of linkage disequilibrium (LD) visualisation using 3,000 SNPs from *Arabidopsis* chromosome 1. (**a**) Scatter plot of LD decay (r^2) against genetic distance (cM). The loess fitting curve (*red curve*) illustrates the LD decay. (**b**) Heat plot for LD extent on chromosome 1. The colours of the squares (*red* being high LD to *white* being less LD) in the heat plot indicate the value of LD (r^2) between the two loci

structure in AM is well known, and several methods have been proposed to combat this problem even in populations with complex genetic structure (Lander and Schork 1994; Price et al. 2006; Pritchard et al. 2000b). However, none of these methods were able to overcome the cryptic population structure inherent of natural populations and are not 100 % efficient in controlling spurious associations, and hence caution should be followed in ascertaining the AM results.

8 Methodology for Association Mapping

The complex breeding history and the restricted gene flow among the gene pools in several important crops have created complex population stratification, which complicates association studies in crop species. The schematic framework for the steps involved in association mapping studies is represented as a flow diagram in Fig. 3. The foremost initiative for AM is the selection of the appropriate germ plasm for the AM studies, which involves several factors like the size of the population, diversity of the population and adaptability of the population. Phenotype in multi-environmental trials to obtain reliable phenotypic data. In some cases, already available robust phenotype information from databases can be directly used for association studies. Genotyping of the population is dependent on the availability of the markers and the cost feasibility of the projects. Apart from the established marker systems like AFLP, RFLP and SSRs, several high-throughput marker platforms like DArTs and SNP arrays are already available for major agricultural crops like in barley, wheat, rice, sorghum and other species (Roy et al. 2010; Wenzl et al. 2004; Marone et al. 2012; Bouchet et al. 2012; Nelson et al. 2011;

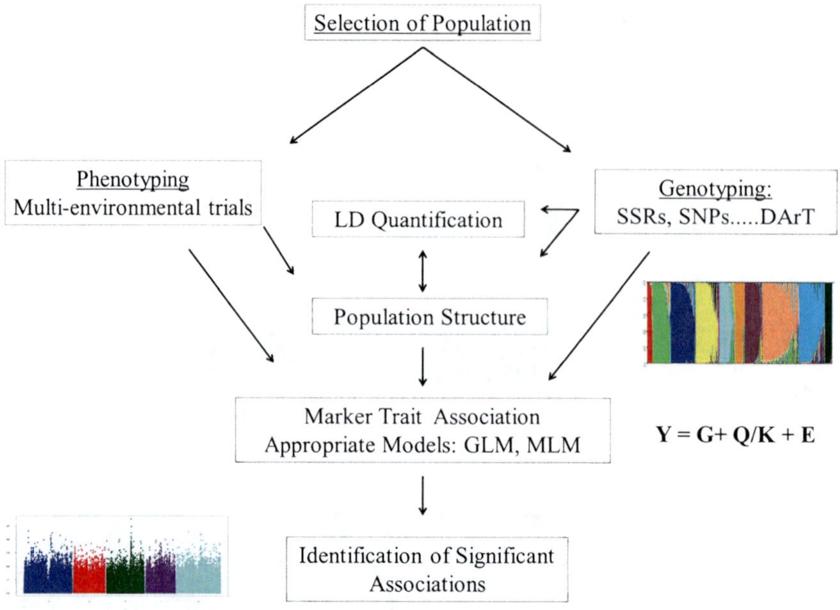

Fig. 3 Schematic framework of the steps involved in genome-wide association studies

Trebbi et al. 2011). Recent years have witnessed amassing of several genetic resources including larger marker databases for so-called neglected or orphan crops (Varshney et al. 2010). These evolving resources can be efficiently utilised in crop breeding in amalgamation with association mapping approaches. The next steps involve the study of LD and population structure. The importance of LD and population structure is illustrated above in detail, and the LD information further helps in the evaluation of results. The final step involves employing an efficient statistical model for association analysis and investigating the outcome for elucidating reliable results. Significant marker-trait association can be due to either (1) direct association of the phenotype with genotyped marker which is the true causal variant, (2) indirect association wherein the phenotype is associated with the genotyped marker which is in LD with the true causal variant, or (3) false-positive association which might be the result of population structure. Differentiating between these three types of associations and determining the true results from false positives while balancing the false-positive and false-negative results with a critical threshold are the crucial tasks in association analysis.

9 Statistical Approaches for Association Mapping

Several statistical methods have been proposed for AM to mitigate spurious associations caused by LD resulting from population structure. There are earlier methods like haplotype relative risk (HRR) approach, 'case-control' approach, transmission disequilibrium test (TDT) and several family-based methods which are of limited use in plant association studies and are currently outdated (Spielman et al. 1993; Falk and Rubinstein 1987; Schulze and McMahon 2002). The two common methods used for adjusting population stratification were genomic control (GC) and structured association (SA) studies (Devlin and Roeder 1999; Pritchard et al. 2000b). In GC approach, a constant factor λ is estimated as the median or mean of test statistics from a series of unlinked random markers genotyped in both cases and controls. In GC approach, it is assumed that population stratification inflates the association test statistics by λ, and hence the test statistics of all markers are divided by λ and compared using chi-square test or F-distribution to test for associations.

GC performs well in many scenarios, but the power of detection is decreased considerably and is conservative in extreme settings (Mackay and Powell 2007).

SA approach is a sophisticated method of AM, wherein each individual involved in the study is attributed to subpopulations and then associations are tested on conditional to the subpopulation allocation. Firstly, population ancestry is inferred using random markers, and based on this ancestry, each individual is allotted to subpopulation. This subpopulation membership is used as control for population stratification in the test for associations. The computer program STRUCTURE is used to allocate individuals to subpopulations (Pritchard et al. 2000a). STRUCTURE runs a Bayesian clustering analysis using model-based approach. This approach uses multilocus information to assign individuals to clusters or subpopulations without prior knowledge of their population affinities and assumes the loci in Hardy-Weinberg equilibrium. The STRUCTURE program is computationally intensive with increasing number of markers as larger numbers of MCMC (Markov chain Monte Carlo) iterations are required for accurate parameter estimation (Pasam et al. 2012). These subpopulation membership coefficients (Q) are incorporated into a general linear model (GLM) for correcting population structure (by logistic regression). In the fitted model, variation attributed to population membership is accounted first from Q and then the presence of any residual association between the markers and phenotype is ascertained. Population structure estimation by STRUCTURE is computationally intensive and sometimes the number of subpopulations is not clear. Furthermore, assignment of individuals to subpopulations is not clearly defined and might result in some discrepancies. Principal component analysis (PCA) method, also termed as EIGENSTRAT method, is proposed to overcome these limitations (Price et al. 2006). This model in effect is similar to SA approach proposed by Pritchard et al. (2000b), but instead of STRUCTURE-derived estimates (Q), this method applies PCA components for adjusting to stratification. PCA method makes it easier to handle a large number of markers and avoids hassles of determining and assigning correct population ancestry to each individual in the analysis. Another approach to reduce false positives in AM studies incorporates two-stage dimension determination approach for both PCA and nonmetric multidimensional scaling (nMDS) to capture major structure pattern in the AM populations (Zhu and Yu 2009).

However, population structure information itself is not sufficient in controlling spurious associations when highly structured populations with closely related individuals are used for AM. A mixed linear model (MLM) approach was introduced by Yu et al. (2006) that combines both the population structure information (Q) and the pairwise relatedness coefficients between the individuals (K matrix). The K estimate tries to estimate the approximate identity by descent between two individuals. In MLM approach, the Q is fitted as fixed effect, whereas K matrix is incorporated as the variance-covariance matrix of random effect for the individuals. MLM incorporate pairwise genetic relatedness between all the individuals in the statistical model, reflecting that the phenotypes of the genetically similar individuals are more likely to be correlated than genetically dissimilar individuals (Kang et al. 2008). The individual random means are constrained by assuming that the phenotypic covariance between individuals is proportional to their relative relatedness (K). The principle of MLM is to use the random effects (K) to explain the phenotype correlations that can be explained by genome-wide relatedness and in addition use either Q or PCA components as fixed effects in the model (Yu et al. 2006; Zhao et al. 2007). Several studies conducted using different models like GLM and MLM concluded that MLM performed better in all cases (Kang et al. 2008; Pasam et al. 2012; Stich and Melchinger 2009; Zhao et al. 2007).

The MLM methods are computationally very intensive; hence to reduce the computational time and speed up the process, a number of procedures like efficient mixed linear model (EMMA) (Kang et al. 2008) and compressed MLM approach called population parameters previously determined (P3D) (Zhang et al. 2010) have been proposed. When inferring the kinship matrix, it is important

Table 1 Recent genome-wide association studies in various plant species

Species	Traits	Reference
Wheat	Kernel quality traits	Breseghello and Sorrells (2006)
Wheat	Rust resistance	Yu et al. (2012)
Rice	Agronomic traits	Huang et al. (2010)
Rice	Agronomic traits	Zhao et al. (2011)
Maize	Leaf architecture	Tian et al. (2011)
Maize	Oil biosynthesis	Li et al. (2013)
Oats	β-Glucan concentration	Newell et al. (2012)
Barley	Growth habit	Rostoks et al. (2006)
Barley	Anthocyanin pigmentation	Cockram et al. (2010)
Barley	Agronomic traits	Pasam et al. (2012)
Rye	Frost tolerance	Li et al. (2011)
Sorghum	Agroclimatic traits	Morris et al. (2013)
Potato	Quality traits	D'hoop et al. (2008)
Tomato	Fruit characters	Ranc et al. (2012)
Soybean	Chlorophyll fluorescence parameters	Hao et al. (2012)
Sugar beet	Agronomic traits	Würschum et al. (2011)
Peach	Fruit characters and flowering time	Cao et al. (2012)
Cotton	Fibre quality traits	Abdurakhmonov et al. (2009)
Arabidopsis	107 Phenotypes	Atwell et al. (2010)

Abbreviations: *AM* association mapping, *DArT* Diversity Arrays Technology, *GC* genomic control, *GLM* general linear model, *GWAS* genome-wide association studies, *LD* linkage disequilibrium, *MAGIC* multiparent advanced generation intercross, *MLM* mixed linear model, *NAM* nested association mapping, *NGP* next-generation population, *NGS* next-generation sequencing, *QTL* quantitative trait locus, *RIL* recombinant inbred line, *SNP* single-nucleotide polymorphism, *SA* structured association

to estimate the matrix in a mathematically correct form. Initially, pedigree information is used to derive the kinship matrix, but with the availability of genome-wide distributed markers, currently, marker-based relative kinship estimates are mostly used. Several software packages like SPAGeDi (Hardy and Vekemans 2002) or TASSEL (Bradbury et al. 2007) can be used to estimate relative or kinship matrices. Association tests can be performed with different options from the models described above and are represented in Table 1. The different models are as follows: (1) naive model: GLM without any correction for population structure; (2) Q model: GLM with Q matrix as correction for population structure; (3) P model: GLM with PCs as correction for population structure; (4) QK model: MLM with Q matrix and K matrix as correction for population structure; (5) PK model: MLM with PCs and K matrix as correction for population structure; and (6) K model: MLM with K matrix as correction for population structure (Kang et al. 2008; Stich and Melchinger 2009; Pritchard et al. 2000a; Yu et al. 2006). Till date, various software programs have been available [like TASSEL (Bradbury et al. 2007), GenStat (Payne et al. 2006), SAS, R and several others] to perform all the above-mentioned analyses. The models PK, QK and K outperformed the other models in several comparison studies. Mostly, QK model and K model showed good fit in different species (Kang et al. 2008; Pasam et al. 2012; Stich and Melchinger 2009; Zhao et al. 2007; Stich et al. 2008).

10 Interpreting Association Mapping Results: What to Expect?

The common method of visualising GWAS is by using Manhattan plot for all the chromosomes. Manhattan plot is a scatter plot with negative logarithm p-values (Y-axis) for the SNP association plotted against the SNP position (X-axis).

The SNPs with most significant association will stand out as prominent peaks in the plot. Due to the quantitative nature of most of the traits and with inevitable false positives, we see several peaks in such plots. Heritability of the trait also affects the outcome of the association results; thus, the trait should be robust and generally replicated trials are used. False positives may arise due to population stratification, statistical analysis methods, error in genotyping and/or multiple testing issues which are difficult to differentiate in simple analyses. In some studies, despite the high heritability and repeatability of the traits, only few associations are reported. This has resulted due to the population structure correction model that captured all the variation, and limited variation is left for the markers for associations. Hence, careful discretion of the results is required before drawing any conclusions.

11 Determining Significance Threshold for GWAS

Setting threshold for the significant p-value to declare true association is challenging for the evaluated tests in a study. The threshold p-value for GWAS determines the stringency of the evaluation. In general, a universal p-value of 0.05 ($-\log10$ (p) = 1.3 on the plot) is too lenient, and several of the tested markers can easily cross this threshold. The GWA analyses should account for the multiplicity of comparisons that are performed as part of the study. This burden of multiple testing constitutes a major challenge for GWAS (Storey et al. 2004). A wide variety of statistical approaches accounting for multiple testing have been proposed, including methods like Bonferroni correction for adjusting the error rate (Bland and Altman 1995), false discovery rate (Benjamini and Hochberg 1995; Storey et al. 2004) and more. Many of these thresholds are already highly conservative, and with the increasing number of tests required with the availability of imputed complete genome information, some might argue for more stringent thresholds. It has been shown by various simulation and theoretical studies that the significance threshold for GWAS may vary for different populations, traits, LD patterns and arrays as well as with different marker systems (Panagiotou et al. 2012). Given the ambiguity and uncertainty regarding the efficient significance threshold in GWAS, various studies use different threshold significances. Due to the reduced power of detection owing to complex association models and the presence of rare alleles, the usage of stringent threshold might overkill other important potential peaks in the analysis. Adjustments for the existing stringent statistical approaches and many new FDR approaches are being proposed for GWAS (Johnson et al. 2010). On the contrary, there are some liberal approaches of determining significance threshold like using the 0.1 percentile distribution of the lower bottom p-values as significant (Chan et al. 2010). While determining significance threshold, it is necessary to strike the balance between selectivity and sensitivity to reduce the problem of false positives and false negatives.

12 Power of Association Mapping

The power of AM is the ability of detecting the true associations within the mapping populations. Several simulation studies and theoretical studies were conducted to determine the power of AM studies. The conclusion of these studies was the power of AM depends on factors like the size of the population, extent of LD in the population, extent of whole-genome marker coverage and experimental designs (Kang et al. 2008; Stich et al. 2008). The power of the QTL detection in GWA studies also depends on the trait analysed and the magnitude of the target allele effects and the number of QTL affecting the trait. The power can be increased by utilising robust data and increasing the population sample size.

13 Genetic Variation Explained in Association Mapping

Significant associations between the markers and phenotype have been found in several plant GWA studies (Abdurakhmonov et al. 2009;

Agrama et al. 2007; Atwell et al. 2010; Breseghello and Sorrells 2006; Rode et al. 2011; Roy et al. 2010). GWAS of the majority of the complex traits show that several loci contribute to the genetic component of variance and most of these associations explain only a limited proportion of phenotypic variability. This leaves geneticists speculating about the missing genetic variation also termed as 'missing heritability' (Brachi et al. 2011). In GWAS, the detected marker-trait associations only explain the fraction of the phenotype variation as compared to family-based studies (Pasam et al. 2012; Roy et al. 2010). The low explained genetic variability raises several questions regarding the methods used to analyse and estimate the genetic variability fraction and genetic architecture of these complex traits. Several plausible reasons were put forth to explain the causes of missing heritability in human GWA studies. Though the number of markers used in plant studies is far less than in human and animal genetic studies, similar reasons also apply to an extent in plant GWAS. In a consorted study for the trait body height, an impressive number of 40 genotypic variants have been identified under a stringent threshold. Together, these variants were able to explain around 5 % of the variation in human body height (Maher 2008; Visscher 2008). Similarly, GWAS in barley showed 34 SNPs significantly associated with the heading date and explained a total of 16 % phenotypic variance (Pasam et al. 2012), whereas in *Arabidopsis*, 45 % of the phenotypic variation for flowering time is explained by the significant markers (Li et al. 2010). But considerable amount of variation is still missing which needs to be accounted. Many explanations for the sources of missing explained variation include (1) insufficient marker coverage, for example, in cases where the causal polymorphism is not in perfect LD with the genotyped SNP, which reduces the power to detect associations and the variation explained by such an SNP marker; (2) rare alleles with a major effect excluded from the analysis and which go undetected; (3) the expression of a character or trait, which depends on a large number of genes/QTL with small individual effects that escape statistical detection; (4) inadequacy of the statistical approaches available to detect epistatic interactions in GWAS; (5) structural variations; (6) gene-environmental interactions; and (7) biased estimates of R^2 for individual SNPs due to the level of population stratification in the panel (Brachi et al. 2011; Frazer et al. 2009; Gibson 2010; Hall et al. 2010; Maher 2008; Manolio et al. 2009). Although the above-mentioned reasons were mainly discussed in the context of GWAS in humans, they also pertain to GWAS in plants and other organisms. In addition to the above-mentioned reasons, the statistical model employed for the analysis will affect the variation explained by the SNPs. As the stringency and threshold of the models increases, the power of detecting small-effect SNPs will be reduced. While using stringent models for GWAS, the larger portion of the trait variation is explained by the model itself and less variation is left to be explained by genetic effects. Reducing the stringency of the model would increase the variation explained by the marker but at the same time would result in more false positives. Determining which combination of these reasons can predict the hidden phenotype variation will have significant implications for future success of association mapping studies.

14 GWAS in Plants

GWAS require three essential elements: (1) sufficiently large populations that provide robust data, (2) adequate whole-genome marker coverage and (3) analytic methods that are statistically powerful and can be employed to identify the genetic associations. Initially, few GWAS with very limited marker coverage were conducted in crops like rice and barley (Virk et al. 1996; Kraakman et al. 2004). Later with the advent of new marker systems, emerging sequencing technologies and genomic resources in several crops, a surge of GWAS were evidenced. Some of the prominent GWAS from the last decade are reported in Table 1. Single-nucleotide polymorphism (SNP) markers are highly used in GWAS due to the ease of parallel genotyping thousands of SNPs at the same time. The evolving SNP arrays and assays for various species like barley, rice, wheat, chickpea, apple and potato with thousands of SNP markers have changed the scenario of GWAS in agricultural crops (Akhunov et al. 2009; Chagné et al. 2012; Hiremath et al. 2012; McCouch et al. 2010; Pasam et al. 2012; Hamilton et al. 2011). The availability of complete de novo genome sequence information and the accessibility of next-generation sequence technologies have brought millions of SNPs per genome into use. In crops like maize and rice, next-generation sequencing (NGS) of all the individuals under low coverage and harnessing of millions of SNPs for GWAS are already in progress (Buckler et al. 2009; Huang et al. 2010). Albeit its caveats, GWAS have proven to be a powerful method to dissect complex traits in crops.

15 Undermining Limitations of Association Mapping

Though AM is proven to be successful in many plant studies, there are several unique challenges in GWA studies pertaining to the causes like the number of false positives, missed rare alleles, low effects of QTL detected and complexities in determining epistatic and environmental effects. Several computational models are helping to reduce false positives and overcome the confounding effects of population structure in GWAS. Statistical approaches for epistatic association mapping techniques are still evolving (Stich and Gebhardt 2011; Lü et al. 2011). The heavy computational requirement of the mixed models might have led to reduced focus on detecting epistatic interactions in association studies. However, there are emerging methods in using different statistical approaches to discover epistatic effects in GWA studies (Wang et al. 2011). Most of the important agronomic traits are controlled by multiple small-effect QTLs and their epistatic interactions and environmental interactions (Bernardo 2008). Hence, it is important to start looking into new statistical methods and interpreting QTL interactions to completely realise the potential of association mapping studies in crop breeding. Another challenge in GWAS is that a large number of loci with small effects contribute to the quantitative trait and only few QTLs with major effect are reported (Buckler et al. 2009). The power of detecting rare alleles and very low-effect QTL can be improved by increasing the population size. In GWA studies, the researchers often tend to ignore or are handicapped in dealing with gene interactions, gene expression, different structural variations in genome like copy number variations (CNV) and balanced weightage of rare alleles. These are some potential areas for future empowerment for GWA studies in dissecting the genetic architecture of complex traits.

16 Future Prospects for Association Mapping in Crop Improvement

The shifting paradigm in plant breeding research in recent years is undoubtedly benefiting from the population genetics framework imputed with linkage mapping, association mapping and comparative genomic approaches. Domestication and plant breeding have resulted in narrow genetic basis of modern crops consequently reducing the genetic variation. In current settings, GWAS is a beacon in the efficient utilisation of natural genetic diversity of worldwide crop germ plasm resources for genetic diversification. The broader germ plasm diversity used in GWAS provides ample scope for allele mining, thereby identifying novel functional variation to be used in crop improvement and broadening the genetic basis of breeding germ plasms (Hamblin et al. 2011). The detection of QTL for economic traits and introgression of superior QTL alleles into breeding germ plasm from exotic materials was proposed as a potential approach for further crop improvement (Prada 2009). Plant breeding emphasises paramount importance on precision of QTL detection for their efficient utilisation. There are some success stories of fine mapping, cloning and characterising QTL/genes, which are discussed elsewhere (Salvi and Tuberosa 2005). Most of these studies demonstrate the importance of using exotic germ plasm and broader diversity towards improvement of cultivated germ plasm. In the near future, efficient utilisation of GWAS results can aid in enhancing the genetic basis of the current crop breeding germ plasms.

GWAS are usually conducted with the object of applying the results to elucidate QTL or for genotype-based selection of superior individuals in breeding or as a step towards positional cloning (Rafalski 2010). The markers that are closely linked to the QTL detected can be used for marker-assisted selection in crop breeding schemes. However, prudence should be shown in selecting the QTL allele for crop improvement. The GWAS result validation by using a different population or from previous studies is crucial before relying on the results. Due to the ever-increasing GWA studies, there is an accumulation of marker-trait associations for several traits, and hence it is recommended to validate the results by replicating the studies. Joint linkage and linkage disequilibrium mapping approaches can also help in simultaneously validating the results and also in overcoming the inherent limitations of linkage mapping and LD mapping (Jung et al. 2005; Lu et al. 2010).

There are also several evolving next-generation populations like NAM and MAGIC which can efficiently capture genetic diversity and also avoid the problem of confounding effect of population structure. The increasing use of these populations in different crops can enhance the opportunities of crop improvement and diversify the genetic basis of breeding germ plasm (Huang et al. 2012a; Buckler et al. 2009; Cook et al. 2012). The NAM population is successfully used to study several traits and map QTL in maize (Cook et al. 2012; Poland et al. 2011; Kump et al. 2011; Buckler et al. 2009). The NAM population in maize is constituted by crossing 25 diverse lines to a single reference parent. From each cross, 200 RILs were generated for a total of 5,000 lines that imparted substantial statistical power for QTL detection (Buckler et al. 2009). However, the optimum design and number of founder lines for establishing population are highly debatable and further depend on the population structure and genetic diversity in the species of interest and also on the genetic architecture of the trait of interest. MAGIC population is another advanced multiparent intercross design, and its success is demonstrated in model species like *Arabidopsis* (Kover et al. 2009) and also recently in wheat (Huang et al. 2012a). However, these resources are not readily available in many of the species and are being developed for efficient utilisation in crop improvement. An alternative approach, genomic selection (GS), is also gaining prominence in crop improvement. GS relies on the breeding values derived from the strength of marker-trait associations for selection, and GS aims at increasing the frequency of desirable alleles across the whole genome rather than focusing on one QTL to improve the trait (Jannink et al. 2010).

Due to the existing and evolving low-cost genotyping and NGS technologies, GWA studies with thousands of markers and large populations are already a reality. Whole-genome re-sequencing (genotyping by sequencing (GBS)) which provides uncompromised genomic polymorphism information is already in use for mapping in crops like rice, wheat and maize (Huang et al. 2010; Buckler et al. 2009; Poland et al. 2012) and will protend to other crops in the near future. Crops with complex genomes (like wheat and barley) encompass a lot of repetitive regions in the genome, and in such situations, reduction representation sequencing or genome enrichment approaches are recommended (Mamanova et al. 2010). With the fast-declining sequencing costs, it will be possible to genotype by sequence large number of accessions in several crops. The advancement of high-throughput gene expression profiling technologies like microarray and RNA-seq in plants gave the scope for genome-wide expression quantitative trait locus (eQTL) studies in plants (Holloway et al. 2011). The advent of new 'omics technologies' like metabolomics and proteomics will concede in new genome-wide profiling and provide platforms for new GWA studies at a different level. The ensuing 'omics' technologies allow global understanding to pathways and intriguing insights into the networks resulting in final quantitative nature of the phenotypes (Adamski and Suhre 2013;

Baerenfaller et al. 2008). These studies in combination with traditional GWA studies will allow a detailed understanding of the gene effects and interactions.

Nonetheless, the costs of GBS and GWAS are still high and are beyond the reach of several common breeders and researchers in developing countries. This restricts the use of GWAS to its full potential in various crop improvement programmes for regional adaptation and in poor resource crops. Today's plant research is not confined to a single problem or purpose but has moved on to cross-border and multidisciplinary approaches. Apparently immense benefits can be harvested by incorporating data warehousing approaches and establishing data integration platforms. The huge data available from various research projects including information about accessions, phenotypic data, omics profiling data, genomics data, sequencing and genotyping data can be integrated into public databases with free accessibility. This also facilitates in easy data sharing and availability of expensive resources across all the plant research communities. As an example, integrated databases including accessions information, phenotype and genotype data along with GWA analysis methods were made available in public databases for *Arabidopsis* research community (Atwell et al. 2010; Huang et al. 2011). Similar public databases with integrated information for the crops could enhance the utilisation of these emerging technologies and realise the complete potential of GWAS in crop improvement.

References

Abdurakhmonov I, Saha S, Jenkins J, Buriev Z, Shermatov S, Scheffler B, Pepper A, Yu J, Kohel R, Abdukarimov A (2009) Linkage disequilibrium based association mapping of fiber quality traits in *G. hirsutum* L. variety germplasm. Genetica 136 (3):401–417

Adamski J, Suhre K (2013) Metabolomics platforms for genome wide association studies linking the genome to the metabolome. Curr Opin Biotechnol 24(1):39–47

Agrama H, Eizenga G, Yan W (2007) Association mapping of yield and its components in rice cultivars. Mol Breed 19(4):341–356

Akhunov E, Nicolet C, Dvorak J (2009) Single nucleotide polymorphism genotyping in polyploid wheat with the Illumina GoldenGate assay. Theor Appl Genet 119(3):507–517

Atwell S, Huang YS, Vilhjalmsson BJ, Willems G, Horton M, Li Y, Meng D, Platt A, Tarone AM, Hu TT, Jiang R, Muliyati NW, Zhang X, Amer MA, Baxter I, Brachi B, Chory J, Dean C, Debieu M, de Meaux J, Ecker JR, Faure N, Kniskern JM, Jones JD, Michael T, Nemri A, Roux F, Salt DE, Tang C, Todesco M, Traw MB, Weigel D, Marjoram P, Borevitz JO, Bergelson J, Nordborg M (2010) Genome-wide association study of 107 phenotypes in *Arabidopsis thaliana* inbred lines. Nature 465 (7298):627–631

Baerenfaller K, Grossmann J, Grobei MA, Hull R, Hirsch-Hoffmann M, Yalovsky S, Zimmermann P, Grossniklaus U, Gruissem W, Baginsky S (2008) Genome-scale proteomics reveals *Arabidopsis thaliana* gene models and proteome dynamics. Science 320(5878):938–941

Barrett JC, Fry B, Maller J, Daly MJ (2005) Haploview: analysis and visualization of LD and haplotype maps. Bioinformatics 21(2):263–265

Beló A, Zheng P, Luck S, Shen B, Meyer D, Li B, Tingey S, Rafalski A (2008) Whole genome scan detects an allelic variant of fad2 associated with increased oleic acid levels in maize. Mol Gen Genomics 279(1):1–10

Benjamini Y, Hochberg Y (1995) Controlling the false discovery rate – a practical and powerful approach to multiple testing. J R Stat Soc Ser B Methodol 57(1):289–300

Bernardo R (2008) Molecular markers and selection for complex traits in plants: learning from the last 20 years. Crop Sci 48(5):1649–1664

Bhosale S, Stich B, Rattunde HF, Weltzien E, Haussmann B, Hash CT, Ramu P, Cuevas H, Paterson A, Melchinger A, Parzies H (2012) Association analysis of photoperiodic flowering time genes in west and central African sorghum [*Sorghum bicolor* (L.) Moench]. BMC Plant Biol 12(1):32

Bland JM, Altman DG (1995) Multiple significance tests – The Bonferroni method. Br Med J 310 (6973):170

Bouchet S, Pot D, Deu M, Rami J-F, Billot C, Perrier X, Rivallan R, Gardes L, Xia L, Wenzl P, Kilian A, Glaszmann J-C (2012) Genetic structure, linkage disequilibrium and signature of selection in Sorghum: lessons from physically anchored DArT markers. PLoS One 7(3):e33470

Brachi B, Morris GP, Borevitz JO (2011) Genome-wide association studies in plants: the missing heritability is in the field. Genome Biol 12(10)

Bradbury PJ, Zhang Z, Kroon DE, Casstevens TM, Ramdoss Y, Buckler ES (2007) TASSEL: software for association mapping of complex traits in diverse samples. Bioinformatics 23(19):2633–2635

Breseghello F, Sorrells ME (2006) Association mapping of kernel size and milling quality in wheat (*Triticum aestivum* L.) cultivars. Genetics 172(2):1165–1177

Brown PJ, Rooney WL, Franks C, Kresovich S (2008) Efficient mapping of plant height quantitative trait loci in a sorghum association population with introgressed dwarfing genes. Genetics 180(1):629–637

Buckler ES, Holland JB, Bradbury PJ, Acharya CB, Brown PJ, Browne C, Ersoz E, Flint-Garcia S, Garcia A, Glaubitz JC, Goodman MM, Harjes C, Guill K, Kroon DE, Larsson S, Lepak NK, Li H, Mitchell SE, Pressoir G, Peiffer JA, Rosas MO, Rocheford TR, Romay MC, Romero S, Salvo S, Sanchez Villeda H, da Silva HS, Sun Q, Tian F, Upadyayula N, Ware D, Yates H, Yu J, Zhang Z, Kresovich S, McMullen MD (2009) The genetic architecture of maize flowering time. Science 325(5941):714–718

Caldwell KS, Russell J, Langridge P, Powell W (2006) Extreme population-dependent linkage disequilibrium detected in an inbreeding plant species, *Hordeum vulgare*. Genetics 172(1):557–567

Cao K, Wang L, Zhu G, Fang W, Chen C, Luo J (2012) Genetic diversity, linkage disequilibrium, and association mapping analyses of peach (*Prunus persica*) landraces in China. Tree Genet Genome 8(5):975–990

Cardon LR, Bell JI (2001) Association study designs for complex diseases. Nat Rev Genet 2(2):91–99

Chagné D, Crowhurst RN, Troggio M, Davey MW, Gilmore B, Lawley C, Vanderzande S, Hellens RP, Kumar S, Cestaro A, Velasco R, Main D, Rees JD, Iezzoni A, Mockler T, Wilhelm L, Van de Weg E, Gardiner SE, Bassil N, Peace C (2012) Genome-wide SNP detection, validation, and development of an 8K SNP array for apple. PLoS One 7(2):e31745

Chan EK, Rowe HC, Kliebenstein DJ (2010) Understanding the evolution of defence metabolites in *Arabidopsis thaliana* using genome-wide association mapping. Genetics 185:991–1007

Cockram J, White J, Zuluaga DL, Smith D, Comadran J, Macaulay M, Luo Z, Kearsey MJ, Werner P, Harrap D, Tapsell C, Liu H, Hedley PE, Stein N, Schulte D, Steuernagel B, Marshall DF, Thomas WTB, Ramsay L, Mackay I, Balding DJ, Consortium TA, Waugh R, O'Sullivan DM (2010) Genome-wide association mapping to candidate polymorphism resolution in the unsequenced barley genome. Proc Natl Acad Sci 107(50):21611–21616

Comadran J, Thomas WT, van Eeuwijk FA, Ceccarelli S, Grando S, Stanca AM, Pecchioni N, Akar T, Al-Yassin A, Benbelkacem A, Ouabbou H, Bort J, Romagosa I, Hackett CA, Russell JR (2009) Patterns of genetic diversity and linkage disequilibrium in a highly structured Hordeum vulgare association-mapping population for the Mediterranean basin. Theor Appl Genet 119:175–187

Cook JP, McMullen MD, Holland JB, Tian F, Bradbury P, Ross-Ibarra J, Buckler ES, Flint-Garcia SA (2012) Genetic architecture of maize kernel composition in the nested association mapping and inbred association panels. Plant Physiol 158(2):824–834

D'hoop B, Paulo M, Mank R, Eck H, Eeuwijk F (2008) Association mapping of quality traits in potato (L.). Euphytica 161:47–60

Devlin B, Roeder K (1999) Genomic control for association studies. Biometrics 55:997–1004

Falk CT, Rubinstein P (1987) Haplotype relative risks: an easy reliable way to construct a proper control sample for risk calculations. Ann Hum Genet 51:227–233

Flint-Garcia SA, Thornsberry JM, Buckler ES 4th (2003) Structure of linkage disequilibrium in plants. Annu Rev Plant Biol 54:357–374

Flint-Garcia SA, Thuillet AC, Yu J, Pressoir G, Romero SM, Mitchell SE, Doebley J, Kresovich S, Goodman MM, Buckler ES (2005) Maize association population: a high-resolution platform for quantitative trait locus dissection. Plant J 44:1054–1064

Frazer KA, Murray SS, Schork NJ, Topol EJ (2009) Human genetic variation and its contribution to complex traits. Nat Rev Genet 10:241–251

Fusari C, Di Rienzo J, Troglia C, Nishinakamasu V, Moreno M, Maringolo C, Quiroz F, Alvarez D, Escande A, Hopp E, Heinz R, Lia V, Paniego N (2012) Association mapping in sunflower for sclerotinia head rot resistance. BMC Plant Biol 12(1):93

Gebhardt C, Ballvora A, Walkemeier B, Oberhagemann P, Schüler K (2004) Assessing genetic potential in germplasm collections of crop plants by marker-trait association: a case study for potatoes with quantitative variation of resistance to late blight and maturity type. Mol Breed 13(1):93–102

Gibson G (2010) Hints of hidden heritability in GWAS. Nat Genet 42(7):558–560

Gupta PK, Rustgi S, Kulwal PL (2005) Linkage disequilibrium and association studies in higher plants: present status and future prospects. Plant Mol Biol 57(4):461–485

Hall D, Tegstrom C, Ingvarsson PK (2010) Using association mapping to dissect the genetic basis of complex traits in plants. Brief Funct Genomic 9:157–165

Hamblin MT, Close TJ, Bhat PR, Chao SM, Kling JG, Abraham KJ, Blake T, Brooks WS, Cooper B, Griffey CA, Hayes PM, Hole DJ, Horsley RD, Obert DE, Smith KP, Ullrich SE, Muehlbauer GJ, Jannink JL (2010) Population structure and linkage disequilibrium in US barley germplasm: implications for association mapping. Crop Sci 50:556–566

Hamblin MT, Buckler ES, Jannink J-L (2011) Population genetics of genomics-based crop improvement methods. Trends Genet 27:98–106

Hamilton J, Hansey C, Whitty B, Stoffel K, Massa A, Van Deynze A, De Jong W, Douches D, Buell CR (2011) Single nucleotide polymorphism discovery in elite North American potato germplasm. BMC Genomics 12:302

Hao D, Chao M, Yin Z, Yu D (2012) Genome-wide association analysis detecting significant single nucleotide polymorphisms for chlorophyll and chlorophyll fluorescence parameters in soybean (*Glycine max*) landraces. Euphytica 186:919–931

Hardy OJ, Vekemans X (2002) SPAGeDi: a versatile computer program to analyse spatial genetic structure at the individual or population levels. Mol Ecol Notes 2:618–620

Hastbacka J, Delachapelle A, Kaitila I, Sistonen P, Weaver A, Lander E (1992) Linkage disequilibrium mapping in isolated founder populations – Diastrophic dysplasia in Finland. Nat Genet 2(3):204–211

Hill WG, Robertson A (1968) The effects of inbreeding at loci with heterozygote advantage. Genetics 60(3):615–628

Hiremath PJ, Kumar A, Penmetsa RV, Farmer A, Schlueter JA, Chamarthi SK, Whaley AM, Carrasquilla-Garcia N, Gaur PM, Upadhyaya HD, Kavi Kishor PB, Shah TM, Cook DR, Varshney RK (2012) Large-scale development of cost-effective SNP marker assays for diversity assessment and genetic mapping in chickpea and comparative mapping in legumes. Plant Biotechnol J 10(6):716–732

Holloway B, Luck S, Beatty M, Rafalski J-A, Li B (2011) Genome-wide expression quantitative trait loci (eQTL) analysis in maize. BMC Genomics 12(1):336

Huang X, Wei X, Sang T, Zhao Q, Feng Q, Zhao Y, Li C, Zhu C, Lu T, Zhang Z, Li M, Fan D, Guo Y, Wang A, Wang L, Deng L, Li W, Lu Y, Weng Q, Liu K, Huang T, Zhou T, Jing Y, Li W, Lin Z, Buckler ES, Qian Q, Zhang Q-F, Li J, Han B (2010) Genome-wide association studies of 14 agronomic traits in rice landraces. Nat Genet 42(11):961–967

Huang YS, Horton M, Vilhjálmsson BJ, Seren Ü, Meng D, Meyer C, Ali Amer M, Borevitz JO, Bergelson J, Nordborg M (2011) Analysis and visualization of *Arabidopsis thaliana* GWAS using web 2.0 technologies. Database 2011

Huang BE, George AW, Forrest KL, Kilian A, Hayden MJ, Morell MK, Cavanagh CR (2012a) A multiparent advanced generation inter-cross population for genetic analysis in wheat. Plant Biotechnol J 10(7):826–839

Huang X, Kurata N, Wei X, Wang Z-X, Wang A, Zhao Q, Zhao Y, Liu K, Lu H, Li W, Guo Y, Lu Y, Zhou C, Fan D, Weng Q, Zhu C, Huang T, Zhang L, Wang Y, Feng L, Furuumi H, Kubo T, Miyabayashi T, Yuan X, Xu Q, Dong G, Zhan Q, Li C, Fujiyama A, Toyoda A, Lu T, Feng Q, Qian Q, Li J, Han B (2012b) A map of rice genome variation reveals the origin of cultivated rice. Nature 490:497–501

Hufford MB, Xu X, van Heerwaarden J, Pyhajarvi T, Chia J-M, Cartwright RA, Elshire RJ, Glaubitz JC, Guill KE, Kaeppler SM, Lai J, Morrell PL, Shannon LM, Song C, Springer NM, Swanson-Wagner RA, Tiffin P, Wang J, Zhang G, Doebley J, McMullen MD, Ware D, Buckler ES, Yang S, Ross-Ibarra J (2012) Comparative population genomics of maize domestication and improvement. Nat Genet 44(7):808–811

Hyten DL, Choi I-Y, Song Q, Shoemaker RC, Nelson RL, Costa JM, Specht JE, Cregan PB (2007) Highly variable patterns of linkage disequilibrium in multiple soybean populations. Genetics 175(4):1937–1944

Jannink J-L, Lorenz AJ, Iwata H (2010) Genomic selection in plant breeding: from theory to practice. Brief Funct Genomic 9(2):166–177

Jennings HS (1917) The numerical results of diverse systems of breeding, with respect to two pairs of characters, linked or independent, with special relation to the effects of linkage. Genetics 2(2):97–154

Johnson R, Nelson G, Troyer J, Lautenberger J, Kessing B, Winkler C, O'Brien S (2010) Accounting for multiple comparisons in a genome-wide association study (GWAS). BMC Genomics 11(1):724

Jun T-H, Van K, Kim M, Lee S-H, Walker D (2008) Association analysis using SSR markers to find QTL for seed protein content in soybean. Euphytica 162(2):179–191

Jung J, Fan R, Jin L (2005) Combined linkage and association mapping of quantitative trait loci by multiple markers. Genetics 170(2):881–898

Kang HM, Zaitlen NA, Wade CM, Kirby A, Heckerman D, Daly MJ, Eskin E (2008) Efficient control of population structure in model organism association mapping. Genetics 178(3):1709–1723

Kover PX, Valdar W, Trakalo J, Scarcelli N, Ehrenreich IM, Purugganan MD, Durrant C, Mott R (2009) A multiparent advanced generation inter-cross to fine-map quantitative traits in *Arabidopsis thaliana*. PLoS Genet 5(7):e1000551

Kraakman AT, Niks RE, Van den Berg PM, Stam P, Van Eeuwijk FA (2004) Linkage disequilibrium mapping of yield and yield stability in modern spring barley cultivars. Genetics 168(1):435–446

Kump KL, Bradbury PJ, Wisser RJ, Buckler ES, Belcher AR, Oropeza-Rosas MA, Zwonitzer JC, Kresovich S, McMullen MD, Ware D, Balint-Kurti PJ, Holland JB (2011) Genome-wide association study of quantitative resistance to southern leaf blight in the maize nested association mapping population. Nat Genet 43(2):163–168

Lander ES, Schork NJ (1994) Genetic dissection of complex traits. Science 265(5181):2037–2048

Lewontin RC, Kojima K-i (1960) The Evolutionary Dynamics of Complex Polymorphisms. Evolution 14(4):458–472

Li YH, Zhang C, Gao ZS, Smulders MJM, Ma ZL, Liu ZX, Nan HY, Chang RZ, Qiu LJ (2009) Development of SNP markers and haplotype analysis of the candidate gene for rhg1, which confers resistance to soybean cyst nematode in soybean. Mol Breed 24(1):63–76

Li Y, Huang Y, Bergelson J, Nordborg M, Borevitz JO (2010) Association mapping of local climate-sensitive quantitative trait loci in *Arabidopsis thaliana*. Proc Natl Acad Sci 107(49):21199–21204

Li Y, Bock A, Haseneyer G, Korzun V, Wilde P, Schon C-C, Ankerst D, Bauer E (2011) Association analysis of frost tolerance in rye using candidate genes and phenotypic data from controlled, semi-controlled, and field phenotyping platforms. BMC Plant Biol 11(1):146

Li H, Peng Z, Yang X, Wang W, Fu J, Wang J, Han Y, Chai Y, Guo T, Yang N, Liu J, Warburton ML, Cheng Y, Hao X, Zhang P, Zhao J, Liu Y, Wang G, Li J, Yan J (2013) Genome-wide association study dissects the genetic architecture of oil biosynthesis in maize kernels. Nat Genet 45(1):43–50

Lu Y, Zhang S, Shah T, Xie C, Hao Z, Li X, Farkhari M, Ribaut J-M, Cao M, Rong T, Xu Y (2010) Joint linkage–linkage disequilibrium mapping is a powerful approach to detecting quantitative trait loci underlying drought tolerance in maize. Proc Natl Acad Sci 107(45):19585–19590

Lü H-Y, Liu X-F, Wei S-P, Zhang Y-M (2011) Epistatic association mapping in homozygous crop cultivars. PLoS One 6(3):e17773

Mackay I, Powell W (2007) Methods for linkage disequilibrium mapping in crops. Trends Plant Sci 12(2):57–63

Maher B (2008) Personal genomes: the case of the missing heritability. Nature 456(7218):18–21

Mamanova L, Coffey AJ, Scott CE, Kozarewa I, Turner EH, Kumar A, Howard E, Shendure J, Turner DJ (2010) Target-enrichment strategies for next-generation sequencing. Nat Methods 7(2):111–118

Manolio TA, Collins FS, Cox NJ, Goldstein DB, Hindorff LA, Hunter DJ, McCarthy MI, Ramos EM, Cardon LR, Chakravarti A, Cho JH, Guttmacher AE, Kong A, Kruglyak L, Mardis E, Rotimi CN, Slatkin M, Valle D, Whittemore AS, Boehnke M, Clark AG, Eichler EE, Gibson G, Haines JL, Mackay TF, McCarroll SA, Visscher PM (2009) Finding the missing heritability of complex diseases. Nature 461(7265):747–753

Marone D, Panio G, Ficco DM, Russo M, Vita P, Papa R, Rubiales D, Cattivelli L, Mastrangelo A (2012) Characterization of wheat DArT markers: genetic and functional features. Mol Gen Genomics 287(9):741–753

Mather KA, Caicedo AL, Polato NR, Olsen KM, McCouch S, Purugganan MD (2007) The extent of linkage disequilibrium in rice (*Oryza sativa* L.). Genetics 177(4):2223–2232

McCouch SR, Zhao KY, Wright M, Tung CW, Ebana K, Thomson M, Reynolds A, Wang D, DeClerck G, Ali ML, McClung A, Eizenga G, Bustamante C (2010) Development of genome-wide SNP assays for rice. Breed Sci 60(5):524–535

Morrell PL, Buckler ES, Ross-Ibarra J (2011) Crop genomics: advances and applications. Nat Rev Genet 13(2):85–96

Morris GP, Ramu P, Deshpande SP, Hash CT, Shah T, Upadhyaya HD, Riera-Lizarazu O, Brown PJ, Acharya CB, Mitchell SE, Harriman J, Glaubitz JC, Buckler ES, Kresovich S (2013) Population genomic and genome-wide association studies of agroclimatic traits in sorghum. Proc Natl Acad Sci 110(2):453–458

Murray SC, Rooney WL, Hamblin MT, Mitchell SE, Kresovich S (2009) Sweet sorghum genetic diversity and association mapping for brix and height. Plant Gen 2(1):48–62

Myles S, Peiffer J, Brown PJ, Ersoz ES, Zhang Z, Costich DE, Buckler ES (2009) Association mapping: critical considerations shift from genotyping to experimental design. Plant Cell Online 21(8):2194–2202

Nelson J, Wang S, Wu Y, Li X, Antony G, White F, Yu J (2011) Single-nucleotide polymorphism discovery by high-throughput sequencing in sorghum. BMC Genomics 12(1):352

Newell M, Asoro F, Scott MP, White P, Beavis W, Jannink J-L (2012) Genome-wide association study for oat (*Avena sativa* L.) beta-glucan concentration using germplasm of worldwide origin. Theor Appl Genet 125(8):1687–1696

Palaisa KA, Morgante M, Williams M, Rafalski A (2003) Contrasting effects of selection on sequence diversity and linkage disequilibrium at two phytoene synthase loci. Plant Cell 15(8):1795–1806

Panagiotou OA, Ioannidis JPA, Genome-Wide Significance Project (2012) What should the genome-wide significance threshold be? Empirical replication of borderline genetic associations. Int J Epidemiol 41 (1):273–286

Pasam R, Sharma R, Malosetti M, van Eeuwijk F, Haseneyer G, Kilian B, Graner A (2012) Genome-wide association studies for agronomical traits in a world-wide spring barley collection. BMC Plant Biol 12(1):16

Payne RW, Murray DA, Harding SA, Baird DB, Soutar DM (2006) GenStat for Windows (9th edition) introduction. VSN International, Hemel Hempstead

Poland JA, Bradbury PJ, Buckler ES, Nelson RJ (2011) Genome-wide nested association mapping of quantitative resistance to northern leaf blight in maize. Proc Natl Acad Sci U S A 108(17):6893–6898

Poland J, Endelman J, Dawson J, Rutkoski J, Wu S, Manes Y, Dreisigacker S, Crossa J, Sánchez-Villeda H, Sorrells M, Jannink J-L (2012) Genomic selection in wheat breeding using genotyping-by-sequencing. Plant Genome 5(3):103–113

Prada D (2009) Molecular population genetics and agronomic alleles in seed banks: searching for a needle in a haystack? J Exp Bot 60(9):2541–2552

Price AL, Patterson NJ, Plenge RM, Weinblatt ME, Shadick NA, Reich D (2006) Principal components analysis corrects for stratification in genome-wide association studies. Nat Genet 38(8):904–909

Pritchard JK, Stephens M, Donnelly P (2000a) Inference of population structure using multilocus genotype data. Genetics 155(2):945–959

Pritchard JK, Stephens M, Rosenberg NA, Donnelly P (2000b) Association mapping in structured populations. Am J Hum Genet 67(1):170–181

Rafalski JA (2010) Association genetics in crop improvement. Curr Opin Plant Biol 13(2):174–180

Ranc N, Muños S, Xu J, Le Paslier M-C, Chauveau A, Bounon R, Rolland S, Bouchet J-P, Brunel D, Causse M (2012) Genome-Wide association mapping in tomato (Solanum lycopersicum) is possible Using Genome Admixture of Solanum lycopersicum var. cerasiforme. G3 Genes Genomes Genet 2(8):853–864

Remington DL, Thornsberry JM, Matsuoka Y, Wilson LM, Whitt SR, Doebley J, Kresovich S, Goodman MM, Buckler ES (2001) Structure of linkage disequilibrium and phenotypic associations in the maize genome. Proc Natl Acad Sci U S A 98 (20):11479–11484

Robbins MD, Sim S-C, Yang W, Van Deynze A, van der Knaap E, Joobeur T, Francis DM (2011) Mapping and linkage disequilibrium analysis with a genome-wide collection of SNPs that detect polymorphism in cultivated tomato. J Exp Bot 62(6):1831–1845

Rode J, Ahlemeyer J, Friedt W, Ordon F (2011) Identification of marker-trait associations in the German winter barley breeding gene pool (*Hordeum vulgare* L.). Mol Breed: 30:1–13

Rostoks N, Ramsay L, MacKenzie K, Cardle L, Bhat PR, Roose ML, Svensson JT, Stein N, Varshney RK, Marshall DF, Graner A, Close TJ, Waugh R (2006) Recent history of artificial outcrossing facilitates whole-genome association mapping in elite inbred crop varieties. Proc Natl Acad Sci 103(49):18656–18661

Roy JK, Smith KP, Muehlbauer GJ, Chao S, Close TJ, Steffenson BJ (2010) Association mapping of spot blotch resistance in wild barley. Mol Breed 26(2):243–256

Salvi S, Tuberosa R (2005) To clone or not to clone plant QTLs: present and future challenges. Trends Plant Sci 10(6):297–304

Sax K (1923) The association of size differences with seed-coat pattern and pigmentation in *Phaseolus vulgaris*. Genetics 8(6):552–560

Schulze TG, McMahon FJ (2002) Genetic association mapping at the crossroads: which test and why? Overview and practical guidelines. Am J Med Genet 114(1):1–11

Semagn K, Bjornstad Å, Xu Y (2010) The genetic dissection of quantitative traits in crops. Electron J Biotechnol 13(5)

Simko I, Pechenick DA, McHale LK, Truco MJ, Ochoa OE, Michelmore RW, Scheffler BE (2009) Association mapping and marker-assisted selection of the lettuce dieback resistance gene Tvr1. BMC Plant Biol 9:135

Singh A, Reimer S, Pozniak CJ, Clarke FR, Clarke JM, Knox RE, Singh AK (2009) Allelic variation at Psy1-A1 and association with yellow pigment in durum wheat grain. Theor Appl Genet 118(8):1539–1548

Spielman RS, Mcginnis RE, Ewens WJ (1993) Transmission test for linkage disequilibrium – The insulin gene region and insulin-dependent diabetes-mellitus (Iddm). Am J Hum Genet 52(3):506–516

Stich B, Gebhardt C (2011) Detection of epistatic interactions in association mapping populations: an example from tetraploid potato. Heredity 107(6):537–547

Stich B, Melchinger AE (2009) Comparison of mixed-model approaches for association mapping in rapeseed, potato, sugar beet, maize, and Arabidopsis. BMC Genomics 10:94

Stich B, Möhring J, Piepho H-P, Heckenberger M, Buckler ES, Melchinger AE (2008) Comparison of mixed-model approaches for association mapping. Genetics 178(3):1745–1754

Storey JD, Taylor JE, Siegmund D (2004) Strong control, conservative point estimation and simultaneous conservative consistency of false discovery rates: a unified approach. J R Stat Soc Ser B (Stat Methodol) 66(1):187–205

Stracke S, Haseneyer G, Veyrieras JB, Geiger HH, Sauer S, Graner A, Piepho HP (2009) Association mapping reveals gene action and interactions in the determination of flowering time in barley. Theor Appl Genet 118(2):259–273

Takeda S, Matsuoka M (2008) Genetic approaches to crop improvement: responding to environmental and population changes. Nat Rev Genet 9(6):444–457

Tanksley SD (1993) Mapping polygenes. Annu Rev Genet 27(1):205–233

Tanksley SD, McCouch SR (1997) Seed banks and molecular maps: unlocking genetic potential from the wild. Science 277(5329):1063–1066

Thornsberry JM, Goodman MM, Doebley J, Kresovich S, Nielsen D, Buckler ES (2001) Dwarf8 polymorphisms associate with variation in flowering time. Nat Genet 28(3):286–289

Tian F, Bradbury PJ, Brown PJ, Hung H, Sun Q, Flint-Garcia S, Rocheford TR, McMullen MD, Holland JB, Buckler ES (2011) Genome-wide association study of leaf architecture in the maize nested association mapping population. Nat Genet 43(2):159–U113

Trebbi D, Maccaferri M, Heer P, Sørensen A, Giuliani S, Salvi S, Sanguineti M, Massi A, Vossen E, Tuberosa R (2011) High-throughput SNP discovery and genotyping in durum wheat (Triticum durum Desf.). Theor Appl Genet 123(4):555–569

Varshney RK, Glaszmann J-C, Leung H, Ribaut J-M (2010) More genomic resources for less-studied crops. Trends Biotechnol 28(9):452–460

Virk PS, Ford-Lloyd BV, Jackson MT, Pooni HS, Clemeno TP, Newbury HJ (1996) Predicting quantitative variation within rice germplasm using molecular markers. Heredity 76(3):296–304

Visscher PM (2008) Sizing up human height variation. Nat Genet 40(5):489–490

Wang D, Eskridge K, Crossa J (2011) Identifying QTLs and epistasis in structured plant populations using adaptive mixed LASSO. J Agric Biol Environ Stat 16 (2):170–184

Waugh R, Jannink JL, Muehlbauer GJ, Ramsay L (2009) The emergence of whole genome association scans in barley. Curr Opin Plant Biol 12(2):218–222

Wenzl P, Carling J, Kudrna D, Jaccoud D, Huttner E, Kleinhofs A, Kilian A (2004) Diversity Arrays Technology (DArT) for whole-genome profiling of barley. Proc Natl Acad Sci U S A 101 (26):9915–9920

Würschum T, Maurer H, Kraft T, Janssen G, Nilsson C, Reif J (2011) Genome-wide association mapping of agronomic traits in sugar beet. Theor Appl Genet 123(7):1121–1131

Yan J, Warburton M, Crouch J (2011) Association mapping for enhancing maize (L.) genetic improvement. Crop Sci 51(2):433–449

Yu J, Pressoir G, Briggs WH, Vroh Bi I, Yamasaki M, Doebley JF, McMullen MD, Gaut BS, Nielsen DM, Holland JB, Kresovich S, Buckler ES (2006) A unified mixed-model method for association mapping that accounts for multiple levels of relatedness. Nat Genet 38(2):203–208

Yu L-X, Morgounov A, Wanyera R, Keser M, Singh S, Sorrells M (2012) Identification of Ug99 stem rust resistance loci in winter wheat germplasm using genome-wide association analysis. Theor Appl Genet 125(4):749–758

Zhang Z, Ersoz E, Lai C-Q, Todhunter RJ, Tiwari HK, Gore MA, Bradbury PJ, Yu J, Arnett DK, Ordovas JM, Buckler ES (2010) Mixed linear model approach adapted for genome-wide association studies. Nat Genet 42(4):355–360

Zhao KY, Aranzana MJ, Kim S, Lister C, Shindo C, Tang CL, Toomajian C, Zheng HG, Dean C, Marjoram P, Nordborg M (2007) An Arabidopsis example of association mapping in structured samples. PLoS Genet 3(1)

Zhao K, Tung C-W, Eizenga GC, Wright MH, Ali ML, Price AH, Norton GJ, Islam MR, Reynolds A, Mezey J, McClung AM, Bustamante CD, McCouch SR (2011) Genome-wide association mapping reveals a rich genetic architecture of complex traits in *Oryza sativa*. Nat Commun 2:467

Zhu C, Yu J (2009) Nonmetric multidimensional scaling corrects for population structure in association mapping with different sample types. Genetics 182(3):875–888

Zhu C, Gore M, Buckler ES, Yu J (2008) Status and Prospects of Association Mapping in Plants. Plant Genome J 1(1):5

The Silent Assassins: Informatics of Plant Viral Silencing Suppressors

Sohini Gupta, Sayak Ganguli, and Abhijit Datta

Abstract

The arms race continues – silencing suppressors have emerged as the silent weapon of counter defense against the indigenous plant RNA silencing machinery. Several reports have identified their presence and in higher plants, these proteins are involved in the inhibition of RNA silencing, which reduces the defense of plants against the pathogen, as well as play important roles for regulation of gene expression during growth and development. These viral silencing suppressors interfere with the various steps of the RNA silencing pathways either by binding to double-stranded RNA or by interfering with the effector proteins and altering their conformation. The informatics challenge lies in the elucidation and unraveling of the diverse structures of these proteins and in understanding and conceptualizing the structure paradigm of these proteins. The basic methods of informatics such as molecular modeling and dynamic simulations would be effective tools for the design and implementation of the above mandates, but advanced techniques such as SVM classification, network analyses, and a systems level approach including phylogenetic analyses should provide a robust framework for such elucidations. This chapter provides a brief overview of the viral silencing suppressors that have been identified so far and provides insights on how they have been studied in the wet lab. Apart from this it focuses on the possibilities of the bioinformatics analyses of the viral silencing suppressors with specific case studies and conceptualizes a future framework for application-oriented use of these silent assassins.

Keywords

Viral silencing suppressor (VSR/VSS) • Bioinformatics • Dynamic simulation • Molecular modeling • Support vector machines (SVM)

S. Gupta • S. Ganguli • A. Datta (✉)
DBT Center for Bioinformatics, Presidency University,
Kolkata, West Bengal, India
e-mail: abhijit_datta21@yahoo.com

K.K. P.B. et al. (eds.), *Agricultural Bioinformatics*,
DOI 10.1007/978-81-322-1880-7_2, © Springer India 2014

1 Introduction

RNA silencing is one of the most potent weapons of the host machinery to fight against the invading viruses (Bies-Etheve et al. 2009). One of the key features of this plant adaptive immune system is the production of virus-derived siRNAs (vsiRNAs) at elevated levels during the phase of infection (Azevedo et al. 2010; Bisaro 2006; Buhler et al. 2006). There have been reports of these vsiRNAs being associated with the AGO1 protein, a key player in the RNA which induces silencing complex (Burgyan 2008; Chao et al. 2005). To equip themselves against these vsiRNAs the viruses have evolved viral silencing suppressors – proteins which serve as counter defense against the vsiRNAs produced by the plants. The importance of these proteins is further exemplified by the fact that if these suppressors are inactivated, then the plant recovers from viral infections probably by using its own defense mechanisms in a much better way (Bies-Etheve et al. 2009; Chen et al. 2008; Csorba et al. 2007).

Three major steps can be identified in the process of antiviral silencing:

(1) Identification and processing of the viral RNAs to viral siRNAs
(2) Increase in expression of these vsiRNAs – a process often referred to as amplification
(3) Targeting of the viral RNAs by incorporating them into an RISC

The key players in the recognition process are plant DICERS which are specialized RNAse III enzymes involved in the sensing of double-stranded or structured RNAs (Aliyari and Ding 2009; Behm-Ansmant et al. 2006; Csorba et al. 2009). Once the viral RNAs have been recognized, DICERS process them to vsiRNAs (Akbergenov et al. 2006; Azevedo et al. 2010; Baumberger et al. 2007; Bisaro 2006; Buhler et al. 2006; Csorba et al. 2010; Cuellar et al. 2009).

Plants have two unique types of these vsiRNAs – primary RNAs which are processed products of an initial trigger RNA by Dicer and secondary siRNAs which are processed by an RDR enzyme (Azevedo et al. 2010; Brosnan and Voinnet 2009; Buhler et al. 2006; Csorba et al. 2010; Cuperus et al. 2010; Deleris et al. 2006; Diaz-Pendon et al. 2007). Studies in the model plant *Arabidopsis* have shown that Dicer-Like 4 (DCL4) and Dicer-Like 2 (DCL2) are the most important DCLs which orchestrate the production of vsiRNAs 21 or 22 nt long by the dicing of ds or viral hairpins (Azevedo et al. 2010; Cuellar et al. 2009; Ding 2010; Ding and Voinnet 2007). RDR1, RDR2, and RDR6 are the enzymes which are responsible for the amplification of vsiRNA production as they are also able to convert aberrant viral ssRNAs lacking quality control to dsRNAs which then serve as substrates for vsiRNA production which are secondary in nature (Brosnan and Voinnet 2009; Buhler et al. 2006; Diaz-Pendon et al. 2007; Donaire et al. 2008; Donaire et al. 2009). Once generated the vsiRNAs are loaded into specialized effector complexes containing Argonaute (AGO) proteins **which** are subsequently guided to their RNA targets (Akbergenov et al. 2006; Dunoyer et al. 2010a, b). Generally the loading of a particular vsiRNA to its AGO partnered complex is controlled by the terminal nucleotides (Eagle et al. 1994). Reports have shown that AGO1 and AGO7 function in unison to ensure the removal of viral RNAs and that AGO7 often acts as a "surrogate slicer" when AGO1 is absent.

2 Types of Viral Silencing Suppressors

The viral silencing suppressors are evolutionarily nascent molecules having great diversity but surprisingly lacking sequence homology. Studies with VSRs have demonstrated that almost all the steps of the RNA silencing mechanisms and its corresponding effector complexes may be targeted by these suppressor proteins. The possible stages and complexes under the VSR attack are as follows (the details are summarized in the figure):

1. Viral RNA recognition
2. Dicing
3. RISC assembly
4. RNA targeting and amplification

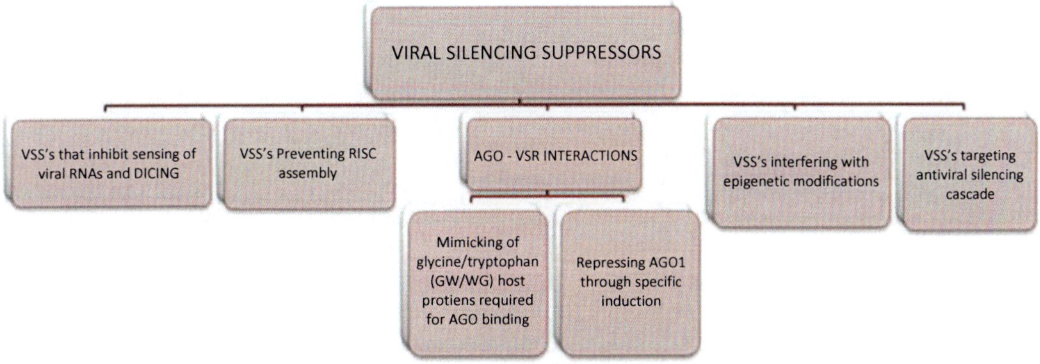

Fig. 1 The various types of identified viral silencing suppressors based on their mode of action

Still to date we are at a very minimum stage of our understanding of VSR biology – production evolution control and action. Many of the VSRs that have been identified are multifunctional – apart from their silencing suppressor activity, they have been reported to perform several structural and functional roles such as functioning as coat proteins, replicases, movement proteins, helper components for viral transmission, proteases, or transcriptional regulators (Fig. 1, Tables 1 and 2).

3 Study of Viral Silencing Suppressors

Over the years as interest has increased regarding the viral silencing suppressors, model assay systems have been developed and formulated so as to identify the expression and provide clues toward the possible modes of action of these proteins. Here five different assay systems would be discussed which have been the most commonly used for studying viral silencing suppressors (Fig. 2, Table 3).

4 The Bioinformatics Approach

Bioinformatics presents a very interesting alternative to the study of viral silencing suppressors. Most of the wet lab experiments have focused on the identification of the basic mechanisms of function and expression levels of the different proteins; structure elucidation and study of interactions of these proteins through simulation techniques offer a feasible alternative toward the understanding of the mode of actions of these suppressors as well as provide insights on the evolutionary history of these unique proteins. The various approaches of bioinformatics and their possible outcomes are summarized in Fig. 3.

Control of viral silencing suppressors also presents an interesting problem for bioinformatics. Designing of siRNAs targeting these suppressor mRNAs can also be utilized using Bayesian and extreme value distributions (Ganguli et al. 2011a). The analyses of regulatory elements in the sequences and prediction of probable functions of such identified elements should provide insights into the basic biology of these suppressors. Such protocols can utilize methods as the ones described in Ganguli et al. (2010 and 2011b).

5 Case Study

Geminiviruses possess much smaller genomes than most of the other viruses. According to their genome organizations, type of insect vectors, and biological properties, i.e., host range, the Geminiviridae family can be classified into four genera (Fauquet et al. 2003). Their genomes comprise of one or two circular single-stranded DNA molecules, each ranging from 2.6 to 2.8 kb, and they are packaged as minichromosomes (Pilartz and Jeske 1992). The two genomic components are DNA-A and DNA-B. Six genes are encoded

Table 1 Host range and geographical distribution of VSRs

Serial No.	Geographical distribution-country	Name of the host plant range	Name of the virus	Name of the VSRs	References
1.	Africa-Cameroon	Cassava (*Manihot esculenta*)	African cassava mosaic virus	AC4	Chellappan et al. (2005)
2.	Africa-Kenya	Cassava (*Manihot esculenta*)	African cassava mosaic virus	AC2	Voinnet et al. (1999)
3.	Sri Lanka	Cassava (*Manihot esculenta*)	Sri Lankan cassava mosaic virus	AC4	Vanitharani et al. (2004)
4.	Brazil	Tomato (*Solanum lycopersicum*)	Tomato golden mosaic virus	AC2	Wang et al. (2005)
5.	Australia	Tomato (*Solanum lycopersicum*)	Tomato leaf curl virus	C2	Ikegami et al. (2011)
6.	Italy, California, and Spain	*Citrus sinensis* and *C. aurantifolia*	Citrus tristeza virus	P23; P20	Ikegami et al. (2011)
7.	Indonesia	*Solanum lycopersicum*, *Nicotiana benthamiana*	Tomato leaf curl Java virus	V2 and C2	Fukunaga et al. (2009) and Glick et al. (2008)
8.	Singapore and Thailand	*Ageratum conyzoides*	Ageratum yellow vein virus	C2	Ikegami et al. (2011)
9.	Israel, Australia	*Solanum lycopersicum*	Tomato yellow leaf curl virus	V2	Fukunaga et al. (2009) Glick et al. (2008)
10.	China	*Solanum lycopersicum*	Tomato yellow leaf curl China betasatellite	βC1	Ikegami et al. (2011)
11.	USA	*Beta vulgaris*	Beet curly top virus	C2	Wang et al. (2005)
12.	Worldwide in temperate regions	*Brassica campestris*, *Arabidopsis thaliana*	Cauliflower mosaic virus	P6	Haas et al. (2008) Love et al. (2007)
13.	Italy	*Epipremnum aureum*	Pothos latent virus	P14	Merai et al. (2005)
14.	Australia, El Salvador, Fiji, the Solomon Islands, Thailand, and the USA	*Hibiscus rosa-sinensis*; *Abelmoschus manihot*	Hibiscus chlorotic ring spot virus	CP	Merai et al. (2005) Deleris et al. (2006)
15.	Denmark, Germany, the Netherlands, and the UK	*Pelargonium zonale*; *Chenopodium quinoa*; *Nicotiana clevelandii*	Pelargonium flower break virus	CP	Martínez-Turiño and Hernández (2009)
16.	Japan	*Pyrus malus*	Apple latent virus	VP20; Vp25; Vp24	Ikegami et al. (2011)
17.	Australia (Tasmania), France, Japan, the Netherlands, the USA (California)	*Beta vulgaris*; *Chenopodium foliosum* or *C. capitatum*; *Claytonia perfoliata*; *Capsella bursa-pastoris*; *Taraxacum officinale*; *Conium maculatum*	Beet yellows virus	P21	Lakatos et al. (2006)
18.	Africa	*Ipomoea batatas*	Sweet potato chlorotic stunt virus	P22	Ikegami et al. (2011)

(continued)

Table 1 (continued)

Serial No.	Geographical distribution-country	Name of the host plant range	Name of the virus	Name of the VSRs	References
19.	Uganda, Tanzania, Peru, Israel	*Ipomoea batatas*	Sweet potato chlorotic stunt virus	RNase3	Ikegami et al. (2011)
20.	Distributed worldwide	*Cucumis sativus*	Cucumber mosaic virus	2b	Goto et al. (2007)
					Guo and Ding (2002)
					Mayers et al. (2000)
21.	Asia, Africa, N. America, C. America, the Caribbean, S. America	*Amaranthus spinosus*, *Chenopodium album*, *Arctium lappa*, *Crotalaria mucronata*, *Solanum nigrum*, and many more	Tomato spotted wilt virus	NSs	Takeda et al. (2005)
22.	Japan, China, Korea	*Oryza sativa*	Rice dwarf virus	Pns10	Chao et al. (2005)
23.	Former Czechoslovakia, Sweden, Poland, UK	*Trifolium praetense*, *Medicago sativa*, *Melilotus officinalis*, *Trifolium repens*, *Chenopodium quinoa*, *Nicotiana clevelandii*	Red clover necrotic mosaic virus	P27; P88	Takeda et al. (2005)
24.	USA, Japan, Italy	*Triticum aestivum*, *Hordeum vulgare*, *Secale cereale*, *Bromus commutatus*	Soilborne wheat mosaic virus	19 K	Ikegami et al. (2011)
25.	Probably distributed worldwide. The Pacific Region, Australia, China, the UK, the USA, and the former USSR	*Hordeum vulgare*, *Triticum aestivum*	Barley stripe mosaic virus	γb	Merai et al. (2006)
26.	Israel, Spain, Jordan, Turkey, Sudan	*Cucumis sativus*, *Citrullus vulgaris*	Cucumber vein yellowing virus	P1b	Kasschau et al. (2003)
27.	Probably distributed worldwide	*Beta vulgaris*, *Lactuca sativa*, *Spinacia oleracea*, *Raphanus sativus*	Beet western yellows virus	P0	Pazhouhandeh et al. (2006)
28.	Probably distributed worldwide especially in Europe, New Zealand, Middle East, North America, and Japan	*Solanum tuberosum*; *Nicotiana tabacum*; *Solanum lycopersicum*	Potato virus Y	HC-Pro	Ebhardt et al. (2005)
29.	Probably distributed worldwide, especially in Western USA and Canada	*Trifolium repens*	Tulip virus X	TGBp1	Ikegami et al. (2011)
30.	Probably distributed worldwide	*Nicotiana tabacum*	Tobacco mosaic virus	122 K	Fukunaga and Doudna (2009)

Table 2 Major viral silencing suppressors in India

Serial No.	Name of the host plant	Name of the virus	Name of the VSRs
1.	Cassava	Indian cassava mosaic virus	AC2
2.	Mung bean (*Vigna radiata*; *Vigna mungo*)	Mung bean yellow mosaic virus	AC2
3.	*Ageratum conyzoides*	Ageratum yellow vein virus	C2
4.	*Abelmoschus esculentus*	Bhendi yellow vein mosaic virus	βC1
5.	*Vigna unguiculata*	Cowpea mosaic virus	S-CP
6.	*Arachis hypogaea*	Peanut clump virus	P15

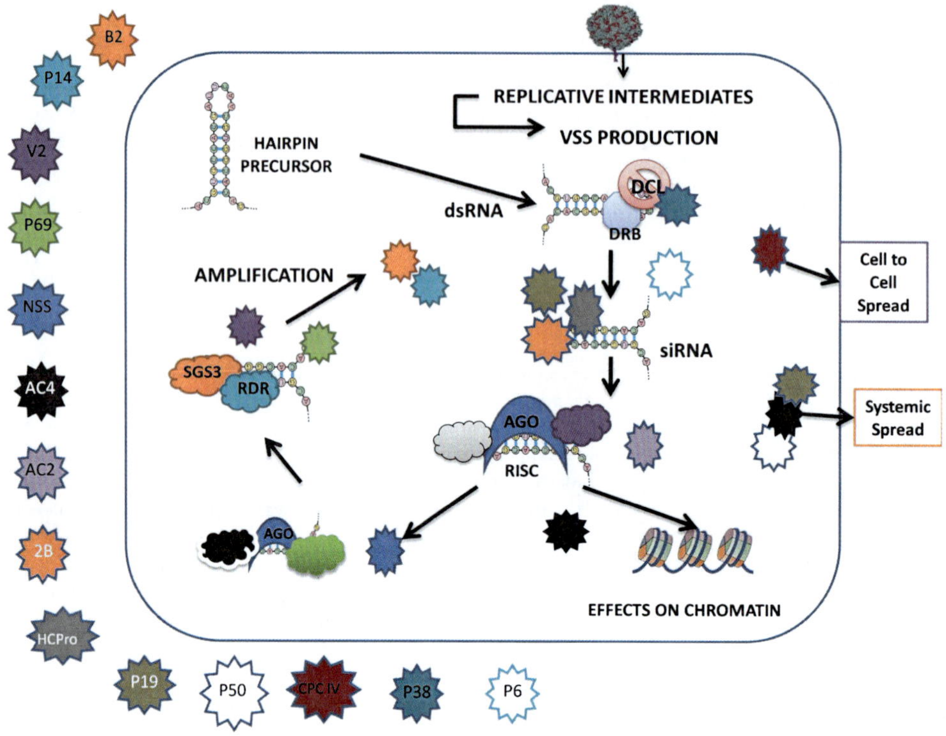

Fig. 2 Schematic representation of the possible modes of action of the most common viral silencing suppressors

by DNA-A: two in the virus sense, *AV1*, which codes for the coat protein (CP), and *AV2*, whose specific role is not yet elucidated, in the complementary sense, rest of the four genes reside, which comprise of *AC1* that codes for replication-associated protein (Rep); *AC2* and *AC3*, which code for the transcription protein and the replication enhancer protein; and AC4. Two genes, BV1 and BC1, are coded by DNA-B, which in turn codes for the movement and the nuclear shuttle proteins, respectively. AC4 is a small protein encoded by bipartite geminiviruses and has been reported to have divergent biological functions. Evidences showed that it plays an important role in the infectivity of several bipartite geminiviruses (Elmer et al. 1988; Etessami et al. 1991; Hoogstraten et al. 1996; Pooma and Petty 1996; Sung and Coutts 1995). According to Vanitharani and associates (2004), African cassava mosaic virus (ACMV) AC4 suppresses posttranscriptional gene silencing (PTGS). Single-stranded DNA genome is comprised by geminiviruses, and during their replication cycle usually they possess no double-stranded RNA phase. It has been reported that they activate PTGS probably by the occurrence of overlapping transcripts with the production of virus-derived small interfering RNA (siRNAs) in infected plants (Chellappan et al. 2004; Vanitharani et al. 2005). One of the common virus-encoded PTGS-suppressor proteins that have been identified is the AC4 protein of African cassava mosaic virus Cameroon Strain and Sri Lankan cassava mosaic virus (SLCMV). The possible mechanism of silencing suppression on the miRNA pathway and the subsequent disease development by the geminivirus-encoded silencing suppressors is still unknown. However, later on it has been identified that AC4 may selectively bind the single-stranded sRNAs including miRNAs (Chellappan et al. 2005).

In this work we have generated a model of the AC4 protein of the African cassava mosaic virus using molecular modeling and threading. Analyses of backbone geometry and conformation taking into consideration the torsion angles of the Ramachandran plot showed that there

Table 3 Assay systems for the identification and study of viral silencing suppressors

Serial No.	Name	Method	Limitation	Related suppressors	References
1.	Agro infiltration assay	The suspect virus-encoded gene is first cloned into an appropriate shuttle vector that is transferred to a strain of *Agrobacterium tumefaciens*	Low	2b gene of *cucumber mosaic virus* (CMV), the coat protein (CP) of *citrus tristeza virus* (CTV)	Brigneti et al. (1998) and Qu et al. (2008)
		Suspensions constructed for delivery of the candidate gene and a reporter gene (green fluorescent protein, GFP, or β-glucuronidase, GUS) into plant cells are usually mixed and pressure-infiltrated into leaves of *Nicotiana benthamiana* plants	Suppression activity		
2.	Silencing reversal	Examines whether the expression of a silenced transgene can be restored upon the infection of a particular virus of interest (silencing reversal)	General problems as there are multiple variables such as plant age and growth conditions which can affect the outcome of the experiments	AC2, AC4	Voinnet et al. (1999) and Schnettler et al. (2009)
					Silhavy and Burgyan (2004)
					Silhavy et al. (2002)
		The initial assay involves observing if the "reporter" transgene silencing is reversed by the viral infection and then it is extended to other genes	In addition, the approach would fail to detect those viral suppressors that are unable to reverse RNA silencing		Chellappan et al. (2005)
3.	Transient expression assays	This involves crossing the transgenic plants expressing the test protein with plants carrying a reporter gene that is regulated by posttranscriptional silencing	Time-consuming	P1/HC-Pro protein of *tobacco etch potyvirus* (TEV)	Anandalakshmi et al. (1998)
					Kasschau et al. (2003)
		Progeny is examined for the expression of the reporter gene as a consequence of being released from the silencing			

(continued)

Table 3 (continued)

Serial No.	Name	Method	Limitation	Related suppressors	References
4.	Movement inhibition by grafting	Plant parts are grafted to evaluate the ability of virus-encoded silencing suppressors to hinder the movement of systemic RNA silencing signals. First used to demonstrate the movement and spread of a systemic silencing signal	Works well only with large plants such as tobacco and *N. benthamiana*	p25 of PVX	Szittya et al. (2010) Takeda et al. (2005) Till and Ladurner (2007)
5.	Fluorescence imaging	Fluorescence imaging of whole intact leaves has been utilized for spatial and quantitative observation of viral suppressor efficiency in plants. This suppressor assay demonstrates that plant viral suppressors greatly enhance transient GFP expression	Control of GFP fluorescence short to long term (range 7–21 dpi)	Beet mild yellowing virus (BMYV-IPP) P0 or plum pox virus (PPV) HC-Pro	Trinks et al. (2005) Vaistij and Jones (2009) Vanitharani et al. (2005)

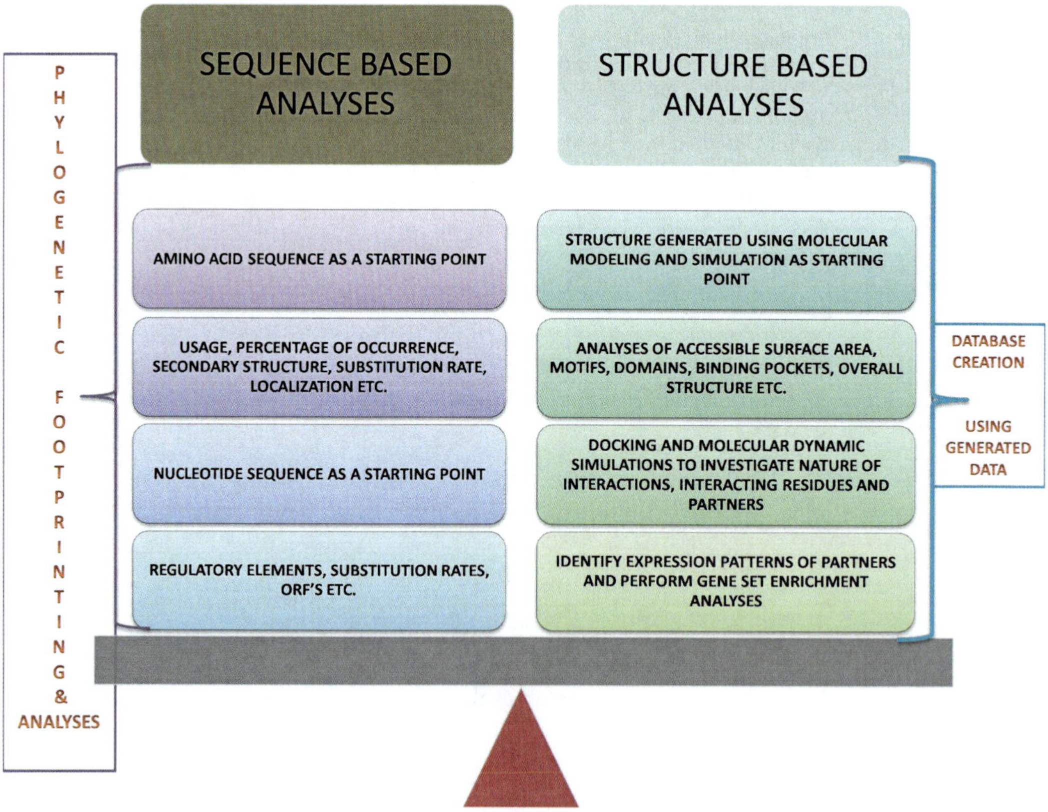

Fig. 3 Bioinformatics approaches for the study of plant viral silencing suppressors

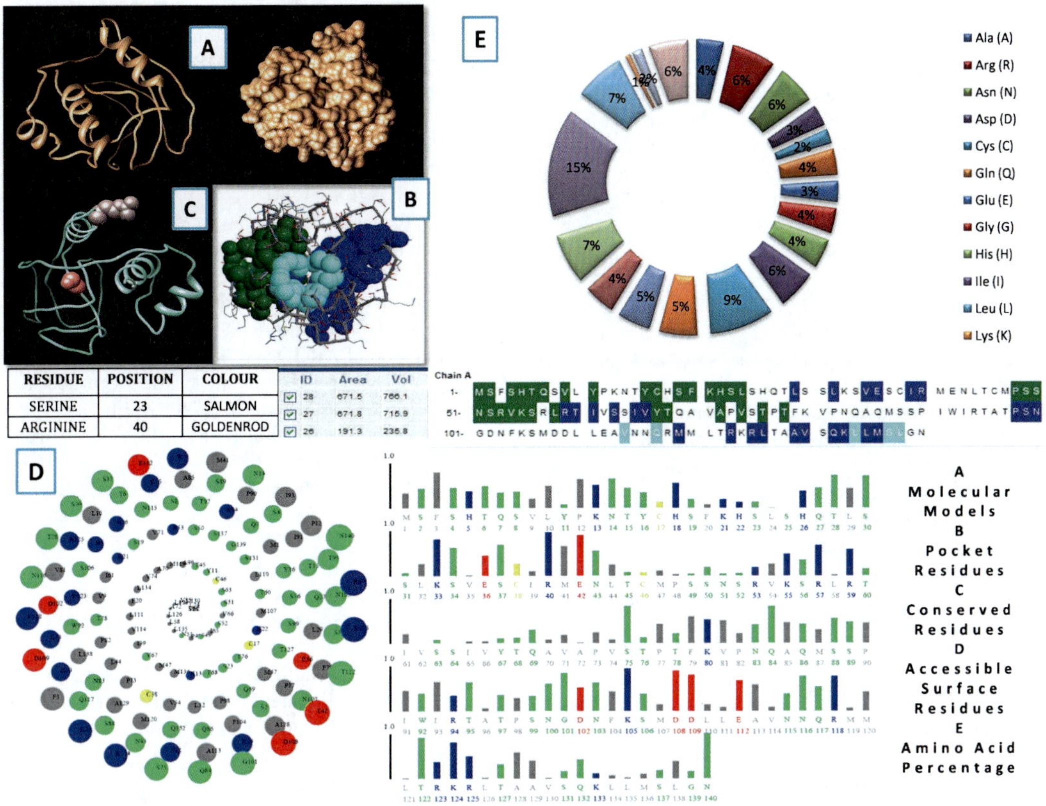

Fig. 4 Analyses of AC4 of African cassava mosaic virus

were no outliers indicative of the fact that the modeling was successful. Once the model was generated, the accessible surface area of the molecule was analyzed, and it was found that hydrophilic residues such as glutamic acid and lysine were present predominantly in the outer layer of the spiral plot. The color scheme used for the analyses is as follows:

1. Blue → Positively charged residues (R, K, H)
2. Red → Negatively charged residues (D, E)
3. Green → Polar uncharged residues (G, N, Y, Q, S, T, W)
4. Yellow → Cysteine
5. Gray → Hydrophobic residues (All others)

The identification of the binding pockets of the protein was performed using the CASTp server which calculates the Delaunay triangulation score and generates the area and volume of the largest pockets (displayed in blue, green, and cyan – of Fig. 4).

6 Conclusion

Viral silencing suppressors are thus resultants of the gene for gene evolution model which has become established in context of plant-pathogen interaction. Further studies need to be performed so as to gain insights on the actual evolutionary plexus of these silent assassins. In silico approaches along with conventional methods of experimental analyses should be explored and utilized in the future so as to elucidate the structure, function, evolution, and interactomics of these proteins which would enable us to utilize and manipulate these molecules to our benefit and help the agriculture fraternity.

Acknowledgment The authors acknowledge the contributions of the BTBI program of the Department of Biotechnology, Government of India.

References

Akbergenov R, Si-Ammour A, Blevins T, Amin I, Kutter C, Vanderschuren H, Zhang P, Gruissem W, Meins F Jr, Hohn T, Pooggin M (2006) Molecular characterization of geminivirus-derived small RNAs in different plant species. Nucleic Acids Res 34:462–471

Aliyari R, Ding SW (2009) RNA-based viral immunity initiated by the Dicer family of host immune receptors. Immunol Rev 227:176–188

Anandalakshmi R, Pruss GJ, Ge X, Marathe R, Mallory AC, Smith TH, Vance VB (1998) A viral suppressor of gene silencing in plants. Proc Natl Acad Sci U S A 95:13079–13084

Azevedo J, Garcia D, Pontier D, Ohnesorge S, Yu A, Garcia S, Braun L, Bergdoll M, Hakimi MA, Lagrange T, Voinnet O (2010) Argonaute quenching and global changes in Dicer homeostasis caused by a pathogen-encoded GW repeat protein. Genes Dev 24:904–915

Baumberger N, Tsai CH, Lie M, Havecker E, Baulcombe DC (2007) The Polerovirus silencing suppressor P0 targets ARGONAUTE proteins for degradation. Curr Biol 17:1609–1614

Behm-Ansmant I, Rehwinkel J, Doerks T, Stark A, Bork P, Izaurralde E (2006) mRNA degradation by miRNAs and GW182 requires both CCR4: NOTdeadenylase and DCP1:DCP2 decapping complexes. Genes Dev 20:1885–1898

Bies-Etheve N, Pontier D, Lahmy S, Picart C, Vega D, Cooke R, Lagrange T (2009) RNA-directed DNA methylation requires an AGO4-interacting member of the SPT5 elongation factor family. EMBO Rep 10:649–654

Bisaro DM (2006) Silencing suppression by geminivirus proteins. Virology 344:158–168

Brigneti G, Voinnet O, Li WX, Ji LH, Ding SW, Baulcombe DC (1998) Viral pathogenicity determinants are suppressors of transgene silencing in *Nicotiana benthamiana*. EMBO J 17:6739–6746

Brosnan CA, Voinnet O (2009) The long and the short of noncoding RNAs. Curr Opin Cell Biol 21:416–425

Bühler M, Verdel A, Moazed D (2006) Tethering RITS to a nascent transcript initiates RNAi- and heterochromatin-dependent gene silencing. Cell 125:873–886

Burgyan J (2008) Role of silencing suppressor proteins. Methods Mol Biol 451:69–79

Chao JA, Lee JH, Chapados BR, Debler EW, Schneemann A, Williamson JR (2005) Dual modes of RNA-silencing suppression by Flock House virus protein B2. Nat Struct Mol Biol 12:952–957

Chapman EJ, Prokhnevsky AI, Gopinath K, Dolja VV, Carrington JC (2004) Viral RNA silencing suppressors inhibit the microRNA pathway at an intermediate step. Genes Dev 18:1179–1186

Chellappan P, Vanitharani R, Fauquet CM (2005) MicroRNA-binding viral protein interferes with Arabidopsis development. Proc Natl Acad Sci U S A 102:10381–10386

Chen HY, Yang J, Lin C, AdamYuan Y (2008) Structural basis for RNA-silencing suppression by Tomato aspermy virus protein 2b. EMBO Rep 9:754–760

Csorba T, Bovi A, Dalmay T, Burgyán J (2007) The p122 subunit of Tobacco Mosaic Virus replicase is a potent silencing suppressor and compromises both small interfering RNA- and microRNA-mediated pathways. J Virol 81:1768–11780

Csorba T, Pantaleo V, Burgyán J (2009) RNA silencing: an antiviral mechanism. Adv Virus Res 75:35–71

Csorba T, Lózsa R, Hutvágner G, Burgyán J (2010) Polerovirus protein P0 prevents the assembly of small RNA-containing RISC complexes and leads to degradation of ARGONAUTE1. Plant J 62:463–472

Cuellar WJ, Kreuze JF, Rajamaki ML, Cruzado KR, Untiveros M, Valkonen JPT (2009) Elimination of antiviral defense by viral RNase III. Proc Natl Acad Sci U S A 106:10354–10358

Cuperus JT, Carbonell A, Fahlgren N, Garcia-Ruiz H, Burke RT, Takeda A, Sullivan CM, Gilbert SD, Montgomery TA, Carrington JC (2010) Unique functionality of 22-nt miRNAs in triggering RDR6-dependent siRNA biogenesis from target transcripts in Arabidopsis. Nat Struct Mol Biol 17:997–1003

Deleris A, Gallego-Bartolome J, Bao J, Kasschau KD, Carrington JC, Voinnet O (2006) Hierarchical action and inhibition of plant Dicer-like proteins in antiviral defense. Science 313:68–71

Diaz-Pendon JA, Li F, Li WX, Ding SW (2007) Suppression of antiviral silencing by cucumber mosaic virus 2b protein in Arabidopsis is associated with drastically reduced accumulation of three classes of viral small interfering RNAs. Plant Cell 19:2053–2063

Ding SW (2010) RNA-based antiviral immunity. Nat Rev Immunol 10:632–644

Ding SW, Voinnet O (2007) Antiviral immunity directed by small RNAs. Cell 130:413–426

Donaire L, Barajas D, Martínez-García B, Martínez-Priego L, Pagán I, Llave C (2008) Structural and genetic requirements for the biogenesis of tobacco rattle virus-derived small interfering RNAs. J Virol 82:5167–5177

Donaire L, Wang Y, Gonzalez-Ibea D, Mayer KF, Aranda MA, Llave C (2009) Deep-sequencing of plant viral small RNAs reveals effective and widespread targeting of viral genomes. Virology 392:203–214

Dunoyer P, Brosnan CA, Schott G, Wang Y, Jay F, Alioua A, Himber C, Voinnet O (2010a) An endogenous, systemic RNAi pathway in plants. EMBO J 29:1699–1712

Dunoyer P, Schott G, Himber C, Meyer D, Takeda A, Carrington JC, Voinnet O (2010b) Small RNA duplexes function as mobile silencing signals between plant cells. Science 328:912–916

Eagle PA, Orozco BM, Hanley-Bowdoin L (1994) A DNA sequence required for geminivirus replication also mediates transcriptional regulation. Plant Cell 6:1157–1170

Ebhardt HA, Thi EP, Wang MB, Unrau PJ (2005) Extensive 3′ modification of plant small RNAs is modulated by helper component-proteinase expression. Proc Natl Acad Sci U S A 102:13398–13403

Elmer JS, Brand L, Sunter G, Gardiner WE, Bisaro DM, Rogers SG (1988) Genetic analysis of the tomato golden mosaic virus. II. The product of the AL1 coding sequence is required for replication. Nucleic Acids Res 16:7043–7060

Etessami P, Saunders K, Watts J, Stanley J (1991) Mutational analysis of complementary-sense genes of African cassava mosaic virus DNA A. J Gen Virol 72:1005–1012

Fauquet CM, Bisaro DM, Briddon RW, Brown JK, Harrison BD, Rybicki EPD, Stenger C, Stanley J (2003) Revision of taxonomic criteria for species demarcation in the family Geminiviridae, and an updated list of begomovirus species. Arch Virol 148:405–421

Fukunaga R, Doudna JA (2009) dsRNA with 5′ overhangs contributes to endogenous and antiviral RNA silencing pathways in plants. EMBO J 28:545–555

Fusaro AF, Matthew L, Smith NA, Curtin SJ, Dedic-Hagan J, Ellacott GA, Watson JM, Wang MB, Brosnan C, Carroll BJ, Waterhouse PM (2006) RNA interference-inducing hairpin RNAs in plants act through the viral defence pathway. EMBO Rep 7:1168–1175

Ganguli S, Dey SK, Dhar P, Basu P, Roy P, Datta A (2010) Catalytic RNA world relics in Dicer RNAs. Int J Genet 2:8–17

Ganguli S, De M, Datta A (2011a) Analyses o argonaute–microrna interactions in Zea mays. Int J Comput Biol 1:32–34.

Ganguli S, Mitra S, Datta A (2011b) Antagomirbase: a putative antagomir database. Bioinformation 7:41–43.

Garcia-Ruiz H, Takeda A, Chapman EJ, Sullivan CM, Fahlgren N, Brempelis KJ, Carrington JC (2010) Arabidopsis RNA-dependent RNA polymerases and dicer-like proteins in antiviral defense and small interfering RNA biogenesis during Turnip Mosaic Virus infection. Plant Cell 22:481–496

Glick E, Zrachya A, Levy Y, Mett A, Gidoni D, Belausov E, Citovsky V, Gafni Y (2008) Interaction with host SGS3 is required for suppression of RNA silencing by tomato yellow leaf curl virus V2 protein. Proc Natl Acad Sci U S A 105:157–161.

Goto K, Kobori T, Kosaka Y, Natsuaki T, Masuta C (2007) Characterization of silencing suppressor 2b of cucumber mosaic virus based on examination of its small RNA binding abilities. Plant Cell Physiol 48:1050–1060

Guo HS, Ding SW (2002) A viral protein inhibits the long range signaling activity of the gene silencing signal. EMBO J 21:398–407

Haas G, Azevedo J, Moissiard G, Geldreich A, Himber C, Bureau M, Fukuhara T, Keller M, Voinnet O (2008) Nuclear import of CaMV P6 is required for infection and suppression of the RNA silencing factor DRB4. EMBO J 27:2102–2112

Hamilton AJ, Baulcombe DC (1999) A species of small antisense RNA in posttranscriptional gene silencing in plants. Science 286:950–952

Hoogstraten RA, Hanson SF, Maxwell DP (1996) Mutational analysis of the putative nicking motif in the replication-associated protein (AC1) of bean golden mosaic geminivirus. Mol Plant-Microbe Interact 9:594–599

Ikegami M, Kon T, Sharma P (2011) RNA silencing and viral encoded silencing suppressors. In: RNAi technology. CRC Press, Enfield, pp 209–240

Kasschau KD, Carrington JC (1998) A counter defensive strategy of plant viruses: suppression of posttranscriptional gene silencing. Cell 95:461–470

Krizan KA, Carrington JC (2003) P1/HC-Pro, a viral suppressor of RNA silencing, interferes with Arabidopsis development and miRNA function. Dev Cell 4:205–217.

Kumakura N, Takeda A, Fujioka Y, Motose H, Takano R, Watanabe Y (2009) SGS3 and RDR6 interact and colocalize in cytoplasmic SGS3/RDR6-bodies. FEBS Lett 583:1261–1266

Lakatos L, Csorba T, Pantaleo V, Chapman EJ, Carrington JC, Liu YP, Dolja VV, Calvino LF, Lo´pez-Moya JJ, Burgya´n J (2006) Small RNA binding is a common strategy to suppress RNA silencing by several viral suppressors. EMBO J 25:2768–2780.

Llave C (2010) Virus-derived small interfering RNAs at the core of plant–virus interactions. Trends Plant Sci 15:701–707

Love AJ, Laird J, Holt J, Hamilton AJ, Sadanandom A, Milner JJ (2007) Cauliflower mosaic virus protein P6 is a suppressor of RNA silencing. J Gen Virol. 88:3439–3444.

Lózsa R, Csorba T, Lakatos L, Burgyán J (2008) Inhibition of 3′ modification of small RNAs in virus-infected plants require spatial and temporal co-expression of small RNAs and viral silencing-suppressor proteins. Nucleic Acids Res 36:4099–4107

Mayers CN, Palukaitis P, Carr JP (2000) Subcellular distribution analysis of the cucumber mosaic virus 2b protein. J Gen Virol 81:219–226

Mayo MA, Ziegler-Graff V (1996) Molecular biology of luteoviruses. Adv Virus Res 46:413–460

Mérai Z, Kerényi Z, Kertész S, Magna M, Lakatos L, Silhavy D (2006) Double-stranded RNA binding may be a general plant RNA viral strategy to suppress RNA silencing. J Virol 80:5747–5756

Mi S, Cai T, Hu Y, Chen Y, Hodges E, Ni F, Wu L, Li S, Zhou H, Long C, Chen S, Hannon GJ, Qi Y (2008) Sorting of small RNAs into Arabidopsis argonaute complexes is directed by the 5′ terminal nucleotide. Cell 133:116–127

Pazhouhandeh M, Dieterle M, Marrocco K, Lechner E, Berry B, Brault V, Hemmer O, Kretsch T, Richards KE, Genschik P, Ziegler-Graff V (2006) F-box-like domain in the polerovirus protein P0 is required for silencing suppressor function. Proc Natl Acad Sci U S A 103:1994–1999.

Phillips JR, Dalmay T, Bartels D (2007) The role of small RNAs in abiotic stress. FEBS Lett 581:3592–3597

Pilartz M, Jeske H (1992) Abutilon mosaic geminivirus double stranded DNA is packed into minichromosomes. Virology 189:800–802

Pooma W, Petty IT (1996) Tomato golden mosaic virus open reading frame AL4 is genetically distinct from its C4 analogue in monopartite geminiviruses. J Gen Virol 77:1947–1951

Qu F, Ye X, Jack Morris T (2008) Arabidopsis DRB4, AGO1, AGO7, and RDR6 participate in a DCL4-initiated antiviral RNA silencing pathway negatively regulated by DCL1. Proc Natl Acad Sci U S A 105:14732–14737

Ratcliff F, Harrison BD, Baulcombe DC (1997) A similarity between viral defense and gene silencing in plants. Science 276:1558–1560

Schnettler E, Vries WD, Hemmes H, Haasnoot J, Kormelink R, Goldbach R, Berkhout B (2009) The NS3 protein of rice hoja blanca virus complements the RNAi suppressor function of HIV-1 Tat. EMBO Rep 10:258–263.

Silhavy D, Burgyan J (2004) Effects and side-effects of viral RNA silencing suppressors on short RNAs. Trends Plant Sci 9:76–83

Silhavy D, Molnár A, Lucioli A, Szittya G, Hornyik C, Tavazza M, Burgyán J (2002) A viral protein suppresses RNA silencing and binds silencing-generated, 21- to 25-nucleotide double-stranded RNAs. EMBO J 21:3070–3080

Sung YK, Coutts RH (1995) Pseudo recombination and complementation between potato yellow mosaic geminivirus and tomato golden mosaic geminivirus. J Gen Virol 76:2809–2815

Szittya G, Moxon S, Pantaleo V, Toth G, Rusholme Pilcher RL, Moulton V, Burgyan J, Dalmay T (2010) Structural and functional analysis of viral siRNAs. PLoS Pathog 6:e1000838.

Takeda A, Tsukuda M, Mizumoto H, Okamoto K, Kaido M, Mise K, Okuno T (2005) A plant RNA virus suppresses RNA silencing through viral RNA replication. EMBO J 24:3147–3157

Till S, Ladurner AG (2007) RNA Pol IV plays catch with Argonaute 4. Cell 131:643–645

Trinks D, Rajeswaran R, Shivaprasad PV, Akbergenov R, Oakeley EJ, Veluthambi K, Hohn T, Pooggin MM (2005) Suppression of RNA silencing by a geminivirus nuclear protein, AC2, correlates with trans activation of host genes. J Virol 79:2517–2527

Vaistij FE, Jones L (2009) Compromised virus-induced gene silencing in RDR6-deficient plants. Plant Physiol 149:1399–1407

Vanitharani R, Chellappan P, Pita JS, Fauquet CM (2004) Differential roles of AC2 and AC4 of cassava geminiviruses in mediating synergism and suppression of posttranscriptional gene silencing. J Virol 78:9487–9498

Vanitharani R, Chellappan P, Fauquet CM (2005) Geminiviruses and RNA silencing. Trends Plant Sci 10(3):144–151

Vaucheret H (2006) Post-transcriptional small RNA pathways in plants: mechanisms and regulations. Genes Dev 20:759–771

Voinnet O, Pinto YM, Baulcombe DC (1999) Suppression of gene silencing: a general strategy used by diverse DNA and RNA viruses of plants. Proc Natl Acad Sci U S A 96:14147–14152.

Wang H, Hao L, Shung CY, Sunter G, Bisaro DM (2003) Adenosine kinase is inactivated by geminivirusAL2 and L2 proteins. Plant Cell 15:3020–3032

Wang H, Buckley KJ, Yang X, Buchmann RC, Bisaro DM (2005) Adenosine kinase inhibition and suppression of RNA silencing by geminivirus AL2 and L2 proteins. J Virol 79:7410–7418.

Wang XB, Wua Q, Itoa T, Cilloc F, Lia WX, Chend X, Yub JL, Ding SW (2010) RNAi-mediated viral immunity requires amplification of virus-derived siRNAs in *Arabidopsis thaliana*. Proc Natl Acad Sci U S A 107:484–489

Tackling the Heat-Stress Tolerance in Crop Plants: A Bioinformatics Approach

Sudhakar Reddy Palakolanu, Vincent Vadez, Sreenivasulu Nese, and Kavi Kishor P.B.

Abstract

Plants are exposed to different types of environmental factors including heat stress that affect negatively various regular activities of the plant. Plants, as sessile organisms, must have developed efficient strategies of response to cope with and adapt to different types of abiotic stresses imposed by the adverse environment. Plant responses to environmental stress are complex and appear to be a difficult task to study in the classical plant-breeding program due to several technical limitations. The current knowledge of the regulatory network governing environmental stress responses is fragmentary, and an understanding of the damage caused by these environmental stresses or the plant's tolerance mechanisms to deal with stress-induced damages is far from complete. The emergence of the novel "omics" technologies from the last few years, such as genomics, proteomics, and metabolomics, is now allowing researchers to enable active analyses of regulatory networks that control abiotic stress

S.R. Palakolanu
International Crops Research Institute for the Semi-Arid Tropics (ICRISAT), Patancheru, Hyderabad 502 324, India

Leibniz Institute of Plant Genetics and Crop Plant Research (IPK), Corrensstrasse 03, D-06466 Gatersleben, Germany

V. Vadez
International Crops Research Institute for the Semi-Arid Tropics (ICRISAT), Patancheru, Hyderabad 502 324, India

S. Nese
Leibniz Institute of Plant Genetics and Crop Plant Research (IPK), Corrensstrasse 03, D-06466 Gatersleben, Germany

K.K. P.B. (✉)
Department of Genetics, Osmania University, Hyderabad 500 007, Andhra Pradesh, India
e-mail: pbkavi@yahoo.com

K.K. P.B. et al. (eds.), *Agricultural Bioinformatics*,
DOI 10.1007/978-81-322-1880-7_3, © Springer India 2014

responses. Recent advances in different omics approaches have been found greatly useful in understanding plant responses to abiotic stress conditions. Such analyses increase our knowledge on plant responses and adaptation to stress conditions and allow improving crop improvement programs including plant breeding. In this chapter, recent progresses on systematic analyses of plant responses to heat stress including genomics, proteomics, metabolomics, and phenomics and transgenic-based approaches to overcome heat stress are summarized.

Keywords
Omics • Phenomics • Molecular regulatory networks • NGS-based transcriptome analysis • Heat-shock response • Heat-shock proteins • Heat-shock element

Abbreviations

HSFs	Heat-Shock Transcription Factors
ROS	Reactive Oxygen Species
GEO	Gene Expression Omnibus
TAIR	The Arabidopsis Information Resource
NGS	Next-Generation Sequencing
GC-MS	Gas Chromatography Coupled to Mass Spectrometry
LC-MS	Liquid Chromatography Coupled to Mass Spectrometry
SGN	Sol Genomics Network

1 Introduction

Since plants are sessile in nature, they develop many physiological and molecular mechanisms to cope with different abiotic stresses. Plants started to emerge 1.5 billion years ago (Lehninger et al. 1993), and the evolutionary pressure has shaped plant responses to environmental fluctuations that minimize damage and ensure protection of cellular homeostasis. Heat stress is one of the main abiotic stresses that can limit the crop productivity drastically in the coming years due to global warming. High temperatures can be detrimental to all phases of plant development. Heat stress causes irreversible damage to plant function and development (Hall 2001). Heat stress affects a broad spectrum of cellular components and metabolism. The timing, duration, and severity of heat stress influence pollen-pistil interactions in crop plants (Snider and Oosterhuis 2011). To counter the effects of heat stress on cellular metabolism, plants and other organisms respond to changes in their ambient temperature by reprogramming the composition of certain transcripts, proteins, and metabolites. Heat stress leads to a series of phenotypical and genetical changes, creates osmotic imbalances, and produces ubiquitous and evolutionarily conserved proteins known as heat-shock proteins (Hsps) (Gupta et al. 2010). Stress responses involving extreme temperature result in excess production of reactive oxygen species (ROS), leading to oxidative damage and thus limiting the growth and productivity of agricultural crops. Genome-wide transcriptional profiles during temperature and oxidative stress revealed coordinated expression patterns and overlapping regulons in crop plants (Mittal et al. 2012). Therefore, understanding plant responses to heat stress is now thought to be one of the hottest topics in agricultural science. Major progress in this research field has come from the application of different bioinformatics/systems biology approaches. These high-throughput techniques have made it possible to analyze thousands of genes in one shot (Smita et al. 2013). With the introduction of bioinformatics tools, many heat-stress-inducible genes were identified from the huge genome databases, their promoter sequences were identified, and the putative functions of the genes were functionally characterized through transgenic approaches. This provides the information to understand the molecular mechanisms for improving heat tolerance in crops. Availability of these data sets publicly has broadened and deepened the view of heat-stress responses and tolerance not only in model plants but also in agricultural crops.

2 Bioinformatics Approaches

Recent advances in functional genomics have allowed us the use of different bioinformatics approaches such as transcriptomics (global gene expression), proteomics (protein profiling/modification), metabolomics (metabolite profiling), and phenomics to understand the complex molecular regulatory networks associated with stress adaptation and tolerance (Cramer et al. 2011). These technologies generate enormous amounts of information which has boosted up the field of bioinformatics, with thousands of new algorithms and software published every year. System-based approaches with a combination of multiple omics analyses have been an efficient tool to determine the global picture of cellular events which would increase our understanding of the complex molecular regulatory networks and find out the interacting partners associated with heat-stress adaptation and tolerance. The data collected from transcriptomics, proteomics, and metabolomics needs to be combined to achieve a better understanding of the plant as a system. In this context, different omics data should contribute greatly to the identification of key regulatory steps to characterize the pathway interactions. The integration of a wide spectrum of omics data sets from various plant species facilitates to promote translational research for future biotechnological applications in crop plants and also in fruit trees. These approaches demonstrate the power of systems biology for understanding the key cellular components underlying plant functions during temperature stress. Thus, cooperation between and beyond disciplines has a role to play in unraveling the intricacies associated with heat-stress adaptation in plants.

2.1 Transcriptomics

Transcriptomics is a powerful approach for studying the responses of plants in relation to their environment. The transcriptome consists of the entire set of transcripts that are expressed within a cell or organism at a particular developmental stage or under various environmental conditions. Recent transcriptomic studies have helped to provide a better understanding of plant response to different abiotic stresses like cold, high salinity, drought, high light intensity, hyperosmolarity, and oxidative stresses (van Baarlen et al. 2008; Deyholos 2010; Wang et al. 2012). The overlap of large number of genes induced by various stress conditions reveals the molecular cross talk of gene regulatory network responses to various abiotic stress conditions (Weston et al. 2011; Friedel et al. 2012). This contribution has enabled the discovery of novel stress-responsive genes on the basis of expression profiles in different developmental stages of the plant under stress conditions (Sreenivasulu et al. 2008; Smita et al. 2013). The availability of complete genome sequences of *Arabidopsis* and *Oryza sativa* model plants and other important crops has provided sufficient genomic information to perform high-throughput genome-wide functional analysis. Compared to other stresses, heat-stress responses in plants have received increasing attention in recent years, and accordingly global transcriptome expression in response to heat stress has been reported in different plant species (Mangelsen et al. 2011; Liu et al. 2012). Lim et al. (2006) found that *Arabidopsis* suspension cells at a moderate heat enhanced the expression profiling of 165 genes, with high quantity of heat-shock proteins (Hsps). Frank et al. (2009) found from the cDNA microarrays and qPCR analysis that Hsp70, Hsp90, and heat-shock transcription factors (HSF) were important to tomato microspore resistance to heat stress. Transcriptomic data from *Triticum aestivum* and other plants show that 5 % of the genes are significantly affected in response to heat stress (Finka et al. 2011). But *Arabidopsis* transcriptomic data show that 11 % of the genes expressed in response to heat stress are encoded for heat-induced chaperones (Qin et al. 2008). The rest of the transcripts encode products involved in calcium signaling, phytohormone signaling, sugar and lipid signaling, and metabolism. Additionally, some studies have identified various transcripts increased during heat treatment, including members of the *DREB2* family of transcription factors, *AsEXP1* encoding an expansion in protein, genes encoding for galactinol synthase and enzymes in the raffinose oligosaccharide pathway, and antioxidant enzymes (Xu et al. 2007). Reports exist which show decrease in transcript levels related to programmed cell death, basic metabolism, and biotic stress responses under heat-stress conditions (Larkindale and Vierling 2008).

Affymetrix Grape Genome Array and qRT-PCR techniques were used to identify the heat-stress- and recovery-regulated genes in the grape and found that about 8 % of total probe sets were responsive to heat stress and subsequent recovery in grape leaves. The responsive genes identified in this study belong to a large number of important factors and biological pathways, including those for cell rescue (i.e., antioxidant enzymes), protein fate (i.e., Hsps), primary and secondary metabolism, transcription factors, and signal transduction and development (Liu et al. 2012). Wheat Genome Array was applied to measure the transcriptome changes in response to heat stress in the contrasting genotypes and identified a total of 6,560 probe sets that responded to heat stress (Qin et al. 2008). A combination of heat and drought stresses had a significantly higher detrimental effect on growth and productivity of maize, barley, sorghum, and different grasses than each of the stresses applied individually (Abraham 2008). Nonetheless, apart from a notable study on the effects of simultaneous drought and heat stress (Rizhsky et al. 2004), the effects of stress combinations have been little studied (Atkinson and Urwin 2012). Transcriptome analysis has been used to investigate the expression in response to heat stress as well as combined stresses in several plant species (Oshino et al. 2007; Rasmussen et al. 2013). Transcriptome profiling of *Arabidopsis* plants during a combination of drought and heat stress influences the changes in the expression pattern of more than 400 transcripts (Rizhsky et al. 2004). Distinct responses were also observed in plants exposed to a combination of heat and high light intensity (Hewezi et al. 2008), heat, and salinity (Keles and Oncel 2002). Their work suggests that some pathways/mechanisms are dependent on genotype, duration, intensity, and type of abiotic stress.

Until now, most of the transcriptome responses have focused on an improved stress tolerance during the vegetative phase of plant growth. Heat stress decreases the duration of developmental phases leading to fewer organs, smaller organs, reduced light perception over the shortened life cycle, and perturbation of the processes related to carbon assimilation. These parameters ultimately contribute to losses in the final yield of plants (Hussain and Mudasser 2007). But the most sensitive developmental stages to heat stress are flowering and grain filling (Wei et al. 2010). High temperature drastically reduced both yield and quality of wheat (Sharma et al. 2012). In this context, recently few attempts were made to reveal the transcriptome alterations in developing seeds to understand the yield stability under heat stress in rice (Dupont et al. 2006; Yamakawa and Hakata 2010) and barley (Mangelsen et al. 2011). Altogether, transcriptome analyses provide novel insight into the plant response to heat stress and have great implications for further studies on gene function annotation and molecular breeding. In the era of post-genomics, large-scale gene expression data are generated by whole-genome transcriptome platforms. There are a few software applications that have been developed to query high-throughput microarray gene expression databases at the genome-wide gene content level for various stress responses. The available expression data are mostly deposited in online repositories such as GEO (Barrett et al. 2007), NASC Arrays (Craigon et al. 2004), PLEXdb (Dash et al. 2012), and ArrayExpress (Kapushesky et al. 2012). In parallel, various online query-oriented tools have been developed such as Genevestigator (Zimmermann et al. 2004), *Arabidopsis* eFP browser (Winter et al. 2007), RiceArrayNet (Lee et al. 2009), or *Arabidopsis* and rice co-expression data mining tools (Ficklin et al. 2010; Movahedi et al. 2011) Gramene (Youens-Clark et al. 2011), TAIR (Swarbreck et al. 2008), and MaizeGDB (Schaeffer et al. 2011) to extract development- and stress-specific regulons by implementing global normalization and clustering algorithms (Sreenivasulu et al. 2010).

2.1.1 NGS-Based Transcriptome Analysis

Next-generation sequencing (NGS)-based transcriptome analysis is superior to other available techniques since sequencing-based method is digital, high throughput, highly accurate, and easy to perform and is capable of identifying allele-specific expression. The principal advantage of NGS is that their throughputs are much higher than that of classical sequencing. In recent years, researchers have developed various platforms such as the Illumina Genome Analyzer, the Roche/454 Genome Sequencer FLX Instrument, and the ABI SOLiD System that have proven to be powerful and cost-effective tools for advanced research in many areas, including genome sequencing, resequencing of the genome, miRNA expression profiling, DNA methylation analysis, and especially the de novo transcriptome sequencing of non-model organisms (Morozova and Marra 2008). NGS transcriptome analysis is fast and simple because it does not require any cloning of cDNAs and generates an extraordinary depth of short reads. It is a more comprehensive and efficient way to measure transcriptome composition, obtain transcriptome sequencing using NGS technologies provides better alternative for the gene expression studies. Application of NGS technology in the plant transcriptome analysis has been very limited, and only a few proof-of-concept studies have been performed to reveal the transcriptional complexity in plants. Here, we provide some examples of the RNA-seq-based gene expression studies performed in plants, which provide novel insights into the various biological aspects. The Illumina sequencing analysis in maize revealed the differential expression of a very high fraction of genes (64.4 %) and provided the evidence for dynamic reprogramming of transcriptome with transcripts for basic cellular metabolism like photosynthesis (Li et al. 2010). An integrated transcriptome atlas of the soybean has been generated, which resulted in the identification of tissue-specific genes (Libault et al. 2010). Further, this expression data has been utilized for comparative analyses of gene expression from other legumes, *Medicago truncatula* and *Lotus japonicus*. In another independent study, Severin et al. (2010) identified more than 177 genes involved in the agronomically important trait, like seed filling process using RNA-seq in soybean. Garg et al. (2011) identified the differentially expressed genes in a tissue-by-tissue comparison and tissue-specific transcripts in the chickpea, using massively parallel pyrosequencing. The transcriptional complexity in rice has also been unraveled via sequencing of mRNA from various tissues in two subspecies and identified more than 15,000 novel transcriptionally active regions and 3,464 differentially expressed genes (Lu et al. 2010). A novel sequence-based approach using Roche 454 technology focused on sequencing unique 3′-UTRs of genes to distinguish highly conserved, related transcripts such as members of same gene family and quantify their expression (Eveland et al. 2008).

RNA-seq is a popular approach in NGS technologies to collect and quantify the large-scale sequences of coding and noncoding RNAs rapidly (Garber et al. 2011). NGS-based RNA-seq has been used for the rapid development of genomic resources in many plants (Gowik et al. 2011). NGS was employed to create transcriptome databases of species without a sequenced genome such as mangroves (Dassanayake et al. 2009), eucalyptus (Novaes et al. 2008), olive (Alagna et al. 2009), and chestnut (Barakat et al. 2009). For this RNA-seq approach, either fragmented mRNA or fragmented cDNA (Wang et al. 2009a) can be used as input, and read lengths ranging from 100 to 250 nts and 500 nts model length can be received depending on the sequencer and sequencing kit employed. A major challenge in the near future for those who like to begin the work with NGS data is retooling for methods to store data. This is due to the short history of the technology and its continuous development, and there are as yet no standard methods available to detect and analyze differentially expressed genes based on NGS data. Such deep sequencing data from crop plants can help to identify the candidate genes associated with final yield, grain quality, disease resistance, and abiotic including heat-stress tolerance. These data are also useful to identify and isolate new genes and promoters involved in agronomical traits of economically important crops. Generation of such bioinformatics data would be useful in crop improvement programs. NGS-based sequencing applications have rapidly expanded in plant genomics by browsing the Sequence Read Archive (SRA) in NCBI (http://www.ncbi.nlm.nih.gov/sra), European Nucleotide Archive (http://www.ebi.ac.uk/ena/home), and DDBJ Sequence Read Archive (http://trace.ddbj.nig.ac.jp/dra/index_e.shtml), all of which store raw sequencing data from NGS platforms; users can determine how thoroughly a given species has been sequenced and retrieve the publicly available sequencing data for further use.

2.2 Proteomics

Proteomics is not only a powerful molecular tool used in describing complete proteomes at the organelle, cell, organ, or tissue levels, but it can also compare the status of protein profiling under different physiological conditions, such as those resulting from the exposure to stressful conditions (Cushman and Bohnert 2000). The proteome reflects the actual state of the cell or the organism and is an essential bridge between the transcriptome and the metabolome (Zhu et al. 2003) and also acts directly on biochemical processes and thus must be closer to the phenotype. In the last decade, proteomics has been shown to be a powerful tool in exploring many biological mechanisms which brought much deeper insight in the abiotic stress-responsive mechanisms in the crop plants (Rinalducci et al. 2011; Yin et al. 2012). However, the proteomic studies of crop plants under heat stress are not well understood (Neilson et al. 2010; Rinalducci et al. 2011). To understand the modulation mechanisms of heat tolerance in plants, a detailed study of the response to high temperature at the proteomics level is essential. Thus far, there have been only a few proteomic studies regarding heat stress in plants (Koussevitzky et al. 2008; Neilson et al. 2010). Recent developments in protein analysis methods have made possible the evaluation and identification of many proteins and to exploit proteomic data in the context of stress response particularly heat stress (Nanjo et al. 2010). Proteome approach has been successfully used to study the effect of heat shock on wheat grain quality and to identify protein markers that enable breeders to produce cultivars with desired characters especially cultivars that tolerate heat-stress conditions (Skylas et al. 2002). The effect of heat stress in the wheat endosperm by MALDI-TOF coupled with 2-DE analysis identified a total of 48 differentially expressed proteins (Majoul et al. 2003). Of these, more than 37 % of the proteins have been identified as Hsps that are involved in protein stability and folding, which suggests that high temperature has severe effects on protein denaturation and regulation. Rice leaf proteomic analysis by 2-DE-MS method in response to heat stress identified 1,000 protein spots, wherein 73 protein spots were differentially expressed at least at one time point. These proteins were further categorized into different classes related to heat-shock proteins, energy and metabolism, redox homeostasis, and regulatory proteins.

Proteomic analysis in barley cultivars under heat stress identified several isoforms of sHsps and S-adenosylmethionine synthetase (SAM-S) and found to be upregulated (Sule et al. 2004). Several studies that analyzed the proteomic response to heat stress have been carried out in *Arabidopsis* and identified 45 spots which were unique to the combined heat and drought stresses. Proteins uniquely regulated by heat in *A. scabra* included sucrose synthase, superoxide dismutase, glutathione S-transferase, and stress-inducible Hsps. This suggests that these proteins may contribute to increased survival of *A. scabra* under high-temperature conditions. Using differential metabolic labeling, Palmblad et al. (2008) identified a number of known Hsps as well as other proteins previously not associated with heat shock in *Arabidopsis*. Polenta et al. (2007) identified the Hsps from tomato pericarp by thermal treatment. They highlighted the importance of class I sHsps that are involved in the process and further characterized them by using monospecific polyclonal antiserum and MS/MS analysis. Thus, the results of this study suggest that plants cope with heat stress in a complex manner, where Hsps play a pivotal role in a complex cellular network. The identification of some novel proteins in the heat-stress response provides new insights that can lead to a better understanding of the molecular basis of heat sensitivity in plants (Lee et al. 2007).

Protein profiling of two ecotypes (low and high elevations) of Norway spruce was investigated in response to high-temperature stress using 2-DE and LC-MS/MS. This analysis showed an accumulation of sHsps during the recovery from heat stress, specifically in the low-elevation ecotype (higher level of thermotolerance) (Valcu et al. 2008). Root protein profiling under heat stress identified 70 protein spots which showed differential accumulation in at least one species. More proteins were downregulated as a result of heat stress, but *A. scabra* exhibited many upregulated protein spots under heat-stress regimes. The two grasses displayed different proteomic profiles. Some of the uniquely regulated genes by heat stress in *A. scabra* included sucrose synthase, superoxide dismutase, glutathione S-transferase, and stress-inducible heat-shock proteins. This suggests that these proteins may contribute to increased survival of *A. scabra* under high-temperature conditions (Xu et al. 2008). Heat-stress treatment in combination with drought resulted in the expression of approximately 650 protein spots in *C. spinarum*. Forty-nine spots changed their expression levels upon heat and drought treatment, and 30 proteins were identified by MS and 2-D Western blot. These proteins were classified into Hsps, photosynthesis-related proteins, RNA-processing proteins, and proteins involved in metabolism and energy production (Zhang et al. 2010). Proteomic profiling of radish leaves in response to high-temperature stress resulted in the identification of 11 differentially expressed protein spots, and they were divided into four categories: Hsps, redox homeostasis-related proteins, energy- and metabolism-related proteins, and signal transduction-associated proteins (Zhang et al. 2012). Such studies provide a good starting point in understanding the overall thermal responses of plants; however, further heat treatments and comparative analyses should be conducted in order to gain a better understanding of the overall thermal responses of plants.

Availability of proteomics data is important to support published results and conclusions. Several proteomics resources and repositories available for plant species were updated (Schneider et al. 2012), e.g., Plant Proteome Database (http://ppdb.tc.cornell.edu/) which provides information on maize and *Arabidopsis* proteomes, RIKEN Plant Phosphoproteome Database (RIPP-DB, https://database.riken.jp/sw/en/Plant_Phosphoproteome_Database/ria102i/) updated with a data set of large-scale identification of rice phosphorylated proteins (Nakagami et al. 2012), and OryzaPG-DB launched as a rice proteome database based on shotgun proteomics (Helmy et al. 2011). Besides those repositories, numerous very valuable resources, each focused on a specific aspect like tandem mass spectra evidences, quantitative information, and localization of phosphorylation sites, are available for plant proteomics such as ProMEX (Wienkoop et al. 2012); PhosPhAt, a plant phosphorylation site database (Arsova and Schulze 2012); PaxDb (Wang et al. 2012), a meta-resource integrating information on absolute protein abundance levels across different organisms, including *A. thaliana*; MASCP Gator (Joshi et al. 2011), an aggregation portal for the visualization of *Arabidopsis* proteomics data; or PPDB, the Plant Proteome Database (Sun et al. 2009) to cite only a few. UniProtKB is cross-linked to several of those proteomics resources, including PRIDE, IntAct, ProMEX, PeptideAtlas, and PhosphoSite. A complete list of the cross-references, with bibliographic references, is available at http://www.uniprot.org/docs/dbxref. These databases help us in identifying and understanding the complex protein networks associated with the heat-stress tolerance and the functions of these proteins during heat stress.

2.3 Metabolomics

Metabolomics is the qualitative and quantitative collection of all low-molecular-weight metabolites present in a cell that participate in general metabolic reactions and are required for the maintenance, growth, and normal function of a cell (Arbona et al. 2009; Jordan et al. 2009). Metabolome directly influences the phenotype when compared to transcriptome or proteome and bridges the gap between genotype and phenotype. The study of the metabolome represents the integration of the genetic background and the influence of the environmental conditions, thus describing more accurately the phenotype of a given plant species. Metabolic regulation during stressful events has been facilitated much in the last decade, and the identification of metabolites has been improved through mass spectrometric studies (Sawada et al. 2009). More comprehensive coverage can only be achieved by using several extraction and detection technologies in parallel and subjecting them to chemical analysis using different analytical methods like gas chromatography coupled to mass spectrometry (GC-MS) and liquid chromatography coupled to mass spectrometry (LC-MS) (De Vos et al. 2007). Other analytical techniques include liquid chromatography (photodiode array detection) coupled to mass spectrometry (LC-PDA/MS) (Huhman and Sumner 2002), capillary electrophoresis coupled to mass spectrometry (CE-MS) (Harada and Fukusaki 2009, Takahashi et al. 2009), Fourier transform ion cyclotron resonance mass spectrometry (FTICR/MS) (Oikawa et al. 2006), and NMR spectroscopy (Krishnan et al. 2005). Among all analyzers that can be used with the separation techniques mentioned above, the most popular in metabolomics are MS and, particularly, those providing accurate mass measures (Arbona et al. 2013). Therefore, the future objective is the standardization and annotation of data from multiple metabolomics technologies in public databases (Castellana and Bafna 2010). The data obtained can then be investigated by multivariate and correlation analyses for functional genomics in order to study the systems biology of plant metabolism and make use of the data for crop improvement (Arbona et al. 2013). From the genome sequence information of the *A. thaliana* and other model plants, it is evident that plants reorganize their metabolic network in order to adapt to such adverse conditions (Kaplan et al. 2004). Many plants respond to different stresses by a progressive adjustment of their metabolism with early and late responsive gene networks. Some metabolic changes are common to salt, drought, and temperature stresses, whereas others are specific (Urano et al. 2009; Lugan et al. 2010). Using metabolic changes as a "map" or "marker," factors regulating metabolic movements were investigated by Saito et al. (2008) in combination with other "omics" approaches. It appears, therefore, metabolomics plays a key role in understanding cellular functions and decoding the functions of genes (Hagel and Facchini 2008).

In plant systems, metabolomics approach has already been used to study metabolomic changes during a variety of stresses, for example, water and salinity (Cramer et al. 2007), sulfur (Nikiforova et al. 2005), phosphorus (Hernandez et al. 2007), oxidative (Baxter et al. 2007), and heavy metals (Le Lay et al. 2006). But, less work has been done in the case of heat stress. A recent metabolome analysis showed common metabolites in response to cold and other stresses and demonstrated a prominent role for the DREB1/CBF transcriptional network in the cold-response pathway (Maruyama et al. 2009). Comparative metabolite analysis has been carried out using GC-MS (Kaplan et al. 2004) and GCTOF-MS (Wienkoop et al. 2008) between *Arabidopsis* plants responding to heat and cold shocks. Many metabolites produced in response to heat shock overlapped with those produced in response to cold shock also. Many metabolite levels changed specifically in response to cold than to heat. This response points out a strong impact of cold stress on plant metabolism. Wang et al. (2004) reported that a combination of drought and heat stress results in decrease of the growth and productivity of the crops when compared with each of the different stresses applied individually. Integrated metabolome and transcriptome results were applied by Yamakawa and Hakata (2010) to analyze rice developing caryopses under high-temperature conditions. Molecular events underlying pollination-induced and pollination-independent fruit sets were carried out by Wang et al. (2009a, b) and also the effects of DE-ETIOLATED1 downregulation in tomato fruits (Enfissi et al. 2010). Heat stress induced the accumulation of key metabolites like alanine, allantoin, arachidic acid, 2-ketoisocaproic acid, myo-inositol, putrescine, and rhamnose, while it decreased fructose-6-phosphate (Luengwilai et al. 2012). Moreover, these results suggested that a metabolic network of compatible solutes including proline, monosaccharides, galactinol, and raffinose has an important role to play in temperature stress tolerance (Alcazar et al. 2010).

Information resources related to metabolic profiling are available and updated and provide data archives for metabolome data sets and analytical platforms such as LC-MS-based metabolome database (http://appliedbioinformatics.wur.nl/moto/) (Moco et al. 2006), KOMICS (Iijima et al. 2008), Plant MetGenMAP (Joung et al. 2009), Metabolome Express (https://www.metabolome-express.org/) (Carroll et al. 2010; Ferry-Dumazet et al. 2011), MeRy-B (http://www.cbib.u-bordeaux2.fr/MERYB/) (Ferry-Dumazet et al. 2011), KaPPA-View4 SOL (Sakurai et al. 2011), MetaCrop 2.0 (http://metacrop.ipk-gatersleben.de) (Schreiber et al. 2012), and PRIMe (http://prime.psc.riken.jp/) (Sakurai et al. 2013). Apart from this, several individual species-wise databases are available at Gramene database like RiceCyc, MaizeCyc, BrachyCyc, SorghumCyc, and Sol Genomics Network (SGN). These databases play crucial roles as information resources and repositories of large-scale data sets and also serve as tools for further integration of metabolic profiles containing comprehensive data acquired from other omics research (Akiyama et al. 2008). Following these successes, several multi-omics-based systems analyses have been used for understanding plant cellular systems.

2.4 Phenomics

Phenomics is the systematic study of the physical and morphological properties of organism as they change in response to genetic mutation and environmental influences. Traditional methods of measuring growth and other morphological features are time consuming and costly and involve many genotypes and the destructive harvest of plants. Phenomics has been considered as one of the important techniques to screen the germplasm and to utilize the available morphological variation in breeding programs aimed at heat-stress tolerance. Therefore, phenomics as a technique remains critical in the post-genomics era. Phenomics approach also enables us to understand the precise molecular mechanism involved in conferring tolerance against different kinds of abiotic stresses. This has stimulated the research in several institutions to invest in developing technologies and platforms able to speed up the phenotyping process. The investments started earlier in the private sectors, and more recently this has been embraced by public research institutions that are developing an international collaboration network (www.plantphenomics.com). There are a large number of initiatives launched (International Plant Phenomics Network, Deutsche Plant Phenomics Network, and European Plant Phenomics Network) to create phenotyping facilities to screen populations using high-throughput methods located in Australia, Germany, France, Canada, Italy, and many more (Furbank 2009; Finkel 2009). Large phenotyping platforms represent technologies that are mainly based on nondestructive image analyses of plant tissues or structural and functional features obtained by advanced technologies (Nagel et al. 2009; Yazdanbakhsh and Fisahn 2009). In other labs, glass houses and greenhouses can be fitted with cameras, and plants may be carried on the conveyor belts to the imaging stations. Such facilities exist in several laboratories around the world (CropDesign, Belgium; The Plant Accelerator, Australia; PhenoPhab, Holland; Metapontum Agrobios, Italy; IPK, Germany) and have the advantage of acquiring 3-D images. Using high-throughput phenomics platforms, various parameters like water-deficit responses can be studied (Sadok et al. 2007; Berger et al. 2010). So far, only a handful of studies have been carried out in the phenomics area in response to heat stress in the crop plants (Sharma et al. 2012; Yeh et al. 2012). However, the application of phenomics will really become useful and important if specific questions are asked to these platforms.

Plants show numerous responses to heat stress regarding carbon metabolism and water balance, but unfortunately no single key physiological trait that relates to a genetic base for heat-stress tolerance has been identified (Allakhverdiev et al. 2008; Wolkovich et al. 2012). It is known that the reproductive processes are the most sensitive to heat stress in many species. Heat stress significantly affects cellular homeostasis including both protein and membrane stability. These responses include basal thermotolerance, short- and long-term acquired thermotolerance, and thermotolerance to moderately high temperatures. High temperatures adversely affect the seed germination, growth, photosynthetic efficiency, core metabolic processes, pollen viability, respiration, water relations, and protein and membrane stability. Different species and cultivars may vary their tolerance to high temperatures with the stage of development, but all vegetative and reproductive stages are highly affected by heat stress (Hall 1992). Different phenological stages of plants differ in their sensitivity to high temperature. During vegetative stage, high day temperature can damage leaf gas-exchange properties. High night temperatures make the pollen sterile. But this depends on species and the genotype under study. Sharma et al. (2012) identified 41 contrasting lines in terms of heat tolerance by mass screening of 1,274 wheat cultivars of diverse origin. This contrasting set of cultivars was then used to compare the ability of chlorophyll fluorescence parameters to detect genetic difference in heat tolerance. This identification may aid future studies to understand the genetic and physiological nature of heat-stress tolerance (Sharma et al. 2012). The temperature and duration of heat-stress treatments resulting in changes in growth and development of seeds, seedlings, mature leaves, panicles or spikes, and fruits have been used in crop thermotolerance studies (Rahman et al. 2007; Seepaul et al. 2011). But, high-throughput phenotyping analyses are necessary for deepening our understanding of the molecular genetics of thermotolerance.

3 Heat-Shock Proteins

Heat-shock response (HSR) is a highly conserved reaction caused by exposure of an organism or tissue or cells to sudden high-temperature stress. High-temperature stress is characterized by rapid induction and transient expression of conserved heat-shock transcripts and other regulators. Among five conserved families of Hsps (Hsp100, Hsp90, Hsp70, Hsp60, and sHsp), the small heat-shock proteins (sHsps) are found to be most prevalent in plants, the expression of which can increase up to 200-fold under heat stress. Different classes of molecular chaperones appear to bind to specific nonnative substrates and states. Molecular Hsps/chaperones are located in the cytoplasm and cell organelles such as the nucleus, mitochondria, chloroplasts, and ER (Wang et al. 2004). The mechanism by which Hsps contribute to heat tolerance is still enigmatic though several roles have been ascribed to them. Many studies assert that Hsps are molecular chaperones ensuring the native configuration and functionality of cell proteins under heat stress. During stress, many enzymes and structural proteins undergo deleterious structural and functional changes. Therefore, it is vital to maintain proteins in their functional conformations, preventing aggregation of nonnative proteins and refolding of denatured proteins. It is also important to remove nonfunctional and harmful polypeptides arising from misfolding, denaturation, or aggregation for cell survival under stress. Thus, the different classes of Hsps/chaperones cooperate and play complementary and sometimes overlapping roles in the protection of proteins from stress (Bowen et al. 2002). When denatured or misfolded proteins form aggregates, they can be resolubilized by Hsp100/Clp followed by refolding or degraded by proteases (Schöffl et al. 1998). Some Hsps/chaperones (Hsp70, Hsp90) accompany the signal transduction and transcription activation that lead to the synthesis of other members of Hsps/chaperones. Similar observations have been reported with plant chaperones. It has been shown that Hsp18.1 from *Pisum sativum* could stably bind to heat-denatured protein and maintained it in a folding-competent state for further refolding by Hsp70/Hsp100 complexes (Mogk et al. 2003). A recent study has shown that Hsp70 and Hsp90 have roles to play in protecting the enzymes at higher temperatures (Reddy et al. 2010, 2011). HSR in plants was investigated in depth, and the presence of multiple signaling pathways was proposed (Kotak et al. 2007; von Koskull-Doring et al. 2007). Many studies noted upregulation of transcripts including high-molecular-weight Hsps like Hsp101, Hsp70s, and small Hsps (Sarkar et al. 2009; Mittal et al. 2009; Chauhan et al. 2011) and also identified additional transcripts such as DREBs, galactinol synthases and other enzymes in the raffinose oligosaccharide pathway, and oxidative stress enzymes (Frank et al. 2009; Suzuki et al. 2011). Genome-wide survey for Hsps and Hsfs using the tools of bioinformatics helps us to find out not only the number of genes present in a genome but also their chromosomal location. Software tools also help us to find their subcellular locations and the upstream promoter sequences along with their predicted functions.

The regulation of heat-shock gene expression in eukaryotes is mediated by the heat-shock transcription factors (Hsfs), which are highly conserved throughout the eukaryotic kingdom (Scharf et al. 2012). Plant *Hsfs* have unique characteristics and the existence of heat-stress-induced *Hsf* genes might have a major role to play in the modulation of transcription during long-term heat-shock response (Chauhan et al. 2011). Temperature stress-response signal transduction pathways and defense mechanisms involving Hsfs and Hsps are thought to be intimately associated with reactive oxygen species (ROS) production (Frank et al. 2009). Heat-shock transcription factor-dependent expression of antioxidant enzymes such as ascorbate peroxidases in *Arabidopsis* (Frank et al. 2009) suggested that Hsfs might be involved not only in Hsp regulation but also in the regulation of oxidative stress (Reddy et al. 2009). Recent research revealed the involvement of noncanonical transcription factors in HSR; for instance, bZip28, a gene-encoding membrane, tethered TF, which was induced by HS, and the bzip28 null mutant became hypersensitive to HS (Gao et al. 2008). Besides Hsps, there are other plant proteins including ubiquitin, LEA proteins, and cytosolic Cu/Zn-SOD and Mn-SOD whose expressions are stimulated upon heat stress. A number of osmotin-like proteins induced by heat and nitrogen stresses, collectively called Pir proteins, have also been found to be overexpressed in many plant cells under heat stress conferring them resistance. Microarray expression data in *Hordeum vulgare* revealed that most of the *sHsp* and *Hsf* genes are differentially regulated during drought and temperature stresses at different plant developmental stages suggesting considerable cross talk between stress and non-stress regulatory networks. In silico *cis*-regulatory motif analysis of *Hsf* promoters showed an enrichment with abscisic acid-responsive *cis*-elements (ABREs), implying regulatory role of ABA in mediating transcriptional response of *HvsHsf* genes.

4 Heat-Shock Promoters

The need for inducible or specific promoters can be a key tool in plant biotechnology, and their need will increase as we attempt to transfer and validate genes associated with abiotic stress tolerance. In the last decade, several candidate genes, pathways, and strategies have been identified by various groups across the globe and provided insights in plant heat-stress adaptation. Strong constitutive promoters are routinely used in plant transformation with a regulated expression of heat-stress-responsive genes. But such use of constitutive promoters is resulting in serious penalties on plant growth and development with overall negative performance of transgenics (Sakuma et al. 2006a). Since constitutive promoters are hampering the final productivity, it is important for us to identify and isolate heat-stress-inducible promoters and use them while developing transgenic crops. However, the regulated expression of transgenes in plants in a tissue-specific manner and at a specific developmental stage remains a challenging task. The isolation and characterization of heat-stress-responsive promoters and their regulatory regions will have more biotechnological applications as these promoters could be used to engineer the target genes to express only at the site of stress. A powerful approach for measuring the activity of any heat-shock promoter is by fusing the promoter of heat-shock gene to reporter genes such as GFP or GUS. This allows measuring the developmental and tissue-specific expression of genes with or without heat stress (Khurana et al. 2013). It has been found that while some transcripts exhibit translational repression, others escape such repression and remain actively translated. But the underlying mechanisms that mediate this control especially the identities of the regulatory RNA elements involved were poorly understood. Using a computational and experimental approach, Matsuura et al. (2013) identified a novel *cis*-regulatory element in the 5′-UTR that affects differential translation and has a *cis*-regulatory signature responsible for heat-stress-regulated mRNA translation in *Arabidopsis*. A comprehensive transcriptome analysis by using microarrays revealed the relationships among stress-regulated transcripts and enabled the prediction of their *cis*-regulatory elements in temperature stress-inducible genes (Weston et al. 2008). In addition, characterization of the transcriptional dynamics during seed development under different stress conditions enabled the prediction of their *cis*-regulatory elements (Weston et al. 2008). Ma and Bohnert (2007) showed a clear correlation between expression profiles and the 5′*cis*-regulatory motifs of stress-regulated genes. These analyses indicated that stress-regulated genes are controlled by a complicated regulatory network and cross talk between pathways. This type of network has been proposed based on transcriptome data using different bioinformatics approaches (Long et al. 2008). The basic structure and promoter recognition of Hsfs are highly conserved throughout the eukaryotic kingdom (Scharf et al. 2012). Expression of *cis*-motifs containing these *Hsf* genes might be regulated by *Hsfs* themselves, via formation of a regulatory network as proposed by Nover et al. (2001). The expression of *Hsps* during stress treatments and different developmental stages will depend on the *cis*-motifs present in the respective *Hsp* and *Hsf* promoters which will subsequently bind to different transcription factors particularly *Hsfs* as demonstrated by transient reporter assays in sunflower embryos (Almoguera et al. 2002). Hsp18.2 promoter fused to the *GUS* gene transgenic *Arabidopsis* plants showed that heat stress induced the *GUS* gene activity in almost all the organs of the plant (Takahashi et al. 1992). Similarly, heat-shock-induced GUS activity was observed in transgenic *Arabidopsis* when the promoter of *Hsp81* gene was used (Yabe et al. 1994). Crone et al. (2001) did a detailed analysis of the expression of the GmHsp17.5E promoter in all the organs and tissues of the flower and found that promoter is differentially expressed after heat shock in different floral tissues. Hsfs bind to heat-stress elements (HSEs) with the core sequence nGAAnnTTCn or nTTCnnGAAn and form trimers, thereby regulate downstream gene expression (Wu 1995). Despite the occurrence of heat-shock elements in the promoters of heat-stress-inducible genes, a set of *Hsp* genes are expressed during seed development (Kotak et al. 2007). Atsp90-1 promoter region contributes in a combinatorial manner to regulate the expression in development and stress conditions (Haralampidis et al. 2002). The heat-stress induction of *Arabidopsis* HsfA3 is regulated directly by DREB2A, a transcription factor functioning in drought stress responses (Sakuma et al. 2006a, b). Accordingly, the DRE has been identified in the promoters of a cluster of heat-inducible genes (Larkindale and Vierling 2008).

Promoters of heat-shock protein (Hsp) genes are good candidates for inducible expression, since they are rapidly and highly induced under heat-stress conditions. Besides, their induction can be accurately controlled by varying the temperature and duration of induction. Several detailed studies have been performed using the reporter gene expression driven by plant small Hsp promoters in different hosts. The AtHsp18.2 promoter has been successfully used in *Arabidopsis* (Takahashi et al. 1992) and in other species, such as *N. plumbaginifolia* (Moriwaki et al. 1999) and *N. tabacum* hairy roots (Lee et al. 2007). Besides, the inducibility of soybean GmHsp17.3B promoter was studied in the moss *Physcomitrella patens* (Saidi et al. 2007). *Arabidopsis* Hsp18.2 promoter was used to drive the expression of GUS gene in N. *tabacum* BY-2 cells, and maximum activity of protein was obtained under the heat stress (Shinmyo et al. 1998). Khurana et al. (2013) studied the wheat sHsp26 promoter activity in transgenic *Arabidopsis* and observed consistently high levels of *GUS* gene expression under different abiotic stress conditions especially in heat stress. However, the mechanisms that regulate Hsp expression during seed maturation remain largely unknown. In addition to their direct functions in acquired stress tolerance and development, Hsps/chaperones function synergistically with other components, thus help in decreasing cellular damage. If the range of promoters is wide, then it is possible to introduce multiple transgenes into plants that are expressed differentially in response to various environmental stresses. Identification of heat-stress-inducible promoters from crop plants would be of immense help in generating transgenic plants with improved agronomic performance.

5 Heat-Stress-Tolerant Transgenic Crops Generated Through Expression of Hsps

Most crops are affected by daily/seasonal fluctuations in day and/or night temperatures. Conventional breeding for high-temperature stress tolerance has not been successful so far. This could be due to lack of our understanding on the genetic mechanisms associated with heat stress, suitable source of genes, and complex nature of the HS trait. This complexity is now being dissected out including features like heat-shock elements (HSEs), heat-shock factors (HSFs), possible receptors of the heat-shock response, signaling components, and chromatin remodeling aspects (Proveniers and van Zanten 2013). Several groups have altered the levels of sHsps in bacterial systems and shown that when overexpressed in bacterial cells, Hsps have a role in conferring thermotolerance. The overexpression of *OsHsp16.9* in *E. coli* conferred thermotolerance. Yeh et al. (2012) constructed deletion mutants of this sHsp to find out the regions associated with heat-stress tolerance. They overexpressed the constructs in *E. coli* (Yeh et al. 2012) and found out that the deletion of amino acid residues 30–36 (PATSDND) in the N-terminal domain or 73–78 (EEGNVL) in the consensus-II domain of *OsHsp16.9* caused the loss of chaperone activities and also rendered the *E. coli* incapable of surviving at 47.5 °C. When three sHsps were introduced into *E. coli*, they acquired thermotolerance and were able to protect malate dehydrogenase (MDH) from in vitro thermal aggregation (Pike et al. 2001). The survivability of *E. coli* Bl21 (DE3) cells transformed with a recombinant plasmid containing different Hsps was compared with the control *E. coli* cells (transformed with the pET28a vector) under heat and different abiotic stresses. The PgHsp transformed cells showed thermotolerance at 47.5 °C, a treatment that was lethal to the untransformed bacterial cells. When the cell lysates from transformed and untransformed were heated at 55 °C, the amount of protein denatured in the PgHsps-Bl21 DE3 cells was 50 % less than that of the PET28a vector (control) cells (Reddy et al. 2010, 2011). Furthermore, genetically modified *E. coli* expressing DcHsp17.7 exhibited a higher salt stress tolerance than control *E. coli* (Song and Ahn 2011). These results suggest that expression of Hsps confers abiotic stress tolerance to *E. coli* cells and may play a role in the plant's adaptation to harsh environments.

The involvement of Hsps in regulating thermotolerance has been further carried out in higher plants by downregulating their expression levels using either antisense or RNAi approach. Mutants of *Zea mays* and *A. thaliana* plants under-expressing their respective Hsp100 proteins were observed to lack both basal and induced thermotolerance (Hong and Vierling 2000, Nieto-Sotelo et al. 2002). Yang et al. (2006) showed that the tomato plants silenced for Hsp100/ClpB protein were impaired in thermotolerance. Acquisition of thermotolerance has been found to be negatively affected in Hsp70 antisense *A. thaliana* plants (Lee and Schoffl 1996). Mutants of *Zea mays* and *Arabidopsis* with low levels of their respective Hsp100 proteins were observed to lack both basal and induced thermotolerance. Plants lacking Hsa32 do not survive HS treatment even after a pretreatment at a sublethal temperature (Charng et al. 2006). Heat-inducible transactivator HsfA2 with low levels of expression results in an increased sensitivity of the mutant plants to heat stress (Charng et al. 2007). Genome-wide transcriptome analysis of HsfA1a, A1b, and A2 knockout mutants in *Arabidopsis* suggests that HsfA1a and A1b play important roles in the initial phase of heat-stress response, but HsfA2 functions under prolonged heat-stress conditions and during the recovery phase (Schramm et al. 2006; Nishizawa et al. 2006). The heat-stress-induced expression of HsfA2 in *Arabidopsis* is not influenced by HsfA1a or HsfA1b (Busch et al. 2005). The *HsfA2* gene is also induced by high light intensity and H_2O_2 (Nishizawa et al. 2006). It is also closely related to the regulation of *ASCORBATE PEROXIDASE 2* (*APX2*) encoding a key enzyme in oxidative stress response, indicating that HSFA2 plays diverse roles under various environmental stresses.

Conversely, upregulation of Hsps has been achieved in a large number of plant species. Transgenic carrot cell lines and plants in which carrot *sHsp17.7* was overexpressed resulted in enhanced survival of transgenic tissues at high temperature (Malik et al. 1999). Transgenic tobacco plants overexpressing tobacco sHsps result in higher cotyledon opening rate (Park and Hong 2002). Similarly, transgenic tomato plants overexpressing tomato *HsfA1* gene showed increased thermotolerance. Tomato mitochondrial *Hsp* gene overexpressed in tobacco showed that transgenics were more thermotolerant at 48 °C than the transgenics produced with the antisense construct of the same gene (Sanmiya et al. 2004). Transgenic rice overexpressing *OsHsp17.7* gene showed increased thermotolerance and greater resistance to UV-B stress than untransformed control plants (Murakami et al. 2004). Constitutive expression of *RcHsp17.8* in transgenic *Arabidopsis* conferred higher thermotolerance and resistance to salt, drought, and osmotic stresses (Jiang et al. 2009). Overexpression of *CaHsp26* in transgenic tobacco protected PSII and PSI during chilling stress under low irradiance (Guo et al. 2007). When transgenic *A. thaliana* plants were generated with overexpression of high-molecular-weight Hsps, transgenics survived at temperatures as high as 45 °C (1 h) and they showed vigorous growth after relieving the plants from stress, while vector-transformed control plants could not regain growth during the post-stress recovery period (Queitsch et al. 2000). Similarly, transgenic rice lines overexpressing *AtHsp100* exhibited regrowth in the post-high-temperature-stress recovery phase, while the untransformed plants could not recover to the similar extents (Katiyar-Agarwal et al. 2003). Recent study in maize demonstrated that small *Hsp* gene, *ZmsHsp*, might have a function in cytokinin response (Cao et al. 2010). Also, *MsHsp23* gene in the tall fescue (*Festuca arundinacea*) transgenics protected the leaves from oxidative damage through chaperon and antioxidant activities. These results suggest that *MsHsp23* confers abiotic stress tolerance in transgenic tall fescue and may be useful in developing stress tolerance in other crops also (Lee et al. 2012). Overexpression of *ZmHSP16.9* in transgenic tobacco conferred tolerance to both heat and oxidative stresses and increased the seed germination rate, root length, and antioxidant enzyme activities when compared with wild-type plants (Sun et al. 2012). Transgenic *Arabidopsis* plants overexpressed with *WsHsp26* were tolerant under continuous high temperature and produced bold seeds under high temperature, having higher germination rate than wild type (Chauhan et al. 2012). The list of transgenic plants raised for high-temperature tolerance using Hsps is shown in the Table 1.

Table 1 Comprehensive details on plant transgenics raised by using Hsp genes for high-temperature tolerance

Gene	Protein	Source	Target plant	Function	Reference
Hsf1	Hsf	*A. thaliana*	*A. thaliana*	Thermotolerance and also constitutive expression of the Hsp genes	Lee et al. (1995)
Hsf3	Hsf	*A. thaliana*	*A. thaliana*	Increase in basal thermotolerance and thermoprotective processes	Prändl et al. (1998)
HsfA1	Hsf	*S. lycopersicon*	*S. lycopersicon*	Advantage for growth and fruit ripening processes under high temperature	Mishra et al. (2002)
Hsf3	Hsf	*A. thaliana*	*A. thaliana*	Lower threshold temperature	Panchuk et al. (2002)
HsfA2	Hsf	*A. thaliana*	*A. thaliana*	Mutants displayed reduced basal and acquired thermotolerance, while the overexpression lines displayed increased tolerance	Li et al. (2005)
HsfA2	Hsf	*A. thaliana*	*A. thaliana*	Increased thermotolerance but also salt/osmotic stress tolerance and enhanced callus growth	Ogawa et al. (2007)
HsfA2e	Hsf	*O. sativa*	*A. thaliana*	Enhances tolerance to environmental stresses	Yokotani et al. (2008)
Hsf7	Hsf	*O. sativa*	*A. thaliana*	Response to high temperature	Liu et al. (2009)
HsfC1b	Hsf	*O. sativa*	*O. Sativa*	Osmotic stress and is required for plant growth under non-stress conditions	Schmidt et al. (2012)
DnaK	Hsp70	*Aphanothece halophytica*	*O. sativa and N. tabacum*	Increased seed yield and total plant biomass in high temperature and salt stress	Uchida et al. (2008)
Hsp70	Hsp70	*Trichoderma harzianum*	*A. thaliana*	Enhanced tolerance to heat stress	Montero-Barrientos et al. (2010)
mtHsp70	Hsp70	*O. sativa*	*O. Sativa*	Suppresses programmed cell death	Qi et al. (2011)
Hsc70-1	Hsp70	*A. thaliana*	*A. thaliana*	More tolerant to heat shock	Sung and Guy (2003)
Hsp101	HSP100	*A. thaliana*	*A. thaliana*	Sudden shifts to extreme temperature better than the controls	Quietsch et al. (2000)
Hsp101	Hsp100	*A. thaliana*	*O. sativa*	Enhanced tolerance to high temperature	Katiyar-Agarwal et al. (2003)
Hsp17.7	Hsp17.7	*D. carota*	*D. carota*	Increased thermotolerance	Malik et al. (1999)
mtsHsp	sHsp	*S. lycopersicon*	*N. tabacum*	Thermotolerance	Sanmiya et al. (2004)
Hsp21	sHsp	*S. lycopersicon*	*S. lycopersicon*	Temperature-dependent oxidative stress	Neta-Sharir et al. (2005)
sHsp17.7	sHsp	*O. sativa*	*O. Sativa*	Drought tolerance in transgenic rice seedlings	Sato and Yokoya (2008)
Hsp16.9	sHsp	*Zea mays L.*	*N. tabacum*	Enhanced tolerance to heat and oxidative stress	Sun et al. (2012)
Hsp17.5	sHsp	*Nelumbo nucifera*	*A. thaliana*	Improved basal thermotolerance	Zhou et al. (2012)
Hsp26	sHsp	*Capsicum annuum*	*N. tabacum*	Protection of PSII and PSI during chilling stress under low irradiance	Guo et al. 2007

(continued)

Table 1 (continued)

Gene	Protein	Source	Target plant	Function	Reference
Hsp17.8	sHsp	*Rosa chinensis*	*A. thaliana*	Increased tolerance to heat, salt, osmotic, and drought stresses	Jiang et al. (2009)
sHsp17.7	sHsp	*O. sativa*	*O. sativa*	Increased thermotolerance	Murakami et al. (2004)
sHsp26	sHsp	*Triticum*	*A. thaliana*	Seed maturation and germination and imparts tolerance to heat stress	Chauhan et al. (2012)
sHsp18	sHsp	*Opuntia streptacantha*	*A. thaliana*	Increased the seed germination rate under salt, osmotic, and ABA treatments	Salas-Muñoz et al. (2012)

6 Conclusions

It appears that a wide range of "omics" studies are currently in progress using numerous methodologies, plant species, and stress conditions. As more results are published, it is becoming increasingly clear that high-temperature stress causes distinct molecular responses in plant tissues. As more data are generated in such studies, it provides suitable candidates for selective breeding programs aimed at enhancing stress tolerance in ecologically and economically important plant species. Plant cells are fundamentally different to those of mammalian species, and these biological differences cause inherent difficulties in plant functional genomics studies. Advances in techniques and approaches will change the way plant heat-stress omics studies are conducted in future. Genomics, transcriptomics, proteomics, and metabolomics investigate different facets of a given scientific issue, such as heat-stress tolerance, but complement each other. Integration of phenotypic, genetic, transcriptomic, proteomic, and metabolomic data will enable accurate and detailed gene network reconstruction. This will ultimately result in the elucidation of the molecular pathways involved in complex phenotypic traits. A better understanding of genetic and cellular mechanisms behind heat-stress tolerance would facilitate generation of transgenic plants with desired traits with little or no undesired/unforeseen effects. Bioinformatics tools are also helping us to obtain genome-wide data on the number of Hsps and Hsfs and their regulations. Taken together, the omics data and the information generated using the tools of bioinformatics would help us to understand better about heat-stress tolerance in crop plants. Future knowledge of tolerance components and the identification of QTLs and cloning of responsible genes may allow transformation of plants with multiple genes and production of highly stress-tolerant transgenic crop plants.

Acknowledgments PSR is thankful to the Department of Science and Technology, Govt. of India, New Delhi, for research funding through the INSPIRE Faculty Fellowship Award Grant No. IFA-11LSPA-06. PBK is thankful to the University Grants Commission, New Delhi, for providing UGC-BSR faculty fellowship.

References

Abraham EM (2008) Differential responses of hybrid bluegrass and Kentucky bluegrass to drought and heat stress. Aristotle University of Thessaloniki, 54006, Thessaloniki, Greece, William A. Meyer, Stacy A. Bonos, and Bingru Huang; Hort Sci 43:2191–2195

Akiyama K, Chikayama E, Yuasa H, Shimada Y, Tohge T, Shinozaki K, Hirai MY, Sakurai T, Kikuchi J, Saito K (2008) PRIMe: a Web site that assembles tools for metabolomics and transcriptomics. In Silico Biol 8:339–345. doi:2008080027

Alagna F, D'Agostino N, Torchia L, Servili M, Rao R, Pietrella M, Giuliano G, Chiusano ML, Baldoni L, Perrotta G (2009) Comparative 454 pyrosequencing of transcripts from two olive genotypes during fruit development. BMC Genomics 10. doi:Artn 399. doi:10.1186/1471-2164-10-399

Alcazar R, Planas J, Saxena T, Zarza X, Bortolotti C, Cuevas J, Bitrian M, Tiburcio AF, Altabella T (2010) Putrescine accumulation confers drought tolerance in transgenic *Arabidopsis* plants over-expressing the homologous arginine decarboxylase 2 gene. Plant Physiol Biochem 48:547–552. doi:S0981-9428(10)00032-X [pii]10.1016/j.plaphy.2010.02.002

Allakhverdiev SI, Kreslavski VD, Klimov VV, Los DA, Carpentier R, Mohanty P (2008) Heat stress: an overview of molecular responses in photosynthesis. Photosynth Res 98:541–550. doi:10.1007/s11120-008-9331-0

Almoguera C, Rojas A, Diaz-Martin J, Prieto-Dapena P, Carranco R, Jordano J (2002) A seed-specific heat-shock transcription factor involved in developmental regulation during embryogenesis in sunflower. J Biol Chem 277:43866–43872. doi:10.1074/jbc.M207330200 M207330200

Arbona V, Iglesias DJ, Talon M, Gomez-Cadenas A (2009) Plant phenotype demarcation using nontargeted LC-MS and GC-MS metabolite profiling. J Agric Food Chem 57:7338–7347. doi:10.1021/jf9009137

Arbona V, Manzi M, Ollas C, Gomez-Cadenas A (2013) Metabolomics as a tool to investigate abiotic stress tolerance in plants. Int J Mol Sci 14:4885–4911. doi:ijms14034885 [pii]10.3390/ijms14034885

Arsova B, Schulze WX (2012) Current status of the plant phosphorylation site database PhosPhAt and its use as a resource for molecular plant physiology. Front Plant Sci 3:132. doi:10.3389/fpls.2012.00132

Atkinson NJ, Urwin PE (2012) The interaction of plant biotic and abiotic stresses: from genes to the field. J Exp Bot 63:3523–3543. doi:ers100 [pii]10.1093/jxb/ers100

Barakat A, DiLoreto DS, Zhang Y, Smith C, Baier K, Powell WA, Wheeler N, Sederoff R, Carlson JE (2009) Comparison of the transcriptomes of American chestnut (*Castanea dentata*) and Chinese chestnut (*Castanea mollissima*) in response to the chestnut blight infection. BMC Plant Biol 9. doi:Artn 51. doi:10.1186/1471-2229-9-51

Barrett T, Troup DB, Wilhite SE, Ledoux P, Rudnev D, Evangelista C, Kim IF, Soboleva A, Tomashevsky M, Edgar R (2007) NCBI GEO: mining tens of millions of expression profiles-database and tools update. Nucleic Acids Res 35:D760–D765. doi:10.1093/nar/gkl887

Baxter CJ, Redestig H, Schauer N, Repsilber D, Patil KR, Nielsen J, Selbig J, Liu J, Fernie AR, Sweetlove LJ (2007) The metabolic response of heterotrophic *Arabidopsis* cells to oxidative stress. Plant Physiol 143:312–325. doi:pp.106.090431 [pii]10.1104/pp.106.090431

Berger B, Parent B, Tester M (2010) High-throughput shoot imaging to study drought responses. J Exp Bot 61:3519–3528. doi:erq201 [pii]10.1093/jxb/erq201

Bowen J, Lay-Yee M, Plummer K, Ferguson I (2002) The heat shock response is involved in thermotolerance in suspension-cultured apple fruit cells. J Plant Physiol 159:599–606

Busch W, Wunderlich M, Schoffl F (2005) Identification of novel heat shock factor-dependent genes and biochemical pathways in *Arabidopsis thaliana*. Plant J 41:1–14. doi:TPJ2272 [pii]10.1111/j.1365-313X.2004.02272.x

Cao Z, Jia Z, Liu Y, Wang M, Zhao J, Zheng J, Wang G (2010) Constitutive expression of ZmsHSP in *Arabidopsis* enhances their cytokinin sensitivity. Mol Biol Rep 37:1089–1097. doi:10.1007/s11033-009-9848-0

Carroll AJ, Badger MR, Harvey Millar A (2010) The Metabolome Express Project: enabling web-based processing, analysis and transparent dissemination of GC/MS metabolomics datasets. BMC Bioinformatics 11:376. doi:1471-2105-11-376 [pii]10.1186/1471-2105-11-376

Castellana N, Bafna V (2010) Proteogenomics to discover the full coding content of genomes: a computational perspective. J Proteomics 73:2124–2135. doi:S1874-3919(10)00185-5 [pii]10.1016/j.jprot.2010.06.007

Charng YY, Liu HC, Liu NY, Hsu FC, Ko SS (2006) *Arabidopsis* Hsa32, a novel heat shock protein, is essential for acquired thermotolerance during long recovery after acclimation. Plant Physiol 140:1297–1305. doi:pp.105.074898 [pii]10.1104/pp.105.074898

Charng YY, Liu HC, Liu NY, Chi WT, Wang CN, Chang SH, Wang TT (2007) A heat-inducible transcription factor, HsfA2, is required for extension of acquired thermotolerance in *Arabidopsis*. Plant Physiol 143:251–262. doi:pp.106.091322 [pii]10.1104/pp.106.091322

Chauhan H, Khurana N, Agarwal P, Khurana P (2011) Heat shock factors in rice (*Oryza sativa* L.): genome-wide expression analysis during reproductive development and abiotic stress. Mol Genet Genomics 286:171–187. doi:10.1007/s00438-011-0638-8

Chauhan H, Khurana N, Nijhavan A, Khurana JP, Khurana P (2012) The wheat chloroplastic small heat shock protein (sHSP26) is involved in seed maturation and germination and imparts tolerance to heat stress. Plant Cell Environ 35:1912–1931. doi:10.1111/j.1365-3040.2012.02525.x

Craigon DJ, James N, Okyere J, Higgins J, Jotham J, May S (2004) NASCArrays: a repository for microarray data generated by NASC's transcriptomics service. Nucleic Acids Res 32:D575–D577

Cramer GR, Ergul A, Grimplet J, Tillett RL, Tattersall EA, Bohlman MC, Vincent D, Sonderegger J, Evans J, Osborne C, Quilici D, Schlauch KA, Schooley DA, Cushman JC (2007) Water and salinity stress in grapevines: early and late changes in transcript and metabolite profiles. Funct Integr Genomics 7:111–134. doi:10.1007/s10142-006-0039-y

Cramer GR, Urano K, Delrot S, Pezzotti M, Shinozaki K (2011) Effects of abiotic stress on plants: a systems biology perspective. BMC Plant Biol 11:163. doi:1471-2229-11-163 [pii]10.1186/1471-2229-11-163

Crone D, Rueda J, Martin KL, Hamilton DA, Mascarenhas JP (2001) The differential expression of a heat shock promoter in floral and reproductive tissues. Plant Cell Environ 24:869–874

Cushman JC, Bohnert HJ (2000) Genomic approaches to plant stress tolerance. Curr Opin Plant Biol 3:117–124. doi:S1369-5266(99)00052-7

Dash S, Van Hemert J, Hong L, Wise RP, Dickerson JA (2012) PLEXdb: gene expression resources for plants and plant pathogens. Nucleic Acids Res 40(Database issue):D1194–D1201. doi:gkr938 [pii]10.1093/nar/gkr938

Dassanayake M, Haas JS, Bohnert HJ, Cheeseman JM (2009) Shedding light on an extremophile lifestyle through transcriptomics. New Phytol 183:764–775. doi:10.1111/j.1469-8137.2009.02913.x

De Vos RC, Moco S, Lommen A, Keurentjes JJ, Bino RJ, Hall RD (2007) Untargeted large-scale plant metabolomics using liquid chromatography coupled to mass spectrometry. Nat Protoc 2:778–791. doi:nprot.2007.95 [pii]10.1038/nprot.2007.95

Deyholos MK (2010) Making the most of drought and salinity transcriptomics. Plant Cell Environ 33:648–654. doi:PCE2092 [pii]10.1111/j.1365-3040.2009.02092.x

Dupont FM, Hurkman WJ, Vensel WH, Tanaka CK, Kothari KM, Chung OK, Altenbach SB (2006) Protein accumulation and composition in wheat grains: effects of mineral nutrients and high temperature. Eur J Agron 25:96–107

Enfissi EM, Barneche F, Ahmed I, Lichtle C, Gerrish C, McQuinn RP, Giovannoni JJ, Lopez-Juez E, Bowler C, Bramley PM, Fraser PD (2010) Integrative transcript and metabolite analysis of nutritionally enhanced DE-ETIOLATED1 downregulated tomato fruit. Plant Cell 22:1190–1215. doi:tpc.110.073866 [pii]10.1105/tpc.110.073866

Eveland AL, McCarty DR, Koch KE (2008) Transcript profiling by 3′-untranslated region sequencing resolves expression of gene families. Plant Physiol 146:32–44. doi:pp.107.108597 [pii]10.1104/pp.107.108597

Ferry-Dumazet H, Gil L, Deborde C, Moing A, Bernillon S, Rolin D, Nikolski M, de Daruvar A, Jacob D (2011) MeRy-B: a web knowledgebase for the storage, visualization, analysis and annotation of plant NMR metabolomic profiles. BMC Plant Biol 11:104. doi:1471-2229-11-104 [pii]10.1186/1471-2229-11-104

Ficklin SP, Luo F, Feltus FA (2010) The association of multiple interacting genes with specific phenotypes in rice using gene coexpression networks. Plant Physiol 154:13–24. doi:pp.110.159459 [pii]10.1104/pp.110.159459

Finka A, Mattoo RU, Goloubinoff P (2011) Meta-analysis of heat and chemically upregulated chaperone genes in plant and human cells. Cell Stress Chaperones 16:15–31. doi:10.1007/s12192-010-0216-8

Finkel E (2009) Imaging. With 'phenomics', plant scientists hope to shift breeding into overdrive. Science 325:380–381. doi:325/5939/380 [pii]10.1126/science.325_380

Frank G, Pressman E, Ophir R, Althan L, Shaked R, Freedman M, Shen S, Firon N (2009) Transcriptional profiling of maturing tomato (*Solanum lycopersicum* L.) microspores reveals the involvement of heat shock proteins, ROS scavengers, hormones, and sugars in the heat stress response. J Exp Bot 60:3891–3908. doi:erp234 [pii]10.1093/jxb/erp234

Friedel S, Usadel B, von Wiren N, Sreenivasulu N (2012) Reverse engineering: a key component of systems biology to unravel global abiotic stress cross-talk. Front Plant Sci 3:294. doi:10.3389/fpls.2012.00294

Furbank RT (2009) Plant phenomics: from gene to form and function. Funct Plant Biol 36:V–VI

Gao H, Brandizzi F, Benning C, Larkin RM (2008) A membrane-tethered transcription factor defines a branch of the heat stress response in *Arabidopsis thaliana*. Proc Natl Acad Sci U S A 105:16398–16403. doi:0808463105 [pii]10.1073/pnas.0808463105

Garber M, Grabherr MG, Guttman M, Trapnell C (2011) Computational methods for transcriptome annotation and quantification using RNA-seq. Nat Methods 8:469–477. doi:10.1038/Nmeth.1613

Garg R, Patel RK, Jhanwar S, Priya P, Bhattacharjee A, Yadav G, Bhatia S, Chattopadhyay D, Tyagi AK, Jain M (2011) Gene discovery and tissue-specific transcriptome analysis in chickpea with massively parallel pyrosequencing and web resource development. Plant Physiol 156:1661–1678. doi:pp.111.178616 [pii]10.1104/pp.111.178616

Gowik U, Brautigam A, Weber KL, Weber AP, Westhoff P (2011) Evolution of C4 photosynthesis in the genus Flaveria: how many and which genes does it take to make C4? Plant Cell 23:2087–2105. doi:tpc.111.086264 [pii]10.1105/tpc.111.086264

Guo SJ, Zhou HY, Zhang XS, Li XG, Meng QW (2007) Overexpression of CaHSP26 in transgenic tobacco alleviates photoinhibition of PSII and PSI during chilling stress under low irradiance. J Plant Physiol 164:126–136. doi:S0176-1617(06)00048-4 [pii]10.1016/j.jplph.2006.01.004

Gupta SC, Sharma A, Mishra M, Mishra R, Chowdhuri DK (2010) Heat shock proteins in toxicology: how close and how far? Life Sci 86:377–384

Hagel J, Facchini PJ (2008) Plant metabolomics: analytical platforms and integration with functional genomics. Phytochem Rev 7:479–497

Hall AE (1992) Breeding for heat tolerance. Plant Breeding Rev 10:129–168

Hall AE (2001) Crop responses to environment. CRC Press LLC, Boca Raton

Haralampidis K, Milioni D, Rigas S, Hatzopoulos P (2002) Combinatorial interaction of cis elements specifies the expression of the *Arabidopsis* AtHsp90-1 gene. Plant Physiol 129:1138–1149. doi:10.1104/pp. 004044

Harada K, Fukusaki E (2009) Profiling of primary metabolite by means of capillary electrophoresis-mass spectrometry and its application for plant science. Plant Biotechnol 26:47–52. doi:10.5511/plantbiotechnology.26.47

Helmy M, Tomita M, Ishihama Y (2011) OryzaPG-DB: rice proteome database based on shotgun proteogenomics. BMC Plant Biol 11:63. doi:1471-2229-11-63 [pii]10.1186/1471-2229-11-63

Hernandez G, Ramirez M, Valdes-Lopez O, Tesfaye M, Graham MA, Czechowski T, Schlereth A, Wandrey M, Erban A, Cheung F, Wu HC, Lara M, Town CD, Kopka J, Udvardi MK, Vance CP (2007) Phosphorus stress in common bean: root transcript and metabolic responses. Plant Physiol 144:752–767. doi:pp.107.096958 [pii]10.1104/pp.107.096958

Hewezi T, Howe P, Maier TR, Hussey RS, Mitchum MG, Davis EL, Baum TJ (2008) Cellulose binding protein from the parasitic nematode *Heterodera schachtii* interacts with *Arabidopsis* pectin methylesterase: cooperative cell wall modification during parasitism. Plant Cell 20:3080–3093. doi:tpc.108.063065 [pii]10.1105/tpc.108.063065

Hong SW, Vierling E (2000) Mutants of *Arabidopsis thaliana* defective in the acquisition of tolerance to high temperature stress. Proc Natl Acad Sci U S A 97:4392–4397. doi:97/8/4392

Huhman DV, Sumner LW (2002) Metabolic profiling of saponins in *Medicago sativa* and *Medicago truncatula* using HPLC coupled to an electrospray ion-trap mass spectrometer. Phytochemistry 59:347–360

Hussain SS, Mudasser M (2007) Prospects for wheat production under changing climate in mountain areas of Pakistan- an econometric analysis. Agric Syst 94:494–501

Iijima Y, Nakamura Y, Ogata Y, Tanaka K, Sakurai N, Suda K, Suzuki T, Suzuki H, Okazaki K, Kitayama M, Kanaya S, Aoki K, Shibata D (2008) Metabolite annotations based on the integration of mass spectral information. Plant J 54:949–962. doi:TPJ3434 [pii]10.1111/j.1365-313X.2008.03434.x

Jiang C, Xu J, Zhang H, Zhang X, Shi J, Li M, Ming F (2009) A cytosolic class I small heat shock protein, RcHSP17.8, of *Rosa chinensis* confers resistance to a variety of stresses to *Escherichia coli*, yeast and *Arabidopsis thaliana*. Plant Cell Environ 32:1046–1059. doi:PCE1987 [pii]10.1111/j.1365-3040.2009.01987.x

Jordan KW, Nordenstam J, Lauwers GY, Rothenberger DA, Alavi K, Garwood M, Cheng LL (2009) Metabolomic characterization of human rectal adenocarcinoma with intact tissue magnetic resonance spectroscopy. Dis Colon Rectum 52:520–525. doi:10.1007/DCR.0b013e31819c9a2c00003453-200903000-00024

Joshi HJ, Hirsch-Hoffmann M, Baerenfaller K, Gruissem W, Baginsky S, Schmidt R, Schulze WX, Sun Q, van Wijk KJ, Egelhofer V, Wienkoop S, Weckwerth W, Bruley C, Rolland N, Toyoda T, Nakagami H, Jones AM, Briggs SP, Castleden I, Tanz SK, Millar AH, Heazlewood JL (2011) MASCP Gator: an aggregation portal for the visualization of *Arabidopsis* proteomics data. Plant Physiol 155:259–270

Joung JG, Corbett AM, Fellman SM, Tieman DM, Klee HJ, Giovannoni JJ, Fei Z (2009) Plant MetGenMAP: an integrative analysis system for plant systems biology. Plant Physiol 151:1758–1768. doi:pp.109.145169 [pii]10.1104/pp.109.145169

Kaplan F, Kopka J, Haskell DW, Zhao W, Schiller KC, Gatzke N, Sung DY, Guy CL (2004) Exploring the temperature-stress metabolome of *Arabidopsis*. Plant Physiol 136:4159–4168. doi:pp.104.052142 [pii]10.1104/pp.104.052142

Kapushesky M, Adamusiak T, Burdett T, Culhane A, Farne A, Filippov A, Holloway E, Klebanov A, Kryvych N, Kurbatova N et al (2012) Gene expression atlas update-a value-added database of microarray and sequencing-based functional genomics experiments. Nucleic Acids Res 40:D1077–D1081

Katiyar-Agarwal S, Agarwal M, Grover A (2003) Heat-tolerant basmati rice engineered by over-expression of hsp101. Plant Mol Biol 51:677–686

Keles Y, Oncel I (2002) Response of antioxidative defence system to temperature and water stress combinations in wheat seedlings. Plant Sci 163:783–790

Khurana N, Chauhan H, Khurana P (2013) Wheat chloroplast targeted sHSP26 promoter confers heat and abiotic stress inducible expression in transgenic *Arabidopsis* Plants. PLoS One 8:e54418. doi:10.1371/journal.pone.0054418PONE-D-12-26816

Kotak S, Vierling E, Baumlein H, von Koskull-Doring P (2007) A novel transcriptional cascade regulating expression of heat stress proteins during seed development of *Arabidopsis*. Plant Cell 19:182–195. doi:tpc.106.048165 [pii]10.1105/tpc.106.048165

Koussevitzky S, Suzuki N, Huntington S, Armijo L, Sha W, Cortes D, Shulaev V, Mittler R (2008) Ascorbate peroxidase 1 plays a key role in the response of *Arabidopsis thaliana* to stress combination. J Biol Chem 283:34197–34203. doi:M806337200 [pii]10.1074/jbc.M806337200

Krishnan P, Kruger NJ, Ratcliffe RG (2005) Metabolite fingerprinting and profiling in plants using NMR. J Exp Bot 56:255–265. doi:eri010 [pii]10.1093/jxb/eri010

Larkindale J, Vierling E (2008) Core genome responses involved in acclimation to high temperature. Plant Physiol 146:748–761. doi:pp.107.112060 [pii]10.1104/pp.107.112060

Le Lay P, Isaure MP, Sarry JE, Kuhn L, Fayard B, Le Bail JL, Bastien O, Garin J, Roby C, Bourguignon J (2006) Metabolomic, proteomic and biophysical analyses of *Arabidopsis thaliana* cells exposed to a caesium stress. Influence of potassium supply. Biochimie 88:1533–1547. doi:S0300-9084(06)00050-2 [pii]10.1016/j.biochi.2006.03.013

Lee JH, Hübel A, Schöffl F (1995) Derepression of the activity of genetically engineered heat shock factor causes constitutive synthesis of heat shock protein and increased thermotolerance in transgenic *Arabidopsis*. Plant J 8:603–612. doi:10.1046/j.1365-313X.1995.8040603.x

Lee JH, Schoffl F (1996) An Hsp70 antisense gene affects the expression of HSP70/HSC70, the regulation of HSF, and the acquisition of thermotolerance in transgenic *Arabidopsis thaliana*. Mol Gen Genet 252:11–19

Lee KP, Kim C, Landgraf F, Apel K (2007) EXECUTER1- and EXECUTER2-dependent transfer of stress-related signals from the plastid to the nucleus of *Arabidopsis thaliana*. Proc Natl Acad Sci U S A 104:10270–10275. doi:0702061104 [pii]10.1073/pnas.0702061104

Lee TH, Kim YK, Pham TT, Song SI, Kim JK, Kang KY, An G, Jung KH, Galbraith DW, Kim M, Yoon UH, Nahm BH (2009) RiceArrayNet: a database for correlating gene expression from transcriptome profiling, and its application to the analysis of coexpressed genes in rice. Plant Physiol 151:16–33. doi:pp.109.139030 [pii]10.1104/pp.109.139030

Lee KW, Cha JY, Kim KH, Kim YG, Lee BH, Lee SH (2012) Overexpression of alfalfa mitochondrial HSP23 in prokaryotic and eukaryotic model systems confers enhanced tolerance to salinity and arsenic stress. Biotechnol Lett 34:167–174. doi:10.1007/s10529-011-0750-1

Lehninger AL, Nelson DL, Cox MM (1993) Principles of biochemistry second edition. Worth, New York

Li C, Chen Q, Gao X, Qi B, Chen N, Xu S, Chen J, Wang X (2005) AtHsfA2 modulates expression of stress responsive genes and enhances tolerance to heat and oxidative stress in *Arabidopsis*. Sci China C Life Sci 48:540–550

Li P, Ponnala L, Gandotra N, Wang L, Si Y, Tausta SL, Kebrom TH, Provart N, Patel R, Myers CR, Reidel EJ, Turgeon R, Liu P, Sun Q, Nelson T, Brutnell TP (2010) The developmental dynamics of the maize leaf transcriptome. Nat Genet 42:1060–1067. doi: ng.703 [pii]10.1038/ng.703

Libault M, Farmer A, Joshi T, Takahashi K, Langley RJ, Franklin LD, He J, Xu D, May G, Stacey G (2010) An integrated transcriptome atlas of the crop model *Glycine max*, and its use in comparative analyses in plants. Plant J 63:86–99. doi:TPJ4222 [pii]10.1111/j.1365-313X.2010.04222.x

Lim CJ, Yang KA, Hong JK, Choi JS, Yun DJ, Hong JC, Chung WS, Lee SY, Cho MJ, Lim CO (2006) Gene expression profiles during heat acclimation in *Arabidopsis thaliana* suspension-culture cells. J Plant Res 119:373–383. doi:10.1007/s10265-006-0285-z

Liu J, Jung C, Xu J, Wang H, Deng S, Bernad L, Arenas-Huertero C, Chua NH (2012) Genome-wide analysis uncovers regulation of long intergenic noncoding RNAs in *Arabidopsis*. Plant Cell 24:4333–4345. doi: tpc.112.102855 [pii]10.1105/tpc.112.102855

Liu JG, Qin QL, Zhang Z, Peng RH, Xiong AS, Chen JM, Yao QH (2009) OsHSF7 gene in rice, Oryza sativa L., encodes a transcription factor that functions as a high temperature receptive and responsive factor. BMB Rep 42:16–21

Long TA, Brady SM, Benfey PN (2008) Systems approaches to identifying gene regulatory networks in plants. Annu Rev Cell Dev Biol 24:81–103. doi:10.1146/annurev.cellbio.24.110707.175408

Lu T, Lu G, Fan D, Zhu C, Li W, Zhao Q, Feng Q, Zhao Y, Guo Y, Huang X, Han B (2010) Function annotation of the rice transcriptome at single-nucleotide resolution by RNA-seq. Genome Res 20:1238–1249. doi: gr.106120.110 [pii]10.1101/gr.106120.110

Luengwilai K, Saltveit M, Beckles DM (2012) Metabolite content of harvested Micro-Tom tomato (*Solanum lycopersicum* L.) fruit is altered by chilling and protective heat-shock treatments as shown by GC–MS metabolic profiling. Postharvest Biol Technol 63:116–122

Lugan R, Niogret MF, Leport L, Guegan JP, Larher FR, Savoure A, Kopka J, Bouchereau A (2010) Metabolome and water homeostasis analysis of *Thellungiella salsuginea* suggests that dehydration tolerance is a key response to osmotic stress in this halophyte. Plant J 64:215–229. doi:10.1111/j.1365-313X.2010.04323.x

Ma S, Bohnert HJ (2007) Integration of *Arabidopsis thaliana* stress-related transcript profiles, promoter structures, and cell-specific expression. Genome Biol 8: R49. doi:gb-2007-8-4-r49 [pii]10.1186/gb-2007-8-4-r49

Majoul T, Bancel E, Triboi E, Ben Hamida J, Branlard G (2003) Proteomic analysis of the effect of heat stress on hexaploid wheat grain: characterization of heat-responsive proteins from total endosperm. Proteomics 3:175–183. doi:10.1002/pmic.200390026

Malik MK, Slovin JP, Hwang CH, Zimmerman JL (1999) Modified expression of a carrot small heat shock protein gene, hsp17.7, results in increased or decreased thermotolerancedouble dagger. Plant J 20:89–99. doi: tpj581

Mangelsen E, Kilian J, Harter K, Jansson C, Wanke D, Sundberg E (2011) Transcriptome analysis of high-temperature stress in developing barley caryopses: early stress responses and effects on storage compound biosynthesis. Mol Plant 4:97–115. doi:ssq058 [pii]10.1093/mp/ssq058

Maruyama K, Takeda M, Kidokoro S, Yamada K, Sakuma Y, Urano K, Fujita M, Yoshiwara K, Matsukura S, Morishita Y, Sasaki R, Suzuki H, Saito K, Shibata D, Shinozaki K, Yamaguchi-Shinozaki K (2009) Metabolic pathways involved in cold acclimation identified by integrated analysis of metabolites and transcripts regulated by DREB1A and DREB2A. Plant Physiol 150:1972–1980

Matsuura H, Takenami S, Kubo Y, Ueda K, Ueda A, Yamaguchi M, Hirata K, Demura T, Kanaya S, Kato K (2013) A Computational and experimental approach reveals that the 5′-proximal region of the 5′-UTR has a cis-regulatory signature responsible for heat stress-regulated mRNA translation in *Arabidopsis*. Plant Cell Physiol 54:474–483

Mishra SK, Tripp J, Winkelhaus S, Tschiersch B, Theres K, Nover L, Scharf KD (2002) In the complex family of heat stress transcription factors, HSfA1 has a unique role as master regulator of thermotolerance in tomato. Genes Dev 16:1555–1567. doi:10.1101/gad.228802

Mittal D, Chakrabarti S, Sarkar A, Singh A, Grover A (2009) Heat shock factor gene family in rice: genomic organization and transcript expression profiling in response to high temperature, low temperature and oxidative stresses. Plant Physiol Biochem 47:785–795. doi:S0981-9428(09)00122-3 [pii]10.1016/j.plaphy.2009.05.003

Mittal D, Madhyastha DA, Grover A (2012) Genome-wide transcriptional profiles during temperature and oxidative stress reveal coordinated expression patterns and overlapping regulons in rice. PLoS One 7:e40899. doi:10.1371/journal.pone.0040899PONE-D-12-02988 [pii]

Moco S, Bino RJ, Vorst O, Verhoeven HA, de Groot J, van Beek TA, Vervoort J, de Vos CH (2006) A liquid chromatography-mass spectrometry-based metabolome database for tomato. Plant Physiol 141:1205–1218. doi:141/4/1205 [pii]10.1104/pp.106.078428

Mogk A, Schlieker C, Friedrich KL, Schonfeld HJ, Vierling E, Bukau B (2003) Refolding of substrates bound to small Hsps relies on a disaggregation reaction mediated most efficiently by ClpB/DnaK. J Biol Chem 278:31033–31042. doi:10.1074/jbc.M303587200M303587200

Montero-Barrientos M, Hermosa R, Cardoza RE, Gutierrez S, Nicolas C, Monte E (2010) Transgenic expression of the Trichoderma harzianum hsp70 gene increases Arabidopsis resistance to heat and other abiotic stresses. J Plant Physiol 167:659–665. doi:10.1016/j.jplph.2009.11.012

Moriwaki M, Yamakawa T, Washino T, Kodama T, Igarashi Y (1999) Delayed recovery of β-glucuronidase activity driven by an *Arabidopsis* heat shock promoter in heat-stressed transgenic *Nicotiana plumbaginifolia*. Plant Cell Rep 19:96–100

Morozova O, Marra MA (2008) Applications of next-generation sequencing technologies in functional genomics. Genomics 92:255–264. doi:10.1016/j.ygeno.2008.07.001

Movahedi S, Van de Peer Y, Vandepoele K (2011) Comparative network analysis reveals that tissue specificity and gene function are important factors influencing the mode of expression evolution in *Arabidopsis* and rice. Plant Physiol 156:1316–1330. doi:pp.111.177865 [pii]10.1104/pp.111.177865

Murakami T, Matsuba S, Funatsuki H, Kawaguchi K, Saruyama H, Tanida M, Sato Y (2004) Over-expression of a small heat shock protein, sHSP17.7, confers both heat tolerance and UV-B resistance to rice plants. Mol Breeding 13:165–175

Nagel KA, Kastenholz B, Jahnke S, van Dusschoten D, Aach T (2009) Temperature responses of roots: impact on growth, root system architecture and implications for phenotyping. Funct Plant Biol 36:947–959. doi:10.1071/FP09184

Nakagami H, Sugiyama N, Ishihama Y, Shirasu K (2012) Shotguns in the front line: phosphoproteomics in plants. Plant Cell Physiol 53:118–124. doi:10.1093/pcp/pcr148

Nanjo Y, Skultety L, Ashraf Y, Komatsu S (2010) Comparative proteomic analysis of early-stage soybean seedlings responses to flooding by using gel and gel-free techniques. J Proteome Res 9:3989–4002. doi:10.1021/pr100179f

Neilson KA, Gammulla CG, Mirzaei M, Imin N, Haynes PA (2010) Proteomic analysis of temperature stress in plants. Proteomics 10:828–845. doi:10.1002/pmic.200900538

Nieto-Sotelo J, Martinez LM, Ponce G, Cassab GI, Alagon A, Meeley RB, Ribaut JM, Yang R (2002) Maize HSP101 plays important roles in both induced and basal thermotolerance and primary root growth. Plant Cell 14:1621–1633

Nikiforova VJ, Daub CO, Hesse H, Willmitzer L, Hoefgen R (2005) Integrative gene-metabolite network with implemented causality deciphers informational fluxes of sulphur stress response. J Exp Bot 56:1887–1896. doi:eri179 [pii]10.1093/jxb/eri179

Nishizawa A, Yabuta Y, Yoshida E, Maruta T, Yoshimura K, Shigeoka S (2006) *Arabidopsis* heat shock transcription factor A2 as a key regulator in response to several types of environmental stress. Plant J 48:535–547. doi:TPJ2889 [pii]10.1111/j.1365-313X.2006.02889.x

Novaes E, Drost DR, Farmerie WG, Pappas GJ, Grattapaglia D, Sederoff RR, Kirst M (2008) High-throughput gene and SNP discovery in *Eucalyptus grandis*, an uncharacterized genome. BMC Genomics 9. doi:Artn 312. doi:10.1186/1471-2164-9-312

Nover L, Bharti K, Doring P, Mishra SK, Ganguli A, Scharf KD (2001) *Arabidopsis* and the heat stress transcription factor world: how many heat stress transcription factors do we need? Cell Stress Chaperones 6:177–189

Ogawa D, Yamaguchi K, Nishiuchi T (2007) High-level overexpression of the Arabidopsis HsfA2 gene confers not only increased themotolerance but also salt/osmotic stress tolerance and enhanced callus growth. J Exp Bot 58:3373–3383

Oikawa A, Nakamura Y, Ogura T, Kimura A, Suzuki H, Sakurai N, Shinbo Y, Shibata D, Kanaya S, Ohta D (2006) Clarification of pathway-specific inhibition by Fourier transform ion cyclotron resonance/mass spectrometry-based metabolic phenotyping studies. Plant Physiol 142:398–413. doi:pp.106.080317 [pii] 10.1104/pp.106.080317

Oshino T, Abiko M, Saito R, Ichiishi E, Endo M, Kawagishi-Kobayashi M, Higashitani A (2007) Premature progression of anther early developmental programs accompanied by comprehensive alterations in transcription during high-temperature injury in barley plants. Mol Genet Genomics 278:31–42. doi:10.1007/s00438-007-0229-x

Palmblad M, Mills DJ, Bindschedler LV (2008) Heat-shock response in *Arabidopsis thaliana* explored by multiplexed quantitative proteomics using differential metabolic labeling. J Proteome Res 7:780–785. doi:10.1021/pr0705340

Panchuk II, Volkov RA, Schoffl F (2002) Heat stress- and heat shock transcription factor-dependent expression and activity of ascorbate peroxidase in Arabidopsis. Plant Physiol 129:838–853

Park SM, Hong CB (2002) Class I small heat shock protein gives thermotolerance in tobacco. J. Plant Physiol 159:25–30

Pike CS, Grieve J, Badger MR, Price GD (2001) Thermoprotective properties of small heat shock proteins from rice, tomato and *Synechocystis* sp PCC6803 overexpressed in, and isolated from, *Escherichia coli*. Aust J Plant Physiol 28:1219–1229

Polenta GA, Calvete JJ, Gonzalez CB (2007) Isolation and characterization of the main small heat shock proteins induced in tomato pericarp by thermal treatment. FEBS J 274:6447–6455. doi:EJB6162 [pii]10.1111/j.1742-4658.2007.06162.x

Prändl R, Hinderhofer K, Eggers-Schumacher G, Schöffl F (1998) HSF3, a new heat shock factor from *Arabidopsis thaliana*, derepresses the heat shock response and confers thermotolerance when overexpressed in transgenic plants. Mol Gen Genet 258:269–278. doi:10.1007/s004380050731

Proveniers MC, van Zanten M (2013) High temperature acclimation through PIF4 signaling. Trends Plant Sci

18:59–64. doi:S1360-1385(12)00209-9 [pii]10.1016/j.tplants.2012.09.002
Qi Y, Wang H, Zou Y, Liu C, Liu Y, Wang Y, Zhang W (2011) Over-expression of mitochondrial heat shock protein 70 suppresses programmed cell death in rice. FEBS Lett 585:231–239. doi:10.1016/j.febslet.2010.11.051
Qin D, Wu H, Peng H, Yao Y, Ni Z, Li Z, Zhou C, Sun Q (2008) Heat stress-responsive transcriptome analysis in heat susceptible and tolerant wheat (*Triticum aestivum* L.) by using Wheat Genome Array. BMC Genomics 9:432. doi:1471-2164-9-432 [pii]10.1186/1471-2164-9-432
Queitsch C, Hong SW, Vierling E, Lindquist S (2000) Heat shock protein 101 plays a crucial role in thermotolerance in *Arabidopsis*. Plant Cell 12:479–492
Rahman H ur, Malik SA, Saleem M, Hussain F (2007) Evaluation of seed physical traits in relation to heat tolerance in upland cotton. Pakistan J Bot 39:475–483
Rasmussen S, Barah P, Suarez-Rodriguez MC, Bressendorff S, Friis P, Costantino P, Bones AM, Nielsen HB, Mundy J (2013) Transcriptome responses to combinations of stresses in *Arabidopsis*. Plant Physiol 161:1783–1794
Reddy RA, Kumar B, Reddy PS, Mishra RN, Mahanty S, Kaul T, Nair S, Sopory SK, Reddy MK (2009) Molecular cloning and characterization of genes encoding *Pennisetum glaucum* ascorbate peroxidase and heat-shock factor: interlinking oxidative and heat-stress responses. J Plant Physiol 166:1646–1659. doi:S0176-1617(09)00134-5 [pii]10.1016/j.jplph.2009.04.007
Reddy PS, Mallikarjuna G, Kaul T, Chakradhar T, Mishra RN, Sopory SK, Reddy MK (2010) Molecular cloning and characterization of gene encoding for cytoplasmic Hsc70 from *Pennisetum glaucum* may play a protective role against abiotic stresses. Mol Genet Genomics 283:243–254. doi:10.1007/s00438-010-0518-7
Reddy PS, Thirulogachandar V, Vaishnavi CS, Aakruti A, Sopory SK, Reddy MK (2011) Molecular characterization and expression of a gene encoding cytosolic Hsp90 from *Pennisetum glaucum* and its role in abiotic stress adaptation. Gene 474:29–38. doi:S0378-1119(10)00459-2 [pii]10.1016/j.gene.2010.12.004
Rinalducci S, Egidi MG, Karimzadeh G, Jazii FR, Zolla L (2011) Proteomic analysis of a spring wheat cultivar in response to prolonged cold stress. Electrophoresis 32:1807–1818. doi:10.1002/elps.201000663
Rizhsky L, Liang H, Shuman J, Shulaev V, Davletova S, Mittler R (2004) When defense pathways collide. The response of *Arabidopsis* to a combination of drought and heat stress. Plant Physiol 134:1683–1696. doi:10.1104/pp. 103.033431pp.103.033431
Sadok W, Naudin P, Boussuge B, Muller B, Welcker C, Tardieu F (2007) Leaf growth rate per unit thermal time follows QTL-dependent daily patterns in hundreds of maize lines under naturally fluctuating conditions. Plant Cell Environ 30:135–146
Saidi Y, Domini M, Choy F, Zryd JP, Schwitzguebel JP, Goloubinoff P (2007) Activation of the heat shock response in plants by chlorophenols: transgenic *Physcomitrella patens* as a sensitive biosensor for organic pollutants. Plant Cell Environ 30:753–763. doi:PCE1664 [pii]10.1111/j.1365-3040.2007.01664.x
Saito K, Hirai MY, Yonekura-Sakakibara K (2008) Decoding genes with coexpression networks and metabolomics – 'majority report by precogs'. Trends Plant Sci 13:36–43
Sakuma Y, Maruyama K, Osakabe Y, Qin F, Seki M, Shinozaki K, Yamaguchi-Shinozaki K (2006a) Functional analysis of an *Arabidopsis* transcription factor, DREB2A, involved in drought-responsive gene expression. Plant Cell 18:1292–1309
Sakuma Y, Maruyama K, Qin F, Osakabe Y, Shinozaki K, Yamaguchi-Shinozaki K (2006b) Dual function of an *Arabidopsis* transcription factor DREB2A in water-stress-responsive and heat-stress-responsive gene expression. Proc Natl Acad Sci U S A 103:18822–18827. doi:0605639103 [pii]10.1073/pnas.0605639103
Sakurai N, Ara T, Ogata Y, Sano R, Ohno T, Sugiyama K (2011) KaPPA-View4: a metabolic pathway database for representation and analysis of correlation networks of gene co-expression and metabolite co-accumulation and omics data. Nucleic Acids Res 39:D677–D684
Sakurai T, Yamada Y, Sawada Y, Matsuda F, Akiyama K, Shinozaki K, Hirai MY, Saito K (2013) PRIMe Update: innovative content for plant metabolomics and integration of gene expression and metabolite accumulation. Plant Cell Physiol 54:e5. doi:pcs184 [pii]10.1093/pcp/pcs184
Sanmiya K, Suzuki K, Egawa Y, Shono M (2004) Mitochondrial small heat-shock protein enhances thermotolerance in tobacco plants. FEBS Lett 557:265–268. doi:S0014579303014947
Sarkar NK, Kim YK, Grover A (2009) Rice sHsp genes: genomic organization and expression profiling under stress and development. BMC Genomics 10:393. doi:1471-2164-10-393 [pii]10.1186/1471-2164-10-393
Sawada Y, Akiyama K, Sakata A, Kuwahara A, Otsuki H, Sakurai T, Saito K, Hirai MY (2009) Widely targeted metabolomics based on large-scale MS/MS data for elucidating metabolite accumulation patterns in plants. Plant Cell Physiol 50:37–47. doi:pcn183 [pii]10.1093/pcp/pcn183
Schaeffer ML, Harper LC, Gardiner JM, Andorf CM, Campbell DA, Cannon EK, Sen TZ, Lawrence CJ (2011) MaizeGDB: curation and outreach go hand-in-hand. Database (Oxford). bar022.doi:bar022 [pii]10.1093/database/bar022
Scharf KD, Berberich T, Ebersberger I, Nover L (2012) The plant heat stress transcription factor (Hsf) family: structure, function and evolution. Biochim Biophys Acta 1819:104–119. doi:S1874-9399(11)00178-7 [pii]10.1016/j.bbagrm.2011.10.002

Schmidt R, Schippers JH, Welker A, Mieulet D, Guiderdoni E, Mueller-Roeber B (2012) Transcription factor OsHsfC1b regulates salt tolerance and development in Oryza sativa ssp. japonica. AoB Plants 2012: pls011. doi: 10.1093/aobpla/pls011

Schneider M, Consortium TU, Poux S (2012) UniProtKB amid the turmoil of plant proteomics research. Front Plant Sci 3:270

Schöffl F, Prändl R, Reindl A (1998) Regulation of the heat-shock response. Plant Physiol 117:1135–1141

Schramm F, Ganguli A, Kiehlmann E, Englich G, Walch D, von Koskull-Doring P (2006) The heat stress transcription factor HsfA2 serves as a regulatory amplifier of a subset of genes in the heat stress response in *Arabidopsis*. Plant Mol Biol 60:759–772. doi:10.1007/s11103-005-5750-x

Schreiber F, Colmsee C, Czauderna T, Grafahrend-Belau E, Hartmann A, Junker A, Junker BH, Klapperstuck M, Scholz U, Weise S (2012) MetaCrop 2.0: managing and exploring information about crop plant metabolism. Nucleic Acids Res 40(Database issue): D1173–D1177. doi:gkr1004 [pii]10.1093/nar/gkr1004

Seepaul R, Macoon B, Reddy KR, Baldwin B (2011) Switchgrass (*Panicum virgatum* L.) intraspecific variation and thermotolerance classification using *in vitro* seed germination assay. Am J Plant Sci 2:134–147

Severin AJ, Woody JL, Bolon YT, Joseph B, Diers BW, Farmer AD, Muehlbauer GJ, Nelson RT, Grant D, Specht JE, Graham MA, Cannon SB, May GD, Vance CP, Shoemaker RC (2010) RNA-Seq Atlas of *Glycine max*: a guide to the soybean transcriptome. BMC Plant Biol 10:160. doi:10.1186/1471-2229-10-160

Sharma RC, Crossa J, Velu G, Huerta-Espino J, Vargas M, Payne TS, Singh RP (2012) Genetic gains for grain yield in CIMMYT spring bread wheat across international environments. Crop Sci 52:1522–1533

Shinmyo A, Shoji T, Bando E, Nagaya S, Nakai Y, Kato K, Sekine M, Yoshida K (1998) Metabolic engineering of cultured tobacco cells. Biotechnol Bioeng 58:329–332. doi:10.1002/(SICI)1097-0290(19980420)58:2/3<329::AID-BIT34>3.0.CO;2-4

Skylas DJ, Cordwell SJ, Hains PG, Larsen MR, Basseal DJ, Walsh BJ, Blumenthal C, Rathmell W, Copeland L, Wrigley CW (2002) Heat shock of wheat during grain filling: proteins associated with heat-tolerance. J Cereal Sci 35:175–188. doi:UNSP jcrs.2001.0410. doi:10.1006/jcrs.2001.0410

Smita S, Katiyar A, Pandey DM, Chinnusamy V, Archak S, Bansal KC (2013) Identification of conserved drought stress responsive gene-network across tissues and developmental stages in rice. Bioinformation 9:72–78. doi:10.6026/97320630009072973206300 09072

Snider JL, Oosterhuis DM (2011) How does timing, duration and severity of heat stress influence pollen-pistil interactions in angiosperms? Plant Signal Behav 6:930–933

Song NH, Ahn YJ (2011) DcHsp17.7, a small heat shock protein in carrot, is tissue-specifically expressed under salt stress and confers tolerance to salinity. Nat Biotechnol 28:698–704. doi:S1871-6784(11)00089-6 [pii]10.1016/j.nbt.2011.04.002

Sreenivasulu N, Usadel B, Winter A, Radchuk V, Scholz U, Stein N, Weschke W, Strickert M, Close TJ, Stitt M, Graner A, Wobus U (2008) Barley grain maturation and germination: metabolic pathway and regulatory network commonalities and differences highlighted by new MapMan/PageMan profiling tools. Plant Physiol 146:1738–1758. doi: pp.107.111781 [pii]10.1104/pp.107.111781

Sreenivasulu N, Sunkar R, Wobus U, Strickert M (2010) Array platforms and bioinformatics tools for the analysis of plant transcriptome in response to abiotic stress. Methods Mol Biol 639:71–93. doi:10.1007/978-1-60761-702-0_5

Sule A, Vanrobaeys F, Hajos G, Van Beeumen J, Devreese B (2004) Proteomic analysis of small heat shock protein isoforms in barley shoots. Phytochemistry 65:1853–1863. doi:10.1016/j.phytochem.2004.03.030S0031942204001414

Sun Q, Zybailov B, Majeran W, Friso G, Olinares PD, van Wijk KJ (2009) PPDB, the Plant Proteomics Database at Cornell. Nucleic Acids Res 37(Database issue): D969–D974. doi:gkn654 [pii]10.1093/nar/gkn654

Sun L, Liu Y, Kong X, Zhang D, Pan J, Zhou Y, Wang L, Li D, Yang X (2012) ZmHSP16.9, a cytosolic class I small heat shock protein in maize (*Zea mays*), confers heat tolerance in transgenic tobacco. Plant Cell Rep 31:1473–1484. doi:10.1007/s00299-012-1262-8

Sung DY, Guy CL (2003) Physiological and molecular assessment of altered expression of Hsc70-1 in Arabidopsis. Evidence for pleiotropic consequences. Plant Physiol 132:979–987

Suzuki N, Koussevitzky S, Mittler R, Miller G (2011) ROS and redox signaling in the response of plants to abiotic stress. Plant Cell Environ 35:259–270

Swarbreck D, Wilks C, Lamesch P, Berardini TZ, Garcia-Hernandez M, Foerster H (2008) The *Arabidopsis* Information Resource (TAIR): gene structure and function annotation. Nucleic Acids Res 36:D1009–D1014

Takahashi T, Naito S, Komeda Y (1992) The *Arabidopsis* HSP18.2 promoter/GUS gene fusion in transgenic *Arabidopsis* plants: a powerful tool for the isolation of regulatory mutants of the heat-shock response. Plant J 2:751–761. doi:10.1111/j.1365-313X.1992.tb00144

Takahashi H, Takahara K, Hashida SN, Hirabayashi T, Fujimori T, Kawai-Yamada M, Yamaya T, Yanagisawa S, Uchimiya H (2009) Pleiotropic modulation of carbon and nitrogen metabolism in *Arabidopsis* plants overexpressing the NAD kinase2 gene. Plant Physiol 151:100–113. doi:pp.109.140665 [pii]10.1104/pp.109.140665

Uchida A, Hibino T, Shimada T, Saigusa M, Takabe T, Araki E, Kajita H, Takabe T (2008) Overexpression of DnaK chaperone from a halotolerant cyanobacterium Aphanothece halophytica increases seed yield in rice and tobacco. Plant Biotechnol 25:141–150

Urano K, Maruyama K, Ogata Y, Morishita Y, Takeda M, Sakurai N, Suzuki H, Saito K, Shibata D, Kobayashi M,

Yamaguchi-Shinozaki K, Shinozaki K (2009) Characterization of the ABA-regulated global responses to dehydration in *Arabidopsis* by metabolomics. Plant J 57:1065–1078. doi:TPJ3748 [pii]10.1111/j.1365-313X.2008.03748.x

Valcu CM, Lalanne C, Plomion C, Schlink K (2008) Heat induced changes in protein expression profiles of Norway spruce (*Picea abies*) ecotypes from different elevations. Proteomics 8(20):4287–4302. doi:10.1002/pmic.200700992

van Baarlen P, van Esse HP, Siezen RJ, Thomma BP (2008) Challenges in plant cellular pathway reconstruction based on gene expression profiling. Trends Plant Sci 13:44–50. doi:S1360-1385(07)00304-4 [pii] 10.1016/j.tplants.2007.11.003

von Koskull-Doring P, Scharf KD, Nover L (2007) The diversity of plant heat stress transcription factors. Trends Plant Sci 12:452–457. doi:S1360-1385(07)00193-8 [pii]10.1016/j.tplants.2007.08.014

Wang W, Vinocur B, Shoseyov O, Altman A (2004) Role of plant heat-shock proteins and molecular chaperones in the abiotic stress response. Trends Plant Sci 9:244–252. doi:10.1016/j.tplants.2004.03.006S1360-1385(04)00060-3

Wang W, Wang YJ, Zhang Q, Qi Y, Guo DJ (2009a) Global characterization of *Artemisia annua* glandular trichome transcriptome using 454 pyrosequencing. BMC Genomics 10. doi:Artn 465. doi:10.1186/1471-2164-10-465

Wang Y, Xiao J, Suzek TO, Zhang J, Wang J, Bryant SH (2009b) PubChem: a public information system for analyzing bioactivities of small molecules. Nucleic Acids Res 37(Web Server issue):w623–w633. doi:gkp456 [pii]10.1093/nar/gkp456

Wang M, Weiss M, Simonovic M, Haertinger G, Schrimpf SP, Hengartner MO (2012) PaxDb, a database of protein abundance averages across all three domains of life. Mol Cell Proteomics 11:492–500

Wei LQ, Xu WY, Deng ZY, Su Z, Xue Y, Wang T (2010) Genome-scale analysis and comparison of gene expression profiles in developing and germinated pollen in *Oryza sativa*. BMC Genomics 11:338. doi:10.1186/1471-2164-11-338

Weston DJ, Gunter LE, Rogers A, Wullschleger SD (2008) Connecting genes, coexpression modules, and molecular signatures to environmental stress phenotypes in plants. BMC Syst Biol 2:16. doi:1752-0509-2-16 [pii]10.1186/1752-0509-2-16

Weston DJ, Karve AA, Gunter LE, Jawdy SS, Yang X, Allen SM, Wullschleger SD (2011) Comparative physiology and transcriptional networks underlying the heat shock response in *Populus trichocarpa*, *Arabidopsis thaliana* and *Glycine max*. Plant Cell Environ 34:1488–1506. doi:10.1111/j.1365-3040.2011.02347.x

Wienkoop S, Morgenthal K, Wolschin F, Scholz M, Selbig J, Weckwerth W (2008) Integration of metabolomic and proteomic phenotypes: analysis of data covariance dissects starch and RFO metabolism from low and high temperature compensation response in *Arabidopsis thaliana*. Mol Cell Proteomics 7:1725–1736

Wienkoop S, Staudinger C, Hoehenwarter W, Weckwerth W, Egelhofer V (2012) ProMEX – a mass spectral reference database for plant proteomics. Front Plant Sci 3:125. doi:10.3389/fpls.2012.00125

Winter D, Vinegar B, Nahal H, Ammar R, Wilson GV (2007) An "Electronic Fluorescent Pictograph" Browser for Exploring and Analyzing Large-Scale Biological Data Sets. PLoS One 2(8):e718

Wolkovich EM, Cook BI, Allen JM, Crimmins TM, Betancourt JL, Travers SE, Pau S, Regetz J, Davies TJ, Kraft NJ, Ault TR, Bolmgren K, Mazer SJ, McCabe GJ, McGill BJ, Parmesan C, Salamin N, Schwartz MD, Cleland EE (2012) Warming experiments underpredict plant phenological responses to climate change. Nature 485:494–497. doi:nature11014 [pii]10.1038/nature11014

Wu C (1995) Heat shock transcription factors: structure and regulation. Annu Rev Cell Dev Biol 11:441–469

Xu J, Tian J, Belanger F, Huang B (2007) Identification and characterization of an expansin gene AsEXP1 associated with heat tolerance in C3 *Agrostis* grass species. J Exp Bot 58:3789–3796

Xu J, Belanger F, Huang B (2008) Differential gene expression in shoots and roots under heat stress for a geothermal and non-thermal *Agrostis* grass species contrasting in heat tolerance. Environ Exp Bot 63:240–247

Yabe N, Takahashi T, Komeda Y (1994) Analysis of tissue-specific expression of *Arabidopsis thaliana* HSP90-family gene HSP81. Plant Cell Physiol 35:1207–1219

Yamakawa H, Hakata M (2010) Atlas of rice grain filling-related metabolism under high temperature: joint analysis of metabolome and transcriptome demonstrated inhibition of starch accumulation and induction of amino acid accumulation. Plant Cell Physiol 51:795–809. doi:pcq034 [pii]10.1093/pcp/pcq034

Yang JY, Sun Y, Sun AQ, Yi SY, Qin J, Li MH, Liu J (2006) The involvement of chloroplast HSP100/ClpB in the acquired thermotolerance in tomato. Plant Mol Biol 62:385–395. doi:10.1007/s11103-006-9027-9

Yazdanbakhsh N, Fisahn J (2009) High throughput phenotyping of root growth dynamics, lateral root formation, root architecture and root hair development enabled by PlaRoM. Funct Plant Biol 36:938–946. doi:10.1071/FP09167

Yeh CH, Kaplinsky NJ, Hu C, Charng YY (2012) Some like it hot, some like it warm: phenotyping to explore thermotolerance diversity. Plant Sci 195:10–23

Yin YL, Yu GJ, Chen YJ, Jiang S, Wang M, Jin YX, Lan XQ, Liang Y, Sun H (2012) Genome-wide transcriptome and proteome analysis on different developmental stages of *Cordycepsmilitaris*. PLoS One 7 (12).doi:ARTN e51853. doi:10.1371/journal.pone.0051853

Youens-Clark K, Buckler E, Casstevens T, Chen C, Declerck G, Derwent P, Dharmawardhana P, Jaiswal P,

Kersey P, Karthikeyan AS, Lu J, McCouch SR, Ren L, Spooner W, Stein JC, Thomason J, Wei S, Ware D (2011) Gramene database in 2010: updates and extensions. Nucleic Acids Res 39(Database issue): D1085–D1094. doi:gkq1148 [pii]10.1093/nar/gkq1148
Yokotani N, Ichikawa T, Kondou Y, Matsui M, Hirochika H, Iwabuchi M, Oda K (2008) Expression of rice heat stress transcription factor OsHsfA2e enhances tolerance to environmental stresses in transgenic Arabidopsis. Planta 227:957–967
Zhang M, Li G, Huang W, Bi T, Chen G, Tang Z, Su W, Sun W (2010) Proteomic study of *Carissa spinarum* in response to combined heat and drought stress. Proteomics 10:3117–3129. doi:10.1002/pmic.200900637
Zhang H, Guo F, Zhou H, Zhu G (2012) Transcriptome analysis reveals unique metabolic features in the *Cryptosporidium parvum* Oocysts associated with environmental survival and stresses. BMC Genomics 13:647. doi:10.1186/1471-2164-13
Zhu H, Bilgin M, Snyder M (2003) Proteomics. Annu Rev Biochem 72:783–812. doi:10.1146/annurev.biochem.72.121801.161511
Zimmermann P, Hirsch-Hoffmann M, Hennig L, Gruissem W (2004) GENEVESTIGATOR. *Arabidopsis* microarray database and analysis toolbox. Plant Physiol 136:2621–2632. doi:10.1104/pp.104.046367136/1/2621

Comparative Genomics of Cereal Crops: Status and Future Prospects

Sujay Rakshit and K.N. Ganapathy

Abstract

Cereals are members of grass family and play an important role in providing food security to billions of people across the globe since the beginning of agriculture. Cereal crops differ considerably from each other in terms of morphology, adaptation and genetic architecture. This has motivated researchers across the world to study their evolution, genetics and development. During the last few decades, phenomenal progress in genomics research has paved the way for comparative genomic studies across crop species, especially the cereals due to their economic importance. These studies together have revealed a good level of conservation across cereals both at macro and micro level. However, most of the comparative studies in cereals prior to genome sequencing projects have been performed at genetic map level. Large-scale genome sequencing projects during the beginning of the twenty-first century, especially in rice, sorghum, maize, barley, wheat and foxtail millet, led to better understanding of the conservations of genes/genomic regions at sequence level. This chapter reviews the status of cereal comparative genomics prior to genome sequencing and progress post-genome sequencing of major cereals. The genomic organization of major cereals has been discussed in detail. The chapter also describes the distinguishing features and the mechanism of evolution of cereal genomes. The advancement in genome sequencing technologies, especially the next-generation sequencing technologies and its effectiveness in performing genomic studies across crops, is also discussed. The various genomic tools, databases and resources for performing comparative genomic studies are reviewed here. The chapter presents the opportunities on how the knowledge gained from comparative genomics of cereals can be used for gene discovery programmes, functional genomics and subsequently genetic improvement of cereal crops.

S. Rakshit (✉) • K.N. Ganapathy
Directorate of Sorghum Research, Rajendranagar, Hyderabad 500 030, India
e-mail: rakshit@sorghum.res.in

K.K. P.B. et al. (eds.), *Agricultural Bioinformatics*,
DOI 10.1007/978-81-322-1880-7_4, © Springer India 2014

Keywords
Comparative genomics • Cereals • NGS • Genomic resources • Collinearity

1 Introduction

Cereal crops (cultivated grass species used for their high-energy edible grains) like wheat (*Triticum aestivum* L.), rice (*Oryza sativa* L.), maize (*Zea mays* L.), barley (*Hordeum vulgare* L.), oat (*Avena sativa* L.), sorghum [*Sorghum bicolor* (L.) Moench] and millets are the main source of calorific supply for human population and represent over 50 % of total crop production across the world. Besides being used as food, the grains and straw are also important component of feedstock, and in recent past, these are gaining importance for production of bioenergy as an alternative to fossil fuel-based energy (Salse and Feuillet 2007). Cereal crops are members of the grass (*Poaceae*) family, which is the fourth largest angiosperm family. Such crops have a relatively recent history of evolution. As compared to flowering plants (angiosperm) lineage which is about 200 million years (MY), major cereals, namely, maize, rice, sorghum and wheat, started diverging from a common ancestor about 50–70 MY (Kellogg 2001). These crops have been domesticated simultaneously and independently with reproductive isolation from each other. Wheat domesticated in SouthWestern Asia in the Fertile Crescent region, maize in Mexico, rice in both Southeast Asia and West Africa, and sorghum in Africa (Harlan 1971, 1992; Zohary and Hopf 2000; Piperno and Flannery 2001). Cereal crops are adapted to wide range of environments making them potent source of resistance against many stresses. These crop plants represent very diverse genome organization and genome size (Feuillet and Keller 2002). Due to their economic importance, broad diversity and relatively recent evolutionary history, grasses in general and cereals in particular have been studied extensively individually and as a group for comparative study. Earlier comparative studies were restricted using isozymes. However, with genomic revolution in the early 1980s, efforts were made to compare genomes using the common language of genetic information, i.e. deoxyribonucleic acid (DNA).

Critical issues of plant growth and development can be better understood using information gathered through genomics research, which may directly be applied for crop improvement (Buell 2009). Recent advances in sequencing of genomes and transcriptomes have led to accumulation of plethora of sequences across crop species. This accumulated information is used in studying the relationship of genome structure and function across different biological species through an emerging area of comparative genomics. Through comparative genomics, the information provided by the signatures of selection is used to understand the function and evolutionary processes acting on genomes. Besides giving insight to the mechanisms of genome evolution and speciation, comparative genomics helps us in densification of DNA markers on the genetic maps and identification of conserved genes and regulatory sequences. It also helps in map-based cloning of important genes. Such study can be carried out at the level of cytology to genetic maps to partial or whole genome sequences. Genomic information developed or generated on model crops may be used for gene discovery in orphan crops to better understand the basis of diversity and adaptation leading to effective utilization of genetic resources for crop improvement.

With the publication of rice genome draft sequence in 2002 (Goff et al. 2002; Yu et al. 2002), comparative genomics in cereals has taken a new dimension. In the recent past, many cereal genomes have been fully or partially sequenced, which has opened up the scope to carry out comparative genomics study in a much systematic manner. Initial results indicated a good level of conservation at genetic map level. The biggest challenges before the comparative genomics are functional validation of over 30,000 annotated genes and identification of the regions responsible for gene function and regulation so that these may be deployed effectively in crop improvement.

2 Cytological Differences Among Cereals

Cytology played an important role in studying the process of evolution. Cytological studies to understand the process of evolution were initiated much before the mechanism of inheritance was established. Cytology principally focused on chromosome number and ploidy level. After recognizing DNA as the genetic material and thereafter understanding the mechanisms of encoding genetic material, it became a major focus for biological research. Much before it was established that the number of chromosomes remains constant in a given species. Till the 1970s, the major focus of research across plant species was to determine the chromosome number, ploidy level and the DNA content (C value). It was found that with advancement in the process of evolution, the DNA content or C value increases. However, this rarely correlates with the organizational complexity, which means that in many cases, an organism at higher level of evolutionary tree has lower DNA than an organism at lower level of evolution. This always puzzles the biologists and this puzzle is referred to as C value paradox or C value enigma. Cereal crops differ drastically from each other in terms of chromosome number, ploidy level as well as DNA content (Table 1). Among the major cereals, barley has the lowest chromosome number (2n = 14) as against bread wheat (2n = 42). Maize and its close relative sorghum have similar chromosome number (2n = 20). Some wild relatives of cultivated maize, like *Z. perennis*, have double the chromosome number than normal maize. Rice has chromosome number higher than sorghum and maize (2n = 24). Among the cultivated cereals, only bread wheat is polyploid (allohexapolyploid), while the remaining are diploid. C values vary drastically among the cereals. Rice has the lowest DNA content (1C = 0.5 pg), while wheat has highest values (1C = 18.1 pg). It is noteworthy that chromosome number does not correlate with the C values. For example, though barley has lower chromosome number (2n = 14) than maize and sorghum (2n = 20), its DNA content is almost two times and seven times higher than maize and sorghum, respectively. In fact, within closely related species like sorghum and maize, though chromosome number is similar, DNA content varies three to four times.

3 Genome Sequencing Methods

Genome sequencing refers to the process of determining the nucleotide sequence of the genome through DNA sequencing. The rapid speed of sequencing attained with modern DNA sequencing technology has been instrumental in the sequencing of complete DNA sequences or genomes of numerous types and species of life. The first breakthrough in DNA sequencing took place in the 1970s with Frederick Sanger from Cambridge, UK, publishing a method for 'DNA sequencing with chain-terminating inhibitors' in 1977, and Walter Gilbert and Allan Maxam from Harvard University developing another DNA sequencing method called 'DNA sequencing by chemical degradation'. Genome of bacteriophage ΦX174 was the first to be sequenced fully (Sanger et al. 1977a). These two methodologies, particularly the chain termination technique, dominated the DNA sequencing programmes almost for two decades. In the middle to late 1990s, several new methods for DNA sequencing were developed, which together is referred to as 'next-generation' sequencing (NGS) tools.

3.1 Next-Generation Sequencing Technologies

Sanger sequencing method prevailed from the 1980s till mid-2000s, when the sequencing

Table 1 Chromosome number, ploidy level, life cycle type and nuclear DNA content of major cereals

Species	Family	2n	Ploidy level	Life cycle types	DNA content (pg) 1C	2C	3C	4C	Present amount	Method
Hordeum vulgare L. cv. Asse	Gramineae	14	2	A	5.4	10.9	16.3	21.8	C^{ar}	Fe
Hordeum vulgare L. cv. Trumpf	Gramineae	14	2	A	5.9	11.7	17.6	23.4	R^{aq}	FC:EB/O
Hordeum vulgare L.	Gramineae	14	2	A	5.1	10.1	15.2	20.2	O	FC:PI
Oryza sativa L. ssp. indica	Gramineae	24	2	P	0.5	1.0	1.4	1.9	O	FC:PI
Oryza sativa L. ssp. indica *Oryza sativa* L. spp. japonica	Gramineae	24	2	P	0.4	0.9	1.3	1.8	O	FC:PI
Oryza sativa L. ssp. javanica	Gramineae	24	2	P	0.5	0.9	1.4	1.8	O	FC:PI
Sorghum bicolor (L. Moench)	Gramineae	20	2	A	0.8	1.6	2.4	3.2	O	FC:PI
Sorghum bicolor (L. Moench) spp. bicolor cv Hegari white	Gramineae	21	2	A	0.9	1.7	2.6	3.5	O	FC:PI
Zea diploperennis	Gramineae	20	2	A	1.8	3.6	5.4	7.1	O	FC:PI
Zea luxurians	Gramineae	20	2	A	4.4	8.8	13.2	17.7	O	Fe
Zea mays L. inbred line B75[h]	Gramineae	20	2	A	2.8	5.6	8.3	11.1	O	FC:DAPI
Zea mays L. spp. mays[h]	Gramineae	20	2	A	2.8	5.6	8.3	11.1	O	Fe
Zea mays L. spp. Mexicana K69-5[h]	Gramineae	20	2	A	3.4	6.8	10.2	13.6	O	Fe
Zea perennis (Hitchc.)	Gramineae	40	4	P	5.7	11.4	17.0	22.7	O	Fe
Triticum aestivum L. cv. Chinese Spring h	Gramineae	42	6	A	18.1	36.1	54.2	72.2	O	FC:PI
Triticum araraticum Jakubz. I	Gramineae	28	4	A	10.1	20.1	30.2	40.2	R^{aq}	FC:EB/O
Triticum duram Desf. var. alexandrinum	Gramineae	28	4	A	13.2	26.3	39.5	52.6	R^{aq}	FC:EB/O
Triticum monococcum L.	Gramineae	14	2	A	6.0	11.9	17.9	23.8	O	FC:PI
Triticum timopheevi	Gramineae	28	4	A	9.6	19.1	28.7	38.2	R^{aq}	FC:EB/O
Triticum turgidum	Gramineae	28	4	A	12.9	25.8	38.6	51.5	C^{ar}	FE
Triticum urartu Thum	Gramineae	14	2	A	5.7	11.4	17.1	22.7	C^{ar}	FE

Modified from Bennett and Leitch (1995)

A annual, *P* perennial, *O* original, *R* recalibrated value, *C* calibrated value, *FC* flow cytometry with one of the fluorochromes (*PI* propidium iodide, *DAPI* 4′,6-diamidinophenylindole, *EB*/*O* ethidium bromide and olivomycin, *M: DAPI* microdensitometry using DAPI)

chemistry took a paradigm shift. Several non-Sanger high-throughput sequencing technologies became commercially available after 2000 which are referred to as 'second-generation' or 'next-generation' sequencing technologies. These include the pyrosequencing (commercialized by 454/Roche), SOLiD sequencing (commercialized by Applied Biosystems) and Sequencing by Synthesis (commercialized by Illumina/Solexa). NGS technologies allow resequencing of large number of plant genomes at greater speed and lower cost compared to traditional sequencing methods (Gupta 2008).

3.1.1 Pyrosequencing

It was introduced commercially in 2004 by 454/Roche. The platform produces on an average read length of 250 bp. The pyrosequencing reaction is carried out with mixture of single-stranded DNA template, sequencing primer, DNA polymerase, ATP sulfurylase, luciferase and apyrase. The two substrates adenosine 5′phosphosulfate (APS) and luciferin are added to the reaction. The first one of the four deoxynucleotides (dNTPs) is added to the sequencing reaction, and the DNA polymerase catalyzes its incorporation into the DNA strand in case there is base complementarity. During incorporation of each base, a phosphodiester bond is formed between the dNTPs, releasing pyrophosphate (PPi) in an equivalent quantity to the amount of incorporated nucleotide. In sequence, the enzyme ATP sulfurylase converts PPi to ATP in the presence of APS. ATP is used for conversion of luciferin to oxyluciferin influenced by the enzyme luciferase. This gives rise to light intensity which is proportional to the amount of ATP used. Light is detected by a charge-coupled device camera and detected as a peak in a pyrogram. The height of each peak is proportional to the number of nucleotides incorporated. The reaction is regenerated with the enzyme apyrase that degrades ATP and unincorporated dNTPs. Then, the next dNTP is added. Addition of dNTPs is performed one at a time. Generation of a signal indicates which nucleotide is occurring next in the sequence. As the process goes on, the complementary DNA strand grows and the nucleotide sequence is determined according to the signal peaks in the programme (Siqueira et al. 2012).

3.1.2 SOLiD

SOLiD stands for Sequencing by Oligonucleotide Ligation and Detection, which is marketed by the Applied Biosystems, USA. In the SOLiD approach, sequencing is carried out by quantifying serial ligation of an oligonucleotide to the sequencing primer by a DNA ligase enzyme (Varshney et al. 2009). As in pyrosequencing, DNA fragments are ligated to oligonucleotide adapters, attached to beads and amplified by emulsion PCR to provide signal for the sequencing reactions. Beads are deposited on a flow cell surface, and the ligase-mediated sequencing begins by annealing of the sequencing primer to the adapter sequences on each amplified fragment (Siqueira et al. 2012). The read lengths for SOLiD platform are in range of 25–35 bp, and each sequencing run produces between 2 and 4 Gb of DNA sequence data. After base calling of the reads with high-quality values, the reads are aligned to a reference genome to enable a second tier of quality evaluation. This process is called two-base encoding (Mardis 2008).

3.1.3 Sequencing by Synthesis

The Sequencing by Synthesis approach is commercialized by the Illumina. In this approach, all the four nucleotides are added to the flow cell channels along with DNA polymerase, for incorporation into the oligo-primed DNA cluster fragments. The nucleotides carry a fluorescent label and the 3′OH group is chemically blocked. This is followed by an imaging step in which each flow cell lane is captured. After the imaging step, the 3′ blocking group is chemically removed to prepare each strand for the next incorporation by DNA polymerase. The series of steps continues for a specific number of cycles producing a read length of 25–35 bases. Unlike pyrosequencing, the DNA chains are extended, one nucleotide at a time, and image capturing can be performed at a delayed moment. This allows very large arrays of DNA clusters to be captured by sequential images taken from a single camera.

3.1.4 Other NGS Technologies

Single molecule real time (SMRT) is another recent sequencing reaction gaining popularity. This has been developed by Pacific Biosciences. This is also based on the Sequencing by Synthesis approach. The DNA is synthesized in zero-mode waveguides (ZMWs), which is a small well-like container with the capturing tools located at the bottom of the well. The sequencing is performed using unmodified polymerase and fluorescently labelled nucleotides. Upon incorporation of the nucleotide in the elongating DNA strand, the fluorescent label is detached leaving an unmodified DNA strand. This approach allows reads of up to 15,000 nucleotides, with mean read lengths of 2,500–2,900 bases. This is often referred to as third-generation sequencing techniques, which is predicted to be much cheaper and faster than the second-generation sequencing technologies (Gupta 2008). HiSeq NGS sequencing combines Illumina's proven and widely adopted reversible terminator-based sequencing by synthesis chemistry with innovative engineering (www.res.illumina.com). With the scanning and imaging technology, clusters on both surfaces of the flow cell can be sequenced increasing the number of reads and sequence output. The Hiseq 2000 platform can provide an output of 600 GB in about 3–8 days. Hiseq 2000 has improved much on GC sequencing, and compared with 454 and SOLiD, Hiseq 2000 is said to be the cheapest in sequencing (Liu et al. 2012). Recently many advanced compact personal genome sequencers are being launched. Ion Personal Genome Machine (PGM) was launched in 2010 by Ion Torrent, a division of life science technologies (Merriman et al. 2012). Ion PGM uses semiconductor-based high-density array of micro-reaction chambers and produces sequence reads of 100–200 bp and up to 1 Gbp data per run. During sequencing process, the four nucleotides are added separately across the micro-reaction chambers. The system then records the nucleotide sequence by sensing the pH change when the hydroxyl group is released during specific base extension (Berkmann et al. 2012). PGM is the first commercial machine that does sequencing without fluorescence and camera scanning, resulting in higher speed, lower cost and smaller instrument size.

MiSeq is another personal genome machine from Illumina which uses Sequencing by Synthesis technology. It integrates cluster generation, SBS and data analysis in a single instrument and can provide sequence information of whole genome in about 8 h. The highest integrity of data and broader range of applications which included amplicon sequencing, clone checking, ChIP-Seq and small genome sequencing are the major advantages of MiSeq. The PGM provides 120 MB–1.25 GB output and up to 2 × 150 bp read length (Liu et al. 2012).

4 Genome Sequencing Progress in Cereals

Major advances have been made in the field of cereal genomics over the last two to three decades. These include availability of high-density genetic maps, physical maps and QTL maps. This was followed by generation of EST databases in cereals. In recent past, large-scale cereal genome sequencing projects in crops like rice, sorghum, maize, wheat and foxtail millet provided extensive information on genome organization of cereals. The extensive information and knowledge gained from the genome sequencing aid in understanding the structural and functional components of the genome for its effective utilization in genetic improvement of cereals.

4.1 Rice

Among cereals, rice has smaller genome (420 Mbp) and predicted to have high gene density as compared to other cereals. Realizing the fact the International Rice Genome Sequencing Project (IRGSP) was initiated involving ten nations to achieve >99.99 % accurate sequence using a mapped clone sequencing strategy. Monsanto announced on April 4, 2000, that the company had completed a working draft of the rice genome, which would be made available to the IRGSP (Barry 2001). In April 2002, Goff et al. from USA and Yu et al. from China published the draft sequence of *japonica* cultivar, Nipponbare, and *indica* cultivar, 93–11, respectively. Both these groups used whole genome shotgun approach to sequence the respective cultivars. Goff et al. (2002) generated over 2.5 billion bases with a 98 % probability of being correct, which suggested more than sixfold coverage of the rice genome. After removal of an estimated 38 Mbp of repetitive DNA, above 5.5 million sequences assembled into 42,109 contiguous sequences (contigs) with a total coverage of 389,809,244 bp (93 % of the rice genome). This sequence is referred to as Syd (Syngenta draft sequence, www.tmri.org). Yu et al. (2002) generated 127,550 contigs, representing total contig length of 361 Mb (86 % coverage). About one simple sequence repeat (SSR) could be detected in every 8,000 bp of rice genome. Di-, tri- and tetranucleotide SSRs account for 24 %, 59 % and 17 % of the SSRs found in rice, respectively. The most common dinucleotide SSR is AG/CT, while CGG/CCG is the most common trinucleotide, and ATCG/CGAT is the most common tetranucleotide repeat unit. Within predicted genes, more than 7,000 SSRs are found, 92 % of which are trinucleotide in nature. Besides SSRs, within rice genome ~38Mbp long repetitive DNA and 150 Mbp of short repetitive DNA are identified. The estimated gene number ranges from 32,000 to 50,000 in Nipponbare and 46,022 to 55,615 in 93–11. Detailed map-based sequence of Nipponbare was published through the initiative of IRGSP by Matsumoto et al. (2005). The study precisely put the rice genome size at 389 Mb and generated 370 Mb finished sequences of less than one error per 10,000 bases. With this 95 % of rice genome is covered. The study identified a total of 37,544 non-transposable-element-related protein-coding sequences, of which 71 % have putative homologue in *Arabidopsis*. Likewise, about 90 % of the *Arabidopsis* proteins are reported to have putative homologue in the predicted rice proteome. Twenty-nine per cent of the 37,544 predicted genes got grouped into gene families. A total of 22,840 genes matched with a rice EST or full-length cDNAs. Out of the identified genes, 2,859 are unique to rice and other cereals. In rice one gene per 9.9 kb has been reported. It is observed that between 0.38 and 0.43 % of the nuclear genome contains organellar DNA fragments, suggesting repeated and ongoing transfer of organellar DNA to the nuclear genome. Between two cultivated rice subspecies, *japonica* and *indica*, 80,127 polymorphic sites were identified, with frequency of SNP ranging from 0.53 to 0.78 %. The centromeres of all rice chromosomes were reported to contain highly repetitive 155–165 bp CentO satellite DNA, together with centromere-specific retrotransposons. The CentO satellites are located within the functional domain of the rice centromere. Except short arms of chromosomes 4, 9 and 10, which are highly heterochromatic, density of expressed genes was observed to be greater on the distal portions compared with the regions around the centromeres. About 14 % of the genes are arranged in tandem repeats. A total of 763 tRNA genes, including 14 tRNA pseudogenes, were detected in the 12 chromosomes. A total of 158 miRNAs were mapped onto the rice pseudomolecules and 215 small nucleolar RNA and 93 spliceosome RNA genes showed nonrandom chromosomal distributions in the rice genome. The rice genome sequence information helped in organizing sequence information of other diverse cereals and together with genetic maps and sequence samples from other cereals yielded new insights into cereal evolution.

4.2 Sorghum

Among cereals, in terms of genome complexity, sorghum is positioned after rice. Its genome size is ~730 Mb. Being a C4 and drought-tolerant species and being closely related to other important cereals like maize and sugarcane, after rice it became a natural choice for genome sequencing. Towards this direction, members of the worldwide sorghum community as well as community representatives from closely related crops like sugarcane and maize met in St. Louis, Missouri, on November 9, 2004, to lay the groundwork for future advances in sorghum genomics and, in particular, to coordinate plans for sequencing of the sorghum genome (Kresovich et al. 2005). Five years later with combined efforts of over 20 laboratories across six countries, the draft sequence of the US cultivar, Btx623, was published using whole genome shotgun approach (Paterson et al. 2009a). It was found that ~75 % larger quantity of DNA in the sorghum genome as compared to rice was mostly heterochromatin. The euchromatin which represents the gene-rich regions is almost 252 Mb in size. One-third of the sorghum genome is recombination-rich regions while the rest portion is assumed to be recombination poor. Net size expansion of the sorghum genome relative to rice largely involved long terminal repeat (LTR) retrotransposons. Nearly 55 % of sorghum genome is represented by retrotransposons, which is intermediate between maize genome (79 %) and rice genome (26 %). Sorghum resembles closely to rice in having higher ratio of *Gypsy*-like to *Copia*-like elements (3.7 to 1 and 4.9 to 1) than maize (1.6 to 1). The CACTA-like elements (4.7 % of the genome) are the predominant class among the sorghum DNA transposons. The miniature inverted repeat elements are present in 1.7 % of the genome while the *helitrons* elements are present in 0.8 %. A total of 34,496 genes were identified in the sorghum genome, of which 27,640 are protein-coding genes. Almost 24 % (3,983) gene families are specific to sorghum and grasses, and 1,153 (7 %) are unique to sorghum. In 1,491 loci, evidences of alternative splicing were noticed. In addition to this, 5,197 hypothetical or uncharacterized 727 processed pseudogenes and 932 models containing transposon-specific domains are also identified in sorghum. More than 98 % concordance in terms of intron position and phases is observed between orthologous genes of sorghum and rice. A total of 5,303 paralogues belonging to 1,947 families are identified in sorghum, which are predominantly proximally duplicated. In sorghum genome, 149 microRNAs (miRNA) have been identified, 5 of which are putative polycistronic miRNA. Nearly 94 % (25,875) of high-confidence sorghum genes have orthologues in rice, *Arabidopsis* and/or poplar. The characteristic adaptation of sorghum to drought is likely due to expansion of one miRNA and several gene families. Rice miRNA 169 g, which is upregulated during drought stress, has five sorghum homologues. Aligning of the nuclear genome with the chloroplast and mitochondria genome revealed several insertions of mitochondria and chloroplast sequence into the nuclear genome. The organellar insertion in sorghum is smaller compared to that of rice. The sizes of most organellar insertion were less than 500 bp while 1.5 % of the insertion was reported to be more than 2 kb. The organellar DNA insertion is 0.085 % of the nuclear genome and is 0.53 % less than that of rice. Nelson et al. (2011) subjected eight diverse sorghum genotypes, namely, BTx623, BTx430, P898012, Segalone, SC35, SC265, PI653737 (*Sorghum propinquum*) and 12–26 (ssp. *verticilliflorum*), to short-read genome sequencing to characterize distribution of SNPs. The study revealed 283,000 SNPs at ≥82 % confirmation probability. Zhang et al. (2011) uncovered 1,057,018 SNPs, 99,948 indels of 1–10 bp in length and 16,487 presence/absence variations as well as 17,111 copy number variations by sequencing of two sweet sorghum lines, Keller and E-Tian, and a Chinese grain sorghum line, Ji2731. Mace and Jordan (2010) integrated the whole genome sequence information with reported 771 QTLs for 161 unique traits from 44 different studies and showed uneven distribution of QTLs and identified gene-rich regions.

4.3 Maize

Maize genome is more complex as compared to rice and sorghum. It has undergone several rounds of genome duplication making it distinguishable from its close relative, sorghum. Schnable et al. (2009) sequenced the genome of one of the most popular maize inbred lines of the USA, B73 using a minimum tiling path of bacterial artificial chromosomes (BACs) and fosmid clones. Shotgun sequencing of clones coupled with automated and manual sequence improvement of the unique regions gave rise to B73 genome sequence of 2.3 Gb. The genome of maize consists mostly the nongenic-rich repetitive fraction punctuated by islands of unique or low-copy DNA that harbour single genes or small groups of genes. The repetitive elements are reported to contribute to the wide range of diversity within the species. These include transposable elements (TEs), ribosomal DNA (rDNA) and high-copy short-tandem repeats. Almost 85 % of the B73 reference sequence consists of TEs. It contains 855 families of DNA TEs. The most complex of these superfamilies is *Mutator*. Maize contains eight families of *Helitrons*. LTR transposons compose >75 % of the B73 reference genome. In maize, *Copia*-like elements are over-represented in gene-rich euchromatic regions, whereas *Gypsy*-like elements are found more in gene-poor heterochromatic regions. A total of 32,540 protein-coding genes belonging to 11,892 families and 150 miRNA genes are recorded in maize. Exon sizes of maize genes are similar to that of their orthologous genes in rice and sorghum. Because of the presence of repetitive elements, maize genes contain larger introns. Out of 11,892 gene families in maize, 8,494 were shared between maize, sorghum, rice and *Arabidopsis*. Tandem CentC satellite repeat and centromeric retrotransposon elements of maize (CRMs) are located in variable amounts in maize centromeres. Hypomethylated genes are dispersed more widely in maize as compared to sorghum in which such genes are largely excluded from the pericentric regions.

4.4 Barley

The fourth most abundant cereal, barley, which is treated as model for the Triticeae tribe, was sequenced by the International Barley Genome Sequencing Consortium in 2012 (Mayer et al. 2012). For this purpose, malting variety, Morex, released by Minnesota AES, USDA, in 1978 was used. A total of 571,000 BAC clones originating from six independent BAC libraries were assembled giving rise to a physical map comprised of 9,265 BAC contigs with a cumulative length of 4.98 Gb. This is represented by a minimum tiling path of 67,000 BAC clones. With this more than 95 % of the barley genome of 5.1 Gb is represented. Whole genome shotgun sequencing strategy was followed using short-read Illumina GAIIx technology. Pericentromeric and centromeric regions of the barley chromosomes exhibit significantly reduced recombination frequency, which is also observed in other grass species. Approximately 1.9 Gb or 48 % of the sequence was assigned to these regions. Nearly 84 % of the barley genome is comprised of mobile elements or other repeat structures. Almost 99.6 % of these are LTR retrotransposons. The non-LTR retrotransposons contribute only 0.31 %. In barley LTR *Gypsy* retrotransposon superfamily is 1.5-fold more abundant than the *Copia* superfamily. Barley contains approximately 30,400 genes. Gene-family-directed comparison with the genome of sorghum, rice, *Brachypodium* and *Arabidopsis* led to identification of 26,159 genes having homology to at least one reference genome. A total of 15,719 high-confidence genes are identified in barley genome. High proportions of functional genes are 'locked' into recombinationally 'inert' pericentromeric regions of each chromosome. In barley 72–84 % of high-confidence genes express in more than one developmental or tissue sample and 36–55 % of the high-confidence genes are differentially regulated. Posttranscriptional processing is key regulatory mechanism in barley. Alternative splicing is recorded in as high as 73 % of high-confidence genes, and only 17 % of alternative splicing transcripts are found in all tissues. In barley genome, as many as 27,009 preferentially single-exon low-confidence genes are classified as putative novel transcriptionally active regions, lacking homology to protein-coding genes or open reading frames (ORFs). Fifteen million nonredundant single-nucleotide variants (SNVs) are detected in barley through resequencing of four diverse barley cultivars, namely, Bowman, Barke, Igri and Haruna Nijo. These SNVs tend to reduce in frequency towards the peri-centromeric regions of all chromosomes.

4.5 Wheat

As compared to the other cereals, wheat genome is of much larger size making it difficult to sequence. Compared to other sequenced cereals which are diploid, bread wheat is complicated as the genome ($2n = 6x = 42$, AABBDD) is allohexaploid and approximately 17 Gb in genome size. Bread wheat originated from hybridization between cultivated tetraploid emmer wheat (AABB, *Triticum dicoccoides*) and diploid goat grass (DD, *Aegilops tauschii*) approximately 8,000 years ago. Using 454 pyrosequencing, Brenchley et al. (2012) developed 17 Gb sequence of hexaploid wheat cultivar, Chinese Spring (CS42). To identify A-, B- and D-genome-derived gene assemblies in the hexaploid sequences, Illumina sequence assemblies of *T. monococcum*, related to the A-genome donor, *Ae. speltoides* complementary DNA (cDNA) assemblies and 454 sequences from the D-genome donor *Ae. tauschii*, respectively, were used. The SOLiD platform was used to generate additional sequence of CS42. In the hexaploid wheat genome, gene number ranges between 94,000 and 96,000. Nearly two-thirds of the identified genes are assigned to the three component genomes (A, B and D). Between the gene sets of *Brachypodium* and wheat, high degree of overlap with regions of lower conservation existed. The hexaploid genome experienced significant loss of gene family members on polyploidization and domestication. Abundance of gene fragments across wheat genome was recorded. Orthologous group assembly between rice, sorghum, *Brachypodium* and barley full-length cDNAs using OrthoMCL clustering generated 20,496 orthologous groups. Nearly 90 % of the metabolic genes in *Arabidopsis* matched to these groups. Ling et al. (2013) reported the genome sequence of *Triticum urartu* accession G1812 (PI428198) using a whole genome shotgun strategy and assembled the genome with 4.48 Gb of filtered high-quality sequence data. The genome size of *T. urartu* to be estimated as 4.94 Gb. Genome annotation predicted 34,879 protein-coding gene models with an average gene size of 3,207 bp and a mean of 4.7 exons per gene. In comparison with the 28,000 genes estimated from A-genome of hexaploid wheat, gene set for *T. urartu* contained 6,800 additional genes indicating more complete representation of genes. The difference could be attributed to the various approaches used for gene prediction and also could be due to extensive gene loss in the hexaploid A-genome compared with its diploid progenitors. About 66.88 % of the *T. urartu* genome sequence was predicted to be repetitive elements which include mainly long terminal repeat retrotransposons (49.07 %), DNA transposons (9.77 %) and unclassified elements (8.04 %). In total, 412 conserved and 24 new microRNAs distributed into 116 families were identified. Comparison with the miRNAs of five monocots and five dicots showed that 73 miRNA families were specific to monocots, of which 23 were unique to *T. urartu*. The investigation predicted 244 target genes for these miRNAs and found that the target gene (TRIUR3_06170) of miRNA MIR5050 responded to cold treatment, which provides a new resource for investigating the regulation of cold adaptation through miRNA. The study revealed 593 genes encoding 'R proteins' which were more abundant than in *B. distachyon* (197), rice (460), maize (106) and sorghum (211), and the study also reported higher number of R genes compared to other cereals. A total of 739,534 insertion-site-based polymorphism markers (ISBP) and 166,309 SSRs were identified. PCR validation resulted in 94.5 % of the SSRs and 87 % of the ISBP markers with the expected products. More than 33 % of the SSRs and more than 10 % of the ISBP markers amplified specific to the A-genome. Resequencing of another *T. urartu* accession (DV2138) discovered 2,989,540 SNPs, which will be useful for the future development of SNP markers.

4.6 Foxtail Millet

Foxtail millet (*Setaria italica*) is also a member of the Poaceae grass family. It is an important nutritious food crop in arid regions and has potential for use as a C4 biofuel. Foxtail millet has a genome size of (~490 Mb). The genome is very rich in repetitive sequences and is consistent with the sequenced genomes of other grass species. The long terminal repeat retrotransposons are the most abundant class of retrotransposons comprising >25 % of the total nuclear genome. Exons of the protein-encoding genes are predicted to comprise ~46 Mb or about 9 % of the genome. A draft genome (~423 Mb) anchored onto nine chromosomes and annotated 38,801 genes is reported (Zhang et al. 2012). The genome assembly of foxtail millet genome covers ~86 % of the estimated genome size and the unassembled part is largely due to the repeat elements. Functional annotation confirmed that 78.8 % of the genes have homologues with known functions in protein databases. Search for conserved genes present in other grasses in the foxtail millet gene set revealed that 99 % of conserved genes have homologues in foxtail millet. The study also identified 1,367 pseudogenes in the genome. A total of 99 ribosomal RNA genes, of which 23 rRNA genes harbour four large clusters on chromosomes 8 and 9, were identified. The genome sequencing also revealed that ~46 % of the genome comprised transposable elements. Both retroelements (class I transposable element; 31.6 %) and DNA transposons (class II transposable element; 9.4 %) were identified.

4.7 Other Grasses

Vogel et al. (2010) through the International *Brachypodium* Initiative published the genome sequence of model grass, *Brachypodium*. The diploid inbred line Bd21 was sequenced using whole genome shotgun sequencing. The effort generated 272 Mb sequence information. A total of 25,532 protein-coding genes is predicted, which is close to rice and sorghum. Between the three grass families, 77–84 % of gene families are shared. DNA transposons comprise 4.77 % of the *Brachypodium* genome. In this retrotransposon, sequences comprise 21.4 % of the genome, as against 26 % in rice, 54 % in sorghum and more than 80 % in wheat. A minimum of 17.4 Mb has been reported to have lost by long terminal repeat (LTR) recombination in *Brachypodium*. This demonstrates that in several grass families, retroelement expansion is countered by removal of these through recombination. Fourteen major syntenic disruptions between *Brachypodium* and rice/sorghum are recorded, which can be explained by nested insertions of entire chromosomes into centromeric regions. Similar nested insertions in sorghum and barley are also identified.

5 Comparative Genomics Resources

5.1 Databases

Several web resources, like Gramene, PlantGDB, Phytozome, GreenPhylDB, CoGE, PLAZA and GRASIUS, are available online for various comparative studies. Main comparative genomic resources are discussed.

5.1.1 Gramene

One of the most popular, open-sourced, curated data resources for comparative genomics analysis in grasses is Gramene (http://www.gramene.org/). This database was put on public domain in 2002 (Ware et al. 2002). In the beginning in this database, the rice sequence was used as base information to facilitate genomics research in other grass families. Subsequently the database was further strengthened integrating the Ensembl technology and including the information on barley, *Brachypodium*, foxtail millet, maize, oats, pearl millet, rye, wheat and sorghum. Information from other plant species, like *Glycine*, *Musa*, *Solanum*, *Brassica*, *Arabidopsis*, *Vitis* and *Populus*, have also been included in recent version. Objectives of the database are to facilitate cross-species homology study using information from public genome and EST sequencing projects, to support protein structure and function analysis, to create genetic and physical maps, to interpret biochemical pathways, to localize gene and QTL and to describe phenotypic characters and mutations. Different modules like *Genome*, *Genetic Diversity*, *Pathways*, *Protein*, *Gene*, *Ontologies*, *Markers*, *Comparative Maps*, *QTL*, *BLAST*, *Gramene Mart* and *Species* are available in the database. '*Genome*' module has detailed information on the plant species covered in the database. Population structure and evolutionary and diversity pattern are covered in '*Genetic Diversity*' module. '*Pathways*' module has *RiceCyc*, *MaizeCyc*, *BrachyCyc* and *SorghumCyc* submodules, in which pathway databases for respective crop plants are described. Mirrors of pathway databases from *Arabidopsis*, tomato, potato, pepper, coffee, *Medicago*, *E. coli*, etc. are also provided, which helps in comparative analysis. Collective information on Swiss-Prot–Trembl protein entries from family *Poaceae* is dealt in the '*Protein*' module. The protein entries are annotated based on the molecular function of the gene product (protein), biological process in which the gene product (protein) is involved and cellular localization of the gene product. In the '*Genes*' or '*Gene and Allele*' module, genes and their alleles associated with morphological, developmental and agronomically important phenotypes, variants of physiological characters, biochemical functions and isozymes are described. The '*Ontologies*' module provides information on *Plant Ontology* (plant anatomy and the stages of plant development), *Trait Ontology* (plant traits and phenotypes), *Gene Ontology* (molecular function, biological process, cellular component), *Environment Ontology* and *Gramene's Taxonomy Ontology*. Basic information on the different markers used for mapping are provided in the '*Markers*' module. The '*SSRIT tool*' available in this module is a very useful tool to identify microsatellite in a given sequence. '*Maps*' module helps to visualize genetic, physical, sequence and QTL maps for the species dealt in the database. *Comparative Map Viewer* (*CMap*) helps in constructing and comparing different maps. Information in *Maps Module* is built from the *Markers Module*. Three sequenced genomes of rice, sorghum, and *Brachypodium* have been compared using syntenic blocks in *CMap*. QTLs identified for agronomic traits in the species are covered in the '*QTL*' module. Information on associated traits and the mapped locus on the genetic map for each QTL is available in this module. This module may be searched by trait name, symbol, category, linkage group, QTL accession ID or species using wild cards. '*BlastView*' is an integrated platform for sequence similarity search against Ensembl Plants database. Species-wise search is possible in both DNA and protein databases using BLASTn (aligns the nucleotide sequences) and BLASTx (aligns translated sequences of any nucleotide sequence in all six reading frames) search tools, respectively. '*Gramene Mart*' module has the following databases, namely, Plant Gene 37, Plant variation 37, Gramene mapping and Gramene QTL 37. Each database can be searched in 10 datasets, of which sorghum is one. '*Species Page*' provides detailed information on all the 11 cereal species dealt in the database with full phylogenetic information.

5.1.2 PlantGDB

PlantGDB (http://www.plantgdb.org) is another comprehensive database for comparative study. The database aims at development of robust genome annotation methods, tools and standard training sets for the number of sequenced or soon to be sequenced plant genomes (Duvick et al. 2007). The database has genome sequence information of 16 dicots and 7 monocots. Plant genome/gene sequences are downloaded from GenBank approximately every 4 months. These are arranged by species and made available for download, search and BLAST analysis. Custom transcript assemblies (*PUTs* or *PlantGDB-derived Unique Transcripts*) are provided for all plant species with >10,000 ESTs. It may be obtained by special request as well. *PUT* datasets can be downloaded and batch BLAST search can be performed. Keyword searches based on Gene Ontology annotations and top UniProt BLAST hits can be performed in this database. Genome browsers (chromosome, scaffold or BAC-based) for 16 plant species are provided here. Genome assembly is splice-aligned to transcripts and protein from similar species and presented in a simple graphical interface (the *xGDB platform*). Powerful search tools to find sequences and retrieve sequence data adjacent to coding regions, as well as BLAST and GeneSeqer, are provided in this portal. PlantGDB has *yrGATE tool* which allows the community to create gene annotations in a xGDB genome browser itself. All splice junctions revealed by transcript evidence are shown in this, and it allows the user to easily create and validate gene models. Gene annotations can be submitted by registered users for curation. Upon approval of the annotation by the administrator, it is incorporated into the genome browser and be viewed publicly on the *yrGATE* track. Potentially mis-annotated genes are classified and displayed in tabular format in the *GAEVAL tool*. Additional databases like *SRGD* (*Splice-Related Gene*) database (detailing splicing-related genes from model species), *ASIP* (*Alternative Splicing in Plants*) database (alternative splicing database from *Arabidopsis*, rice, *Medicago truncatula* and *Lotus japonicas*), Ac/Ds Transposons database (particularly for maize), etc. are available in PlantGDB. A variety of tools for sequence analysis like *BLAST*, *GeneSeqer Spliced Alignment*,

GenomeThreader Spliced Alignment, *MuSeqBox*, *PatternSearch*, *Tracembler* and *TE nest* are available in the database. *TableMaker* and *BioExtract* are two additional tools available in PlantGDB. *TableMaker* accesses GenBank tables at PlantGDB using MySQL queries. With *BioExtract* in a single environment, users can query sequence databases, analyse data with web-based or local bioinformatics tools, save results and create and manage workflows. The *Plant Genome Research Outreach (PGROP) Portal* allows a centralized access to locate different outreach activities, programmes and resources.

5.1.3 Phytozome

Phytozome (http://www.phytozome.net) is another online resource to facilitate comparative genomic studies among plants. Objectives of this resource are to enable users with varying degrees of computational sophistication to access annotated plant gene families, to navigate the evolutionary history of gene families and individual genes, to examine plant genes in their genomic context, to assign putative function to uncharacterized user sequences and to provide uniform access to plant genomics datasets consisting of complete genomes, gene and related (e.g. homologous) sequences and alignments, gene functional information and gene families, either in bulk or as the result of on-the-fly complex queries' (Goodstein et al. 2012). Information on 25 plant genomes, 18 of which were sequenced, assembled and partially or completely annotated at the Joint Genomics Institute (JGI), is available in Phytozome database. Here genes and gene families can be retrieved by both keyword and sequence similarity searches. From proteome and gene family consensus sequences, genomic regions, gene transcripts, peptides and gene families most similar to a given query sequence may be obtained. Relevant attributes of gene and gene family like names, symbols, synonyms, external database identifiers and functional annotation IDs can also be searched. In this database, family of uncurated genes can be identified easily. Genes and gene families derived by keyword or sequence similarity searches can be viewed individually or be combined 'on the fly' to produce composite families. Information on each family and its constituent members is provided in the *Gene Family view*. Using the '*Family History*' tab, evolutionary history of the families can be viewed. There is a *Gene Page* in Phytozome, which shows single gene functional annotations and evolutionary history, as well as it links to alternatively spliced transcripts (if they exist), and genomic, transcript, coding and peptide sequences associated with this gene locus (exon–intron and UTR boundaries are colour-coded), and a graphical view of all other Phytozome peptides aligned against this gene's peptide. *GBrowse* tool of Phytozome allows genome-centric views for all the 25 genomes. This can be accessed from the Phytozome home page directly or from individual member gene links on the Gene Family or Gene Page and from the BLAST/BLAT results page of a query search.

5.1.4 GreenPhylDB

GreePhylDB (http://www.greenphyl.org/cgi-bin/index.cgi) is a comparative genomics database, which has genome information of 22 species (Conte et al. 2008). It comprises complete proteome sequences from the major plant phylum. The proteome information is clustered into consistent homeomorphic plant families. Once a group of sequences (cluster) is validated, phylogenetic analyses are performed to predict homologue relationships such as orthologs and ultraparalogs. The resource is particularly useful for functional genomics and identification of candidate genes for traits of interests. The web resource has 3,371 annotated gene families. Genomics tools in GreenPhylDB include *BLAST*, *family classification of given sequence*, *Get homologues and/or similar sequences* (this tool provides sequence ID of predicted homologue sequences inferred from phylogeny and best blast mutual hit matching sequences upon submitting a list of sequence ID), *InterPro Domain Distribution* (domain distribution is displayed by sequence and by species; the IPR domains can be selected using special operators to identify different combinations) and *export sequences* (provides a list of sequence ID used in database and one can export them in the selected format).

5.1.5 CoGE

Accelerating Comparative Genomics or CoGE resource (http://genomevolution.org/CoGe/) is a unique web resource having many interconnected tools capable to create open-ended analysis network (Lyons and Freeling 2008; Lyons et al. 2008). The main functions of CoGE are to store multiple versions of multiple genomes from multiple organism in a single platform, to quickly find sequences of interest in genomes of interest (with associated information), to compare multiple genomic regions using any algorithms and to visualize the results of analyses in such a way as to make the identification of 'interesting' patterns quick and easy. CoGe has the following options: *OrganismView* (gives an overview of an organism and its genomic information), *CoGeBlast* (allows BLAST search against any number of organisms), *FeatView* (finds genomic features by name or description), *SynMap* (generates syntenic dot plots of any two genomes) and *GEvo* (allows to compare multiple genomic regions using a variety of sequence comparison algorithms). *Orthology Viewer* integrates information from different orthology prediction tools. For example, if sorghum genome is compared with maize genome using *SynMap* and a region with an inversion is identified, breakpoints of that region in high detail may be compared using *GEvo* and extracted out the maize sequence using *SeqView*. Subsequently all the protein-coding regions can be found out using *FeatView*, and generated information can be used to find homologues in other plant genomes (e.g. rice) using *CoGeBlast*. Putative syntenic regions may be validated using *GEvo*. If say a gene with extra copy number is identified in a syntenic region, its sequence may be obtained using *FeatView* once again and putative intra- and interspecific homologues of it may be obtained using *CoGeBlast*, which will generate a FASTA file of those putative homologues using *FastaView*. This can be aligned using *CoGeAlign* and then be used to build a phylogenetic tree using *TreeView* or be exported to more expansive phylogenetic tool sets such as *CIPRES*. At the same time, the codon and protein usage variation of the genes may be checked using *FeatList*. If some interesting variation in some genes is observed, their overall GC content and wobble-position GC content may be checked in *FeatView*. Horizontal transfer of any sequence from the mitochondria can be checked using *CoGeBlast* or *GEvo* to search mitochondrial genomes.

5.1.6 PLAZA

PLAZA (http://bioinformatics.psb.ugent.be/plaza/) is also a centralized genomic database integrating information from different genome sequencing initiatives. Evolutionary analyses and data mining within green plant lineage (*Viridiplantae*) can be performed through an integrated module of plant sequence data mining and comparative genomics methodologies. It has integrated structural and functional annotation of 25 green plant species representing 909,850 genes. Above 85 % of these genes are protein coding, which are clustered in 32,294 multi-gene families. This has resulted into 18,547 phylogenetic trees. Precomputed datasets cover homologous gene families, multiple sequence alignments, phylogenetic trees, intraspecies whole genome dot plots and genomic collinearity between species. Through the integration of high-confidence gene ontology annotations and tree-based orthology between related species, thousands of genes lacking any functional description are functionally annotated. Advanced query systems, as well as multiple interactive visualization tools, are available through a user-friendly and intuitive web interface. In addition, detailed documentation and tutorials introduce the different tools, while the workbench provides an efficient means to analyse user-defined gene sets through PLAZA's interface.

5.1.7 GRASSIUS

GRASSIUS (Grass Regulatory Information Services) available at http://grassius.org/commcontrib.html is a public web resource composed of databases, computational and experimental resources related to regulation of gene expression in the grasses and their relationship with agronomic traits. Here information on the interactions of transcription factors (TFs) and *cis*-regulatory elements in the promoters of the genes in regulating gene expression, which are contributed either by the community or by literature analysis, are compiled. Its objective is to provide a "one-stop" resource to facilitate research and communication within the plant community with regards to genome-wide regulation of gene expression processes. The resource currently contains regulatory information on maize, rice, sorghum, sugarcane and *Brachypodium*. It has integrated information from three large databases: *GrassTFDB*, *GrassPROMDB* and *GrassCoRegDB*. *GrassTFDB* deals with information on TFs, their DNA-binding properties and the genes that interact with these TFs. Based on unique structural characteristics, the TFs are grouped into 50 families. Information on TFs can be accessed species-wise by accessing *MaizeTFDB*, *RiceTFDB*, *SorghumTFDB* and *CaneTFDB* or be accessed TF family-wise like *MYB*. This database also has compiled information on the availability of clones for particular TFs in the *TFome Collection*. *GrassPROMDB* is dedicated on promoter sequences across the four grasses. *GrassCoRegDB* is a collection of transcriptional regulatory factors that do not bind DNA in a sequence-specific fashion. These either interact with transcription factors or act as chromatin modifiers controlling accessibility of DNA. Specific role of many of the proteins listed in this database is still unknown. Currently GRASSIUS contains information on 9,028 TFs, 419 coregulators, 149,075 promoters and 180 TFomes, which have been grouped crop-wise.

5.2 Tools

With the rapid progress in sequencing techniques, the biggest challenge before the genomic scientists is to complement the accumulated information with appropriate analysis tools. In this regard, bioinformatics play a very prominent role. Available sequence information needs to be annotated appropriately using right annotation algorithm. Without annotating regulatory elements (both *cis*- and *trans*-acting) appropriately generated information cannot be correlated with any biological role. Generated sequence information from different genotypes needs to be aligned to understand the common features among them, which ultimately can be used in comparative genomic studies.

5.2.1 Annotation Tools/Resources

Annotation tools may be grouped as genome annotation tools, gene structural annotation tools and gene prediction tools. TriAnnot is one of the versatile and high-throughput pipeline for automated annotation of plant genomes. The pipeline is parallelized on a 712 CPU computing cluster and can run about 1Gb sequence annotation in less than 5 days. It can be used through a web interface for small-scale analysis or using server for large-scale annotations. The TriAnnot pipeline has four main panels. Panel I performs transposable elements annotation and masking. Panel II is used for structural and functional annotation of protein-coding genes. Panel III is used for identification of ncRNA genes and conserved non-coding sequences, and Panel IV is used for the development of molecular markers. The TriAnnot pipeline was initially used for annotation of 21 bread wheat chromosomes under the International Wheat Genome Sequencing Consortium. TriAnnot is operational and in use for the 3BSEQ project which would assist in the continuing improvement of TriAnnot pipeline for additional wheat chromosomes and plant genome annotation projects. TriAnnot pipeline is believed to easily get adapted to other plant species with minor modifications (Leroy et al. 2012). Other tools and resources for gene and genome annotations and their features and web resources are provided in Table 2.

5.2.2 Promoter and Regulatory Element Identification Tools/Resources

A significant fraction of the genome of any organism is responsible to specify when, where and how much a gene product needs to be produced. This regulatory information is hardwared in the genome and largely constant over time and generations. These regulatory sequences may be in close proximity with the genes they control (*cis*-regulatory element) or their products (mostly

Table 2 Common resources for gene and genome annotation

S. no	Tool/resource	Particulars	Source
Genome annotation pipeline resources			
1	PASA	Annotation pipeline that aligns EST and protein sequences to the genome and produces evidence-driven consensus gene models	www.pasa.sourceforge.net/
2	MAKER	Identifies repeats, aligns ESTs and proteins to a genome, produces ab initio gene predictions and automatically synthesizes these data into gene annotations with evidence-based quality values	http://gmod.org/wiki/MAKER
3	NCBI	The genome annotation pipeline from the US National Center for Biotechnology Information (NCBI). Uses BLAST alignments together with predictions from Gnomon and GenomeScan to produce gene models	www.ncbi.nlm.nih.gov
4	Ensembl	Ensembl's genome annotation pipeline. Uses species-specific and cross-species alignments to build gene models. Also annotates non-coding RNAs	www.ensembl.org
Gene structural annotation tools			
1.	RepeatMasker	Analysis DNA sequences for interspersed repeats and low complexity DNA sequences	www.repeatmasker.org/
2.	GENEMARK	Gene prediction programmes	www.exon.gatech.edu
3.	TSSP-TCM	Plant promoter identification	http://www.arabidopsis.org
4.	WISE2	Compares a protein sequence to a genomic DNA sequence, allowing for introns and frameshifting errors	www.ebi.ac.uk/Tools/psa/genewise/
5.	GrailEXP	Software package that predicts exons, genes, promoters, polyas, CpG islands, EST similarities and repetitive elements within DNA sequence	www.compbio.ornl.gov/grailexp/
6.	GeneScan	MIT's new web server for GeneScan. GeneScan is used to predict the location and intron–exon boundaries in a genomic sequence	http://genes.mit.edu/GENSCAN.html
7.	yrGATE	A web-based gene structure annotation tool for the identification and dissemination of eukaryotic genes	www.plantgdb.org/prj/yrGATE/
Gene prediction tools			
1.	mGene	Computational tool for the genome-wide prediction of protein-coding genes from eukaryotic DNA sequences	www.mgene.org/
2.	SNAP	SNAP calculates synonymous and non-synonymous substitution rates based on a set of codon-aligned nucleotide sequences	www.snap.cs.berkeley.edu/
3.	FGENESH	Predicting multiple genes in genomic DNA sequences	www.softberry.com
4.	Twinscan	System for predicting gene structure in eukaryotic genomic sequences	www.bioinformatics.ca/
5.	GenomeScan	Predicting the locations and exon–intron structures of genes in genomic sequences from a variety of organisms	www.genes.mit.edu/genomescan.html
6.	GeneSeqer@PlantGDB	Web server provides a gene structure prediction tool tailored for applications to plant genomic sequences	www.plantgdb.org/cgi-bin/GeneSeqer/

proteins) may act in *trans* as transcription factor (TFs). The TFs interpret 'the sequence code hardwared in the *cis*-regulatory apparatus and execute it in the form of a signal to the basal transcription machinery that will result in RNA production. TFs are organized into hierarchial gene regulatory networks in which one TF, often in cooperation with other proteins, positively or

Table 3 Features of promoter and regulatory element resources

S. no	Tool/resource	Particulars	Source
1.	GRASSIUS	TF resource dealing specifically grass family	http://grassius.org/commcontrib.html
2.	PlantProm	A database of plant promoter sequences	http://linux1.softberry.com/berry.phtml?topic=plantprom&group=data&subgroup=plantprom
3.	AthaMap	A genome-wide map of putative transcription factor binding sites in *Arabidopsis thaliana*	www.athamap.de/
4.	PlantTFDB	Plant transcription factor database from about 49 species	planttfdb.cbi.edu.cn/
5.	PlantPromoterdb 3.0	Plant promoter database of Arabidopsis, rice, etc.	ppdb.agr.gifu-u.ac.jp/
6.	PLACE	A database of plant *cis*-acting regulatory DNA elements	www.dna.affrc.go.jp/PLACE/
7.	Transfac	Database on eukaryotic transcription factors, their genomic binding sites and DNA-binding profiles	www.gene-regulation.com/pub/databases.html
8.	PlantCare	Database of plant *cis*-acting regulatory elements and a portal to tools for in silico analysis of promoter sequences	bioinformatics.psb.ugent.be/webtools/plantcare/html/

negatively regulates the expression of another TF' (Yilmaz et al. 2009). Many tools to identify these regulatory elements are available, and several resources harbour these information for use by the communities. Features of common tools and resources are given in Table 3.

5.2.3 Multiple Sequence Alignment Resources

Sequence alignment helps to detect similarity or differences between given biological sequences, which may be protein, DNA or RNA. The sequence alignment may be pairwise or multiple sequence. Using pairwise sequence alignment methods, best-matching piecewise (local) or global alignments of two query sequences can be found out. It can be used between only two sequences at a time. Most commonly this method is used under situation where extreme precision is not required like searching of database for sequences with high similarity to a query. Dot-matrix methods, dynamic programming, and word methods are mostly used for pairwise alignments. As against pairwise alignment, the multiple sequence alignment can incorporate more than two sequences at a time. Such methodologies try to align all of the sequences in a given query set. It often identifies conserved sequences across a group of sequences, which is likely evolutionarily related. Such alignments are computationally demanding. Common resources for multiple sequence alignment are given in Table 4. Among various sequence alignment tools, Clustal is a widely used multiple sequence alignment programme (Chenna et al. 2003). There are three main variations: ClustalW (command line interface, Larkin et al. 2007), ClustalX (this version has a graphical user interface, Thompson et al. 1997) and Clustal Omega (allows hundreds of thousands of sequences to be aligned in only a few hours. It also makes use of multiple processors. In addition, the quality of alignments is superior to previous versions, Sievers et al. 2011). A wide range of input formats, including NBRF/PIR, FASTA, EMBL/Swiss-Prot, Clustal, GCC/MSF, GCG9 RSF and GDE, are acceptable in this programme. The output format can be one or many of the following: Clustal, NBRF/PIR, GCG/MSF, PHYLIP, GDE or NEXUS. There are three main steps in the alignment process, i.e. pairwise alignment, followed by creation of a guide tree (or use a user-defined tree) and finally use the guide tree to carry out a multiple alignment. The 'Do Complete Alignment' option when selected, all these steps are done automatically or else the task may

Table 4 List of multiple sequence alignment resources

Name	Description	Sequence type[a]	Alignment type[b]	Link
ABA	A-Bruijn alignment	Protein	Global	http://nbcr.sdsc.edu/euler/
ALE	Manual alignment; some software assistance	Nucleotide	Local	http://www.red-bean.com/ale/
AMAP	Sequence annealing	Both	Global	http://baboon.math.berkeley.edu/mavid/
anon.	Fast, optimal alignment of three sequences using linear gap costs	Nucleotide	Global	http://www.csse.monash.edu.au/~lloyd/tildeStrings/
BAli-Phy	Tree + multi alignment; probabilistic/Bayesian; joint estimation	Both	Global	http://www.biomath.ucla.edu/msuchard/bali-phy
Base-By-Base	Java-based multiple sequence alignment editor with integrated analysis tools	Both	Local or global	http://athena.bioc.uvic.ca/virology-ca-tools/base-by-base/
ClustalW	Progressive alignment	Both	Local or global	Available at EBI, DDBHPBIL, EMBNet, GenomeNet
CodonCodeAligner	Multi alignment; ClustalW and Phrap support	Nucleotide	Local or global	http://www.codoncode.com/aligner/
Compass	Comparison of multiple protein sequence alignments with assessment of statistical significance	Protein	Global	http://prodata.swmed.edu/compass/compass_advanced.php
DIALIGN-TX and DIALIGN-T	Segment-based method	Both	Local (preferred) or global	http://dialign-tx.gobics.de/
DNA Alignment	Segment-based method for intraspecific alignments	Both	Local (preferred) or global	http://www.fluxus-engineering.com/align.htm
DNA Baser Sequence Assembler	Multi alignment; automatic batch alignment	Nucleotide	Local or global	www.DnaBaser.com
MARNA	Multiple alignment of RNAs	RNA	Local	http://biwww2.informatik.uni-freiburg.de/Software/MARNA/index.html
MAVID	Progressive alignment	Both	Global	http://baboon.math.berkeley.edu/mavid/
MULTALIN	Dynamic programming/clustering	Both	Local or global	http://prodes.toulouse.inra.fr/multalin/multalin.html
Multi-LAGAN	Progressive dynamic programming alignment	Both	Global	http://genome.lbl.gov/vista/lagan/submit.shtml
Praline	Progressive/iterative/consistency/homology-extended alignment with pre-profiling and secondary structure prediction	Protein	Global	http://www.ibi.vu.nl/programs/pralinewww/
RevTrans	Combines DNA and protein alignment, by back translating the protein alignment to DNA	DNA/Protein (special)	Local or global	http://www.cbs.dtu.dk/services/RevTrans/
SAGA	Sequence alignment by genetic algorithm	Protein	Local or global	http://www.tcoffee.org/Projects_home_page/saga_home_page.html

(continued)

Table 4 (continued)

Name	Description	Sequence type[a]	Alignment type[b]	Link
SAM	Hidden Markov model	Protein	Local or global	http://www.cse.ucsc.edu/compbio/HMM-apps/T02-query.html
Se-Al	Manual alignment	Both	Local	http://tree.bio.ed.ac.uk/software/seal/
StatAlign	Bayesian co-estimation of alignment and phylogeny (MCMC)	Both	Global	http://phylogeny-cafe.elte.hu/StatAlign/
Stemloc	Multiple alignment and secondary structure prediction	RNA	Local or global	http://biowiki.org/StemLoc
UGENE	Supports multiple alignment with MUSCLE, KAlign, Clustal and MAFFT plugins	Both	Local or global	http://ugene.unipro.ru/download.html

Modified from http://en.wikipedia.org/wiki/Sequence_alignment_software
[a]Sequence type: protein or nucleotide
[b]Alignment type: local of global

be carried out by the following options, namely, 'Do Alignment from guide tree' and 'Produce guide tree only'. There is option for default setting or customized settings.

5.2.4 Whole Genome Alignment and Comparative Genomics Tools

Alignment of nucleic or amino acid sequences has been one of the most important tools in sequence analysis, and with much dedicated research, many sophisticated algorithms are available for aligning sequences with similar regions (Chain et al. 2003). Until very recently, most of these algorithms were primarily designed for comparing single protein sequences or DNA sequences containing a single gene or operon. There are several problems associated with aligning long genomic sequences or entire genomes. In recent past several programmes have been developed to address this requirement (Table 5). Accelerated Search for SImilar Regions in Chromosomes (ASSIRC) tool is used for locating similarities of nucleic sequences in large genomes (Vincens et al. 2002). The new version of D-ASSIRC uses two different strategies for performing the above task: (i) distributed search using splitting of the target sequence into several large overlapping sequences and (ii) distributed searches for repeated exact motifs of fixed size either managed by a central processor (strategy AGD) or locally managed by numerous processors (strategy ALD). A Comparative Genomic Tool (ACGT) is another comparative sequence analysis tool which can analyse, compare as well as provide graphical view of the compared sequences (Xie and Hood 2003). The tool reads pair of DNA sequences in GenBank, Embl or FASTA formats, with or without a comparison file, and provides users with many options to view and analyse the similarities between the input sequences. Generic Model Organism Database (GMOD) is another comprehensive comparative genomic tool which can be used for managing and visualizing comparative genomics and synteny data. GMOD has several components for managing and visualizing comparative genomics data. The components incorporated into GMOD are CMAP, GBrowse_syn, SynView and Sybil. CMAP tool allows researchers to view comparison of genetic maps, physical maps, sequence assemblies, QTL and deletion maps. Unlike other comparative genomic tools, CMAP does not require sequence data. GBrowse_syn displays view of multiple sequence alignment data and synteny from other sources against genome annotations provided by GBrowse. Sybil provides whole genome comparisons, regional comparisons (synteny) and orthologous gene comparisons. SynView displays synteny at the

Table 5 Sequence alignment and comparative genomic tools

Tool	Website
ASSIRC	ftp://ftp.biologie.ens.fr/pub/molbio/
DIALIGN	http://bibiserv.TechFak.Uni-Bielefeld.DE/dialign/
MUMmer	http://www.tigr.org/software/mummer/
PipMaker/BlastZ	http://bio.cse.psu.edu/pipmaker/
GLASS	http://crossspecies.lcs.mit.edu/
WABA	http://www.soe.ucsc.edu/&#3;kent/xenoAli/
LSH-ALL-PAIRS	Contact author at: jbuhler@cs.washington.edu
Vmatch	http://www.vmatch.de
MGA	http://bibiserv.techfak.uni-bielefeld.de/mga/
PipMaker/BlastZ	http://bio.cse.psu.edu/pipmaker/
Alfresco	http://www.sanger.ac.uk/Software/Alfresco/
Intronerator	http://www.cse.ucsc.edu/&#3;kent/intronerator/
VISTA	http://www-gsd.lbl.gov/vista/
SynPlot	http://www.sanger.ac.uk/Users/jgrg/SynPlot/
ACT	http://www.sanger.ac.uk/Software/ACT/
DisplayMUMS	http://www.tigr.org/software/displaymums/
ACGT	http://db.systemsbiology.net/projects/local/mhc/acgt/
GMOD	http://ccg.murdoch.edu.au/index.php/GMOD
DIALIGN	http://dialign.gobics.de/
GATA	http://gata.sourceforge.net/

region and/or gene level. DIALIGN is a software program for multiple sequence alignment programme. It constructs pairwise and multiple alignments by comparing entire segments of the sequences. DIALIGN is used for both global and local alignment, but it is particularly successful in situations where sequences share only local homologies (Morgenstern et al. 1998). *Wobble Aware Bulk Aligner* (WABA) is another tool for sequence comparison between large genomes (Baillie and Rose 2000). Graphic Alignment Tool for Comparative Sequence Analysis (GATA) is one major tool for fine-grained alignment and visualization tool suited for non-coding 0–200 kb pairwise sequence analysis (Nix and Eisen 2005). GATA uses NCBI-BLASTN programme along with post-processing features to exhaustively align two DNA sequences. The tool visualizes both large and small sequence inversions, duplications and segment shuffling. Since the alignment is visual and does not contain gaps, gene annotation can be added to both sequences to create a thoroughly descriptive picture of DNA conservation that is well suited for comparative sequence analysis.

6 Progress in Comparative Genomics in Cereals

Comparative genome mapping in cereals was carried out extensively during the 1990s using genetic markers (Gale and Devos 1998; Bennetzen 2000). Subsequently development of EST and genome segments sequencing projects improved the resolution of the comparative mapping. The availability of whole genome sequences of major cereals facilitates the validation of the earlier results on comparative genomics of cereals. Moreover, the wealth of information available provides a major resource for genomic studies in many other minor cereals where the genomes are yet to be sequenced.

6.1 Macro and Micro Level Collinearity in Cereals

In the sense of comparative genomics, collinearity refers to conservation of gene/marker order across genomes separated from each other in the process of evolution. This collinearity at genetic map level is referred to as macro-collinearity, while that at sequence level is referred to as micro-collinearity. Though large differences exist between cereals in terms of ploidy level, chromosome number and haploid DNA content (Table 1), marker orders are largely conserved among cereal crops even with million years of evolutionary history (Feuillet and Keller 2002). QTLs underlying important agronomic traits like shattering and dwarfing found to be collinear between cereals and other grass species (Paterson et al. 1995; Pereira and Lee 1995). Moore et al. (1995) were first to develop grass consensus map aligning seven different grass species using rice as a reference genome. Further refinement of this consensus map by Gale and Devos (1998) and Devos and Gale (2000) has shown that 10 grass genomes can be described using less than 30 rice linkage blocks. This gave rise to the famous 'crop circles' model, representing the relationships between orthologous chromosomes in eight species, namely, rice, foxtail millet, sugarcane, sorghum, pearl millet, maize, triticeae and oats. However, further analysis of collated data on comparative genomics in cereals indicated that the average probability of one marker found in the vicinity of one species to be in a collinear region of another species is low (on an average 50 % even between closely related species like maize and sorghum) (Gaut 2002). Such results clearly indicated extensive rearrangements between cereal genomes and questioned the concept of using small genomes like rice and sorghum as a proxy for more complex genomes like maize or wheat.

With progress in sequence information and analysis tools using 2,600 mapped sequence markers, Salse et al. (2004) identified 656 putative orthologous genes in the rice genomes. They could identify six new collinear regions between maize chromosomes 1, 4, 5 and 6 and rice chromosomes 9–12, 6–8, 6 and 1, respectively. Similar studies in wheat using 4,485 ESTs showed that among the wheat chromosomes, chromosome 3 is the most conserved to rice and chromosome 5 to be the least conserved (La Rota and Sorrells 2004). However, Guyot et al. (2004) observed mosaic conservation of genes between short arm of wheat chromosome 1A and rice chromosome 5S. Further use of ESTs and low-pass BAC sequences could identify new regions of collinearity as well as structural aberrations like inversions within single chromosome groups of cereal genomes (Klein et al. 2003; Singh et al. 2004; Buell et al. 2005).

First micro-collinearity studies involving two maize loci (*sh2*/*a1* and *Adh1*) and the homologous regions in sorghum and rice demonstrated conservation of gene order between maize, sorghum and rice (Chen et al. 1997). Similar micro-collinearity was demonstrated between reception-like kinase orthologs of wheat and barley (Feuillet and Keller 1999). However, other studies involving *Adh1* locus between maize and sorghum (Tikhonov et al. 1999), and rice (Tarchini et al. 2000; Ilic et al. 2003); rust resistance gene *rpg1* in barley and rice (Kilian et al. 1997); *Rph7* between barley and rice (Scherrer et al. 2005); *r1*/*b1* between maize, sorghum and rice (Swigonova et al. 2005); *Orp1*/*Orp2* in rice and sorghum (Ma et al. 2005); *Bz* between maize and rice (Lai et al. 2005) and many more revealed significant gene rearrangements within otherwise macro-collinear regions. Comparing 454 pyrosequences of 20 sugarcane BACs with sorghum sequences revealed that the two genomes are mostly collinear in the genic regions, and sorghum genomes can be used as a template for assembling much of the genic DNA of the autopolyploid sugarcane genome (Wang et al. 2010). However, they also detected 54 events of chromosomal rearrangements (translocation, inverted translocation, inversion, genome-specific duplications) between the two genomes. Within macro-collinear regions, different types of small rearrangements are likely to disrupt the micro-collinearity, which are unlikely to be reflected at the genetic map level. Some rearrangements, like small inversions or gene duplications, have little effect on macro-collinearity, whereas deletions and translocations may make the analysis much complicated (Feuillet and Keller 2002). Schnable et al. (2009), upon comparison of genome sequence of rice and sorghum with maize, demonstrated considerable collinearity between these genomes. They observed that exons of rice, sorghum and maize orthologs are of similar size but they differed for intron lengths because of insertion of repetitive elements. Paterson et al. (2009a) reported that a total of 19,929 sorghum gene models are in blocks collinear with rice. A very close conservation of gene has been demonstrated between wheat and *Brachypodium* upon comparison of the two genome sequences (Brenchley et al. 2012). Earlier Vogel et al. (2010) demonstrated that 77–84 % of gene families identified in *Brachypodium* are shared among the three grass subfamilies represented by rice, sorghum and *Brachypodium.*

6.2 Euchromatin Versus Heterochromatin in Cereal Genomes

Despite the large differences among the cereals in terms of the genome sizes, the gene-rich regions in these genomes remain quite similar in size and organization (Feuillet and Keller 1999; Bowers et al. 2005). Bowers et al. (2005) reported close correlation between euchromatic and heterochromatic regions of the chromosome with gene-rich versus repeat-abundant regions in rice/sorghum synteny map. Frequency of recombination is observed more in the euchromatic regions. They further observed highest synteny and lowest occurrence of retroelements at the distal regions of the chromosomes. These regions are highly recombinogenic as against pericentromeric regions. Pericentromeric regions are high in repetitive DNA content and low in genic sequences (Akhunov et al. 2003; Bowers et al. 2005; Schnable et al. 2009). However, preferential preservation of microsynteny in recombinogenic regions is of common occurrence among cereals (Bowers et al. 2005). This suggests that gene rearrangements in cereals are generally somewhat deleterious but the reason behind this is still not clear (Paterson et al. 2009b).

Euchromatin and heterochromatin segments have been largely preserved in the process of evolution (Paterson et al. 2009b). It is observed that sorghum has ~75 % larger DNA than rice, most of which are heterochromatin (Paterson et al. 2009a). Alignment of genetic (Bowers et al. 2003) and cytological (Kim et al. 2005) maps of sorghum and rice shows that these two genomes have similar quantities of euchromatin. These regions account for 97–98 % of recombination and 75.4–94.2 % of genes with high collinearity (Bowers et al. 2005; Paterson et al. 2009a). It is established in *Arabidopsis* that TEs and related tandem repeats under the control of chromatin remodelling ATPase DDM1 being guided by siRNAs determine the heterochromatic state (Lippman et al. 2004). This preservation of 'genomes within a genome' in the cereals needs to be understood more precisely in the future (Paterson et al. 2009b). Genome sequences of rice, sorghum, maize, wheat, barley and others have revealed the predominance of TEs and other repetitive sequences in the heterochromatic regions (Matsumoto et al. 2005; Paterson et al. 2009a; Schnable et al. 2009; Brenchley et al. 2012; Mayer et al. 2012). Differences between species in terms of the predominance of TE families are widely reported. For example, LTR *Gypsy* retrotransposon superfamily is 1.5-fold more abundant than the *Copia* superfamily in the highly repetitive regions of barley (Mayer et al. 2012) as against *Brachypodium* (Vogel et al. 2010) and rice (Matsumoto et al. 2005). Suppression of recombination along with high tolerance for repetitive DNA in pericentromeric regions is suggested to create a sort of genomic environment that favours the evolution of 'co-adapted gene complexes', predicted to favour the process of domestication (Paterson et al. 2010).

6.3 Distinguishing Features of Cereal Genomes

Comparison of cereal genomes with that of other dicot genomes has identified good number of genes specific to cereal lineages (Campbell et al. 2007; Schnable et al. 2009). Schnable et al. (2009) compared the genome sequences of maize, rice and sorghum to that of *Arabidopsis.* A core set of 8,494 gene families is shared among all the four species. Gene family numbers for maize, sorghum and rice are 11,892, 12,353 and 13,055, respectively. Out of these 2,077 gene families are common among the three species, 405 between maize and sorghum, 229 between maize and rice and 661 between sorghum and rice. Unique gene families identified for maize, sorghum and rice are 465, 265 and 1,110, respectively. Identification of such species-specific genes is not unprecedented (Paterson et al. 2009b). For example, genes for major seed storage proteins of maize (zein) and sorghum (kafirin), which are absent in rice, appear to have evolved since the divergence of the panicoids from the oryzoids (Song and Messing 2003). Juxtapositioning of existing gene fragments are the predominant mechanism of creation of new genes. Predominant gene-transducing elements in rice and maize are reported to be *Mutator* like TE, *Pack-MULE* (Jiang et al. 2004) and *Helitrons* (Brunner et al. 2005), respectively. Maize is reported to contain eight families of *Helitrons* totaling ~20,000 in number, predominating in the gene-rich regions (Schnable et al. 2009).

It is intriguing to note that similarity of gene order is higher between sorghum and rice, which diverged 40–50 MY (Paterson et al. 2004; Bowers et al. 2005), than sorghum and maize, which diverged 10–15 MY back (Bowers et al. 2003). Study of Schnable et al. (2009) also demonstrated that more numbers of gene families are conserved between sorghum and rice (661) than sorghum and maize (405). The specific reason explaining this enigma is still unknown. Occurrence of bimodal GC content distribution is another striking compositional feature of cereal genes (Paterson et al. 2009b). It is also observed that often the cereal genes show a unique negative GC gradient from the 5′ untranslated region and extend nearly 1 kb into the coding region (Wong et al. 2002). Again causes behind such distinct feature of cereal genes are largely elusive.

6.4 Mechanism of Genome Evolution in Cereals

Five primary mechanisms, namely, polyploidy, amplification and movement of TEs, chromosome breakage, unequal homologous recombination and illegitimate recombination, are responsible for origin of genomic complexity (Bennetzen 2007). Sequence comparison of angiosperm genomes to date suggests predominance of ancient genome duplication playing an important role in shaping respective genomes (Paterson et al. 2009b). Earlier studies with RFLP maps hinted occurrence of large-scale duplications in the process of evolution even among diploid texa. This is further confirmed using the rice genome sequence (Paterson et al. 2004; Tian et al. 2005; Yu et al. 2005). Based on the divergence of gene sequences, Paterson et al. (2004) estimated that this duplication occurred about 70 MY back, following which with further evolution of 20 MY the panicoid, pooid and oryzoid lineages diverged. Within this time period, most of the genome reorganizations in terms of gene loss, neofunctionalization or subfunctionalization and others have occurred (Lynch et al. 2001). It is suggested that recent duplications than the shared ancient duplication must have contributed more towards evolution of cereal genomes (Paterson et al. 2009b). Paterson et al. (2009a) identified 5,303 paralogues belonging to 1,947 families, which are predominantly proximally duplicated.

Major mechanisms of genome rearrangements are not the same in all cereal lineages. However, it is evident from generated genome sequences that retrotransposons and particularly the LTR retrotransposons are responsible for plant genome expansion, often referred to as 'genomic obesity' (Vitte and Bennetzen 2006). For example, at least 35 % of rice genome is represented by TEs (Matsumoto et al. 2005), 55 % in sorghum (Paterson et al. 2009a), 85 % in maize (Schnable et al. 2009) and 84 % in barley (Mayer et al. 2012). It is evident that the TE families vary significantly from species to species, which are likely responsible for unique genome organization and functioning of respective genomes. Chromosomal structural changes, often associated with movement of TEs, also play a crucial role in evolution of different genomes (Feuillet and Keller 2002). TEs often lead to illegitimate recombination leading to unequal homologous recombination resulting into segment removal or addition. Rates of segment removal also appear to be quite different from one plant lineage to another (Kirik et al. 2000; Vitte and Bennetzen 2006). Most of the micro-collinearity studies have shown conservation of exons between homologous grass genes, while intron positions are also conserved except their size. Among cereals with bigger genome like in maize, intronic regions are reported to be populated with TEs leading to increase in their sizes (Schnable et al. 2009). Comparative genomic studies to date suggest that retroelements have played a major role in shaping and remodelling the cereal genomes during evolution. The most evident example is almost doubling of maize genome after its divergence from sorghum within a period of 3 MY with the invasion of retrotransposons. Retrotransposons are generally transcriptionally more active and more conserved in grasses than dicots (Vicient et al. 2001).

6.5 Gene Discovery and Marker Development Using Comparative Genomics

Cereal comparative genomics has revealed overall conservation of genes and gene orders across cereals. Progress in bioinformatics has also provided tools that can be used for gene and marker isolation across genomes.

6.5.1 Comparative Genomics-Based Gene Cloning

Much before availability of genome sequence studies with RFLP maps showed collinearity of genes among cereals, and rice genome data was considered to support positional cloning of genes from other genomes through the so-called cross genome map-based cloning (Kilian et al. 1997). Collinearity exists among the 'green revolution' genes across cereals, which helped in direct cloning of dwarfing genes, *Sd1* in rice (Monna et al. 2002), *Rht-1* in wheat and *D8* in maize (Peng et al. 1999). Wherever micro-collinearity is high, using the rice sequence candidate genes can be identified easily even in the absence of noncollinear map of the target traits (Salse and Feuillet 2007). Following this approach in barley, powdery mildew resistance gene *Ror2* was isolated by Collins et al. (2003), and dwarfing gene *sw3* by Gottwald et al. (2004). Genes and QTL involved in developmental process and those selected during the process of domestication are suggested to be a better candidate for direct gene isolation (Salse and Feuillet 2007). Disease resistance (R) genes show little collinearity among cereal genomes and are reported not to be a good candidate for gene isolation strategy using rice genome information.

It has been observed that even in the absence of orthology in rice, flanking genes are often conserved between two genomes and markers from these regions may be used to saturate the target regions. Using the rice ESTs, the genetic interval around R loci, *Rpg1* and *Rph7* in barley could be reduced to initiate chromosome walking (Brueggeman et al. 2002; Brunner et al. 2003). Griffiths et al. (2006) combining the information from *Brachypodium* and rice-isolated pairing controlling gene, *Ph1*, in wheat. In case of very low collinearity, alternative strategy like transposon tagging can be used as has been reported by Vollbrecht et al. (2005) in isolating tassel architecture gene, *Ramosa1*, in maize. Following comparative genomics and physical mapping approaches, Shinozuka et al. (2010) cloned self-incompatibility gene in ryegrass using sequence information from rice, sorghum and *Brachypodium*.

6.5.2 Comparative Genomics-Based Gene Annotation and Marker Development

High amount of conservation of genes and genic sequences across cereals opens up the possibility for rapid and accurate gene annotation in new species. For this purpose, genome sequence of the model crops like rice and sorghum can be used as reference. Annotated EST sequences can also be used as reference for this purpose. Annotated intro–exon junction information helps in annotation of new sequences. In the process new markers are identified across genomes, which can be deployed in marker-assisted selection. Varshney et al. (2005) demonstrated across taxa transferability of barley EST-SSRs, in which more than 78 % of the barley markers showed amplification in wheat followed by 75.2 % in rye and 42.4 % in rice. Furthermore, in silico comparison of EST-SSRs against 1,369,182 publicly available cereal ESTs showed significant homology with ESTs of wheat (93.5 %), rye (37.3 %), rice (57.3 %), sorghum (51.9 %) and maize (51.9 %). The study indicated utility of comparative genomics for development of robust marker system. In a similar study, Srinivas et al. (2008) could develop 50 genic-SSR markers for the four major sorghum stay-green QTLs, *Stg1*, *Stg2*, *Stg3* and *Stg4*, using rice genome information. Recently Kumari et al. (2013) developed 447 eSSRs from 24,828 nonredundant foxtail millet ESTs, of which 327 were mapped physically onto nine chromosomes. Out of these SSRs, 106 showed high level of cross-genera amplification at an average of ~88 % in eight millets and four non-millet species. The large-scale SSR markers developed in these studies demonstrate the utility of comparative genomics in germplasm characterization and subsequently for its utilization in molecular breeding programmes.

7 Future Prospects

The whole genome sequence information available in major cereals like rice, sorghum, maize, foxtail millet and wheat has revolutionized the understanding of the mechanisms underlying genome evolution in these important cereal crops as well as in unravelling the genes governing important mechanisms in plant growth and development, cellular processes and tolerance to various biotic and abiotic stresses. Comparative mapping using large-scale DNA markers revealed the conservation of genes across members of cereal family. The whole genome sequences of rice and sorghum serve as validation tool for comparative genomics at sequence level in other cereals. This also provides an excellent tool for comparative genomes in orphan crops where genome has not yet been sequenced.

Functional genomics across cereals has become more practical due to whole genome sequencing and large-scale EST sequencing projects. The translation of genome DNA sequence data into protein structures and predicted functions will provide important link between the genotype of an organism and its phenotype. The practical applications of the genome sequence and high-throughput sequencing projects are best realized only when allelic diversity patterns existing among the diverse germplasms are better understood. Information on the allelic variation patterns may contribute to functional analysis of crop-specific genes for genetic improvement of cereals for traits of agronomic importance. Eventually integrating information on structural and functional genomics will provide a better understanding about network of genes involved in complex biological responses. In orphan crops where sequence information is not available, gene synteny and collinearity play an important role in comparative genome mapping for studying ancestral genome evolution. With the availability of genome sequence in model crops and understanding synteny relationships between these crop species, the development of cross-amplification markers in the related crop species will now be a practical option. These approaches will enhance the development of genomic tools required for marker-assisted breeding in orphan crop species that would otherwise be unlikely to benefit from the current genomic revolution.

Improved sequencing is accompanied by many challenges as well as new developments. NGS technologies create a vast amount of data in a much shorter time compared to conventional Sanger sequencing techniques, presenting many challenges to computational biologists, bioinformaticians and more importantly to the crop breeders for its utilization in crop improvement programmes. One of the major goals of modern NGS tools is to resequence traditional landraces and to understand the molecular basis for phenotype–genotype relationships. Diversity panels of thousands of individuals selected to sample the extent of diversity with reference genome sequences using NGS technologies will provide a platform for understanding existing genetic diversity, associating gene(s) with phenotypes and exploiting natural genetic diversity to help develop superior genotypes using association mapping approaches. The major challenge of the modern genome tools remains to convert this mass of data into knowledge that can be applied in crop-breeding programmes. Collecting accurate phenotypic information is one of the major hindrances for effective utilization of genomics technologies in advanced crop improvement. Development of precise phenotyping techniques has not kept pace with advances in genomics. Breeders are required to apply the genomic tools and precise phenotyping techniques to truly advance the crop improvement process and take advantage of the potential of genomics. The most effective effort to fulfil the gap is to integrate various research disciplines which form core components of molecular plant breeding. The integration of various approaches required knowledge of whole genome organization, strong statistical knowledge to estimate the gene/genetic effects and good experience in molecular biology techniques and traditional breeding methodologies. These integrated approaches will revolutionize the crop improvement in future.

References

Akhunov ED, Goodyear AW, Geng S et al (2003) The organization and rate of evolution of wheat genomes are correlated with recombination rates along chromosome arms. Genome Res 13:753–763

Baillie DL, Rose AM (2000) WABA success: a tool for sequence comparison between large genomes. Genome Res 10:1071–1073

Barry GF (2001) The use of the Monsanto draft rice genome sequence in research. Plant Physiol 125:1164–1165

Bennett MD, Leitch IJ (1995) Nuclear DNA amounts in angiosperms. Ann Bot 76:113–176

Bennetzen JL (2000) Comparative sequence analysis of plant nuclear genomes: microcolinearity and its many exceptions. Plant Cell 12:1021–1029

Bennetzen JL (2007) Patterns in grass genome evolution. Plant Biol 10:176–181

Berkmann PJ, Lai K, Lorenc MT et al (2012) Next-generation sequencing applications for wheat crop improvement. Am J Bot 99(2):365–371

Bowers JE, Abbey C, Anderson S et al (2003) A high-density genetic recombination map of sequence-tagged sites for Sorghum, as a framework for comparative structural and evolutionary genomics of tropical grains and grasses. Genetics 165:367–386

Bowers JE, Arias MA, Asher R et al (2005) Comparative physical mapping links conservation of microsynteny to chromosome structure and recombination in grasses. Proc Natl Acad Sci U S A 102:13206–13211

Brenchley R, Spannag M, Pfeifer M et al (2012) Analysis of the bread wheat genome using whole-genome shotgun sequencing. Nature 491:705–710

Brueggeman R, Rostoks N, Kudrna D et al (2002) The barley stem rust-resistance gene Rpgl is a novel disease-resistance gene with homology to receptor kinases. Proc Natl Acad Sci U S A 99:9328–9333

Brunner S, Keller B, Feuillet C (2003) A large rearrangement involving genes and low copy DNA interrupts the microcolinearity between rice and barley at the Rph7 locus. Genetics 164:673–683

Brunner S, Fengler K, Morgante M et al (2005) Evolution of DNA sequence nonhomologies among maize inbreds. Plant Cell 17:343–360

Buell CR (2009) Poaceae genomes: going from unattainable to becoming a model clade for comparative plant genomics. Plant Physiol 149:111–116

Buell CR, Yuan Q, Ouyang S et al (2005) Sequence, annotation, and analysis of synteny between rice chromosome 3 and diverged grass species. Genome Res 15:1284–1291

Campbell MA, Zhu W, Jiang N et al (2007) Identification and characterization of lineage-specific genes within the Poaceae. Plant Physiol 145:1311–1322

Chain P, Kurtz S, Ohlebusch E et al (2003) An applications-focused review of comparative genomics tools: capabilities, limitations and future challenges. Brief Bioinform 4:105–123

Chen M, SanMiguel P, De Oliveira AC et al (1997) Microcollinearity in sh-2 homologous regions of the maize, rice and sorghum genomes. Proc Natl Acad Sci U S A 94:3431–3435

Chenna R, Sugawara H, Koike T et al (2003) Multiple sequence alignment with the Clustal series of programs. Nucleic Acids Res 31:3497–3500

Collins NC, Thordal-Christensen H, Lipka V et al (2003) SNARE-protein –mediated disease resistance at the plant cell wall. Nature 425:973–977

Conte MG, Gaillard S, Lanau N et al (2008) GreenPhylDB: a database for plant comparative genomics. Nucleic Acids Res 36:991–998

Devos KM, Gale MD (2000) Genome relationships: the grass model in current research. Plant Cell 12:637–646

Duvick J, Fu A, Muppirala U et al (2007) PlantGDB: a resource for comparative plant genomics. Nucleic Acids Res 36:959–965

Feuillet C, Keller B (1999) High gene density is conserved at syntenic loci of small and large grass genomes. Proc Natl Acad Sci U S A 96:8265–8270

Feuillet C, Keller B (2002) Comparative genomics in the grass family: molecular characterization of grass genome structure and evolution. Ann Bot 89:3–10

Gale MD, Devos KM (1998) Comparative genetics in the grasses. Proc Natl Acad Sci U S A 95:1971–1974
Gaut BS (2002) Evolutionary dynamics of grass genomes. New Phytol 154:15–28
Goff SA, Ricke D, Lan TH et al (2002) A draft sequence of the rice genome (Oryza sativa L. ssp. japonica). Science 296:92–100
Goodstein DM, Shu S, Howson R et al (2012) Phytozome: a comparative platform for green plant genomics. Nucleic Acids Res 40:1178–1186
Gottwald S, Stein N, Borner A et al (2004) The gibberellic-acid insensitive dwarfing gene sdw3 of barley is located on chromosome 2HS in a region that shows high colinearity with rice chromosome 7L. Mol Genet Genomics 271:426–436
Griffiths S, Sharp R, Foote TN et al (2006) Molecular characterization of Ph1 as a major chromosome pairing locus in polyploid wheat. Nature 439:749–752
Gupta PK (2008) Single molecule DNA sequencing technologies for future genomics research. Trends Biotechnol 26:602–611
Guyot R, Yahiaoui N, Feuillet C et al (2004) In silico comparative analysis reveals a mosaic conservation of genes within a novel colinear region in wheat chromosome 1AS and rice chromosome 5S. Funct Integr Genomics 4:47–58
Harlan JR (1971) Agricultural origins: centers and non-centers. Science 174:568–574
Harlan JR (1992) Origins and process of domestication. In: Chapman GP (ed) Grass evolution and domestication. Cambridge University Press, Cambridge, pp 159–175
Ilic K, SanMiguel PJ, Bennetzen JL (2003) A complex history of rearrangement in an orthologous region of the maize, sorghum, and rice genomes. Proc Natl Acad Sci U S A 100:12265–12270
Jiang N, Bao ZR, Zhang XY et al (2004) Pack-MULE transposable elements mediate gene evolution in plants. Nature 431:569–573
Kellogg EA (2001) Evolutionary history of the grasses. Plant Physiol 125:1198–1205
Kilian A, Chen J, Han F et al (1997) Towards map-based cloning of the barley stem rust resistance gene Rpg1 and rpg4 using rice as a intergenomic cloning vehicle. Plant Mol Biol 35:187–195
Kim JS, Klein PE, Klein RR et al (2005) Chromosome identification and nomenclature of Sorghum bicolor. Genetics 169:1169–1173
Kirik A, Salomon S, Puchta H (2000) Species-specific double-strand break repair and genome evolution in plants. EMBO J 19:5562–5566
Klein PE, Klein RR, Vrebalov J, Mullet JE (2003) Sequence-based alignment of sorghum chromosome 3 and rice chromosome 1 reveals extensive conservation of gene order and major chromosomal rearrangement. Plant J 34:605–621
Kresovich S, Barbazuk B, Bedell JA et al (2005) Toward sequencing the sorghum genome. A U.S. National Science Foundation-sponsored workshop report[w]. Plant Physiol 138:1898–1902
Kumari K, Muthamilarasan M, Misra G et al (2013) Development of eSSR-markers in Setaria italica and their applicability in studying genetic diversity, cross-transferability and comparative mapping in millet and non-millet species. PLoS One 8(6):e67742. doi:10.1371/journal.pone.0067742
La Rota M, Sorrells ME (2004) Comparative DNA sequence analysis of mapped wheat ESTs reveals the complexity of genome relationships between rice and wheat. Funct Integr Genomics 4:34–46
Lai J, Li Y, Messing J, Dooner HK (2005) Gene movement by helitron transposons contributes to the haplotype variability of maize. Proc Natl Acad Sci U S A 102:9068–9073
Larkin MA, Blackshields G, Brown NP et al (2007) Clustal W and Clustal X version 2.0. Bioinformatics 23:2947–2948
Leroy P, Guilhot N, Sakai H et al (2012) TriAnnot: a versatile and high performance pipeline for the automated annotation of plant genomes. Front Plant Sci 3:5. doi:10.3389/fpls.2012.00005
Ling HQ, Zhao S, Liu D et al (2013) Draft genome of the wheat A genome progenitor Triticum urartu. Nature 496:87–90
Lippman Z, Gendrel AV, Black M et al (2004) Role of transposable elements in heterochromatin and epigenetic control. Nature 430:471–476
Liu L, Li Y, Li S et al (2012) Comparison of next-generation sequencing systems. J Biomed Biotechnol. doi:10.1155/2012/251364
Lynch M, O'Hely M, Walsh B, Force A (2001) The probability of preservation of a newly arisen gene duplicate. Genetics 159:1789–1804
Lyons E, Freeling M (2008) How to usefully compare homologous plant genes and chromosomes as DNA sequences. Plant J 53:661–673
Lyons E, Pedersen B, Kane J, Freeling M (2008) The value of nonmodel genomes and an example using SynMap within CoGe to dissect the hexaploidy that predates the rosids. Trop Plant Biol 1:181–190
Ma J, SanMiguel P, Lai J et al (2005) DNA rearrangement in orthologous orp regions of the maize, rice and sorghum genomes. Genetics 170:1209–1220
Mace ES, Jordan DR (2010) Location of major effect genes in sorghum (Sorghum bicolor (L.)). Theor Appl Genet 121:1339–1356
Mardis ER (2008) Next generation DNA sequencing methods. Annu Rev Genomics Hum Genet 9:387–402
Matsumoto T, Wu J, Kanamori H et al (2005) The map-based sequence of the rice genome. Nature 436:793–800
Mayer KFX, Waugh R, Langridge P et al (2012) The International Barley Genome Sequencing Consortium. Nature 491:711–716
Merriman B, Ion Torrent R&D Team, Rothberg JM (2012) Progress in Ion Torrent semiconductor chip

based sequencing. Electrophoresis 33(23):3397–3417. doi:10.1002/elps.201200424
Monna L, Kitazawa N, Yoshino R et al (2002) Positional cloning of rice semidwarfing gene, sd-1: rice 'green revolution gene' encodes a mutant enzyme involved in gibberellin synthesis. DNA Res 9:11–17
Moore G, Devos KM, Wang Z, Gale MD (1995) Cereal genome evolution: grasses, line up and form a circle. Curr Biol 5:737–739
Morgenstern B, Frech K, Dress A et al (1998) DIALIGN: finding local similarities by multiple sequence alignment. Bioinformatics 14:290–294
Nelson JC, Wang S, Wu Y et al (2011) Single nucleotide polymorphism discovery by high-throughput sequencing in sorghum. BMC Genomics 12:352–364
Nix DA, Eisen MB (2005) GATA: a graphic alignment tool for comparative sequence analysis. BMC Bioinformatics 6:9
Paterson AH, Lin YR, Li ZK et al (1995) Convergent domestication of cereal crops by independent mutations at corresponding genetic loci. Science 269:1714–1718
Paterson AH, Bowers JE, Chapman BA (2004) Ancient polyploidization predating divergence of the cereals, and its consequences for comparative genomics. Proc Natl Acad Sci U S A 101:9903–9908
Paterson AH, Bowers JE, Bruggmann R et al (2009a) The sorghum bicolor genome and the diversification of grasses. Nature 457:551–556
Paterson AH, Bowers JE, Feltus FA et al (2009b) Comparative genomics of grasses promises a bountiful harvest. Plant Physiol 149:125–131
Paterson AH, Freeling M, Sasaki T (2010) Grains of knowledge: genomics of model cereals. Genome 15:1643–1650
Peng JR, Richards DE, Hartley NM et al (1999) 'Green revolution' genes encode mutant gibberellin response modulators. Nature 400:256–261
Pereira MG, Lee M (1995) Identification of genomic regions affecting plant height in sorghum and maize. Theor Appl Genet 90:380–388
Piperno DR, Flannery KV (2001) The earliest archaeological maize (Zea mays L.) from highland Mexico: new accelerator mass spectrometry dates and their implications. Proc Natl Acad Sci U S A 98: 2101–2103
Salse J, Feuillet C (2007) Comparative genomics of cereals. In: Varsheny RK, Tuberosa R (eds) Genomic assisted crop improvement, vol 1, Genomics approaches and platforms. Springer, The Netherlands, pp 177–205
Salse J, Piegu B, Cooke R, Delseny M (2004) New in silico insight into the synteny between rice (Oryza sativa L.) and maize (Zea mays L.) highlights reshuffling and identifies new duplications in the rice genome. Plant J 38:396–409
Sanger F, Air GM, Barrell BG et al (1977a) Nucleotide sequence of bacteriophage phi X174 DNA. Nature 265:687–695
Scherrer B, Isidore E, Klein P et al (2005) Large intraspecific haplotype variability at the Rph7 locus results from rapid and recent divergence in the barley genome. Plant Cell 17:361–374
Schnable PS, Ware D, Fulton RS et al (2009) The B73 maize genome: complexity, diversity, and dynamics. Science 326:1112–1115
Shinozuka H, Cogan NOI, Smith KF et al (2010) Fine-scale comparative genetic and physical mapping supports map-based cloning strategies for the self-incompatibility loci of perennial ryegrass (*Lolium perenne* L.). Plant Mol Biol 72:343–355
Sievers F, Wilm A, Dineen D et al (2011) Fast, scalable generation of high-quality protein multiple sequence alignments using Clustal Omega. Mol Syst Biol 7:539
Singh NK, Raghuvanshi S, Srivastava SK et al (2004) Sequence analysis of the long arm of rice chromosome 11 for rice-wheat synteny. Funct Integr Genomics 4:102–117
Siqueira JF Jr, Fouad AF, Rocas IN (2012) Pyrosequencing as a tool for better understanding of human microbiomes. J Oral Microbiol. doi:10.3402/jom.v4i0.10743
Song R, Messing J (2003) Gene expression of a gene family in maize based on noncollinear haplotypes. Proc Natl Acad Sci U S A 100:9055–9060
Srinivas G, Satish K, Murali Mohan S et al (2008) Development of genic-microsatellite markers for sorghum staygreen QTL using a comparative genomic approach with rice. Theor Appl Genet 117:283–296
Swigonova Z, Bennetzen JL, Messing J (2005) Structure and evolution of the r/b chromosomal regions in rice, maize and sorghum. Genetics 169:891–906
Tarchini R, Biddle P, Wineland R et al (2000) The complete sequence of 340 kb of DNA around the rice Adh1- Adh2 region reveals interrupted colinearity with maize chromosome 4. Plant Cell 12:381–391
Thompson JD, Gibson TJ, Plewniak F et al (1997) The CLUSTAL_X windows interface: flexible strategies for multiple sequence alignment aided by quality analysis tools. Nucleic Acids Res 25:4876–4882
Tian CG, Xiong YQ, Liu TY et al (2005) Evidence for an ancient whole genome duplication event in rice and other cereals. Yi Chuan Xue Bao 32:519–527
Tikhonov AP, SanMiguel PJ, Nakajima Y et al (1999) Colinearity and its exceptions in orthologous adh regions of maize and sorghum. Proc Natl Acad Sci U S A 96:7409–7414
Varshney RK, Sigmund R, Borner A et al (2005) Interspecific transferability and comparative mapping of barley EST-SSR markers in wheat, rye and rice. Plant Sci 168:195–202
Varshney RK, Nayak SN, May GD et al (2009) Next generation sequencing technologies and their implications for crop genetics and breeding. Trends Biotechnol 27:522–530
Vicient CM, Jaaskelainen MJ, Kalendar R, Schulman AH (2001) Active retrotransposons are common feature of grass genomes. Plant Physiol 125:1283–1292

Vincens P, Badel-Chagnon A, Andre C et al (2002) D-ASSIRC: distributed program for finding sequence similarities in genomes. Bioinformatics 18:446–451

Vitte C, Bennetzen JL (2006) Analysis of retrotransposon diversity uncovers properties and propensities in angiosperm genome evolution. Proc Natl Acad Sci U S A 103:17638–17643

Vogel JP, Garvin DF, Mockler TC et al (2010) Genome sequencing and analysis of the model grass Brachypodium distachyon. Nature 463:763–768

Vollbrecht E, Springer PS, Goh L et al (2005) Architecture of floral branch systems in maize and related grasses. Nature 436:1119–1126

Wang J, Roe B, Macmil S et al (2010) Microcollinearity between autopolyploid sugarcane and diploid sorghum genomes. BMC Genomics 11:261

Ware D, Jaiswal P, Ni J et al (2002) Gramene: a resource for comparative grass genomics. Nucleic Acids Res 30:103–105

Wong GKS, Wang J, Tao L et al (2002) Compositional gradients in Gramineae genes. Genome Res 12:851–856

Xie T, Hood L (2003) ACGT-a comparative genomics tool. Bioinformatics 19:1039–1040

Yilmaz A, Nishiyama MY Jr, Fuentes BG et al (2009) GRASSIUS: a platform for comparative regulatory genomics across the grasses. Plant Physiol 149:171–180

Yu J, Hu S, Wang J et al (2002) A draft sequence of the rice genome (Oryza sativa L. ssp. indica). Science 296:79–92

Yu J, Wang J, Lin W et al (2005) The genomes of Oryza sativa: a history of duplications. PLoS Biol 3:e38

Zhang LY, Guo XS, He B et al (2011) Genome-wide patterns of genetic variation in sweet and grain sorghum (Sorghum bicolor). Genome Biol 12:R114

Zhang G, Liu X, Quan Z et al (2012) Genome sequence of foxtail millet (Setaria italica) provides insights into grass evolution and biofuel potential. Nat Biotechnol 30:549–554

Zohary D, Hopf M (2000) Domestication of plants in the old world, 3rd edn. Oxford University Press, Oxford

A Comprehensive Overview on Application of Bioinformatics and Computational Statistics in Rice Genomics Toward an Amalgamated Approach for Improving Acquaintance Base

Jahangir Imam, Mukesh Nitin, Neha Nancy Toppo, Nimai Prasad Mandal, Yogesh Kumar, Mukund Variar, Rajib Bandopadhyay, and Pratyoosh Shukla

Abstract

Rice (*Oryza sativa* L.) is a major crop in the world and provides the staple food for over half of the world population. From thousands of years of cultivation and breeding to recent genomics and systems biology approach, rice has been the focus of agriculture and plant research. Modern scientific research depends on computer technology to organize and analyze large datasets. Rice informatics – a relatively new discipline – has been developing rapidly as a subdiscipline of bioinformatics. Rice informatics devotes to leveraging the power of nature's experiment of breeding and evolution to extract key findings from sequence and experimental data. Recent advances in high-throughput genotyping and sequencing technologies have changed the landscape of data collection

J. Imam
Biotechnology Laboratory, Central Rainfed Upland Rice Research Station (CRRI), Hazaribagh 825301, Jharkhand, India

Enzyme Technology and Protein Bioinformatics Laboratory, Department of Microbiology, Maharshi Dayanand University, Rohtak 124001, Haryana, India

M. Nitin • N.N. Toppo • N.P. Mandal • Y. Kumar
M. Variar (✉)
Biotechnology Laboratory, Central Rainfed Upland Rice Research Station (CRRI), Hazaribagh 825301, Jharkhand, India
e-mail: mukund.variar@gmail.com

R. Bandopadhyay
Department of Biotechnology, Birla Institute of Technology, Mesra, Ranchi 835215, Jharkhand, India

P. Shukla (✉)
Enzyme Technology and Protein Bioinformatics Laboratory, Department of Microbiology, Maharshi Dayanand University, Rohtak 124001, Haryana, India
e-mail: pratyoosh.shukla@gmail.com

K.K. P.B. et al. (eds.), *Agricultural Bioinformatics*,
DOI 10.1007/978-81-322-1880-7_5, © Springer India 2014

and its analysis by using friendly database access and information retrieval. It focuses on developing and applying database tools and computationally intensive techniques and statistical software (e.g., pattern recognition, data mining, machine learning algorithms, R-statistical, MATLAB, and visualization) which give the opportunity to quickly and efficiently study heap of genomics information, chemical structure, and model generation study. Over recent years, various newly emerged diseases to rice varieties have an increasing concern to agriculturists and pathologists. The establishments of International Rice Information System, Rice Genome Research Project, Integrated Rice Genome Explorer, and Rice Proteome Databases are important initiatives for rice improvement using in silico software (e.g., homology modeling using SWISS Model, Modeler, and Autodock); the recent ongoing research on rice protein and its role in metabolic pathways works is being done around the world. Rice informatics has already started showing its profound impact on agricultural research and developments.

Keywords
Rice informatics • IRIS • Rice genomics resources • Rice proteomic resources • Databases • Rice bioinformatics software

Abbreviations

RAP-DB	Rice Annotation Project Database
ESTs	Expressed Sequence Tag
SAS	Statistical Analysis Software
IRRI	International Rice Research Institute
ICIS	International Crop Information System
WWW	World Wide Web
IRIS	International Rice Information System
GMS	Genealogy Management System
NCBI	National Center for Biotechnology Information
EMBL	European Molecular Biology Laboratory
EBI	European Bioinformatics Institute
DDBJ	DNA Data Bank of Japan
TIGR	The Institute of Genomic Research
IRFGC	International Rice Functional Genomics Consortium
GWAS	Genome Wide Association Study
OMAP	Oryza Mapping Alignment Project
MOsDB	MIPS Rice (*Oryza sativa*) database
RiceGAAS	Rice Genome Automated Annotation System
RED	Rice Expression Databases
RMOS	Rice Microarray Opening Site
RAD	Rice Array Database
CREP	Collection of Rice Expression Profiles
RAN	RiceArrayNet
RGKbase	Rice Genome Knowledgebase
MATLAB	MATrixLABoratory
GRNN	Generalized Regression Neural Network
BPNN	Back Propagation Neural Network
PDB	Protein Data Bank

1 Introduction

Rice is a major staple food crop for almost half of the world population. Among several agricultural crops, rice is considered as one of the most important crop plants for bioinformatics and computational biology research as it has become the model monocot plant having a number of biological characteristics and recent research advancement in the field of genetics, breeding, genomics, germplasm collection and maintenance, systems biology, and functional genomics. During the last three decades, advancement in biotechnology led to the acceleration in many rice research programs particularly in breeding, selection of superior genotypes, large-scale cDNA analysis, genetic mapping, and genome sequencing (Khush and Brar 1998; Sasaki and Burr 2000). At the same time, this progression in biotechnology research paved the way for a new era, i.e., rice informatics in rice research, and opened new opportunities and direction for the improvement of rice crop which will address the issues concerning global problems on food security. The *japonica* rice cultivar Nipponbare genome sequencing project was completed in 2005 by consortium research of 10 countries, and Rice Annotation Project Database (RAP-DB) was developed to provide an accurate annotation of the rice genome through HTTP access (IRGSP 2005; Itoh et al. 2007). Parallel with rice genome sequence work and its related genomics resources, advancement in rice breeding research and development of molecular marker resources has helped the researchers to accelerate the identification, isolation, and incorporation of agronomically important genes and QTLs (Ashikari et al. 2005; Konishi et al. 2006; Ma et al. 2006, 2007; Kurakawa et al. 2007).

Recent advances in rice research are associated with the emergence of high-throughput data from large-scale sequencing, expression profiling of thousands of genes, phenotyping, and strategies on transcriptomics, proteomics, and metabolomics (Nagamura and Antonio 2010). In addition, large-scale collections of bioresources, such as mass-produced mutant lines and clones of full-length cDNAs and their integrative relevant databases, are now available (Brady and Provart 2009; Kuromori et al. 2009; Seki and Shinozaki 2009). The vast accumulation of genomics data from these strategies has culminated the need and importance of transforming these data into easily accessible and understandable form to the researcher, which can be ultimately studied and interpreted into useful biological information (Lewis et al. 2000). For this robust infrastructure for organizing data, computational methods for analysis and interfaces for integration and retrieval of various types of data through user-friendly databases have been developed (Nagamura and Antonio 2010).

The application of bioinformatics has triggered the research in rice sciences with speed. This contributed to the easy and convenient way of data handling and data analysis much faster than traditional approach. The potential of the Internet in access of most up-to-date information on scholarly content, communication with colleagues, engaging two-way process of communication between researchers, and publishing materials more easily has been visualized in the advancement of the rice-related information and technologies (Ram and Rao 2012). Many bioinformatics resources are now available to the researchers around the world through the World Wide Web. Nowadays, researchers can easily post their research findings on the Web or compare their discoveries with previous results. The easy access and sharing of data between institutions has increased the opportunity for collaboration and thus dramatically fastened the research work and the development in the field of rice science. These developments are highlighted through the availability of databases, web servers, articles, and research organizations working in this area.

In the field of agriculture, the main focus is rice research, and the last 20 years belong to advancement of sequencing technology. In every stage of rice research, global problem of food security has been the burning issue, and every research activity from morphological and physiological to application of biotechnology to marker-assisted rice improvement to the development in bioinformatics technologies has been focused and addressed. The advancement in bioinformatics rice research played a pivotal role in this. Later, large-scale DNA analysis, genetic mapping, and genome sequencing have resulted in a tremendous increase in computer-generated information on rice genome (Sasaki and Burr 2000). During the same period, the advancement of sequencing technologies such as Expressed Sequence Tags (EST) research changed the path of genetic expression of rice. EST project collectively represents about 1,251,304 entries of GenBank available on NCBI dbEST (http://ncbi.nlm.nih.gov/dbEST/dbEST_summary.html) [GenBank Release 1.3.2011] (Ram and Rao 2012). In recent years, with the development of new software and statistical analyses, the physiology experiments of rice (*Oryza sativa* L.) were performed by analyzing ANOVA and Tukey's HSD mean comparison using Rv. 2.8.0 (Swamy et al. 2013). Rv. 2.8.0 is a free statistical software for computing and graphics. It compiles and runs on a wide variety of UNIX platforms, Windows, and MacOS. MATLAB is a data analysis and visualization package software. Agriculture scientists use it for climate change analysis and in probabilistic model designing as it can cope up with large gridded dataset quite easily. Statistical Analysis Software (SAS) is also one of the widely used statistical packages for analyzing statistical data in crop science. Rice plants respond to different stresses via a number of mechanisms. Availability of rice genome sequences, large amount of information were generated from genomics and proteomics studies and in silico computational bioinformatics tools set a new platform for the management of environmental stresses in rice. In silico docking between the two proteins showed a significant protein–protein interaction between rice EDS1 and PAD4, suggesting that they form a dimeric protein complex, which, similar to that in *Arabidopsis*, is perhaps also important for triggering the salicylic acid signaling pathway in plants (Singh and Shah 2012).

2 Rice (*Oryza sativa*): Model Species for Monocot Plants

Poaceae family is one of the most important among the monocots which mostly includes agricultural crop species like maize, wheat, barley, sugarcane, sorghum, and rice. These species share extensive synteny across their genomes, allowing for one of the species to serve as the base for comparative genomics and bioinformatics study within the family (Moore et al. 1995). Rice is one of the major monocot plants among them and used as a model species for the various biotechnological and bioinformatics research. Rice represents the most suitable species for genomics and bioinformatics research, which is the main reason for selecting it as a model species for its small genome size (~431 Mb). The second reason is the availability of genetic and molecular resources. After the complete genome sequencing of *japonica* rice variety Nipponbare, the research in rice has been revolutionized.

3 Rice Information System: The Rice Informatics

The computer and information technology has revolutionized the research in biotechnology and bioinformatics. Recent technologies in molecular biology and germplasm conservation have speeded up the sequence, genetic, and phenotypic information analysis (Ram and Rao 2012). The IRRI (International Rice Research Institute) in collaboration with CIMMYT (International Maize and Wheat Improvement Center) established the International Crop Information System (ICIS) project. In this project, scientists work for varietal improvement with the use of new and advanced bioinformatics software and also work for the development of software to facilitate and fasten the research and establishing the links between information from different crops like rice and maize. By this, the ICIS is maintaining huge datasets of rice which is publicly available and easily accessible to the scientists.

4 Web Tools and Resources on Rice

The results of huge data and information generated through acquiring knowledge of genomics resources in rice improvement has posed the problem in front of agricultural scientists to maintain data for future use and manipulate the data. Such activities have created the provision for the maintenance of data in the form of databases. Rice research in global world and organizations working in the area of rice have developed and maintained such databases globally. Databases, software tools, web servers, etc. are available for data management and are used for solving problems related to rice, whether it would be the handling of molecular level activities or the production of a disease-resistant variety of rice (Ram and Rao 2012). The World Wide Web (www) provides a mechanism for extraordinary information sharing among the researchers as many bioinformatics resources are now available all over the world through www.

4.1 The International Rice Information System (IRIS)

The International Rice Information System (IRIS) is the rice implementation of the International Crop Information System (ICIS), a database system for the management and integration of global information on genetic resources and germplasm improvement (Bruskiewich et al. 2003). In 1995, the international agricultural research centers CIMMYT and IRRI partnered with other CGIAR centers to establish a project to develop an International Crop Information System (ICIS; Fox and Skovmand 1996) to overcome these deficiencies in crop data management for a wide range of crops. Several CGIAR centers, national agricultural research systems, and advanced research institutes are collaborating to develop ICIS as a generic system that will accommodate all data sources for any crop and breeding system. There are basically two objectives of ICIS: first, to integrate different data types in both private and public datasets into a single information system and, second, to provide specialist views and applications that operate on this integrated platform. After successful completion, ICIS will support a range of activities from germplasm conservation, evaluation, functional genomics, allele mining, breeding, testing, and release. The Genealogy Management System (GMS) of ICIS is the core database which ensures unique identification of germplasm, management of nomenclature (including homonyms and synonyms), and retention of all germplasm development information. The ICIS system is fast, user-friendly, PC based, and is available on CD-ROM and also available online (http://www.iris.irri.org/). Mainly the ICIS (IRIS) is designed to allow biologists to manage local data and query and view their own data fully integrated with global public information. One of the innovative features of ICIS is that it permits independent users to integrate their own local data with public central data. The IRIS is being developed under the open-source ICIS project. Code is freely available to anyone, and the latest information about the ICIS project can be accessed at http://www.icis.cgiar.org.

4.2 Database Development

Computer-based databases are new innovations in the field of molecular biology, biotechnology, and bioinformatics, finding its scope in usage and online accessibility of information. The three major database sources at GenBank at the National Center for Biotechnology Information (NCBI) (http://ncbi.nlm.nih.gov), the European Molecular Biology Laboratory (EMBL) at European Bioinformatics Institute [(EBI) http://www.ebi.ac.uk/], and DNA Data Bank of Japan (DDBJ) (http://ddbj.nig.ac.jp) are imparting major roles in the management of biological information. The basic sequenced data submitted to these institutions can be mirrored to other institutions automatically on a routine basis to suffice for basic data. Beyond sequence data, the range of pertinent functional genomics, proteomics, structured experimental data, and associated data are being utilized by various organizations in order to develop different kinds of databases in various categories. A list of such databases in rice is shown in Table 1.

4.3 Structural and Functional Genomics Databases

Rice genomics databases are the major sources of information that could be used in understanding the genetic and molecular basis of all biological

Table 1 Major rice structural and functional genomics databases available

Sr. no.	Database	Description/major features	URL
1	DTRF	Database of Rice Transcription Factors	http://drtf.cbi.pku.edu.cn/
2	RAP-DB	*Oryza sativa* ssp. *Japonica* cv. Nipponbare genome sequence and annotation	http://rapdb.dna.affrc.go.jp/
3	Rice Genome Annotation	*Oryza sativa* ssp. *Japonica* cv. Nipponbare genome sequence and annotation	http://rice.plantbiology.msu.edu/
4	RISe	*Oryza sativa* ssp. *Indica* (93-11) and *japonica* (Nipponbare) sequence contigs and annotation	http://rice.genomics.org.cn/
5	OryzaSNP	SNP data from *Oryza* species and cultivars	http://oryzasnp.org/
6	Rice Haplotype Map	SNP data from 517 rice landraces	http://www.ncgr.ac.cn/RiceHapMap
7	Koshihikari genome	*Oryza sativa* ssp. *Japonica* cv. Koshihikari genome sequence	http://koshigenome.dna.affrc.go.jp/
8	OMAP	*Oryza* Map Alignment Project	http://www.omap.org/
9	dbEST	Rice ESTs	http://www.ncbi.nlm.nih.gov/
10	KOME	Japonica rice full-length cDNA sequences	http://cdna01.dna.affrc.go.jp/cDNA/
11	RICD	Indica rice full-length cDNA sequences	http://www.ncgr.ac.cn/ricd
12	Rice MPSS	Rice small RNA MPSS data	http://mpss.udel.edu/rice/
13	Rice PARE	Parallel Analysis of RNA Ends	http://mpss.udel.edu/rice/
14	Rice SBS	Rice small RNA SBS data	http://mpss.udel.edu/rice/
15	Indica MPSS	Indica Rice small RNA MPSS data	http://mpss.udel.edu/rice/
16	Tos17 Insertion Mutant Database	Tos17 insertion mutant panel	http://tos.nias.affrc.go.jp/
17	Rice Mutant Database	T-DNA insertion lines	http://rmd.ncpgr.cn/
18	TRIM	Rice insertional mutant lines	http://trim.sinica.edu.tw/index.php
19	SHIP	Shanghai T-DNA Insertion Population	http://ship.plantsignal.cn/index.do
20	Oryza Tag Line	T-DNA and Ds flanking sequence tags	http://urgi.versailles.inra.fr/OryzaTagLine/
21	IR64 Rice Mutant Database	Mutant lines for mutagenesis of IR64	http://irfgc.irri.org/cgi-bin/gbrowse/IR64_deletion_mutants/
22	Rice Array DB	NSF Rice Oligonucleotide Array Project	http://www.ricearray.org/
23	Rice Atlas	Rice cellular expression profile database	www.yalescientific.org/2010/02/the-rice-atlas/
24	OryzaExpress	Geo-based rice gene expression profiles	http:// bioinf.mind.meiji.ac.jp/OryzaExpress/
25	CREP	Collection of rice gene expression profiles	http://crep.ncpgr.cn/crep-cgi/home.pl
26	RiceXpro	Rice Expression Profile Database	http://ricexpro.dna.affrc.go.jp/
27	NIAS Genebank	30,000 rice accessions	http://www.gene.affrc.go.jp/
28	National Plant Germplasm	USDA germplasm database	http://www.ars-grin.gov/npgs/
29	IRIS	International Rice Information System	http://www.iris.irri.org/germplasm/
30	Gramene	Grass genome data anchored on rice	http://www.gramene.org/
31	Oryzabase	Oryza genetics database	http://www.shigen.nig.ac.jp/rice/oryzabase/
32	GreenPhyl	Phylogenomics database	http://greenphyl.cirad.fr/v2/cgi-bin/index.cgi

(continued)

Table 1 (continued)

Sr. no.	Database	Description/major features	URL
33	Rice Phylomics	Rice phylomics analysis data	http://phylomics.ucdavis.edu/
34	RKD	Rice Kinase Database	http://phylomics.ucdavis.edu/kinase/
35	Rice GT	Rice Glycosyltransferase Database	http://phylomics.ucdavis.edu/cellwalls/gt/
36	SALAD	Clustering of conserved amino acid sequences	http://salad.dna.affrc.go.jp/
37	Q-TARO	Rice QTL mapping information	http://qtaro.abr.affrc.go.jp/
38	Rice TOGO Browser	Integrated rice genomics browser	http://agri-trait.dna.affrc.go.jp/
39	INE	Integrated Rice Genome Explorer	http://ine.dna.affrc.go.jp/giot/
40	MOsDB	MIPS *Oryza sativa* database	http://mips.gsf.de/proj/plant/jsf/rice/index.jsp
41	OryGenesDB	Rice genes, T-DNA, and Ds flanking sequence tags	http://orygenesdb.cirad.fr/
42	Rice Annotation DB	Contig data for manual annotation of rice genome	http://ricedb.plantenergy.uwa.edu.au/
43	Rice pipeline	Unification tool for rice databases	http://cdna01.dna.affrc.go.jp/PIPE
44	Rice Proteome DB	Rice Proteome Database	http://gene64.dna.affrc.go.jp/RPD/main_en.html
45	RiceGAAS	Rice Genome Automated Annotation System	http://ricegaas.dna.affrc.go.jp/
46	Rice Genome Project/IRGSP	The International Rice Genome Sequencing Project (IRGSP), a consortium of publicly funded laboratories	http://rgp.dna.affrc.go.jp/E/IRGSP/
47	TIGR Rice	The TIGR Rice Genome Project BLAST, a server for collection of databases for use in searching with BLAST programs	http://blast.jevi.org/euk-blast/index.cgi/project=osa1
48	IRFGC	International Rice Functional Genomics Consortium	http://irfgc.irri.org/
49	RED	Rice Expression Database	http://red.dna.affrc.go.jp/cDNA/
50	Yale Plant Genomics	Yale plant genomics	http://plantgenomics.biology.yale.edu/
51	Genevestigator	Multi-organism microarray database and expression meta-analysis tool	http://www.genevestigator.com/
52	AgriTOGO	A database system for information and resources generated from various genome projects of the Ministry of Agriculture, Forestry and Fisheries, Japan	http://togo.dna.affrc.go.jp/
53	VanshanuDhan	Vanshanu rice gene database	http://www.nrcpb.org/

processes including many economically important traits, which are the main concern of breeders. The basic information that can be deciphered from the genome sequence is indispensable in the development of new cultivars with target traits such as high yield, biotic/abiotic stress resistance, good eating quality, etc. The availability of wide range of genetic and molecular markers made rice as an important species for genomics analysis. With the completion of the rice genome sequence in 2004, a standard annotation is necessary so that the information from the genome sequence can be fully utilized and understood. This led to the establishment of a platform for structural and functional characterization of the rice genome. The Rice Annotation Project Database (RAP-DB) provides sequence and annotation data for rice genome. RAP-DB is a hub for *Oryza sativa* ssp. *japonica* genome information. This web-based tool provides information on rice genome sequence from the Nipponbare subspecies of rice and annotation of the 12 rice chromosomes. The RAP-DB contains the IRGSP genome sequence (build 3 assembly) (IRGSP 2005) and the RAP loci with corresponding locus IDs representing the annotated genes. The primary concept of the RAP-DB is to provide simple access for the IRGSP genome sequence and the RAP annotation. RAP-DB has two different types of annotation viewers, BLAST and BLAT search, and other useful features. The Institute of Genomic Research (TIGR) works in the area of rice genome sequence with new data to improve the quality of the annotation. The TIGR Rice Genome Project BLAST server has a collection of databases for use in searching with the BLAST programs blastn, blastx, tblastn, and tblastx. The TIGR Rice Pseudomolecules database allows user to search against the latest version of the 12 TIGR rice pseudomolecules. These databases have been conceptualized with the aim of providing a comprehensive analysis of the rice genome and include both structural annotations to identify the genomics elements and functional annotations to attach biological meaning to the sequence data.

The International Rice Functional Genomics Consortium (IRFGC) has re-sequenced the 100 Mb of gene-rich genomics sequences to determine SNP variation from 20 diverse rice varieties and Landraces commonly used in breeding programs internationally and has the impressive genotypic and phenotypic diversity of domesticated rice (McNally et al. 2009). Zhao et al. (2011) showed the detailed results of a Genome Wide Association Study (GWAS) based on 44,100 SNP variants across 413 diverse accessions of *O. sativa* collected from 82 countries, which has been systematically phenotyped for 34 traits. SNP variation has been analyzed in 177 Japanese rice accessions, which are categorized into three groups: landraces, improved cultivars developed from 1931 to 1974 (the early breeding phase), and improved cultivars developed from 1975 to 2005 (the late breeding phase) (Yonemaru et al. 2012). Oryza Mapping Alignment Project (OMAP, Wing et al. 2005) has been initiated to characterize the genome of wild rice species and already released the genome sequence of *Oryza glaberrima*, an African species of domesticated rice. Due to genome sequencing advancement during recent years, the researchers are able to access huge information on genetic variation present within the genus *Oryza*, within the two major subspecies, and among diverse rice cultivars and landraces that have great potential for improvement of cultivated rice (Nagamura and Antonio 2010). Along with complete genome sequencing, short ESTs (Expressed Sequence Tags), full-length cDNA sequence databases in dbEST, and KOME are also useful for evaluation of gene expression and variation. Furthermore, short read sequences including microRNAs (miRNAs), small interfering RNAs (siRNAs), transacting siRNAs, and heterochromatic siRNAs can be accessed via the rice databases at the University of Delaware (Simon et al. 2008). In particular, the Rice MPSS database is a repository of small RNA sequences with detailed information on sense and antisense expression of rice annotated genes (Kan et al. 2007). The genome sequences and all short transcript resources immensely augment the information on rice and will help in understanding the genetic control of agronomically important traits.

The MIPS Rice (*Oryza sativa*) database (MOsDB) provides a comprehensive data collection dedicated to the genome information of rice. MOsDB integrates data from two publicly available rice genomics sequences: *O. sativa* L. ssp. *indica* and *O. sativa* L. ssp. *japonica*. MOsDB provides an integrated resource for associated data analysis like internal and external annotation information as well as a complex characterization of all annotated rice genes. It includes an up-to-date access to publicly available rice genomics sequences and various search options. MOsDB is continuously expanding to include increasing range of data type and the growing amount of information on the rice genome. The RiceGAAS (Rice Genome Automated Annotation System) is also extensively used to identify various structural and functional components (Sakata et al. 2002). It has been developed to execute a reliable and up-to-date analysis of the genome sequence as well as to store and retrieve the results of annotation. The RiceGAAS system does the functional analysis by collecting rice genome sequences from GenBank and then executing gene prediction, analysis of exons, splice sites, repeats, and transfer RNA based on algorithm which combines multiple gene prediction programs with homology search results. RiceGAAS system consists of 14 analysis programs. These include BLAST for homology search against protein database and rice EST database, GENSCAN and RiceHMM for gene domain prediction, MZEF for exon prediction, SplicePredictor for splice site prediction, and more (Sakata et al. 2002). Thus, RiceGAAS provides a systematic and comprehensive annotation of accumulated sequence data.

OryGenDB is another database for rice functional genomics. The database is an interactive tool for rice reverse genetics. Insertion mutants of rice genes are cataloged by flanking sequence tag (FST) information that can be readily accessed by this database. Oryzabase is an integrated database of rice genome resources. Oryzabase has been created for a comprehensive view of rice (*Oryza sativa*) as a model monocot plant by integrating biological data with molecular genomics information (http://www.shigen.nig.ac.jp/rice/oryzabase/top/top.jsp). The database contains information about rice development and anatomy, rice mutants, and genetic resources, especially for wild species of rice. Several genetic, physical, and expression maps with full genome and cDNA sequences are also combined with biological data in Oryzabase. This provides a useful tool for gaining greater knowledge about the life cycle of rice, the relationship between phenotype and gene function, and rice genetic diversity (Kurata and Yamazaki 2006).

Database has also been developed to screen and phenotype mutant lines in rice which can be used for selecting lines with specific morphological and physiological features. Some of these mutant databases are Rice Tos17 Insertion Mutant Database with about 50,000 Tos17 insertion mutant lines from *japonica* rice cultivar Nipponbare (Miyao et al. 2007), Rice Mutant Database (RMD) containing 134,346 rice T-DNA insertion lines (Zhang et al. 2006), Taiwan Rice Insertional Mutants (TRIM) with 55,000 T-DNA insertion lines (Chern et al. 2007), Shanghai T-DNA Insertion Population Database (SHIP) containing 65,000 T-DNA insertion lines, Oryza Tag Line with 46,000 T-DNA and Ds insertion lines (Larmande et al. 2008), and IR64 Rice Mutant Database with phenotype information for irradiation and chemical mutants of IR64 cultivar (Wu et al. 2005). These databases mostly provide information on flanking sequences of the disrupted genes and thus allow users to screen the available mutants.

Another major part in bioinformatics is the analysis of huge amount of genome expression data, which is generated by microarray and SAGE (Gerstein and Jansen 2000). For analysis of such vast amount of data, many analysis tools and databases have been developed like Rice Expression Databases (RED) and Rice Microarray Opening Site (RMOS), Rice Array Database (RAD), Rice Atlas, OryzaExpress, Collection of Rice Expression Profiles (CREP), RiceArrayNet (RAN), RiceChip.Org, Rice MPSS, RicePLEX, MGOS (*Magnaporthe grisea, Oryza sativa*) database, and Rice Gene Expression Profile database (RiceXPro). These databases provide analysis tools that allow comparison of expression profiles from different samples based on specific criteria. Moreover, these databases provide Gene Expression Networks and various kinds of omics information including genome annotation, metabolic pathways, and gene expression.

For the better analysis of genome assemblies and annotation, a database, namely, Rice Genome Knowledgebase (RGKbase) – an annotation database for rice comparative genomics and evolutionary biology – has been introduced (Wang et al. 2012). RGKbase has three major components: (1) integrated data curation for rice genomics and molecular biology, (2) user-friendly viewers, and (3) bioinformatics tools for compositional and synteny analyses. Currently RGKbase includes data from five rice cultivars and species, and new datasets are continuously introduced in it. A very important genomics database, known as Rice TOGO

Browser, a component database of AgriTOGO, is an integrated database on rice functional and applied genomics. Rice TOGO Browser can be accessed through a user-friendly web interface that provides three search options, namely, keyword search, region search, and trait search, to retrieve information on specific genes, sequences, genetic markers, and phenotypes associated with a specific region of the genome (Nagamura et al. 2010).

4.4 Rice Proteomics Databases

As genome sequencing of rice has been completed, proteome analysis, which is the detailed investigation of the functions, functional networks, and 3-D structures of proteins, has gained increasing attention. Many rice proteomics databases are also available that are important resources for better understanding of protein functions in cellular system, protein–protein interaction, and downstream protein functions. Today, there are lists of proteome-related databases and tools available for rice proteomics analysis. This includes ExPASy Proteomics tools, Compute pI/Mw tool in ExPASy, ProteinProspector, pIans MW calculation service of aBi, SWISS-PROT and TrEMBL, ProteinProspector (UCSF), Rockefeller Univ Prowl (search engine), Mascot, EMBL, PeptideSearch, Swiss-protExPaSy, YPD (Proteome Inc.), and Sherpa. OryzaPG-DB, a Rice Proteome Database based on shotgun proteogenomics, incorporates the genomics features of experimental shotgun proteomics data. This version of the database was created from the results of 27 nanoLC-MS/MS and runs on a hybrid ion trap-orbitrap mass spectrometer, which offers high accuracy for analyzing tryptic digests from undifferentiated cultured rice cells. Approximately 3,200 genes were covered by these peptides and 40 of them contained novel genomics features. OryzaPG is the first proteogenomics-based database of the rice proteome, providing peptide-based expression profiles, together with the corresponding genomics origin, including the annotation of novelty for each peptide (Helmy et al. 2011).

The Rice Proteome Database is the first detailed database to describe the proteome of rice. The Rice Proteome Database contains 23 reference maps based on 2D-PAGE of proteins from various rice tissues and subcellular compartments. These reference maps comprise 13,129 identified proteins, and the amino acid sequences of 5,092 proteins are entered in the database. Major proteins involved in growth or stress responses were identified using the proteome approach. Rice Proteome Database contains the calculated properties of each protein such as molecular weight, isoelectric point, and expression, experimentally determined properties such as amino acid sequences obtained using protein sequencers and mass spectrometry, and the results of database searches such as sequence homologies. The database is searchable by keyword, accession number, protein name, isoelectric point, molecular weight, and amino acid sequence or by selection of a spot on one of the 2D-PAGE reference maps (Komatsu et al. 2004). The information obtained from the Rice Proteome Database will aid in cloning the genes for and predicting the function of unknown proteins.

The important proteome databases and web resources popular in the rice research are listed in Table 2. These given databases are classified into different categories that have been used for rice proteomics databases comparison.

5 Statistical Rice Informatics Using R-Software

The new science of statistical rice bioinformatics aims to modify available classical and nonclassical statistical methods, develop new methodologies, and analyze the databases to improve the understanding of the complex biological phenomena. This interdisciplinary science requires understanding of mathematical and statistical knowledge in the biological sciences and its applications (Mathur 2010). Statistical programming language R was developed by R Development Core Team (2009), specifically for our purposes as shown in Fig. 1. This will solve the practical issues to follow the stream of reasoning. R can be used as a simple calculator and also for complex statistical analysis.

Table 2 Major rice proteomics databases available worldwide

Sr. no.	Database	Description/major features	URL
1	WORLD-2DPAGE	Index to 2-D PAGE databases and services	http://www.expasy.ch/ch2d/2d-index.html
2	SWISS-2DPAGE	Annotated databases of 2-D PAGE and SDS-PAGE	http://www.expasy.ch/ch2d/2d-top.html
3	YPM	Yeast Proteome Map	http://www.ibgc.u-bordeaux2.fr/YPM
4	YEAST 2D-PAGE	Database of Yeast 2D-PAGE	www.world-2dpage.expasy.org/list/
5	ECO2DBASE	2-D database of *E. Coli*	http://pcsf.brcf.med.umich.edu/eco2dbase
6	Sub2D	2-D database of *Bacillus subtilis*	http://pc13mi.biologie.uni-greifswald.de/
7	Cyano2Dbase	2-D database of cyanobacteria	http://www.kazusa.or.jp/tech/sazuka/cyano/proteome.html
8	Abeerden 2-D db	*Haemophilus* 2-D database	http://www.abdn.ac.uk/-mmb023/2dhome.html
9	Fly 2-D db	Database of Drosophila genes and genomes	http://ty.cmb.ki.se/
10	ECO2DBASE	2-D database of *E. coli*	www.expasy.org/cgi-bin/dbxref?ECO2DBASE
11	HSC-2DPAGE	2-DE gel protein database at Harefield hospital	www.doc.ic.ac.uk/vip/hsc-2dpage/
12	SIENA-2DPAGE	2D-PAGE database from University of Siena, Italy	http://www.bio-mol.unisi.it/2d/2d.html
13	PHCI-2DPAGE	2D-PAGE database of parasite–host cell interaction	http://www.gram.au.dk/
14	Rice Proteome Database	Rice 2D-PAGE and proteomics database	http://gene64.dna.affrc.go.jp/RPD

R is free software and comes with absolutely no warranty. It also generates codes which are actively in use by researchers for crop science to evaluate their experimental data. A rice strip-plot experiment with three replications, variety as the horizontal strip, and nitrogen fertilizer as the vertical strip was calculated using R-Software as shown in Fig. 2 (Gomez and Gomez 1984).

```
dat<- gomez.stripplot

# Gomez figure 3.7
desplot(gen~x*y, data=dat,
out1=rep, num=nitro, cex=1)

# Gomez table 3.12
tapply(dat$yield, dat$rep,sum)
tapply(dat$yield, dat$gen, sum)
tapply(dat$yield, dat$nitro,
sum)

# Gomez table 3.15. Anova table
for strip-plot
dat<-transform(dat,
nf=factor(nitro))
m1 <- aov(yield ~ gen * nf +
Error(rep + rep:gen + rep:nf),
data=dat)
summary(m1)

>library(agridat)
>png(filename="gomez.
 stripplot_%03d_large.png",
 width=1000, height=800)
> ### Name: gomez.stripplot
> ### Title: Rice strip-plot
  experiment
> ### Aliases: gomez.stripplot
>
> ### ** Examples
>
>
>dat<- gomez.stripplot
>
> # Gomez figure 3.7
>desplot(gen~x*y, data=dat,
 out1=rep, num=nitro, cex=1)
>
```

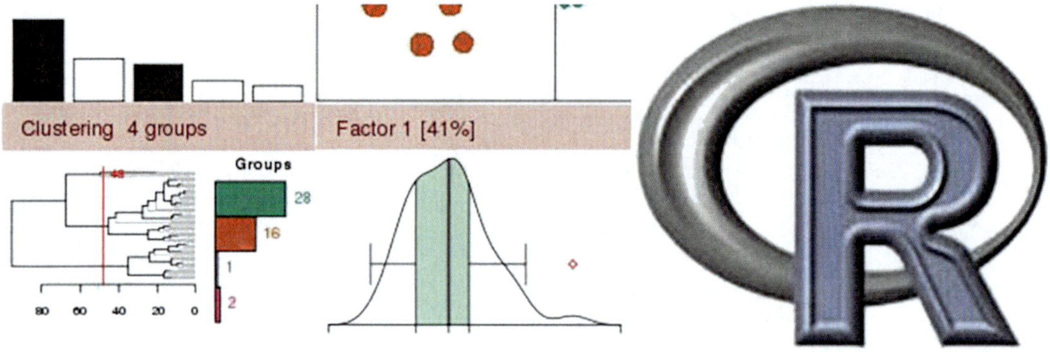

Fig. 1 R version 2.15.1 (2012-06-22). "Roasted Marshmallows." Copyright (C) 2012. The R Foundation for Statistical Computing ISBN 3-900051-07-0. Platform: x86_64-unknown-linux-gnu (64-bit)

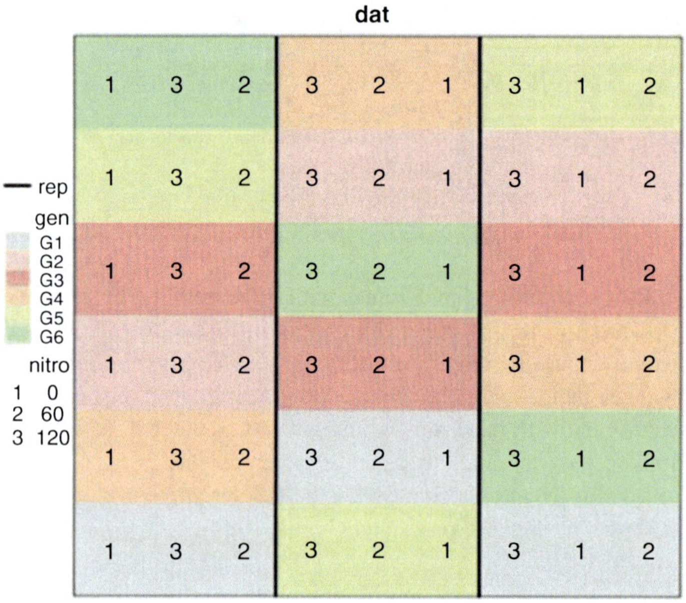

Fig. 2 Rice strip-plot experiment (Gomez and Gomez 1984)

```
> # Gomez table 3.12
>tapply(dat$yield, dat$rep,
 sum)
   R1    R2    R3
 84700 100438 100519
>tapply(dat$yield, dat$gen,
 sum)
   G1   G2   G3   G4   G5   G6
 48755 56578 54721 50121 47241
 28241
>tapply(dat$yield, dat$nitro,
 sum)
    0    60   120
72371  98608 114678
>
> # Gomez table 3.15.Anova table
  for strip-plot
>dat<-transform(dat,
 nf=factor(nitro))
> m1 <- aov(yield ~ gen * nf +
  Error(rep + rep:gen + rep:
  nf), data=dat)
>summary(m1)
```

```
Error: rep
      Df SumSq Mean Sq F value Pr
      (>F)
Residuals 2 9220962 4610481

Error: rep:gen
      Df  Sum Sq  MeanSq F value
      Pr(>F)
gen   5 57100201 11420040 7.653
0.00337 **
Residuals 10 14922619 1492262
---
Signif.codes:   0 '***' 0.001
'**' 0.01 '*' 0.05 '.' 0.1 ' ' 1

Error: rep:nf
Df   Sum Sq  MeanSq F value  Pr
(>F)
nf   2 50676061 25338031  34.07
0.00307 **
Residuals 4 2974908  743727
---
Signif.codes:   0 '***' 0.001
'**' 0.01 '*' 0.05 '.' 0.1 ' ' 1

Error: Within
Df   Sum Sq Mean Sq F value   Pr
(>F)
gen:nf     10 23877979 2387798
5.801 0.000427 ***
Residuals 20 8232917 411646
---
Signif.codes:   0 '***' 0.001
'**' 0.01 '*' 0.05 '.' 0.1 ' ' 1
>dev.off()
null device
        1
>
Result
```

Thiel et al. (2009) used R-statistical package to test three regression methods: (i) linear least squares regression, (ii) local polynomial regression fitting using *loess* with the *degree* = *1* (the degree of polynomials) and *span* = *0.7* (the smoothing parameter), and (iii) local polynomial regression fitting with the same set of parameters as (ii) to normalize the observed map length differences of up to 24 % among chromosomes of both datasets using the previously defined anchor points in order to support the evidence and evolutionary analysis of ancient whole-genome duplication in barley predating the divergence from rice as shown in Fig. 3.

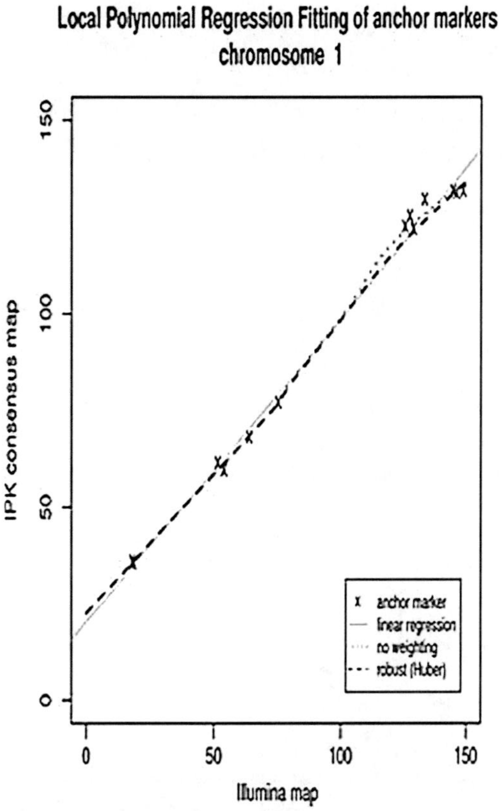

Fig. 3 Local polynomial regressions fitting of anchor markers of chromosome 1 using R-statistical package (Thiel et al. 2009)

6 MATLAB Computing Tool in Rice Science

MATLAB stands for **MAT**rix**LAB**oratory, which is a state-of-the-art mathematical software package that integrates computation, visualization, and programming in an easy-to-use environment where problems and solutions are expressed in familiar mathematical notation. Unlike other mathematical packages, such as MAPLE or MATHEMATICA, MATLAB cannot perform symbolic manipulations without the use of additional toolboxes. It remains, however, one of the leading software packages for numerical computation. MATLAB application is widely used in

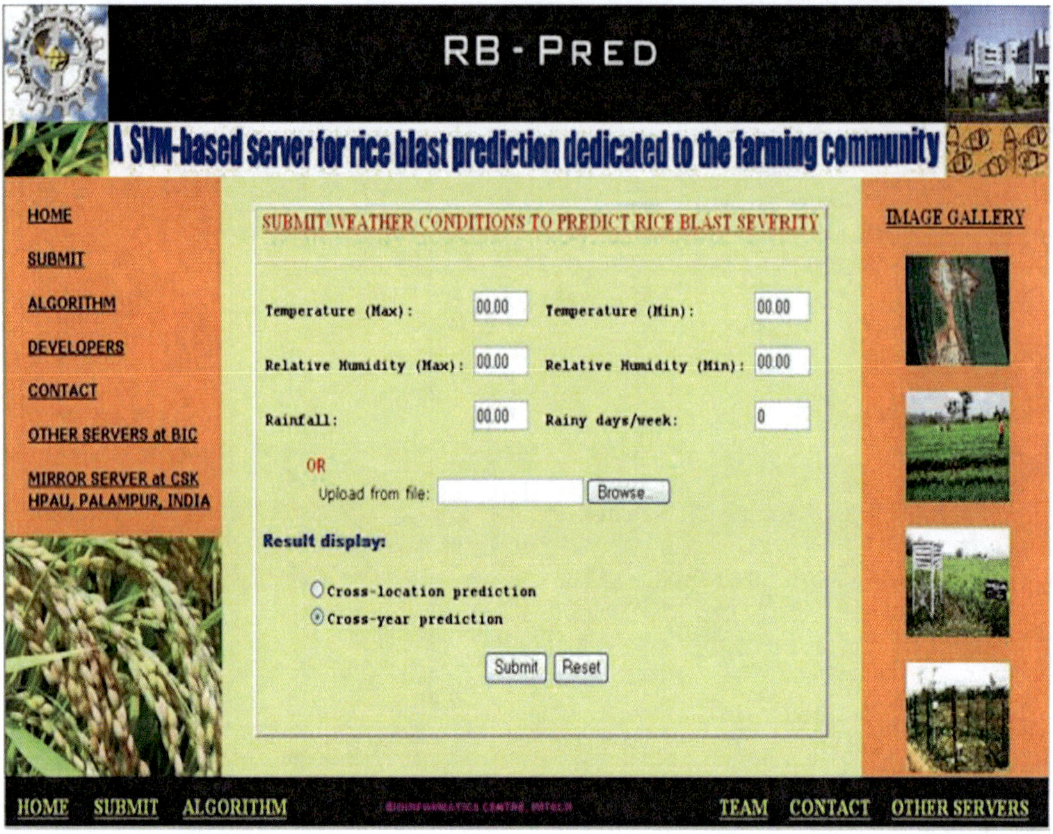

Fig. 4 An overview of submission form for online prediction of rice BLAST severity with "RB-Pred" web server (Kaundal et al. 2006)

the field of rice sciences for designing probabilistic model to predict the time window of various rice diseases. The Generalized Regression Neural Network (GRNN) models were developed using the MATLAB (the MathWorks Inc., Natick, MA) software for designing a second type of neural network which do not require the optimization of multiple parameters that is required in feed-forward Back Propagation Neural Network (BPNN) and help in designing probabilistic model web server for forecasting rice BLAST disease prediction (Kaundal et al. 2006) as shown in Fig. 4.

7 SAS (Statistical Analysis Software) Application in Rice

SAS is an integrated system of software solutions that enables us to perform the various tasks like data entry, retrieval, management, report writing, graphics design, statistical and mathematical analysis, business forecasting, decision support, operations research, project management, and applications development. SAS runs on IBM mainframes, Unix, Linux, OpenVMS Alpha, and Microsoft Windows. Code is "almost" transparently moved between these environments. Older versions have supported PC DOS, the Apple Macintosh, VMS, VM/CMS, PrimeOS, Data General AOS, and OS/2 as shown in Fig. 5.

SAS is also one of the widely used statistical software for analyzing statistical data in crop sciences. The statistical analysis that was carried out using SAS obtained the first three places, respectively, based on Version 8.2. A sample SAS program that can be used to superiority along with a high stability level (Table 3) while carrying out the analysis for three-month maturity group for Bg2845, Bg2834, and Bg300 obtained the first three places 2001/02 wet season

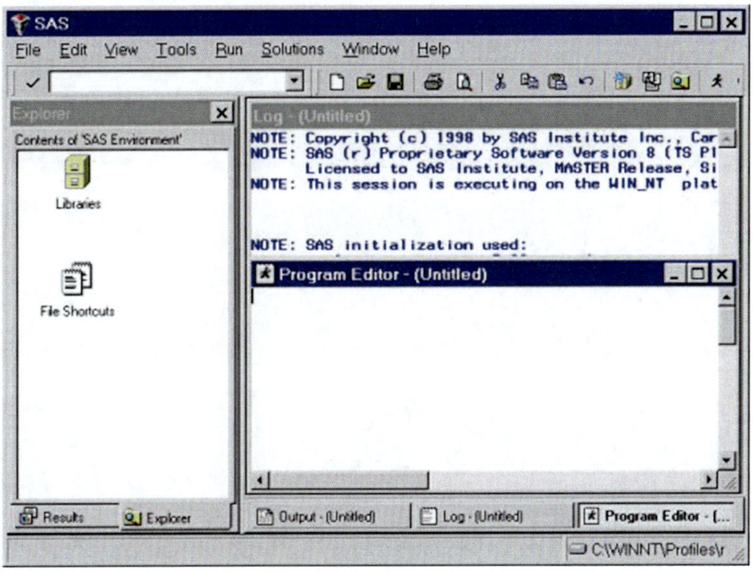

Fig. 5 SAS Windowing Environment (Cary 2001)

Table 3 ANOVA of split plot design for AVT-2 (BT) trial (Cyprien and Kumar 2012)

Source of variation	D.f.	ANOVA SS	Mean square	F value	Pr > F
N	2	0.08	0.04	2.54	0.19

(Samita et al. 2010). Varieties and their interactions were tested by using SAS software. ANOVA of SAS was used to display the results. Table 3 shows the results of ANOVA of split plot design for grain yield (kg/plot) in AVT-2 (BT) trial and observed significant variation in grain yield (Cyprien and Kumar 2012).

8 Computational Modeling in Rice Science

Protein–protein interaction is one of the crucial ways to decipher the functions of proteins and to understand their role in complex pathways at cellular level. The most accurate structural characterization of proteins is provided by X-ray crystallography and NMR spectroscopy. Due to certain technical difficulties and labor intensiveness of these methods, the number of protein structures solved by experimental methods lags far behind the accumulation of protein sequences. By the end of 2007, there were 44,272 protein structures deposited in the Protein Data Bank (PDB) (www.rcsb.org) (Berman et al. 2000) – accounting for just one percent of sequences in the UniProtKB database (http://www.ebi.ac.uk/swissprot). The advancements in the field of bioinformatics have given us efficient tools to understand several biological processes at the molecular level. In recent times, significant progress has been made in computational modeling of protein structures and molecular docking, which holds great promise in prediction of protein–protein interactions. Docking is the computational scheme that attempts to find the best matching between two molecules: a receptor and ligand (Halperin et al. 2002). Protein–protein docking is one of the potential means to study the structure of protein–protein complexes such as antibody-antigen complexes (Gray 2006; Sivasubramanian et al. 2006; Sharma 2008). There are several docking software now introduced into the market such as GOLD, Autodock, Hex etc. Certain online server resources like SWISS Model (http://swissmodel.expasy.org/) and Modbase (http://modbase.compbio.ucsf.edu/modbase-cgi/index.cgi) are also being used

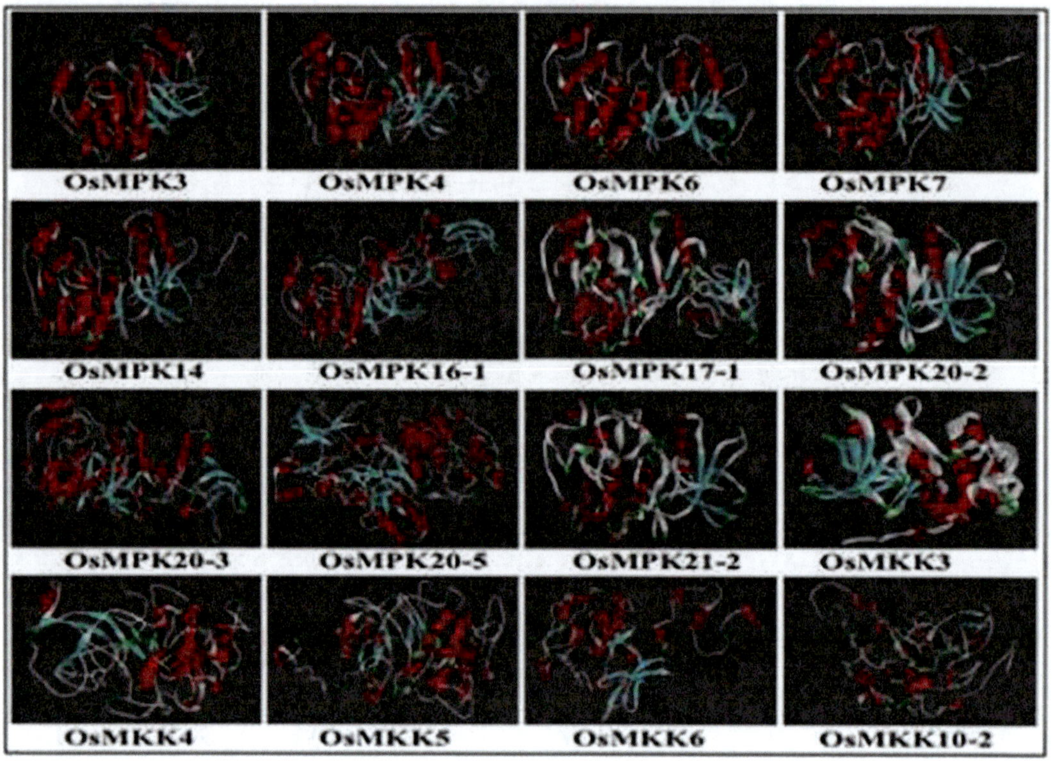

Fig. 6 Theoretical 3-D models of rice MAPKKs and MAPKs built by homology modeling. Structure of 11 rice MAP kinases (OsMPK3, OsMPK4, OsMPK6, OsMPK7, OsMPK14, OsMPK16-1, OsMPK17-1, OsMPK20-2, OsMPK20-3, OsMPK20-5, and OsMPK21-2) and 5 MAP kinase kinases (OsMKK3, OsMKK4, OsMKK5, OsMKK6, OsMKK10-2) are shown. The *red* region represents the alpha helices and *sky blue* regions the beta sheets, *green* colored regions depict the turns, whereas the *gray* color represents the loops (Wankhede et al. 2013)

in homology modeling. Other structure validation programs such as Procheck, WHAT IF, VERIFY3D, and ERRAT were employed for further validation of models.

In the absence of crystallographic structures for rice MAPKs and MAPKK, the homology modeling approach was employed to determine a reasonable 3-D structure of these proteins based on the known structure of the template proteins. The 3-D structures were further used as an input for protein–protein docking using ZDOCK and RDOCK programs, to predict MAPKK–MAPK interactions. Simultaneously, Y2H analyses were used to study rice MAPKK–MAPK protein–protein interaction networks. A direct comparison of computational prediction and Y2H analyses of MAPKK and MAPK was made to assess the reliability of computational docking for prediction of protein–protein interactions (Wankhede et al. 2013). All the 3-D models were refined with the help of loop refinement (Modeler and looper algorithm based) and side chain refinement protocols. The modeled 3-D structure of each of eleven MAPKs and five MAPKKs are shown in Fig. 6.

The overall stereochemical quality of the modeled 3-D structure of proteins was evaluated by using Ramachandran plot which is based on psi (Ca-C bond) and phi (N-Ca bond) angles of the protein and provides information about the number of amino acid residues present in allowed and disallowed regions. All the modeled proteins showed maximum residues in the most favored region followed by allowed region and least in the generously allowed regions in the Ramachandran plot (Fig. 7).

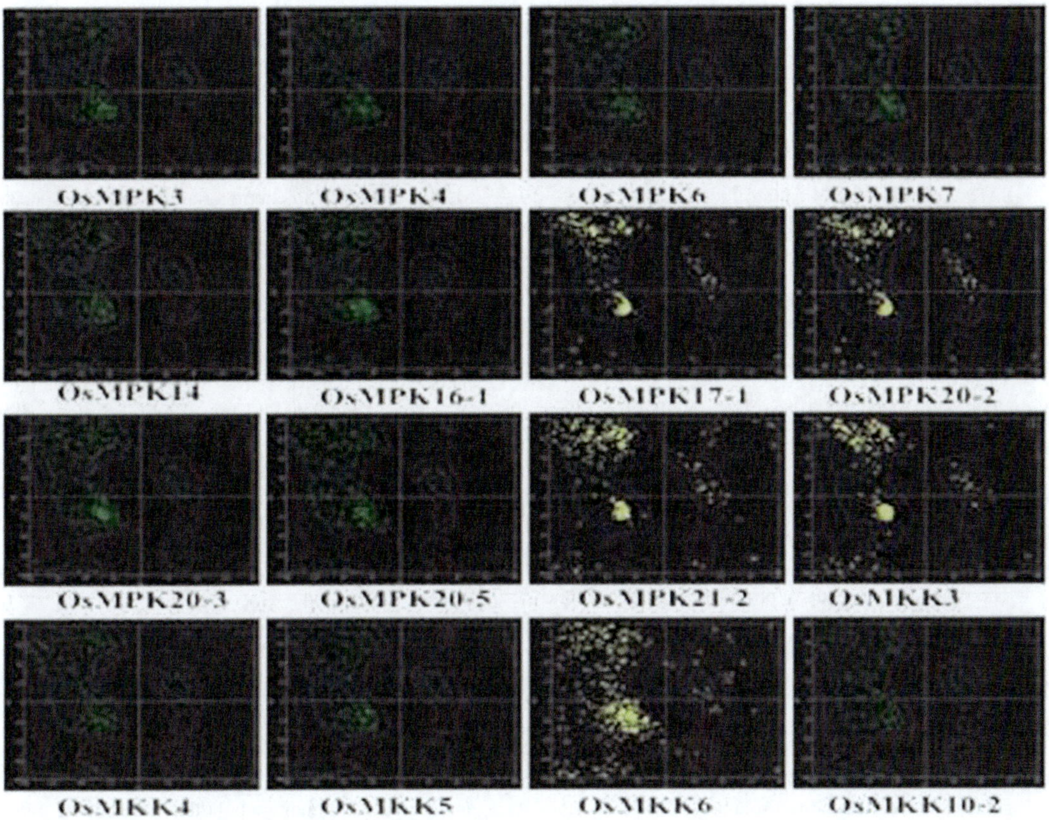

Fig. 7 Ramachandran plot analysis of theoretical 3-D structure of rice MAPKKs and MAPKs. The 3-D structures of 11 rice MAP kinases (OsMPK3, OsMPK4, OsMPK6, OsMPK7, OsMPK14, OsMPK16-1, OsMPK17-1, OsMPK20-2, OsMPK20-3, OsMPK20-5, and OsMPK21-2) and 5 MAP kinase kinases (OsMKK3, OsMKK4, OsMKK5, OsMKK6, OsMKK10-2) were validated using Ramachandran plot. The *green dots/ yellow dots* show the amino acids that are in the most favored regions and additionally allowed region *while red dots* show the amino acids that are in generously allowed region or disallowed regions. The regions covered by *sky blue line* show most favored regions, while the regions covered by *pink line* show additionally allowed regions. Other regions of the plot show the generously allowed or disallowed region loops (Wankhede et al. 2013)

9 Conclusion

Rice is one of the economically important monocot crops in the world. The field of bioinformatics fulfills the needs and provides information on demand to the researcher in all fields of rice research. Bioinformatics is generally applicable to all branches of agriculture and is a boon for varietal improvement. It helps in understanding many agronomic traits involved in crop productivity. With the advancement in structural and functional genomics and proteomics rice databases, the research at both molecular and phenotypic level will advance further. Integrated Bioinformatics Information Resource Access (iBIRA) is an initiative to associate bioinformatics researchers with bioinformatics resources at a single platform. Every day new databases, web server, and software tools are coming up to fulfill the need of the researchers. The challenge for rice informatics is how to translate it into a logical end.

References

Ashikari M, Sakakibara H, Lin S, Yamamoto T et al (2005) Cytokinin oxidase regulates rice grain production. Science 309:741–745

Berman HM, Westbrook J, Feng Z, Gilliland G et al (2000) The protein data bank. Nucleic Acids Res 28:235–242

Brady SM, Provart NJ (2009) Web-queryable large-scale data sets for hypothesis generation in plant biology. Plant Cell 21:1034–1051

Bruskiewich RM, Cosico AB, Eusebio W et al (2003) Linking genotype to phenotype: the International Rice Information System (IRIS). Bioinformatics 19:163–165

Cary NC (2001) Step-by-step programming with base SAS® software. SAS Institute Inc., Cary

Chern C, Fan M, Yu S, Hour S et al (2007) A rice phenomics study-phenotype scoring and seed propagation of a T-DNA insertion-induced rice mutant population. Plant Mol Biol 265:427–438

Cyprien M, Kumar V (2012) A comparative statistical analysis of rice cultivars data. J Reliab Stat Stud 5:143–161

Fox PN, Skovmand B (1996) The International Crop Information System (ICIS)-connects genebank to breeder to farmer's field. In: Cooper M, Hammer GL (eds) Plant adaptation and crop improvement. CAB International, Wallingford

Gerstein M, Jansen R (2000) The current excitement in bio- informatics-analysis of whole-genome expression data: how does it relate to protein structure and function. Curr Opin Struct Biol 10:574–584

Gomez KA, Gomez AA (1984) Statistical procedures for agricultural research. Wiley Interscience, New York

Gray JJ (2006) High resolution protein-protein docking. Curr Opin Struct Biol 16:183–193

Halperin I, Ma B, Wolfson H, Nussinov R (2002) Principles of docking: an overview of search algorithms and a guide to scoring functions. Proteins 47:409–443

Helmy M, Tomita M, Ishihama Y (2011) OryzaPG-DB: rice proteome database based on shotgun proteogenomics. BMC Plant Biol 11:63

International Rice Genome Sequencing Project (2005) The map-based sequence of the rice genome. Nature 436:793–800

Itoh T, Tanaka T, Barrero RA, Yamasaki C et al (2007) Curated genome annotation of *Oryza sativa* ssp. *japonica* and comparative genome analysis with *Arabidopsis thaliana*. Genome Res 17:175–183

Kan N, Venu RC, Cheng L, Belo A et al (2007) An expression atlas of rice mRNAs and small RNAs. Nat Biotechnol 25:473–477

Kaundal R, Kapoor AS, Raghava GPS (2006) Machine learning techniques in disease forecasting: a case study on rice blast prediction. BMC Bioinformatics 7:485. doi:10.1186/1471-2105-7-485

Khush GS, Brar DS (1998) The application of biotechnology to rice. In: Ives C, Bedford B (eds) Agricultural biotechnology in international development. CAB International, Wallingford

Komatsu S, Kojima K, Suzuki K, Ozaki K, Higo K (2004) Rice Proteome Database based on two-dimensional polyacrylamide gel electrophoresis: its status in 2003. Nucleic Acids Res 32:388–392

Konishi S, Izawa T, Lin SY, Ebana K, Fukuta Y, Sasaki T, Yano M (2006) An SNP caused loss of seed shattering during rice domestication. Science 312:1392–1396

Kurakawa T, Ueda N, Maekawa M, Kobayashi K et al (2007) Direct control of shoot meristem activity by a cytokinin-activating enzyme. Nature 445:652–655

Kurata N, Yamazaki Y (2006) Oryzabase. An integrated biological and genome information database for rice. Plant Physiol 140:12–17

Kuromori T, Takahashi S, Kondou Y, Shinozaki K, Matsui M (2009) Phenome analysis in plant species using loss-of-function and gain-of-function mutants. Plant Cell Physiol 50:1215–1231

Larmande P, Gay C, Lorieux M, Perin C et al (2008) Oryza Tag Line, a phenotypic mutant database for the Genoplante rice insertion line library. Nucleic Acids Res 36:1022–1027

Lewis S, Ashburner M, Reese MG (2000) Annotating eukaryote genomes. Curr Opin Struct Biol 10:349–354

Ma JF, Tamai K, Yamaji N, Mitani N et al (2006) A silicon transporter in rice. Nature 440:688–691

Ma JF, Yamaji N, Mitani N et al (2007) An efflux transporter of silicon in rice. Nature 448:209–212

Mathur SK (2010) Statistical bioinformatics: with R. Elsevier, Boston

McNally KL, Childs KL, Bohnert R et al (2009) Genome wide SNP variation reveals relationships among landraces and modern varieties of rice. Proc Natl Acad Sci U S A 106:12273–12278

Miyao A, Iwasaki Y, Kitano H, Itoh JI, Maekawa M, Murata K, Yatou O, Nagato Y, Hirochika H (2007) A large-scale collection of phenotypic data describing an insertional mutant population to facilitate functional analysis of rice genes. Plant Mol Biol 63:625–635

Moore G, Devos KM, Wang Z, Gale MD (1995) Grasses, line up and form a circle. Curr Biol 5:737–739

Nagamura Y, Antonio BA (2010) Current status of rice informatics resources and breeding applications. Breed Sci 60:549–555

Nagamura Y, Antonio BA, Sato Y, Miyao A, Namiki N, Yonemaru J, Minami H, Kamatsuki K, Shimura K, Shimizu Y, Hirochika H (2010) Rice TOGO Browser: a platform to retrieve integrated information on rice functional and applied genomics. Plant Cell Physiol 52:230–237

Ram S, Rao LN (2012) Global information resources on rice for research and development. Rice Sci 19:327–334

Sakata K, Nagamura Y, Numa H, Antonio BA et al (2002) RiceGAAS: an automated annotation system and database for rice genome sequence. Nucleic Acids Res 30:98–102

Samita S, Anputhas M, De DS (2010) Selection of rice varieties for recommendation in Sri Lanka: a complex-free approach. World J of Agric Sci 6:189–194

Sasaki T, Burr B (2000) International Rice Genome Sequencing Project: the effort to completely sequence the rice genome. Curr Opin Plant Biol 3:138–141

Seki M, Shinozaki K (2009) Functional genomics using RIKEN *Arabidopsis thaliana* full-length cDNAs. J Plant Res 122:355–366

Sharma B (2008) Structure and mechanism of a transmission blocking vaccine candidate protein Pfs25 from *P falciparum*: a molecular modeling and docking study. In Silico Biol 8:193–206

Simon SA, Zhai J, Zeng J, Meyers BC (2008) The cornucopia of small RNAs in plant genomes. Rice 1:52–62

Singh I, Shah K (2012) In silico study of interaction between rice proteins enhanced disease susceptibility and phytoalexin deficient, the regulators of salicylic acid signalling pathway. J Biosci 37:563–571

Sivasubramanian A, Chao G, Pressler HM, Wittrup KD, Gray JJ (2006) Structural model of the mAb 806-EGFR complex using computational docking followed by computational and experimental mutagenesis. Structure 14:401–414

Swamy BPM, Ahmed HU, Henry A, Mauleon R, Dixit S et al (2013) Genetic, physiological, and gene expression analyses reveal that multiple QTL enhance yield of rice mega-variety IR64 under drought. PLoS ONE 8:e62795. doi:10.1371/journal.pone.0062795

Thiel T, Graner A, Waugh R, Grosse I, Close TJ, Stein N (2009) Evidence and evolutionary analysis of ancient whole-genome duplication in barley predating the divergence from rice. BMC Evol Biol 9:209. doi:10.1186/1471-2148-9-209

Wang D, Xia Y, Li X, Hou L, Yu J (2012) The Rice Genome Knowledgebase (RGKbase): an annotation database for rice comparative genomics and evolutionary biology. Nucleic Acids Res 41:1199–1205

Wankhede DP, Misra M, Singh P, Sinha AK (2013) Rice mitogen activated protein kinase kinase and mitogen activated protein kinase interaction network revealed by in-silico docking and yeast two-hybrid approaches. PLoS ONE 8:e65011. doi:10.1371/journal.pone.0065011

Wing RA, Ammiraju JS, Luo M, Kim H et al (2005) The *oryza* map alignment project: the golden path to unlocking the genetic potential of wild rice species. Plant Mol Biol 59:53–62

Wu J, Wu C, Lei C, Baraoidan M, Boredos A et al (2005) Chemical and irradiation induced mutants of *indica* rice IR64 for forward and reverse genetics. Plant Mol Biol 59:85–97

Yonemaru J, Yamamoto T, Ebana K, Yamamoto E, Nagasaki H, Shibaya T, Yano M (2012) Genome-wide haplotype changes produced by artificial selection during modern rice breeding in Japan. PLoS ONE 7:e32982. doi:10.1371/journal.pone.0032982

Zhang J, Li C, Wu C, Xiong L, Chen G, Zhang Q, Wang S (2006) RMD: a rice mutant database for functional analysis of the rice genome. Nucleic Acids Res 34:745–748

Zhao K, Tung CW, Eizenga GC et al (2011) Genome-wide association mapping reveals a rice genetic architecture of complex traits in *oryza sativa*. Nat Commun 2:467. doi:10.1038/ncomms1467

Contribution of Bioinformatics to Gene Discovery in Salt Stress Responses in Plants

P. Hima Kumari*, S. Anil Kumar*, Prashanth Suravajhala, N. Jalaja, P. Rathna Giri, and Kavi Kishor P.B.

Abstract

Salinity is the major abiotic stress leading to huge losses in crop productivity. Therefore, understanding the regulatory mechanisms and subsequently improving salinity tolerance in plants are important goals for plant biologists. Salinity tolerance depends upon the ability of plants to exclude salts, the compartmentalization of sodium (Na^+) into vacuoles and acquisition of potassium (K^+) to cope with osmotic stress and to maintain ion homeostasis. The availability of complete genome sequences coupled with effective and high-throughput methods has helped us in identifying many genes associated with salt stress. The tools of bioinformatics have allowed us to identify stress-associated gene families across species based on homology and gene synteny. Besides, whole genome sequencing, cDNA libraries related to stress tolerance, and genome-wide association studies have facilitated the discovery of stress-related genes. This vital information from both halophytic and glycophytic species coupled with the isolation of potential target genes associated with salt stress tolerance helps in crop breeding programs aimed at generating salt stress-tolerant crop plants.

Keywords

Comparative genomics • Protein-protein interactions • Salt stress • Transporters

*Contributed equally

P. Hima Kumari • S. Anil Kumar
N. Jalaja • K.K. P.B. (✉)
Department of Genetics, Osmania University, Hyderabad 500 007, Andhra Pradesh, India
e-mail: pbkavi@yahoo.com

P. Suravajhala
Bioclues Organization, IKP Knowledge Park, Picket, Secunderabad 500 009, Andhra Pradesh, India

P. Rathna Giri
Genomix Molecular Diagnostics Pvt. Ltd., Prashanth Nagar, Kukatpally, Hyderabad 500 072, India

1 Introduction

Plants need optimal environmental conditions for their growth and development. However, they face many adverse conditions due to their sedentary nature. World population mainly relies on cereal crops, which constitute more than 60 % of total worldwide agricultural production (Harlan 1995). A number of abiotic stress factors (Fig. 1)

K.K. P.B. et al. (eds.), *Agricultural Bioinformatics*,
DOI 10.1007/978-81-322-1880-7_6, © Springer India 2014

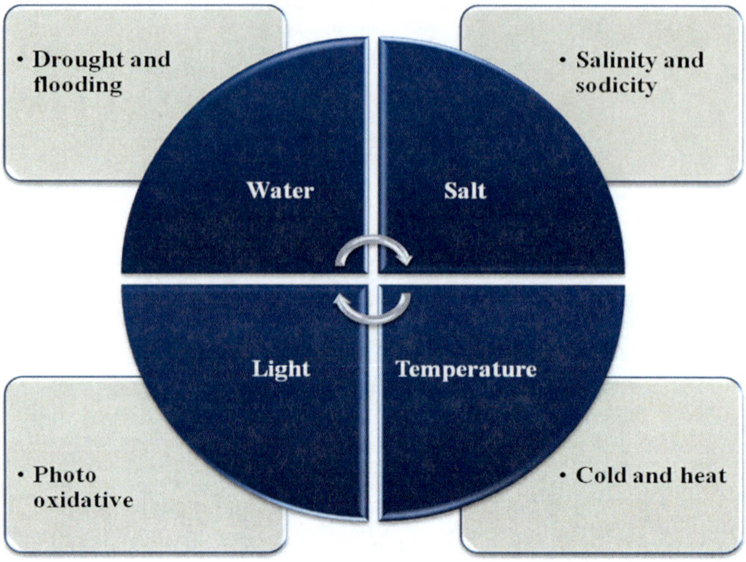

Fig. 1 Abiotic stresses affecting the plant growth

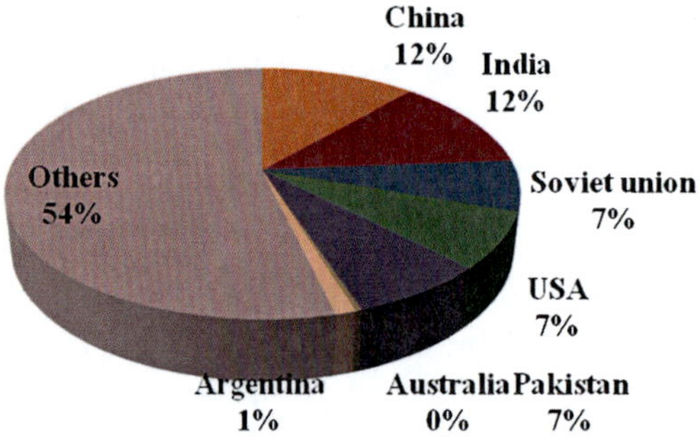

Fig. 2 Percent of irrigated lands affected by soil salinity

like salt, drought, flooding, heat, cold, and oxidative stress together impose severe metabolic alterations leading to the production of reactive oxygen species (ROS). Due to various abiotic stresses, 5 % of world's cultivated land and 50 % of crop productivity are affected every year (Fig. 2). If the abiotic stresses are severe, plants become sterile, and productivity is completely lost. These stress factors are interconnected and lead to a series of biological, morphological, and physiological changes (Table 1) leading to the death of plants (Rodriguez et al. 2005). Salt and drought are the two major stress factors which cause huge crop losses. But, plants have developed multiple mechanisms in response to various stresses, which are complex and integrative (Ciarmiello et al. 2011).

1.1 Impact of Salinity Stress on Plants

Salinity limits the plant growth and decreases the final productivity. The ability of plants to survive under saline conditions varies among different species of glycophytes and halophytes. Most of the dicotyledonous halophytes live at the concentrations of 200–300 mM NaCl by adapting various mechanisms. The growth response of glycophytes occurs in two phases: (a) a rapid response to increase in external salt called as osmotic phase and (b) slower response with accumulation of Na^+ ions in vacuoles called as ionic phase. With an increase in salinity, leaf cells loose water and turgor, but are transient, and they gain osmotic adjustment due to complex mechanisms developed by plants. Salinity affects the plants by closure of stomatal apertures and subsequently reducing the photosynthetic rate. This situation increases the formation of ROS creating an oxidative stress. The strategies and mechanisms for sodium transport include sodium exclusion, sodium inclusion into vacuole, and intracellular compartmentation at different stages of plant. In many crops, salt tolerance is achieved by sodium exclusion from leaves and shoots to minimize the damage caused by accumulated sodium ions (Munns 2002; Tester and Davenport 2003).

Table 1 Effect of salinity on plant metabolism

Morphological effects	Physiological and biochemical effects
White leaf tip followed by tip burning (salinity)	High Na^+ transport to shoot
Leaf browning and necrosis (sodicity)	Preferential accumulation of Na^+ in older leaves
Poor root growth	High Cl^- uptake
Leaf rolling	Lower K^+ uptake
Change in flowering duration	Increase in the compatible solutes and polyamine levels
Stunted plant growth	Change in esterase isozyme pattern
Low grain yield and quality	Lower fresh and dry weights of shoots and roots

1.1.1 Mechanisms of Na^+ Transport at Root and Plasma Membrane Levels

Na^+ enters passively into epidermal and cortical cells by concentration gradient from soil through root hairs. It may also enter into endodermis which does not have any Casparian bands and suberin lamellae (White 2001; Moore et al. 2002). Na^+ influx is unidirectional at plasma membrane level, triggered and controlled by a complex set of signal molecules like Ca^{2+} and many nonselective cation channels (NSCCs) which include cyclic nucleotide-gated channels (CNGCs) and glutamate receptors. Of the 20 CNGCs, AtCNGC1, AtCNGC3, AtCNGC4, and AtCNGC10 are known to be involved in Na^+ uptake (Gobert et al. 2006; Guo et al. 2008), and AtGLR2 and AtGLR3 in Na^+ and K^+ symport. The exchange of K^+ or Na^+ for proton (H^+) is known to occur in plants (Chanroj et al. 2012) and is mediated by a family of transporters known as Na^+/H^+ antiporters (NHXs). Na^+ efflux from plasma membrane to the apoplast is achieved by the expression of salt overly sensitive (*SOS1*) gene, a sodium proton antiporter in root epidermal cells of *Arabidopsis*. With an increase in the cytoplasmic salt concentration, an intracellular Ca^{2+} signal is triggered which recruits an SOS3 (a calcium-binding protein) and forms a complex with serine–threonine protein kinase SOS2, which in turn activates SOS1 for the efflux of Na^+ to maintain pH homeostasis and ionic balance (Fig. 3) that are important regulators of cellular processes associated with plant growth.

1.1.2 Transport of Na^+ in the Plants

In both halophytes and glycophytes, cytosolic Na^+ concentration is maintained at nontoxic levels in spite of considerable influx of Na^+ into the cytosol. This is due to exclusion mechanisms or compartmentalization of Na^+ (and Cl^-) into the vacuoles. The latter allows the plants to use NaCl as an osmoticum. Thus, plants use inorganic ions to maintain cellular osmotic potential that drives water into the cells from soil (Blumwald 2000). Na^+ influx into the cytoplasm of epidermal and cortical cells occurs radially *via* plasmodesmata along a concentration gradient. Not much information is available about radial transport in symplast cells. A set of Na^+ transporters (NHX) mediate the regulation of Na^+ in plasma membrane and vacuolar cells of root hairs. Cryoscanning electron microscopy and X-ray microanalysis have been used to detect Na^+ concentrations in the vacuoles (Lauchli et al. 2008; Moller et al. 2009). Also, a cation exchanger *AtCHX21* gene is implicated in radial transport in stelar cells. The root apoplast cells have affinity for ions like Na^+, K^+, Ca^{2+}, and Mg^{2+} as it has a cation-binding matrix. The apoplastic transport of Na^+ is limited only up to endodermal cells as it has a barrier of Casparian bands (Peterson et al. 1993). Generally, Na^+ first enters from apoplast to symplast via endodermal cells and passes through plasma membrane of cortical cells. The loading of Na^+ into xylem is a crucial process for plant salinity tolerance. This process leads to increased Na^+ concentrations in leaves (Shi et al. 2002). The main site for Na^+ toxicity appears to be leaf blade than the roots

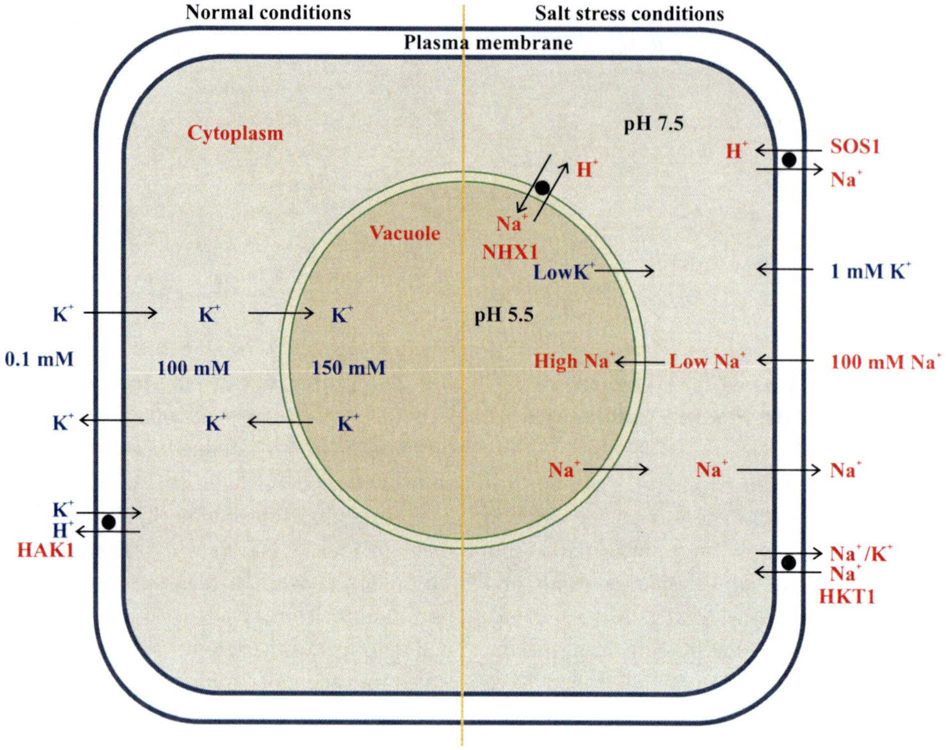

Fig. 3 Ion homeostasis of Na^+ and K^+ under normal and salt stress conditions

(Munns and Tester 2008). High-affinity K^+ transporter1 (*AtHKT1*) is involved in the transport of Na^+ from the shoot into the phloem and also in the unloading of Na^+ into stelar cells. Na^+ accumulation was found to be more in older leaves than younger because of differential distribution of various nonselective cation channels in different cell types (Karley et al. 2000). The entry of Na^+ from roots to the leaves must be restricted and compartmentalized into the vacuoles in order to decrease the accumulation of Na^+ that is toxic to the proteins in the leaf cytoplasm. Low Na^+ concentrations in the cytoplasm are generally achieved because of Na^+/H^+ exchangers located at the plasma membrane. Multiple Na^+/H^+ exchangers have been detected in plants with complex functions. Different Na^+/H^+ exchangers (NHXs) identified and isolated from *Arabidopsis thaliana* so far and their functions are shown in Table 2.

1.2 Na^+ Channels

Plants have different types of membrane transport systems for movement of ions and molecules across a biological membrane barrier. The exchange of ions usually occurs by secondary active transporters like symporters and antiporters (Dahl et al. 2004). Na^+ transporters belong to the monovalent cation proton antiporter-1 (CPA1) family, derived from bacteria, yeast, plants and animals (Fig. 4). Their primary physiological functions may be (1) cytoplasmic pH regulation, extruding the H^+ generated during metabolism and (2) salt tolerance (in plants) due to Na^+ influx into vacuoles (Waditee et al. 2001). In addition to Na^+ transport, this family also involves in pH regulation and K^+ homeostasis (An et al. 2007). Plasma membrane Na^+ transporters (PM-NHE) are found extensively in vertebrates, whereas

Table 2 List of Na^+/H^+ exchangers isolated from different plants

Sl. No.	Protein names	Length of amino acids	Function/expression	Reference
1	Na^+/H^+ exchanger 1	538	Necessary for cell volume regulation and cytoplasmic Na^+ detoxification	Gaxiola et al. (1999)
2	Na^+/H^+ exchanger 2	546	Expressed in roots and shoots. Induced by ABA and NaCl	Yokoi et al. (2002)
3	Na^+/H^+ exchanger 3	503	Expressed in roots	Yokoi et al. (2002)
4	Na^+/H^+ exchanger 4	529	Expressed at very low levels in roots and shoots	Yokoi et al. (2002)
5	Na^+/H^+ exchanger 5	521	Expressed in roots, leaves, stems, flowers, and siliques. Detected at low levels in roots and shoots	Bassil et al. (2011)
6	Na^+/H^+ exchanger 6	535	Expressed in roots, leaves, stems, flowers, and siliques. Detected at low levels in roots and shoots	Bassil et al. (2011)
7	Na^+/H^+ exchanger 7 (NHE-7) (SOS1)	1,146	More expressed in roots than shoots. Mostly localized in parenchyma cells at the xylem/symplast boundary in roots, hypocotyls, stems, and leaves	Shi et al. (2000)
8	Na^+/H^+ exchanger 8 (NHE-8) (protein SALT OVERLY SENSITIVE 1B)	756	Expressed more in leaves than roots and flowers	An et al. (2007)
9	Inactive poly [ADP-ribose] polymerase RCD1 (RADICAL-INDUCED CELL DEATH 1 protein)	589	Expressed in young developing tissues and vasculature of leaves and roots and guard cells	Jaspers et al. (2009)

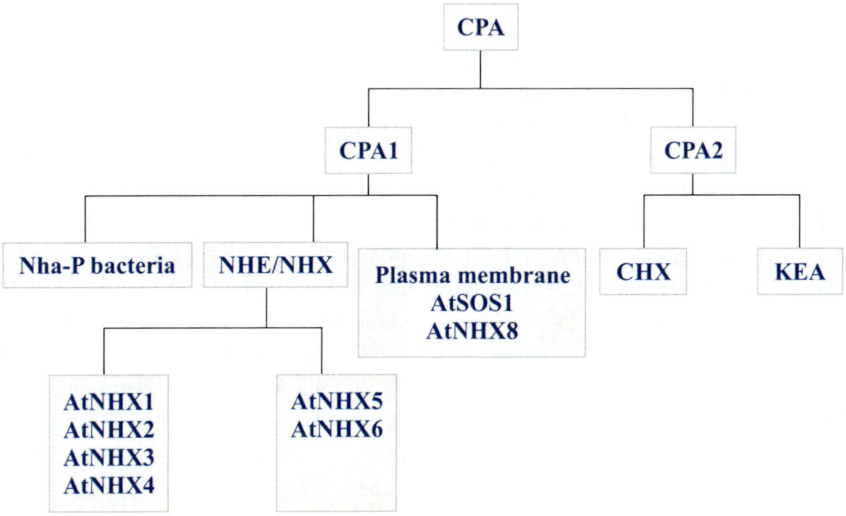

Fig. 4 Classification of monovalent cation transporters

intracellular Na^+ transporter (NHX) genes are found to be evolved from PM-NHE with 20–25 % identity only in fungi, plants, and animals (Rodríguez-Rosales et al. 2009). To date, nine NHE isoforms (*NHE1–NHE9*) have been identified in animals and eight NHX in plants (*NHX1–NHX8*). These are modulated by agents that target primarily tyrosine kinases and also by agonists of Ser/Thr kinases including protein kinases A (PKA) and C (PKC). Besides, they are sensitive to increase in cytosolic Ca^{2+} and to changes in cell volume (Orlowski and Grinstein 1997). Genome-wide survey of *Arabidopsis* for CPA1 family members revealed eight members, *AtNHX1–8*. *AtNHX1* and *AtNHX7/SOS1* have been well characterized as tonoplast and plasma membrane Na^+/H^+ antiporters. The proteins *AtNHX2–6* are phylogenetically similar to *AtNHX1*, while *AtNHX8* appears to be related to *AtNHX7/SOS1*.

1.2.1 Structure of Na^+-H^+ Antiporter

Intracellular pH, cell osmotic balance, and cellular Na^+ levels are controlled by Na^+/H^+ antiporter genes. They are integral membrane proteins and use proton electrochemical gradient for exclusion of Na^+ ions. Structural determination of NHX shows that it is tightly regulated by pH, downregulated, and forms crystals at acidic pH. NHX consists of 350–400 amino acid residues in different plants and measures about 40–45 angstrom units. Its activity is mainly altered in between pH 7 and 8 and fully downregulated below pH 6.5. The Na^+ transporters have 12 transmembrane domains with a conserved domain LLFIYLLPPI, a drug inhibitor binding site for amiloride. The pH sensor responsible for alteration of protonation site is present at the transmembrane IX of the cytoplasmic side.

1.3 Salt Stress Response in Crop Species

Studies were carried out by taking advantage of the nucleotide sequence identity between *Thellungiella* (a salt-tolerant species) and *Arabidopsis* (a salt-susceptible plant) (Gong et al. 2005). Under both standard and stressed conditions, *Thellungiella* metabolome was found to be more soluble, polar, reduced, and massive than the *Arabidopsis*. But, osmotic and salinity stresses changed the metabolic and biophysical properties more in *Arabidopsis* (Lugan et al. 2010). To assess the Na^+/H^+ exchange function, Na^+-dependent H^+ movement was measured in intact vacuoles isolated from both wild type and transgenics. The Na^+/H^+ exchange rates were high in transgenic plants overexpressing Na^+/H^+ than wild-type plants when grown in the presence of 200 mM NaCl (Apse et al. 1999). *AtNHX1* mutant plant leaves had markedly decreased Na^+/H^+ and K^+/H^+ exchange activity at the vacuolar membrane and altered the development (Apse et al. 2003), suggesting a role of *AtNHX1* in mediating K^+ and pH homeostasis.

1.4 K^+ Channels

K^+ is one of the most abundant and key elements and accounts up to 10 % of the plant dry weight. Its major role is the maintenance of ionic balance of the cell, membrane potential, photosynthesis, enzyme activation, etc. The stable cytoplasmic K^+ concentration estimated in plant cell is around 100 mM which is needed for various enzyme activities. The channels transporting K^+ from external environment into a cell are mostly located at the plasma membrane regions. In many plant species, genes encoding K^+ transporters were identified. Plants initially sense external K^+ from root hairs and epidermal cells. There is a linear relationship between plant cell and K^+ external supply. K^+ influx into the cell is inhibited under salt stress conditions since these ions compete with Na^+ ions for the binding site. If there is a decrease in K^+ supply, hyperpolarization of membrane potential is activated by several magnitudes within few minutes. Contrarily, depolarization occurs by excess K^+ concentrations. More than 70 genes encoding K^+ transporters were identified in *Arabidopsis* and rice by genome-wide scanning (Amrutha et al. 2007). Depending on the early and long-term K^+ transport, genes encoding

K^+ channels and transporters were categorized into six families (KUP/HAK/KT, HKT, Shaker, TPK, Kir-like, and CPA).

1.4.1 KUP/HAK/KT Transporters

This family has been identified with a homology to K^+ uptake permeases (KUP) from bacteria, high-affinity K^+ transporters (HAK) from fungi, and K^+ transporter (KT) from plants. These proteins are involved in cell expansion growth. HAK/KUP/KT transporters form a large family, comprising 13 members in *Arabidopsis* and 26 members in rice. They are functionally more diverse than their fungal and bacterial homologues (Grabov 2007; Banuelos et al. 2002). All proteins from this family were isolated and cloned from *Arabidopsis*, rice, and barley with 10–14 transmembrane helices, except OsHAK3 (Gierth and Maser 2007).

1.4.2 HKT Transporters

High-affinity K^+ transporters (HKT) belong to Trk superfamily and share analogy with voltage-gated cation channels. The transporters are monomers and thought to resemble shaker-type transporter tetramers. This family seems to have been lost from animal kingdom and only present in bacteria, fungi and plants. Based on the HKT amino acid sequences, this family is categorized into two subfamilies. Subfamily one contains longer introns and is present in monocots and dicots, whereas subfamily two is present in monocots only. The first *HKT* genes were isolated from wheat (*TaHKT1*) and expressed in *Xenopus laevis* oocytes. The K^+ channels P-loop ends in filter residues called Gly-Tyr-Gly (GYG), and this triad serves to attract a substrate cation into the hydrophobic domain of plasma membrane. This triad is highly conserved in all K^+ channels. Subfamily one has serine residue in place of glycine at first position (Gierth and Maser 2007).

1.4.3 Cyclic Nucleotide-Gated Cation Channels (CNGC)

Plant CNGCs are nonselective cation-conducting channels, functionally characterized and expressed. Till date, 56 coding cation-conducting channel sequences were identified and of that 26 are from CNGC family (Ward et al. 2009). They were first identified by the presence of cyclic nucleotide-binding domain (CNBD) having a conserved phosphate-binding cassette which binds sugar and phosphate moieties of cyclic nucleotide ligand (Cukkemane et al. 2011). These genes are mainly localized to plasma membrane except CNGC20, which is targeted to chloroplast. These channels are members of "P-loop" superfamily of cation channels; evolutionarily their structure provides a basic form capable of conducting various cations (Zhorov and Tikhonov 2004). Biochemical studies with a CNGC ortholog from tobacco and CNGC1 from *A. thaliana* have demonstrated that the CNBD overlaps with that of calmodulin (Arazi et al. 2000; Koehler and Neuhaus 2000), thereby suggesting that cyclic nucleotides and calmodulin interact in the regulation of channel activity.

1.4.4 Shaker-Type Channels

Shaker channels have high selectivity for K^+ ions and are expressed more in plasma membrane regions. Channel gating here mainly depends on the changes in membrane potential, and the family is further classified into K^+ outward rectifying channels (KORC) and K^+ inward rectifying channels (KIRC). KORCs mediate the influx of Na^+ into plant cells (Wegner and Raschke 1994). KORC channels open during the depolarization of the plasma membrane and mediate the efflux of K^+ and the influx of Na^+ (Maathuis and Sanders 1997). The opening of KORC channels in stelar cells of maize results in an increase of cytosolic Ca^{2+}. In plant plasma membranes (Tyerman and Skerrett 1999; White 1999), voltage-independent cation channels (VIC) have a relatively high Na^+/K^+ selectivity in contrast to the voltage-dependent channels (KIRC and KORC).

1.4.5 Two-Pore K^+ (TPK) Channel Family

The members of this family are similar to shaker-type channels, having four transmembrane domains and two pore regions with GYGD motif. Voltage sensor is not present and therefore shows weak substrate affinity at the membrane level (Table 3). Four of five channels isolated

Table 3 List of K^+ transporter proteins isolated from different plants

Sl. No.	Protein name	Organism	Length of amino acids	Function and expression	Reference
1	K^+ transporter 10 (OsHAK10)	*Oryza sativa* subsp. Japonica	843	Expressed in roots, shoots, and panicle at flowering stage	Banuelos et al. (2002)
2	Probable K^+ transporter 11 (OsHAK11)	*O. sativa* subsp. Japonica	791	Involved in pollen development	Banuelos et al. (2002)
3	Putative K^+ transporter 12 (OsHAK12)	*O. sativa* subsp. Japonica	793	K^+ ion transport	Banuelos et al. (2002)
4	Probable K^+ transporter 13 (OsHAK13)	*O. sativa* subsp. Japonica	778	K^+ ion transport	Banuelos et al. (2002)
5	Probable K^+ transporter 14 (OsHAK14)	*O. sativa* subsp. Japonica	859	K^+ ion transport	Banuelos et al. (2002)
6	Probable K^+ transporter 15 (OsHAK15)	*O. sativa* subsp. Japonica	867	K^+ ion transport	Banuelos et al. (2002)
7	Probable K^+ transporter 16 (OsHAK16)	*O. sativa* subsp. Japonica	811	K^+ ion transport	Banuelos et al. (2002)
8	Probable K^+ transporter 17 (OsHAK17)	*O. sativa* subsp. Japonica	707	K^+ ion transport	Banuelos et al. (2002)
9	K^+ transporter 18 (OsHAK18)	*O. sativa* subsp. Japonica	793	K^+ ion transport	Yang et al. (2009)
10	K^+ transporter 19 (OsHAK19)	*O. sativa* subsp. Japonica	742	K^+ ion transport	Yang et al. (2009)
11	K^+ transporter 1 (OsHAK1)	*O. sativa* subsp. Japonica	801	Expressed mainly in roots	Banuelos et al. (2002)
12	K^+ transporter 20 (OsHAK20)	*O. sativa* subsp. Japonica	747	K^+ ion transport	Yang et al. (2009)
13	K^+ transporter 21 (OsHAK21)	*O. sativa* subsp. Japonica	799	K^+ ion transport	Yang et al. (2009)
14	K^+ transporter 22 (OsHAK22)	*O. sativa* subsp. Japonica	790	K^+ ion transport	Yang et al. (2009)
15	K^+ transporter 23 (OsHAK23)	*O. sativa* subsp. Japonica	877	Potassium ion transport	Yang et al. (2009)
16	K^+ transporter 24 (OsHAK24)	*O. sativa* subsp. Japonica	772	K^+ ion transport	Yang et al. (2009)
17	K^+ transporter 25 (OsHAK25)	*O. sativa* subsp. Japonica	770	K^+ ion transport	Yang et al. (2009)

(continued)

Table 3 (continued)

Sl. No.	Protein name	Organism	Length of amino acids	Function and expression	Reference
18	K^+ transporter 26 (OsHAK26)	*O. sativa* subsp. Japonica	739	K^+ ion transport	Yang et al. (2009)
19	K^+ transporter 27 (OsHAK27)	*O. sativa* subsp. Japonica	811	K^+ ion transport	Yang et al. (2009)
20	Probable K^+ transporter 2 (OsHAK2)	*O. sativa* subsp. Japonica	783	K^+ ion transport	Yang et al. (2009)
21	Probable K^+ transporter 3 (OsHAK3)	*O. sativa* subsp. Japonica	808	K^+ ion transport	Yang et al. (2009)
22	HvHAK4	*Hordeum vulgare*	697	Expressed in growing leaf tissue	Boscari et al. (2009)
23	K^+ transporter 5 (OsHAK5)	*O. sativa* subsp. Japonica	781	K^+ uptake in epidermis of main and lateral roots	Gierth et al. (2005)
24	K^+ transporter 6 (OsHAK6)	*O. sativa* subsp. Japonica	748	K^+ ion transport	Yang et al. (2009)
25	K^+ transporter 7 (OsHAK7)	*O. sativa* subsp. Japonica (Rice)	811	Expressed mainly in shoots and roots	Banuelos et al. (2002)
26	Putative K^+ transporter 8 (OsHAK8)	*O. sativa* subsp. Japonica	793	K^+ ion transport	Yang et al. (2009)
27	Probable transporter 9 (OsHAK9)	*O. sativa* subsp. Japonica	788	K^+ ion transport	Yang et al. (2009)
28	K^+ transporter 3 (AtKT3) (AtKUP4) (AtPOT3) (tiny root hair 1 protein)	*Arabidopsis thaliana*	775	Required for tip growth of root hairs	Rigas et al. (2001)
29	AtKUP1	*A. thaliana*	721	Detected at very low levels in roots, stems, leaves, and flowers of mature plant	Kim et al. (1998)
30	Transporter 4 (AtKT4) (AtKUP3) (AtPOT4)	*A. thaliana*	789	Strongly expressed in the roots of K^+-starved plants	Kim et al. (1998)
31	K^+ transporter 5 (AtHAK1) (AtHAK5) (AtPOT5)	*A. thaliana*	785	Induced in both roots and shoots by K^+ starvation	Rubio et al. (2000)
32	K^+ transporter 6 (AtHAK6) (AtPOT6)	*A. thaliana*	782	Expressed in roots	Rubio et al. (2000)
33	K^+ transporter 7 (AtHAK7) (AtPOT7)	*A. thaliana*	858	Expressed in roots	Rubio et al. (2000)
34	K^+ transporter 8 (AtHAK8) (AtPOT8)	*A. thaliana*	781	Expressed in roots	Rubio et al. (2000)
35	Na^+ transporter HKT1 (AtHKT1)	*A. thaliana*	506	Expressed in the vascular tissues of every organs. In roots, leaves, and flower peduncles	Uozumi et al. (2000)

(continued)

Table 3 (continued)

Sl. No.	Protein name	Organism	Length of amino acids	Function and expression	Reference
36	Cation transporter HKT2 (OsHKT2) (Po-OsHKT2)	*O. sativa* subsp. Indica	530	Involved in the regulation of K^+/Na^+ homeostasis	Horie et al. (2001)
37	Cation transporter HKT4 (OsHKT4)	*O. sativa* subsp. Japonica	552	Expressed in shoots	Garciadeblas et al. (2003)
38	Probable cation transporter HKT6 (OsHKT6)	*O. sativa* subsp. japonica	530	Involved in the regulation of K^+/Na^+ homeostasis and weakly expressed	(Garciadeblas et al. 2003)
39	Probable cation transporter HKT7 (OsHKT7)	*O. sativa* subsp. Japonica	500	Involved in the regulation of K^+/Na^+ homeostasis	Garciadeblas et al. (2003)
40	Probable cation transporter HKT9 (OsHKT9)	*O. sativa* subsp. Japonica	509	Involved in the regulation of K^+/Na^+ homeostasis	Garciadeblas et al. (2003)
41	Putative Na^+ transporter HKT7-A2	*Triticum monococcum*	554	Control Na^+ unloading from xylem in roots and sheaths	Huang et al. (2006)
42	CNGC11 (cyclic nucleotide- and calmodulin-regulated ion channel 11)	*A. thaliana*	621	Involved in pathogen resistance response	Yoshioka et al. (2006)
43	Probable CNGC12 (cyclic nucleotide- and calmodulin-regulated ion channel 12)	*A. thaliana*	649	Involved in pathogen resistance response	Yoshioka et al. (2006)
44	AtCNGC1 (cyclic nucleotide- and calmodulin-regulated ion channel 1)	*A. thaliana*	716	Plants exhibit an improved tolerance to Pb^{2+}	Sunkar et al. (2008)
45	AtCNGC2 (cyclic nucleotide- and calmodulin-regulated ion channel 2) (protein DEFENSE NO DEATH 1)	*A. thaliana*	726	Strongly expressed in the expanded cotyledons	Kohler et al. (2001)
46	AtCNGC4 (cyclic nucleotide- and calmodulin-regulated ion channel 4) (AtHLM1)	*A. thaliana*	694	Induced by both ethylene and methyl jasmonate treatments	Balague et al. (2003)
47	AtKT2/AtKUP2	*A. thaliana*		Mediates K^+-dependent cell expansion in growing leaf tissues	Elumalai et al. (2002)
48	HvAKT1	*H. vulgare*		Expressed in leaf and root	Boscari et al. (2009)
49	HvAKT2	*H. vulgare*		Contribute to K^+ uptake into mesophyll cells of growing leaf tissue	Boscari et al. (2009)

from *Arabidopsis* contain "EF hand" domains involving calcium ion binding, indicating a role in K^+ ion homeostasis (Maathuis 2007).

1.5 Genomics, Database Resources, and In Silico Analysis

Genome sequences from *Arabidopsis* and *Arabidopsis*-relative model systems (halophytes) are important for understanding the complex traits like salt stress tolerance. This will ultimately help us in genetic and bioinformatics comparative analyses of the species for better understanding the mechanisms responsible for the tolerance (Bressan et al. 2013). To check the gene expression at the transcriptional level under salt stress, transcriptomic studies have been made in several species. Transcriptomic data from salt-tolerant Pokkali and sensitive

IR29 rice varieties revealed that in both lines salt stress induced an identical set of genes. The expression of these genes in Pokkali was only for few hours and later switched on different sets of genes. On the other hand, IR29 (salt-susceptible line) continued to express such stress-inducible genes, but the plants eventually died (Kawasaki et al. 2001). Similarly, Gong et al. (2005) discovered a large overlap in the genes that are salt induced, repressed, or unaltered in both *Arabidopsis* and *T. salsuginea*. The salt-tolerant *T. salsuginea* showed responses only at higher salt concentrations than *Arabidopsis*. Thus, some genes appear to be common between the lines though they are distinct in terms of salt stress tolerance.

Bioinformatics helps in the analysis of various "omics" like genomics, transcriptomics, proteomics, and metabolomics (Pérez-Clemente et al. 2013). Genes associated with stress responses and their functional analysis with an insight into stress-responsive network can be identified using genome-wide analysis approach (Vij and Tyagi 2007). Genomes of many plants have been sequenced, and web-based databases of these genomes (e.g., Gramene, GrainGenes, KEGGPLANT, NCBI, Phytozome, PlantGDB, TAIR, VISTA, etc.) are available for analysis. Information also exists on coding, noncoding sequences, promoter sequences, gene families, molecular makers, and genetic variability, among different species. The abovementioned databases along with specified databases like Gene Expression Database Resources (AREX LITE, Genevestigator, NOBLE, PlantCARE, PLACE, TriFLDB, etc.), Proteomics-based Database Resources (Proteomics database, RICE PROTEOME DATABASE, SUBA, etc.), Mutants-based Database Resources (GABI-KAT, OTL, RAPID, SALK, SHIP, SIGnAL, TRIM, etc.), and Transcription factor-based Database Resources (DRTF, GrameneTFDB, Grassius, LegumeTFDB, PlantTFDB, etc.) may also serve as excellent sources for the identification of salt stress-responsive genes. Table 4 shows the tools of bioinformatics used for the analysis of transporters in plants.

1.5.1 Next-Generation Sequencing Technology and Comparative Genomics

Next-generation sequencing (NGS) technologies have helped us to produce whole genome sequences from diverse taxa including salt-tolerant (*Thellungiella salsuginea* and *T. parvula*, relatives of *Arabidopsis*) and susceptible (*Arabidopsis*) species. NGS is a robust tool and will facilitate the development of comparative genomics and pinpoint the genetic differences that exist across species. *Thellungiella* can certainly serve as the genetic basis to understand mechanisms associated with salt tolerance in halophytic species and the genetic differences between salt-tolerant and susceptible lines. The availability of several plant genomic sequences and also stress-related cDNA libraries is highly valuable resources for us for studying comparative genomics and the relative differences in stress-associated genes and pathways. If datasets of orthologs are available, comparative genomics would help to transfer gene annotations from model organisms like *Arabidopsis* and *Thellungiella* to more important crop species like rice, wheat, maize, and others (Ma et al. 2012). In the gene ontology (GO) class, "biological processes" and subcategories "response to abiotic or biotic stimulus" and "developmental processes" genes were enriched in *T. parvula*. Similarly, in the GO class "molecular function" and subcategories "transporter activity" and "receptor binding or activity" were found different between *T. parvula* and *Arabidopsis*. Among the genes annotated, the numbers of cation, ATPase, nucleotide, and sugar transporters have been found higher in *T. parvula* when compared to *Arabidopsis* (Dassanayake et al. 2011). It has been pointed out that differences in gene copy numbers for cation transporters (other than Na^+ and K^+ transporters) reflect the adaptation of *T. parvula* to salt-contaminated soils and also soils that are imbalanced in other ions (Amtmann 2009).

In the genome sequence of *T. parvula*, a number of tandem duplications have been identified unlike in *Arabidopsis*. Such gene duplications suggest a possible basis for the extremophile

Table 4 Tools of bioinformatics used for the analysis of transporters in plants

Sl. No.	Tool name	Tool used for the prediction of	Link	Reference
1	TMHMM	Membrane topology	http://www.cbs.dtu.dk/services/TMHMM/	Krogh et al. (2001)
2	GSDS	Gene structure	http://gsds.cbi.pku.edu.cn	Guo et al. (2007)
3	GENSCAN	Finding genes in genome	http://genes.mit.edu/GENSCAN.html	Burge and Karlin (1998)
4	DIVEIN	Divergence	http://indra.mullins.microbiol.washington.edu/DIVEIN/	Deng et al. (2010)
5	MEGA 5	Reconstruction of phylogeny	http://www.megasoftware.net.	Tamura et al. (2011)
6	PhyML	Phylogenetic tree	http://www.atgc-montpellier.fr/phyml/	Guindon et al. (2010)
7	Clustal W/X	Multiple sequence alignment	http://www.clustal.org/clustal2/	Larkin et al. (2007)
8	cMAP	Comparison of genetic maps	http://www.agron.missouri.edu/cMapDB/cMap.html	Fang et al. (2003)
9	GeneMANIA	Predicts the function of gene	http://www.genemania.org/	Warde-Farley et al. (2010)
10	Cytoscape	Biological network visualization	http://www.cytoscape.org/	Smoot et al. (2011)
11	STRING	Protein–protein interaction networks	http://string-db.org/	Franceschini et al. (2013)
12	Circos	Comparison of genomes	http://mkweb.bcgsc.ca/tableviewer/	Krzywinski et al. (2009)
13	SyMAP	Syntenic relationships	http://www.agcol.arizona.edu/software/symap	Soderlund et al. (2011)
14	Plaza	Comparative genomics	http://bioinformatics.psb.ugent.be/plaza/	Van Bel et al. (2011)
15	WoLF PSORT	Protein subcellular localization	http://wolfpsort.org/	Horton et al. (2007)
16	TargetP	Subcellular location of eukaryotic proteins	http://www.cbs.dtu.dk/services/TargetP/	Emanuelsson et al. (2000)
17	MEME	Motif analysis	http://meme.nbcr.net/	Bailey et al. (2009)

lifestyle of *T. parvula* and as a vehicle for evolution (DeBolt 2010; Dassanayake et al. 2011). Cannon et al. (2004) and Dassanayake et al. (2011) compared the genomes of *Arabidopsis* and *T. salsuginea* and *T. parvula* and found out that *Thellungiella* added or retained duplication genes that are associated with abiotic stress. In contrast, *Arabidopsis* retained genes that are relevant to biotic stress. Clear differences between *T. parvula* and *Arabidopsis* existed in gene copy number. It is proposed that gene copy number variation is an important mechanism of phenotypic differentiation in *T. parvula*. Such a gene copy number variation reflects evolutionary adaptation to the environment (Hastings et al. 2009; Sudmant et al. 2010). *T. parvula* genome showed higher copy numbers of orthologous genes related to stress adaptation. It has been found that gene copy number of orthologous genes like *AVP1*, *HKT1*, *NHX8*, *CBL10*, and *MYB47* is more in *T. parvula* (Dassanayake et al. 2011). Several transcription factors were predicted in soybean, maize, sorghum, and barley using a comparative analysis of known stress-responsive transcription factors identified in model plants like *Arabidopsis* and rice (Mochida et al. 2009; Mochida and Shinozaki 2011). Also, in newly sequenced species, comparative genomics will help to predict and understand the functions of genes. Further, comparative genomics helps us not only to analyze the expression profiles of genes related to salt stress in crop plants where the salt stress

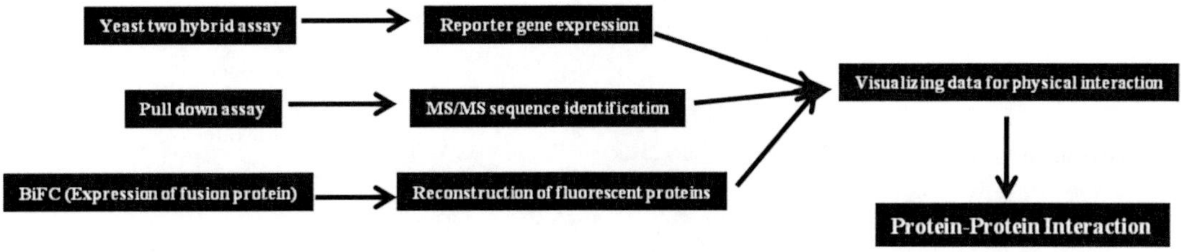

Fig. 5 Three major methods for mapping protein–protein interaction network schemes

mechanisms are less known but also to find out species-specific (model verses crops) differences in stress-responsive genes. Thus, bioinformatics helps us in many ways to understand the salt stress mechanisms.

1.5.2 Target Identification by Proteomics Approach

The analysis of entire genome content for Na^+ and K^+ transporters along with their expression under stress can be achieved by proteomics approach (Agrawal and Rakwal 2006). It gives the information of protein repositories like protein interactions and posttranslational modifications (Peck 2005). UGPase, Cox6b-1, GS root isozyme, α-NAC, putative splicing factor-like protein, and putative ABP novel proteins were identified in rice under salt stress conditions by mass spectrometric analysis (Yan et al. 2005). Similarly, *Arabidopsis* and tobacco leaf apoplast proteome analysis was performed using this technique (Dani et al. 2005). An integrated approach of proteome and metabolome analysis was used to study the differences between rice and wheat coleoptiles under anoxic condition (Shingaki-Wells et al. 2011). Thus, proteomics and bioinformatics are good tools for genome-wide scanning and identification of proteins associated with salt stress.

1.5.3 Protein–Protein Interactions

Protein–protein interaction studies are important to find out the protein networks associated with salt stress. In protein–protein interaction studies, transcription can be activated only when DNA-binding domain of transcriptional activator physically interacts with activation domain leading to protein–protein interactions. Although this process can be studied by using yeast two-hybrid and co-immunoprecipitation techniques (Fields and Song 1989), it is often time consuming and cumbersome (Fig. 5). Hence, several bioinformatics tools are being utilized now to predict such protein networks. For plants whose genome is not yet sequenced, proteins can be identified through cross-species identification approach and by comparing peptides of interest to orthologous peptides of species that were already characterized (Witters et al. 2003). Protein interaction analysis has been used to characterize plant Ca^{2+}-mediated signaling and membrane signaling in *Arabidopsis* (Reddy et al. 2011; Vernoux et al. 2011). A protein–protein interaction map generated in our lab helped us to unravel the interaction of Na^+/H^+ antiporters with other Na^+, K^+, and Ca^{2+} exchangers and also their interactions with radical-induced cell death (RCD) proteins. Such vital information generated through the help of bioinformatics can help us to understand salt stress mechanisms better and also to create crop plants that can cope up well to salt stress conditions. Thus, bioinformatics helps in the identification of key protein networks and their associated genes under multiple stress conditions not only in model plants but also in crop species.

1.5.4 OMICS and Phylogenetic Analysis

Voluminous data are accumulated in relation to individual genes associated with abiotic stress tolerance in literature. But such genes and their functions and the tolerance mechanisms could not be merged into a coherent picture. We need to predict, analyze, interpret, cluster, compare, and functionally annotate many genes associated with salt stress tolerance in crop species. This needs integration of data with biological databases such as the nucleotide databases, protein interaction

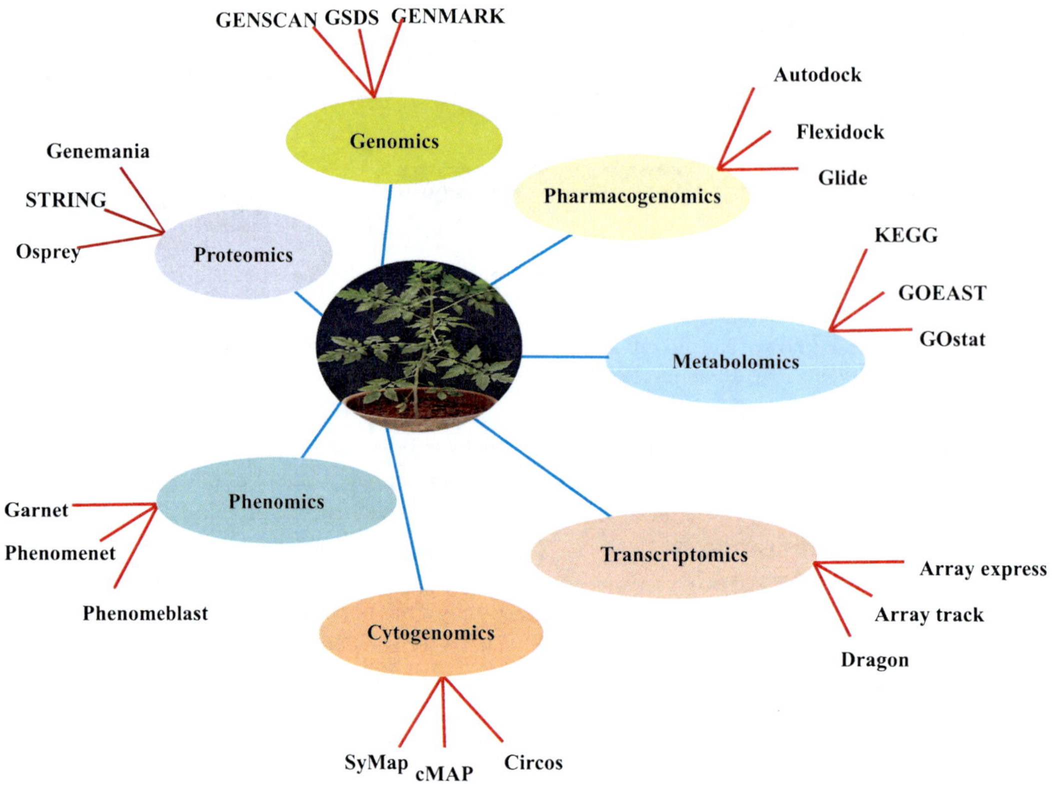

Fig. 6 Different tools used in systems biology of plants

databases, metabolic pathway databases, molecular annotation, as well as data mining with the available literature. We then need to gaze the genome sequences to get meaningful information or mechanisms associated with salinity tolerance. A large number of systems biology tools are available today to study the proteins, their networks, protein cross talks, and their functions associated with salt stress in a simpler way (Fig. 6). To determine the complete picture of plant cellular responses in an efficient way, we need systems analysis with a combination of multiple omics approach (Mochida and Shinozaki 2011). For finding a solution to a complex trait such as salt stress tolerance, we need to combine the information on gene-by-gene interactions with data from protein–protein or transcription factor networks.

With the availability of genomic information in public databases, it is now possible to study the expansion of gene families. It has been recognized that while some genes lost their functions in the course of evolution, others like paralogs must have gained new functions as has been pointed out by Demuth and Hahn (2009). Le et al. (2011) made a genome-wide survey and expression analysis of the plant-specific NAC transcription factor family in *Glycine max* (soybean) during development and dehydration stress. Their phylogenetic analysis using stress-related NAC transcription factors from *Arabidopsis* and rice identified 58 of the 152 *GmNACs* as putative stress-responsive genes. Further, a systematic analysis has identified tissue-specific and/or dehydration-responsive candidate NAC genes for in-depth characterization and development of drought-tolerant transgenic soybean lines. Thus, a bioinformatics approach has provided valuable information by identifying tissue-specific and/or dehydration-responsive NAC genes in soybean.

2 Conclusion

Halophytes thrive in hostile environments and some of them appear to be close relatives of glycophytes such as *Arabidopsis*. But halophytes and glycophytes respond differently for multiple stress conditions, and thus it is highly complex to understand polygenic traits such as salinity tolerance. Halophytes have structural variations such as leaf succulence, salt glands, and salt bladders which are lacking in glycophytic species. Genome sequence analysis in higher plants revealed that nearly 25,000–30,000 genes are common or omnipresent in both halophytes and glycophytes. But species-, or line-, genus, family-, and order-specific differences exist in gene number. The availability of genome sequences from both halophytes and glycophytes can help us in obtaining vital information related to global expression of salt stress-associated genes. Genome sequences of halophytic species such as *Thellungiella* revealed that genes are duplicated, especially with relevance to stress defenses. Further, gene expression alterations or copy number variation is important which is at the basis of salt tolerance (Kvitek et al. 2008; Dassanayake et al. 2011; Waterhouse et al. 2011). This genetic makeup perhaps reflects stress-tolerant way of life in halophytes. The availability of genome sequences of plants that grow in extreme environments and the crop plant genomes and their comparative analysis along with genome-wide association studies can greatly facilitate us both forward and reverse genetic approaches to the study of stress tolerance (Huang et al. 2011). This ultimately may pave the way for crop breeding with improved resistance to salinity. As aforementioned, protein–protein interaction analysis is used to expand our knowledge on molecular mechanisms for abiotic stress responses, signal transduction, and many other biological events (Morsy et al. 2008). The tools of bioinformatics along with the available databases and emerging technologies like next-generation sequencing technologies help us in many ways to unravel the mechanisms and improve the salt stress tolerance in crop plants for sustaining agricultural production under adverse environmental conditions.

References

Agrawal GK, Rakwal R (2006) Rice proteomics, a cornerstone for cereal food crop proteomes. Mass Spectrom Rev 25:1–53. doi:10.1002/mas.20056

Amrutha RN, Jogeswar G, Srilakshmi P, Kavi Kishor PB (2007) Rubidium chloride tolerant callus cultures of rice (*Oryza sativa* L.) accumulate more potassium and cross tolerate to other salts. Plant Cell Rep 26:1647–1662. doi:10.1007/s00299-007-0353-4

Amtmann A (2009) Learning from evolution: *Thellungiella* generates new knowledge on essential and critical components of abiotic stress tolerance in plants. Mol Plant 2:3–13. doi:10.1093/mp/ssn094

An R, Chen QJ, Chai MF, Lu PL, Su Z, Qin ZX, Chen J, Wang XC (2007) AtNHX8, a member of the monovalent cation:proton antiporter-1 family in *Arabidopsis thaliana*, encodes a putative Li^+/H^+ antiporter. Plant J 49:718–728. doi:10.1111/j.1365-313X.2006.02990.x

Apse MP, Aharon GS, Snedden WA, Blumwald E (1999) Salt tolerance conferred by overexpression of a vacuolar Na^+/H^+ antiport in *Arabidopsis*. Science 285:1256–1258. doi:10.1126/science.285.5431.1256

Apse MP, Sottosanto JB, Blumwald E (2003) Vacuolar cation/H^+ exchange, ion homeostasis, and leaf development are altered in a T-DNA insertional mutant of AtNHX1, the Arabidopsis vacuolar Na^+/H^+ antiporter. Plant J 36:229–239. doi:10.1046/j.1365-313X.2003.01871.x

Arazi T, Kaplan B, Formm H (2000) A high affinity Calmodulin-binding site in a tobacco plasma-membrane channel protein coincide with a characteristic element of cyclic nucleotide binding domains. Plant Mol Biol 42:591–601. doi:10.1023/A:1006345302589

Bailey TL, Boden M, Buske FA, Frith M, Grant CE, Clementi L, Ren J, Li WW, Noble WS (2009) MEME SUITE: tools for motif discovery and searching. Nucleic Acids Res 37:W202–W208. doi:10.1093/nar/gkp335

Balague C, Lin B, Alcon C, Flottes G, Malmstrom S, Kohler C, Neuhaus G, Pelletier G, Gaymard F, Roby D (2003) HLM1, an essential signalling component in the hypersensitive response, is a member of the cyclic nucleotide-gated channel ion channel family. Plant Cell 15:365–379. doi:10.1105/tpc.006999

Banuelos MA, Garciadeblas B, Cubero B, Rodriguez-Navarro A (2002) Inventory and functional characterization of the HAK potassium transporters of rice. Plant Physiol 130:784–795. doi:10.1104/pp. 007781

Bassil E, Ohto MA, Esumi T, Tajima H, Zhu Z, Cagnac O, Belmonte M, Peleg Z, Yamaguchi T, Blumwald E (2011) The *Arabidopsis* intracellular Na^+/H^+ antiporters NHX5 and NHX6 are endosome associated and necessary for plant growth and development. Plant Cell 23:224–239. doi:10.1105/tpc.110.079426

Blumwald E (2000) Sodium transport and salt tolerance in plants. Curr Opin Cell Biol 12:431–434. doi:http://dx.doi.org/10.1016/S0955-0674(00)00112-5

Boscari A, Clement M, Volkov V, Golldack D, Hybiak J, Miller AJ, Amtmann A, Fricke W (2009) Potassium channels in barley: cloning, functional characterization and expression analyses in relation to leaf growth and development. Plant Cell Environ 32:1761–1777. doi:10.1111/j.1365-3040.2009.02033.x

Bressan RA, Park HC, Orsini F et al (2013) Biotechnology for mechanisms that counteract salt stress in extremophile species: a genome-based view. Plant Biotechnol Rep 7:27–37. doi:10.1007/s11816-012-0249-9

Burge CB, Karlin S (1998) Finding the genes in genomic DNA. Curr Opin Struct Biol 8:346–354. doi:10.1016/S0959-440X(98)80069-9

Cannon SB, Mitra A, Baumgarten A, Young ND, May G (2004) The roles of segmental and tandem gene duplication in the evolution of large gene families in *Arabidopsis thaliana*. BMC Plant Biol 4:10. doi:10.1186/1471-2229-4-10

Chanroj S, Wang G, Venema K, Zhang MW, Delwiche CF, Sze H (2012) Conserved and diversified gene families of monovalent cation/H^+ antiporters from algae to flowering plants. Front Plant Sci 3:25. doi:10.3389/fpls.2012.00025

Ciarmiello LF, Woodrow P, Fuggi A, Pontecorvo G, Carillo P (2011) Plant genes for abiotic stress. In: Shanker A, Venkateswarlu B (eds) Abiotic stress in plants-mechanisms and adaptations. Venkateswarlu Intech, Rijeka

Cukkemane A, Seifert R, Kaupp UB (2011) Cooperative and uncooperative cyclic-nucleotide-gated ion channels. Trends Biochem Sci 36:55–64. doi:10.1016/j.tibs.2010.07.004

Dahl S, Sylte I, Ravna A (2004) Structures and models of transporter proteins. J Pharmacol Exp Ther 309:853–860. doi:10.1124/jpet.103.059972

Dani V, Simon WJ, Duranti M, Croy RR (2005) Changes in the tobacco leaf apoplast proteome in response to salt stress. Proteomics 5:737–745. doi:10.1002/pmic.200401119

Dassanayake M, Oh D, Haas JS, Hernandez A, Hong H, Ali S, Yun D, Bressan RA, Zhu J, Bohnert HJ, Cheesman JM (2011) The genome of the extremophile crucifer *Thellungiella parvula*. Nat Genet 43:913–918. doi:10.1038/ng.889

DeBolt S (2010) Copy number variation shapes genome diversity in Arabidopsis over immediate family generational scales. Genome Biol Evol 2:441–453. doi:10.1093/gbe/evq033

Demuth JP, Hahn MW (2009) The life and death of gene families. Bioessays 31:29–39. doi:10.1002/bies.080085

Deng W, Maust BS, Nickle DC, Learn GH, Liu Y, Heath L, Kosakovsky Pond SL, Mulins JI (2010) DIVEIN: a web server to analyse phylogenies, sequence divergence, diversity and informative sites. Biotechniques 48:405–408. doi:10.2144/000113370

Elumalai RP, Nagpal P, Reed JW (2002) A mutation in the *Arabidopsis* KT2/KUP2 potassium transporter gene affects shoot cell expansion. Plant Cell 14:119–131. doi:10.1105/tpc.010322

Emanuelsson O, Nielsen H, Brunak S, von Heijne G (2000) Predicting subcellular localization of proteins based on their N-terminal amino acid sequence. J Mol Biol 300:1005–1016. doi:10.1006/jmbi.2000.3903

Fang Z, Polacco M, Chen S, Schroeder S, Hancock D, Sanchez H, Coe E (2003) cMap: the comparative genetic map viewer. Bioinformatics 19:416–417. doi:10.1093/bioinformatics/btg012

Fields S, Song O (1989) A novel genetic system to detect protein–protein interactions. Nature 340:245–246. doi:10.1038/340245a0

Franceschini A, Szklarczyk D, Frankild S, Kuhn M, Simonovic M, Roth A, Lin J, Minguez P, Bork P, von Mering C, Jensen LJ (2013) STRING v9.1: protein-protein interaction networks, with increased coverage and integration. Nucleic Acids Res 41: D808–D815. doi:10.1093/nar/gks1094

Garciadeblas B, Senn ME, Banuelos MA, Rodríguez-Navarro A (2003) Sodium transport and HKT transporters: the rice model. Plant J 34:788–801. doi:10.1046/j.1365-313X.2003.01764.x

Gaxiola RA, Rao R, Sherman A, Grisafi P, Alper SL, Fink GR (1999) The *Arabidopsis thaliana* proton transporters, AtNhx1 and Avp1, can function in cation detoxification in yeast. Proc Natl Acad Sci USA 96:1480–1485. doi:10.1073/pnas.96.4.1480

Gierth M, Maser P (2007) Potassium transporters in plants-involvement in K^+ acquisition, redistribution and homeostasis. FEBS Lett 581:2348–2356. doi:10.1016/j.febslet.2007.03.035

Gierth M, Maser P, Schroeder JI (2005) The potassium transporter AtHAK5 functions in K^+ deprivation-induced high-affinity K^+ uptake and AKT1 K^+ channel contribution to K1 uptake kinetics in *Arabidopsis* roots. Plant Physiol 137:1105–1114. doi:10.1104/pp.104.057216

Gobert A, Park G, Amtmann A, Sanders D, Maathuis FJM (2006) *Arabidopsis thaliana* cyclic nucleotide gated channel 3 forms a non-selective ion transporter involved in germination and cation transport. J Exp Bot 57:791–800. doi:10.1093/jxb/erj064

Gong Q, Li P, Ma S, Indu Rupassara S, Bohnert HJ (2005) Salinity stress adaptation competence in the extremophile *Thellungiella halophila* in comparison with its relative *Arabidopsis thaliana*. Plant J 44:826–839. doi:10.1111/j.1365-313X.2005.02587.x

Grabov A (2007) Plant KT/KUP/HAK potassium transporters: single family-multiple functions. Ann Bot 99:1035–1041. doi:10.1093/aob/mcm066

Guindon S, Dufayard JF, Lefort V, Anisimova M, Hordijk W, Gascuel O (2010) New algorithms and methods to estimate maximum-likelihood phylogenies: assessing the performance of PhyML 3.0. Syst Biol 59:307–321

Guo AY, Zhu QH, Chen X, Luo JC (2007) GSDS: a gene structure display server. Yi Chuan 29:1023–1026. doi:10.1360/yc-007-1023

Guo KM, Babourina O, Christopher DA, Borsics T, Rebgel Z (2008) The cyclic nucleotide-gated channel, AtCNGC10, influences salt tolerance in *Arabidopsis*. Physiol Plant 134:499–507. doi:10.1111/j.1399-3054.2008.01157.x

Harlan JR (1995) The living fields: our agricultural heritage. Cambridge University Press, New York

Hastings PJ, Lupski JR, Rosenberg SM, Ira G (2009) Mechanisms of change in gene copy number. Nat Rev Genet 10:551–564. doi:10.1038/nrg2593

Horie T, Yoshida K, Nakayama H, Yamada K, Oiki S, Shinmyo A (2001) Two types of HKT transporters with different properties of Na^+ and K^+ transport in *Oryza sativa*. Plant J 27:129–138. doi:10.1046/j.1365-313x.2001.01077.x

Horton P, Park KJ, Obayashi T, Fujita N, Harada H, Adams-Collier CJ, Nakai K (2007) WoLF PSORT: protein localization predictor. Nucleic Acids Res 35: W585–W587. doi:10.1093/nar/gkm259

Huang S, Spielmeyer W, Lagudah ES, James RA, Platten JD, Dennis ES, Munns R (2006) A sodium transporter (HKT7) is a candidate for Nax1, a gene for salt tolerance in durum wheat. Plant Physiol 142:1718–1727. doi:10.1104/pp. 106.088864

Huang YS, Horton M, Vilhjálmsson BJ, Seren U, Meng D, Meyer C, Amer MA, Borevitz JO, Bergelson J, Nordborg M (2011) Analysis and visualization of *Arabidopsis thaliana* GWAS using web 2.0 technologies. Database 2011:1–17. doi:10.1093/database/bar014

Jaspers P, Blomster T, Brosché M, Salojarvi J, Ahlfors R, Vainonen JP, Reddy RA, Immink R, Angenent G, Turck F, Overmyer K, Kangasjarvi J (2009) Unequally redundant RCD1 and SRO1 mediate stress and developmental responses and interact with transcription factors. Plant J 60:268–279. doi:10.1111/j.1365-313X.2009.03951.x

Karley AJ, Leigh RA, Sanders D (2000) Differential ion accumulation and ion fluxes in the mesophyll and epidermis of barley. Plant Physiol 122:835–844. doi:10.1104/pp. 122.3.835

Kawasaki S, Borchert C, Deyholos M, Wang H, Brazille S, Kawai K, Galbraith D, Bohnert HJ (2001) Gene expression profiles during the initial phase of salt stress in rice. Plant Cell 13:889–905. doi:http://dx.doi.org/10.1105/tpc.13.4.889

Kim EJ, Kwak JM, Uozumi N, Schroeder JI (1998) AtKUP1: an *Arabidopsis* gene encoding high-affinity potassium transport activity. Plant Cell 10:51–62. doi:10.1105/tpc.10.1.51

Koehler C, Neuhaus G (2000) Characterization of calmodulin binding to cyclic nucleotide-gated ion channels from *Arabidopsis thaliana*. FEBS Lett 471:133–136. doi:10.1016/S0014-5793(00)01383-1

Kohler C, Merkle T, Roby D, Neuhaus G (2001) Developmentally regulated expression of a cyclic nucleotide-gated ion channel from *Arabidopsis* indicates its involvement in programmed cell death. Planta 213:327–332. doi:10.1007/s004250000510

Krogh A, Larsson B, von Heijne G, Sonnhammer EL (2001) Predicting transmembrane protein topology with a hidden Markov model: application to complete genomes. J Mol Biol 305:567–580. doi:10.1006/jmbi.2000.4315

Krzywinski M, Schein J, Birol I, Connors J, Gascoyne R, Horsman D, Jones S, Marra MA (2009) Circos: an information aesthetic for comparative genomics. Genome Res 19:1639–1645. doi:10.1101/gr.092759.109

Kvitek DJ, Will JL, Gasch AP (2008) Variations in stress sensitivity and genomic expression in diverse *Saccharomyces cerevisiae* isolates. PLoS Genet 4:1–18. doi:10.1371/journal.pgen.1000223

Larkin MA, Blackshields G, Brown NP, Chenna R, McGettigan PA, McWilliam H, Valentin F, Wallace IM, Wilm A, Lopez R, Thompson JD, Gibson TJ, Higgins DG (2007) Clustal W and Clustal X version 2.0. Bioinformatics 23:2947–2948. doi:10.1093/bioinformatics/btm404

Lauchli A, James RA, Huang CX, McCully M, Munns R (2008) Cell-specific localization of Na^+ in roots of durum wheat and possible control points for salt exclusion. Plant Cell Environ 31:1565–1574. doi:10.1111/j.1365-3040.2008.01864.x

Le DT, Nishiyama R, Watanabe Y, Mochida K, Yamaguchi-Shinozaki K, Shinozaki K, Phan Tran L (2011) Genome-wide survey and expression analysis of the plant-specific NAC transcription factor family in soybean during development and dehydration stress. DNA Res 18:263–276. doi:10.1093/dnares/dsr015

Lugan R, Niogret MF, Leport L, Guégan JP, Larher FR, Savouré A, Kopka J, Bouchereau A (2010) Metabolome and water homeostasis analysis of *Thellungiella salsuginea* suggests that dehydration tolerance is a key response to osmotic stress in this halophyte. Plant J 64:215–229. doi:10.1111/j.1365-313X.2010.04323.x

Ma Y, Qin F, Phan Tran L (2012) Contribution of genomics to gene discovery in plant abiotic stress responses. Mol Plant 5:1176–1178. doi:10.1093/mp/sss085

Maathuis FJM (2007) Monovalent cation transporters; establishing a link between bioinformatics and physiology. Plant Soil 301:1–15. doi:10.1007/s11104-007-9429-8

Maathuis FJM, Sanders D (1997) Regulation of K^+ absorption in plant root cells by external K^+: interplay of different plasma membrane K^+ transporters. J Exp Bot 48:451–458. doi:10.1093/jxb/48

Mochida K, Shinozaki K (2011) Advances in omics and bioinformatics tools for systems analyses of plant functions. Plant Cell Physiol 52:2017–2038. doi:10.1093/pcp/pcr153

Mochida K, Yoshida T, Sakurai T, Yamaguchi-Shinozaki K, Shinozaki K, Tran LS (2009) In silico analysis of transcription factor repertoire and prediction of stress responsive transcription factors in soybean. DNA Res 16:353–369. doi:10.1093/dnares/dsp023

Moller IS, Gilliham M, Jha D, Mayo GM, Roy SJ, Coates JC, Haseloff J, Tester M (2009) Shoot Na^+ exclusion and increased salinity tolerance engineered by cell type-specific alteration of Na^+ transport in *Arabidopsis*. Plant Cell 21:2163–2178. doi:10.1105/tpc.108.064568

Moore CA, Bowen HC, Scrase-Field S, Knight MR, White PJ (2002) The deposition of suberin lamellae determines the magnitude of cytosolic Ca^{2+} elevations in root endodermal cells subjected to cooling. Plant J 30:457–465. doi:10.1046/j.1365-313X.2002.01306.x

Morsy M, Gouthu S, Orchard S, Thorneycroft D, Harper JF, Mittler R, Cushman JC (2008) Charting plant interactomes: possibilities and challenges. Trends Plant Sci 13:183–191. doi:10.1016/j.tplants.2008.01.006

Munns R (2002) Comparative physiology of salt and water stress. Plant Cell Environ 25:239–250. doi:10.1046/j.0016-8025.2001.00808.x

Munns R, Tester M (2008) Mechanisms of salinity tolerance. Annu Rev Plant Biol 59:651–681. doi:10.1146/annurev.arplant.59.032607.092911

Orlowski J, Grinstein S (1997) Na^+/H^+ exchangers of mammalian cells. J Biol Chem 272:22373–22376. doi:10.1074/jbc.272.36.22373

Peck SC (2005) Update on proteomics in *Arabidopsis*. Where do we go from here? Plant Physiol 138:591–599. doi:10.1105/tpc.107.050989

Pérez-Clemente RM, Vives V, Zandalinas SI, López-Climent MF, Muñoz V, Gómez-Cadenas A (2013) Biotechnological approaches to study plant responses to stress. BioMed Res Int 2013:1–10. doi:10.1155/2013/654120

Peterson CA, Murrmann M, Steudle E (1993) Location of the major barriers to water and ion movement in young roots of *Zea-mays* L. Planta 190:127–136. doi:10.1007/BF00195684

Reddy AS, Ben-Hur A, Day IS (2011) Experimental and computational approaches for the study of calmodulin interactions. Phytochemistry 72:1007–1019. doi:10.1016/j.phytochem.2010.12.022

Rigas S, Debrosses G, Haralampidis K, Vicente-Agullo F, Feldmann KA, Grabov A, Dolan L, Hatzopoulos P (2001) TRH1 encodes a potassium transporter required for tip growth in *Arabidopsis* root hairs. Plant Cell 13:139–151. doi:10.1105/tpc.13.1.139

Rodriguez M, Canales E, Borras-Hidalgo O (2005) Molecular aspects of abiotic stress in plants. Biotecnol Apl 22:1–10

Rodríguez-Rosales MP, Gálvez FJ, Huertas R, Aranda MN, Baghour M, Cagnac O, Venema K (2009) Plant NHX cation/proton antiporters. Plant Signal Behav 4:265–276. doi:10.4161/psb.4.4.7919

Rubio F, Santa-Maria GE, Rodrıguez-Navarro A (2000) Cloning of *Arabidopsis* and barley cDNAs encoding HAK potassium transporters in root and shoot cells. Physiol Plant 109:34–43. doi:10.1034/j.1399-3054.2000.100106.x

Shi H, Ishitani M, Kim C, Zhu JK (2000) The *Arabidopsis thaliana* salt tolerance gene SOS1 encodes a putative Na^+/H^+ antiporter. Proc Natl Acad Sci U S A 97:6896–6901. doi:10.1073/pnas.120170197

Shi H, Quintero FJ, Pardo JM, Zhu JK (2002) The putative plasma membrane Na^+/H^+ antiporter SOS1 controls long distance Na^+ transport in plants. Plant Cell 14:465–477. doi:10.1105/tpc.010371

Shingaki-Wells RN, Huang S, Taylor NL, Carroll AJ, Zhou W, Millar AH (2011) Differential molecular responses of rice and wheat coleoptiles to anoxia reveal novel metabolic adaptations in amino acid metabolism for tissue tolerance. Plant Physiol 156:1706–1724. doi:10.1104/pp. 111.175570

Smoot ME, Ono K, Ruscheinski J, Wang PL, Ideker T (2011) Cytoscape 2.8: new features for data integration and network visualization. Bioinformatics 27:431–432. doi:10.1093/bioinformatics/btq675

Soderlund C, Bomhoff M, Nelson WM (2011) SyMAP v3.4: a turnkey synteny system with application to plant genomes. Nucleic Acids Res 39:1–9. doi:10.1093/nar/gkr123

Sudmant PH, Kitzman JO, Antonacci F, Alkan C, Malig M, Tsalenko A, Sampas N, Bruhn L, Shendure J, Eichler EE (2010) Diversity of human copy number variation and multicopy genes. Science 330:641–646. doi:10.1126/science.1197005

Sunkar R, Kaplan B, Bouche N, Arazi T, Dolev D, Talke IN, Maathuis FJM, Sanders D, Bouchez D, Fromm H (2008) Expression of a truncated tobacco NtCBP4 channel in transgenic plants and disruption of the homologous *Arabidopsis* CNGC1 gene confer Pb^{2+} tolerance. Plant J 24:533–542. doi:10.1111/j.1365-313X.2000.00901.x

Tamura K, Peterson D, Peterson N, Stecher G, Nei M, Kumar S (2011) MEGA5: molecular evolutionary genetics analysis using maximum likelihood, evolutionary distance, and maximum parsimony methods. Mol Biol Evol 28:2731–2739. doi:10.1093/molbev/msr121

Tester M, Davenport RJ (2003) Na^+ tolerance and Na^+ transport in higher plants. Ann Bot 91:503–527. doi:10.1093/aob/mcg058

Tyerman SD, Skerrett IM (1999) Root ion channels and salinity. Sci Hortic 78:175–235. doi:10.1016/S0304-4238(98)00194-0

Uozumi N, Kim EJ, Rubio F, Yamaguchi T, Muto S, Akio T, Bakker EP, Nakamura T, Schroeder JI (2000) The *Arabidopsis* HKT1 gene homolog mediates inward Na^+ currents in *Xenopus laevis* oocytes and Na^+ uptake in *Saccharomyces cerevisiae*. Plant Physiol 122:1249–1260. doi:10.1104/pp. 122.4.1249

Van Bel M, Proost S, Wischnitzki E, Mohavedi S, Scheerlinck C, Van De Peer Y, Vandepoele K (2011) Dissecting plant genomes with the PLAZA comparative genomics platform. Plant Physiol 158:590–600. doi:10.1104/pp. 111.189514

Vernoux T, Brunoud G, Farcot E, Morin V, Van den Daele H, Legrand J et al (2011) The auxin signalling network translates dynamic input into robust patterning at the shoot apex. Mol Syst Biol 7:1–15. doi:10.1038/msb.2011.39

Vij S, Tyagi AK (2007) Emerging trends in the functional genomics of the abiotic stress response in crop plants. Plant Biotechnol J 5:361–380. doi:10.1111/j.1467-7652.2007.00239.x

Waditee R, Hibino T, Tanaka Y, Nakamura T, Incharoensakdi A, Takabe T (2001) Halotolerant cyanobacterium *Aphanothece halophytica* contains an Na^+/H^+ antiporter, homologous to eukaryotic ones, with novel ion specificity affected by C-terminal tail. J Biol Chem 276:36931–36938. doi:10.1074/jbc. M103650200

Ward JM, Maser P, Schroeder JI (2009) Plant ion channels: gene families, physiology, and functional genomics analyses. Annu Rev Physiol 71:59–82. doi:10.1146/annurev.physiol.010908.163204

Warde-Farley D, Donaldson SL, Comes O, Zuberi K, Badrawi R, Chao P, Franz M, Grouios C, Kazi F, Lopes CT, Maitland A, Mostafavi S, Montojo J, Shao Q, Wright G, Bader GD, Morris Q (2010) The GeneMANIA prediction server: biological network integration for gene prioritization and predicting gene function. Nucleic Acids Res 38:W214–W220. doi:10.1093/nar/gkq537

Waterhouse RM, Zdobnov EM, Kriventseva EV (2011) Correlating traits of gene retention, sequence divergence, duplicability and essentiality in vertebrates, arthropods and fungi. Genome Biol Evol 3:75–86. doi:10.1093/gbe/evq083

Wegner LH, Raschke K (1994) Ion channels in the xylem parenchyma of barley roots. Plant Physiol 105:799–813. doi:10.1104/pp. 105.3.799

White PJ (1999) The molecular mechanism of sodium influx to root cells. Trends Plant Sci 4:245–246. doi:10.1016/S1360-1385(99)01435-1

White PJ (2001) The pathways of calcium movement to the xylem. J Exp Bot 52:891–899. doi:10.1093/jexbot/52.358.891

Witters E, Laukens K, Deckers P, Van Dongen W, Esmans E, Van Onckelen H (2003) Fast liquid chromatography coupled to electrospray tandem mass spectrometry peptide sequencing for cross-species protein identification. Rapid Commun Mass Spectrom 17:2188–2194. doi:10.1002/rcm.1173

Yan S, Tang Z, Su W, Sun W (2005) Proteomic analysis of salt stress-responsive proteins in rice root. Proteomics 5:235–244. doi:10.1002/pmic.200400853

Yang Z, Gao Q, Sun C, Li W, Gu S, Xu C (2009) Molecular evolution and functional divergence of HAK potassium transporter gene family in rice (*Oryza sativa* L.). J Genet Genomics 36:161–172. doi:10.1016/S1673-8527(08)60103-4

Yokoi S, Quintero FJ, Cubero B, Ruiz MT, Bressan RA, Hasegawa PM, Pardo JM (2002) Differential expression and function of *Arabidopsis thaliana* NHX Na^+/H^+ antiporters in the salt stress response. Plant J 30:529–539. doi:10.1046/j.1365-313X.2002.01309.x

Yoshioka K, Moeder W, Kang HG, Kachroo P, Masmoudi K, Berkowitz G, Klessig DF (2006) The chimeric *Arabidopsis* cyclic nucleotide-gated ion channel11/12 activates multiple pathogen resistance responses. Plant Cell 18:747–763. doi:10.1105/tpc.105.038786

Zhorov BS, Tikhonov DB (2004) Potassium, sodium, calcium and glutamate-gated channels: pore architecture and ligand action. J Neurochem 88:782–799. doi:10.1111/j.1471-4159.2004.02261.x

Peanut Bioinformatics: Tools and Applications for Developing More Effective Immunotherapies for Peanut Allergy and Improving Food Safety

Venkatesh Kandula, Virginia A. Gottschalk, Ramesh Katam, and Roja Rani Anupalli

Abstract

Advanced tools of bioinformatics have been employed to assess the features critically required for allergenicity and cross-reactivity. A tremendous accumulation of data on plant proteins in recent years has made it possible to classify allergens in different protein families, with most food allergens grouped into four protein families. These families can be grouped together into superfamilies by comparing sequences and related structures. This information makes it possible to identify a wide range of related proteins that may result in the development of multiple food allergies that initiate the development of cross-reactive antibodies in susceptible individuals. Since peanut allergies are responsible for most episodes of food-induced anaphylaxis, a detailed immunological and molecular characterization of these allergenic components is essential to develop suitable immunotherapies. This would also allow us to screen transgenic plants for the possible development of allergens similar to those allergenic components in peanuts. Homology modeling in combination with residue-wise solvent accessibility of monomers and biological assemblies of allergens certainly gives valuable information about antigenic determinants on protein allergens. Through this review, we discuss the applications of bioinformatics tools toward the mitigation of peanut allergy.

Keywords

Peanut • Allergens • Allergen database • Ara h and transgene allergenicity

V. Kandula • R.R. Anupalli
Department of Genetics and Biotechnology, Osmania University, Hyderabad 500007, AP, India

V.A. Gottschalk • R. Katam (✉)
Department of Biological Sciences, College of Science and Technology, Florida A&M University, Tallahassee, FL 32307, USA
e-mail: ramesh.katam@gmail.com

K.K. P.B. et al. (eds.), *Agricultural Bioinformatics*,
DOI 10.1007/978-81-322-1880-7_7, © Springer India 2014

Abbreviations

BLAST	Basic Local Alignment Search Tool
DNA	Deoxyribonucleic acid
FARRP	Food Allergy Research and Resource Program
FASTA	Fast alignment
GM crops	Genetically modified crops
IgE	Immunoglobulin E
IUIS	International Union of Immunological Societies
kDa	Kilodalton
MW	Molecular weight
ORF	Open reading frames
PR	Pathogenesis-related genes
SDAP	Structural Database of Allergenic Proteins
URL	Uniform resource locator

1 Introduction

Food allergy can be described as adverse reactions to certain foods because of immunological mechanisms (Matsuda and Nakamura 1993), and it seems to be increasing when it comes to legumes. Peanut allergy affects approximately 1 % of the population (Burks 2003; Sampson 2004) and up to 8 % of children with IgE mediation (Ortolani et al. 2001; Woods et al. 2002; Mills et al. 2004).

The vast majority of anaphylactic reactions from adults and children can be traced back to the seed storage proteins, especially those from peanuts (Sicherer et al. 2003). There is no cure for peanut allergy, and management is therefore focused on allergen avoidance and prompt treatment (Simons 2008). Peanut allergy is one of the major prevalent food allergies triggered by several proteins known as allergens. About 10–47 % of food-induced anaphylactic reactions and more than 50 % of food allergy fatalities are accounted for peanut allergy (Bock et al. 2001).

The computational and bioinformatics tools in relevance to the study of allergy that helps to probe databases and data repositories belong to different categories: sequence analysis and comparison for classification, characterization of genes and proteins of confirmed and putative allergens, and prediction of function and structural analysis of genes and proteins (Fig. 1) (Midoro-Horiuti et al. 2001; Yang et al. 2000).

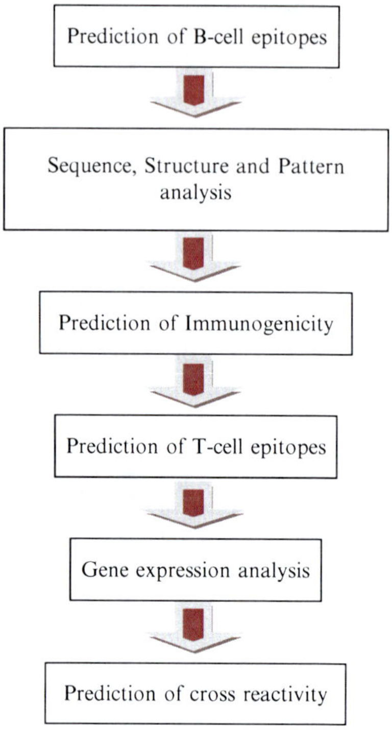

Fig. 1 Different steps of allergic cascade with the scope of bioinformatics tools applications (Brusic et al. 2003)

DNA sequence analysis focuses largely on finding conserved patterns of sequence, prediction of protein secondary structure, 3D structure, and function of whole molecules or their domains (Table 1).

The three major peanut allergenic proteins responsible for the allergenic response in susceptible individuals are Ara h 1 (vicilin), assembled in di- and trimeric complexes (Buschmann et al. 1996; Shin et al. 1998), two of its genes encoding 626 and 614 amino acids (Burks et al. 1995a, b; Wichers et al. 2004); Ara h 2 (2S albumin conglutin), two isoforms with masses of ca. 16 and 17 kDa (Chatel et al. 2003; Hales et al. 2004); and Ara h 3/Arah 4 (glycinin) – consisting of an acidic (40–45 kDa) and a basic (25 kDa) subunit (Piersma et al. 2005; Koppelman et al. 2003; Restani et al. 2005) linked by an intermolecular disulfide bridge (Table 2).

Table 1 Bioinformatics tools used in the identification of structural motifs and 3D structure analysis

Bioinformatics tools	Purpose	URL	References
Predict protein	Secondary structure prediction of a selection of pollen, fruit, and vegetable allergens. Identification of similar structural elements like P-loop region as a common domain of pollen and related food allergens	http://www.predictprotein.org/#	Cuff and Barton (2000) and Scheurer et al. (1999)
Jpred2	Predicts more than 75 % of amino acids that form α-helices, β-sheets, or coils in allergen groups	http://www.compbio.dundee.ac.uk/www-jpred/	Cuff and Barton (2000)
PAIRCOIL and COILS	Prediction of coil structure	http://groups.csail.mit.edu/cb/paircoil/cgi-bin/ http://embnet.vital-it.ch/software/COILS_form.html	Berger et al. (1995) and Lupas (1996)

Table 2 Peanut allergens identified until January 2013

Allergen	Protein family	MW (kDa)	Isoallergens	References
Ara h 1	Cupin: 7S vicilin-like globulins	64.0	Ara h 1.0101	Burks et al. (1995a)
Ara h 2	Prolamin: 2S albumin, conglutin	17.0	Ara h 2.0101, Ara h 2.0201	Kleber-Janke et al. (1999)
Ara h 3	Cupin: legumin-like (11S globin, glycinin)	60.0	Ara h 3.0101, Ara h 3.0201	Rabjohn et al. (1999)
Ara h 4	Cupin: legumin-like (11S globin, glycinin)	37.0	Ara h 4.0101	Kleber-Janke et al. (1999)
Ara h 5	Profilin	15.0	Ara h 5.0101	Kleber-Janke et al. (1999)
Ara h 6	Prolamin: 2S albumin, conglutin	15.0	Ara h 6.0101	Kleber-Janke et al. (1999)
Ara h7	Prolamin: 2S albumin, conglutin	15.0	Ara h 7.0101, Ara h 7.0201, Ara h 7.0202	Kleber-Janke et al. (1999)
Ara h 8	Pathogenesis-related protein, PR-10	17.0	Ara h 8.0101, Ara h 8.0101	Mittag et al. (2004)
Ara h 9	Prolamin: nonspecific lipid-transfer protein	9.8	Ara h 9.0101, Ara h 9.0201	Krause et al. (2009)
Ara h10	Oleosin	16.0	Ara h 10.0101, Ara h 10.0102	Pons et al. (1998, 2002)
Ara h 11	Oleosin	14.0	Ara h 11.0101	IUIS/Allergome: UniProtKB/TrEMB: Q45W87
Ara h 12	Defensin (PR-10)	8.0	Ara h 12.0101	Allergome: GenBank Acc: EY396089 EST name: 5J9
Ara h 13	Defensin (PR-10)	8.0	Ara h 13.0101	Allergome: GenBank Acc: EY396019 EST name: 1N15

Isoallergens are variants which may be recognized differently by patient IgE which either increases or decreases the severity of the allergenic response (Christensen, et al. 2010). Ara h 3 and Ara h 4 are 91 % homologous and are regarded as isoforms of each other and are considered to be the same allergen (Allergome, Boldt et al. 2005 and Koppelman et al. 2003). Ara h 8 is homologous to *Betula verrucosa* (birch) allergen Bet v 1 (Mittag et al. 2004). With regard to Ara h 11, 7 patients with peanut ingestion-related symptoms recognized the 14 kDa band (IUIS). Ara h 12 and Ara h 13 were isolated by analysis of expressed sequence tags (GenBank). Databases used were Allergome (http://www.allergome.org) and IUIS (http://www.allergen.org/index.php)

Abbreviations: *MW* molecular weight, *kDa* kilodalton, *IUIS* International Union of Immunological Societies

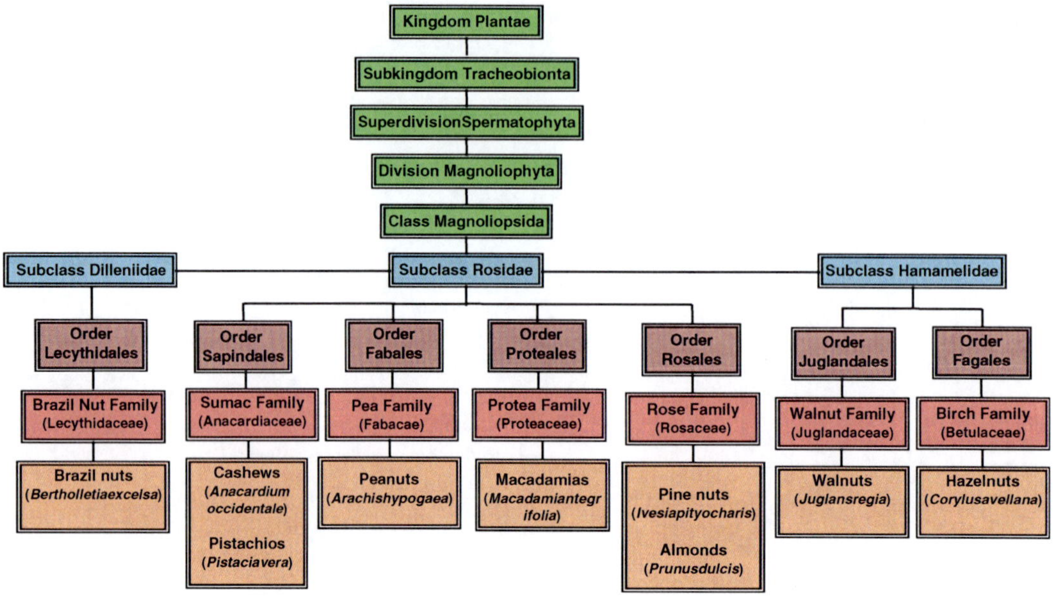

Fig. 2 Taxonomic classification of tree nuts and peanuts. Although they may belong to different families, peanuts and a number of tree nuts arise from the same class, Magnoliopsida, and for this reason may contain homologous proteins responsible for allergenic cross-reactivity. Therefore, it is not surprising that many people suffering from peanut allergy are also allergic to a number of tree nuts as well

The allergen Ara h 1 is the most abundant (20 % of the total protein) followed by Ara h 2 (~10 %) and Ara h 3/4 (Burks et al. 1998; Koppelman et al. 2001).

To elucidate molecular features related to allergenicity and cross-reactivity and for designing hypoallergenic derivatives, allergen structures are necessary. Theoretical methods and computational techniques are used in molecular modeling to mimic the behavior of molecules (Sonika and Anil 2013).

Furthermore, many of the people with allergies to peanuts also have allergies to similar proteins in other tree nuts (Schein et al. 2007). This is not surprising considering the degree of relatedness between these plants as seen in Fig. 2 (Oezguen et al. 2008).

2 ALLERDB Database and Integrated Bioinformatics Tools

Bioinformatics has proven to play a key role in the management of allergen data, analysis of the properties of allergens, protein allergenicity, and cross-reactivity (Aalberse 2000; Hileman et al. 2002). Sequences of allergenic proteins are spread out across a variety of databases, including general-purpose protein databases and specialist allergen databases. Each of these databases has a different focus, and they contain different subsets of allergen sequences (Table 3).

The most authoritative data source in relation to content is the IUIS Nomenclature Subcommittee Allergen Database (IUIS allergens), which defines allergens according to a set of well-defined criteria. Bioinformatics search and software tools have been developed to define the characteristics which distinguish allergens from nonallergens (Zhang et al. 2006; Power et al. 2013). This kind of categorization helps to organize allergenic proteins so that they can be better with further research in facilitating specific immunotherapies that could possibly be used to help allergic individuals to develop immunological tolerance to peanuts (Stadler and Stadler 2003; Hoffman et al. 2005).

Identification of allergic potential is a key issue whenever new proteins are brought in contact with humans, either through food or other modes of exposure. The entire ALLERDB

Table 3 List of different websites with information about allergen database and allergenicity prediction servers (Schein et al. 2007)

Website name	URL
Nomenclature and general information database	
AllFam	http://www.meduniwien.ac.at/allergens/allfam/
All allergy	http://allallergy.net/
IUIS (International Union of Immunological Societies)	http://www.allergen.org
Allergome	http://www.allergome.org
CSL (Central Science Laboratory, UK)	http://www.csl.gov.uk/allergen/index.htm
National Center for Food Safety and Technology	Information on Allergens: ftp://ftp.embl-heidelberg.de/ftp.ebi.ac.uk/pub/delete_databases/swissprot/release/allergen.txt
Protall	http://www.ifr.ac.uk/Protall/
Inform all	http://foodallergens.ifr.ac.uk/
Cross-referenced databases with tools to compare sequences	
FARRP	Allergen Online at Nebraska: http://www.allergenonline.org/
ADFS (Allergen Database for Food Safety)	http://allergen.nihs.go.jp/ADFS/
The Allergen Database	http://allergen.csl.gov.uk//index.htm
Cross-referenced database with tools for prediction of allergenicity	
SDAP (Structural Database of Allergenic Proteins) and SDAP-Food	http://fermi.utmb.edu/SDAP
ALLERDB	Zhang et al. (2007)
Web servers for allergenicity prediction	
WebAllergen	Riaz et al. (2005)
AllerTool	Zhang et al. (2007)
Allermatch	www.allermatch.org/
Algpred	http://www.imtech.res.in/raghava/algpred/

database has been created and maintained using Bioware data warehousing system (Koh et al. 2004). To assist allergen data analysis and retrieval, ALLERDB has five integrated tools, BLAST, Search, XRblast, AllerPredict, and XRgraph.

ALLERDB (Zhang et al. 2007) is a database of allergenic proteins with integrated bioinformatics tools for the assessment of allergenicity and allergic cross-reactivity (Schein et al. 2007). It facilitates the assessment of allergenic potential of anonymous proteins that might be introduced in food chain or potential close contact with allergic individuals. This database is, along with SDAP, a representative of a new brand of allergen databases which include a rich set of tools, including those for sequence comparison, pattern identification, and visualization of results (Ivanciuc et al. 2003, 2011). The progress in clinical allergology, genomics, and proteomics generates huge amounts of data for the study of allergens. The molecular basis for allergic cross-reactivity is in structural similarity among proteins from diverse sources, and the clinical relevance appears to be influenced by several factors including the IgE response against the allergen, route of exposure, and the sequence and structure of the allergen (Ferreira et al. 2004). The main purpose of ALLERDB is the support of molecular analysis of allergenicity, assessment of allergic responses, and allergic cross-reactivity of clinically relevant protein allergens. This coded program proved to be accurate, but it still needs more programs to be created to help reduce the allergenicity of human foods.

3 Molecular Modeling of Peanut Allergen

Molecular modeling of Ara h 1, Len c 1, and Pis s 1 was carried out on a Silicon Graphics O2 R10000 workstation, using the programs Insight II, Homology, and Discover 3 (Accelrys, San

Diego CA, USA) (Viquez et al. 2003). It is an X-ray method used to look at the percentage of identity and homology of the vicilin monomers of peanut, lentil, and pea share with β-conglycinin allowed to build accurate 3D models for the vicilin monomers (Barre et al. 2008). We looked at the homology-based molecular modeling to characterize the structural features responsible for the epitopic community among the major vicilin allergens from peanuts (Ara h 1), lentils (Len c 1), and peas (Pis s 1). The B-cell epitopes share sequence similarity and conformation, suggesting they could be recognized by the same IgE antibodies. The B-cell epitopes most probably account for the IgE-binding cross-reactivity toward different seed storage vicilins and readily participate in the allergic response of sensitized individuals to multiple legume seeds (Mills et al. 2002). In addition, the resistance of the homotrimeric arrangement of the cupin monomers to both thermal denaturation and proteolysis preserved the most binding sites. This greatly increased the propensity of allergenicity.

The remaining B-cell epitopes characterized in the C-terminal region of Ara h 1 share a high degree of homology and most probably correspond to B-cell epitopes common to the seed storage protein family. In this respect, most of these homologous regions were predicted as being exposed B-cell epitopes along the amino acid sequences of Len c 1 and Pis s 1 using the classical hydrophilicity/ exposure/ flexibility-based predictive methods. Ara h 1 and other vicilin allergens were tentatively assigned as homotrimers displaying an arrangement of the cupin motifs similar to that found in β-conglycinin, phaseolin, or canavalin (Burks et al. 1995a, b). In vicilin allergens most of the amino acid residues are highly conserved at the interfaces of the cupin motifs of the β-conglycinin homotrimer (Singh and Raghava 2001). The cross-reactivity implication has proven to be a suppressant to legume seed allergy. Research today still resumes, trying to find better methods on shutting down allergenicity completely.

4 Food Allergenicity Associated with Hypothetical Open Reading Frames

Genetically modified crops have successfully been grown and consumed across the world from the past 16 years. Water scarcity, energy resources, and food have become a matter of concern with the growing population (Bruinsma 2009; Gregory and George 2011). To avoid confusion with expressed or predicted protein-coding sequence, open reading frames (ORFs) are referred to as novel polypeptides, even though at times the novel polypeptide sequence may overlap with an expressed or protein-coding sequence (Vrtala 2008; Silvanovich et al. 2009). Translation of stop-to-stop frames, alignments to the allergen online database using FASTA, and search for eight residue identical matches were used (Silvanovich et al. 2009). The novel polypeptide sequences were created by scanning all six DNA reading frames (starting at base 1, 2, or 3 of each DNA sequence and translating each non-overlapping nucleotide triplet into an amino acid using the standard genetic codon table and then reversing the sequence and repeating the process).

5 Assessment of the Potential Allergenicity of Transgenes

GM foods are developed using recombinant DNA technology by the transfer of foreign genes from one organism to another for human or animal consumption. The expression of proteins corresponding to foreign gene may cause allergenic reactions, and this potential has been seized upon by critics and regulatory bodies of GM crops. Gene suppression technology as an alternative GM approach can reduce or eliminate allergens. In silico approach to assess the allergenic potential of transgenes is gaining importance nowadays (Mishra et al. 2012). Most of the novel gene products showing cross-reactivity with allergens have demonstrable homology. To evaluate the allergenicity of the transgene, there is a need to perform an initial homology search of the transgene against the database and final validation with relevant experiments.

6 Results

Gene	Length of the protein sequence (amino acids)	Full-length FASTA alignment in three databases (% sequence identity)	Sequence identity with
Manganese superoxide dismutase	228	85.09	*Hevea brasiliensis*
		86.207	
		86.20	
Rice chitinase	175	68.57 (Hev b 1.0102)	Known allergens
		63.102 (Pers a 1)	
		63.40 (chitinase Ib)	
Glycine betaine aldehyde dehydrogenase	490	40.00	Known allergens
		41.02	
		43.70	
Alfalfa beta-1, 3-glucanase	507	34.32	Known allergens
		38.96	
		40.86	
		40.90	
Wheat beta-1, 3-glucanase	399	33.83	Known allergens
		36.33	
		41.20	
Osmotin gene	246	91.87 (Lyc e NP24)	Known allergens
		88.21 (Cap a 1w)	
		66.50 (Act d 2)	

In many cases inhalation or ingestion of food allergens has resulted in severe systemic reactions (Helm and Burks 2000). In Asia there is a gradual increase in food allergy, particularly in North America and Europe; peanut allergy is exhibiting a rising trend (Sharma et al. 2011; Singh et al. 2006; Chandra 2003; Grundy et al. 2002; Sicherer et al. 2003).

Acknowledgment Venkatesh K acknowledges CSIR, New Delhi, for the financial assistance.

References

Aalberse RC (2000) Structural biology of allergens. J Allergy Clin Immunol 106:228–238

Barre A, Sordet C, Culerrier R, Rance F, Didier A, Rouge P (2008) Vicilin allergens of peanut and tree nuts (walnut, hazelnut and cashew nut) share structurally related IgE-binding epitopes. Mol Immunol 45(5):1231–1240

Berger B, Wilson DB, Wolf E, Tonchev T, Milla M, Kim PS (1995) Predicting coiled coils by use of pairwise residue correlations. Proc Natl Acad Sci U S A 92 (18):8259–8263

Bock SA, Muñoz-Furlong A, Sampson HA (2001) Fatalities due to anaphylactic reactions to foods. J Allergy Clin Immunol 107(1):191–193

Boldt A, Fortunato D, Conti A, Petersen A, Ballmer-Weber B, Lepp U, Reese G, Becker WM (2005) Analysis of the composition of an immunoglobulin E reactive high molecular weight protein complex of peanut extract containing Ara h1 and Ara h3/4. Proteomics 5(3):675–686

Bruinsma J (2009) The resource outlook to 2050: by how much do land, water, and crop yields need to increase by 2050? FAO expert meeting on 'How to feed the world in 2050'. FAO, Rome, pp 24–26

Brusic V, Petrovsky N, Gendel SM, Millot M, Gigonzac O, Stelman SJ (2003) Computational tools for the study of allergens. Allergy 58(11):1083–1092

Burks W (2003) Peanut allergy: a growing phenomenon. J Clin Invest 111(7):950–952. doi:10.1172/JCI18233

Burks AW, Cockrell G, Stanley JS, Helm RM, Bannon GA (1995a) Recombinant peanut allergen Ara h I expression and IgE binding in patients with peanut hypersensitivity. J Clin Invest 96(4):1715–1721

Burks AW, Cockrell G, Stanley JS, Helm RM, Bannon GA (1995b) Recombinant peanut allergen Ara h I expression and IgE binding in patients with peanut hypersensitivity. J Clin Invest 96(4):1715–1721. doi:10.1172/JCI118216

Burks AW, Sampson HA, Bannon GA (1998) Peanut allergens. Allergy 53(8):725–730. doi:10.1111/j.1398-9995.1998.tb03967.x

Buschmann L, Petersen A, Schlaak M, Becker WM (1996) Reinvestigation of the major peanut allergen Ara h 1 on molecular level. Monogr Allergy 32:92–98

Chandra RK (2003) Food hypersensitivity and allergic disease: a new threat in India. Indian Pediatr 40 (2):99–101

Chatel JM, Bernard H, Orson FM (2003) Isolation and characterization of two complete Ara h 2 isoforms cDNA. Int Arch Allergy Immunol 131(1):14–18

Christensen LH, Riise E, Bang L, Zhang C, Lund K (2010) Isoallergen variations contribute to the overall complexity of effector cell degranulation: effect mediated through differentiated IgE affinity. J Immunol 184(9):4966–4972. doi:10.4049/jimmunol.0904038

Cuff JA, Barton GJ (2000) Application of multiple sequence alignment profiles to improve protein secondary structure prediction. Proteins 40(3):502–511

Ferreira F, Hawranek T, Gruber P, Wopfner N, Mari A (2004) Allergic cross-reactivity: from gene to the clinic. Allergy 59(3):243–267. doi:10.1046/j.1398-9995.2003.00407.x

Gregory PJ, George TS (2011) Feeding nine billion: the challenge to sustainable crop production. J Exp Bot 62(15):5233–5239

Grundy J, Matthews S, Bateman B, Dean T, Arshad SH (2002) Rising prevalence of allergy to peanut in children: data from 2 sequential cohorts. J Allergy Clin Immunol 110(5):784–789

Hales BJ, Bosco A, Mills KL, Hazell LA, Loh R, Holt PG, Thomas WR (2004) Isoforms of the major peanut allergen ara h 2: IgE binding in children with peanut allergy. Int Arch Allergy Immunol 135(2):101–107

Helm RM, Burks AW (2000) Mechanisms of food allergy. Curr Opin Immunol 12(6):647–653

Hileman RE, Silvanovich A, Goodman RE, Rice EA, Holleschak G, Astwood JD, Hefle SL (2002) Bioinformatic methods for allergenicity assessment using a comprehensive allergen database. Int Arch Allergy Immunol 128(4):280–291. doi:10.1159/000063861

Hoffman M, Arnoldi C, Chuang I (2005) The clinical bioinformatics ontology: a curated semantic network utilizing RefSeq information. Pac Symp Biocomput:139–150

Ivanciuc O, Mathura V, Midoro-Horiuti T, Braun W, Goldblum RM, Schein CH (2003) Detecting potential IgE-reactive sites on food proteins using a sequence and structure database, SDAP-food. J Agric Food Chem 51(16):4830–4837

Ivanciuc O, Gendel SM, Power TD, Schein CH, Braun W (2011) AllerML: markup language for allergens. Regul Toxicol Pharmacol 60(1):151–160. doi:10.1016/j.yrtph.2011.03.006

Kleber-Janke T, Crameri R, Appenzeller U, Schlaak M, Becker WM (1999) Selective cloning of peanut allergens, including profilin and 2S albumins, by phage display technology. Int Arch Allergy Immunol 119(4):265–274

Koh LYJ, Brusic V, Krishnan SPT, Seah SH, Tan PTJ, Khan AM, Li ML (2004) BioWare: a framework for bioinformatics data retrieval, annotation, and publishing. In: Proceedings of the symposium model analysis and simulation of computer and telecommunication systems. The University of Sheffield, Sheffield, 25–29 July 2004

Koppelman SJ, Vlooswijk RA, Knippels LM, Hessing M, Knol EF, van Reijsen FC, Bruijnzeel-Koomen CA (2001) Quantification of major peanut allergens Ara h 1 and Ara h 2 in the peanut varieties Runner, Spanish, Virginia and Valencia, bred in different parts of the world. Allergy 56(2):132–137

Koppelman SJ, Knol EF, Vlooswijk RA, Wensing M, Knulst AC, Hefle SL, Gruppen H, Piersma S (2003) Peanut allergens Ara h 3: isolation from peanut and biochemical characterisation. Allergy 58 (11):1144–1151

Krause S, Reese G, Randow S, Zennaro D, Quaratino D, Palazzo P, Ciardiello MA, Petersen A, Becker WM, Mari A (2009) Lipid transfer protein (Ara h 9) as a new peanut allergen relevant for a Mediterranean allergic population. J Allergy Clin Immunol 124(4):771–778

Lupas A (1996) Prediction and the analysis of coiled-coil structures. Methods Enzymol 266:513–525

Matsuda T, Nakamura R (1993) Molecular structure and immunological properties of food allergens. Trends Food Sci Technol 4(9):289–293. doi:10.1016/0924-2244(93)90072-I

Midoro-Horiuti T, Goldblum RM, Brooks EG (2001) Identification of mutations in the genes for the pollen allergens of eastern red cedar (*Juniperus virginiana*). Clin Exp Allergy 31(5):771–778

Mills EN, Jenkins J, Marigheto N, Belton PS, Gunning AP, Morris VJ (2002) Allergens of the cupin superfamily. Biochem Soc Trans 30(Pt 6):925–929

Mills EN, Jenkins JA, Alcocer MJ, Shewry PR (2004) Structural, biological, and evolutionary relationships of plant food allergens sensitizing via the gastrointestinal tract. Crit Rev Food Sci Nutr 44(5):379–407

Mishra A, Gaur S, Singh BP, Arora N (2012) In silico assessment of the potential allergenicity of transgenes used for the development of GM food crops. Food Chem Toxicol 50(5):1334–1339

Mittag D, Akkerdaas J, Ballmer-Weber BK, Vogel L, Wensing M, Becker WM, Koppelman SJ, Knulst AC, Helbling A, Hefle SL, Van Ree R, Vieths S (2004) Ara h 8, a Bet v 1-homologous allergen from peanut, is a major allergen in patients with combined birch pollen and peanut allergy. J Allergy Clin Immunol 114(6):1410–1417

Oezguen N, Zhou B, Negi SS, Ivanciuc O, Schein CH, Labesse G, Braun W (2008) Comprehensive 3D-modeling of allergenic proteins and amino acid composition of potential conformational IgE epitopes. Mol Immunol 45(14):3740–3747. doi:10.1016/j.molimm.2008.05.026

Ortolani C, Ispano M, Scibilia J, Pastorello EA (2001) Introducing chemists to food allergy. Allergy 56(67):5–8

Piersma SR, Gaspari M, Hefle SL, Koppelman SJ (2005) Proteolytic processing of the peanut allergen Ara h 3. Mol Nutr Food Res 49(8):744–755

Pons L, Olszewski A, Gueant JL (1998) Characterization of the oligomeric behavior of a 16.5 kDa peanut oleosin by chromatography and electrophoresis of the iodinated form. J Chromatogr B Biomed Sci Appl 706(1):131–140

Pons L, Chery C, Romano A, Namour F, Artesan MC, Guéant JL (2002) The 18 kDa peanut oleosin is a candidate allergen for IgE-mediated reactions to peanuts. Allergy 57(Suppl 72):88–93

Power TD, Ivanciuc O, Schein CH, Braun W (2013) Assessment of 3D models for allergen research. Proteins 81(4):545–554. doi:10.1002/prot.24239

Rabjohn P, Helm EM, Stanle JS, West CM, Sampson HA, Burks AW, Bannon GA (1999) Molecular cloning and epitope analysis of the peanut allergen Ara h 3. J Clin Invest 103(4):535–542

Restani P, Ballabio C, Corsini E, Fiocchi A, Isoardi P, Magni C, Poiesi C, Terracciano L, Duranti M (2005) Identification of the basic subunit of Ara h 3 as the major allergen in a group of children allergic to peanuts. Ann Allergy Asthma Immunol 94(2):262–266

Riaz T, Hor HL, Krishnan A, Tang F, Li KB (2005) WebAllergen: a web server for predicting allergenic proteins. Bioinformatics. 21(10):2570–2571

Sampson HA (2004) Update on food allergy. J Allergy Clin Immunol 113(5):805–819

Schein CH, Ivanciuc O, Braun W (2007) Bioinformatics approaches to classifying allergens and predicting cross-reactivity. Immunol Allergy Clin North Am 27(1):1–27

Scheurer S, Son DY, Boehm M, Karamloo F, Franke S, Hoffmann A, Haustein D, Vieths S (1999) Cross-reactivity and epitope analysis of Pru a 1, the major cherry allergen. Mol Immunol 36(3):155–167

Sharma P, Singh AK, Singh BP, Gaur SN, Arora N (2011) Allergenicity assessment of osmotin, a pathogenesis-related protein, used for transgenic crops. J Agric Food Chem 59(18):9990–9995. doi:10.1021/jf202265d

Shin DS, Compadre CM, Maleki SJ, Kopper RA, Sampson H, Huang SK, Burks AW, Bannon GA (1998) Biochemical and structural analysis of the IgE binding sites on ara h1, an abundant and highly allergenic peanut protein. J Biol Chem 273 (22):13753–13759

Sicherer SH, Munoz-Furlong A, Sampson HA (2003) Prevalence of peanut and tree nut allergy in the United States determined by means of a random digit dial telephone survey: a 5-year follow up study. J Allergy Clin Immunol 112(6):1203–1207

Silvanovich A, Bannon G, McClain S (2009) The use of E-scores to determine the quality of protein alignments. Regul Toxicol Pharmacol 54(3l): S26–S31. doi:10.1016/j.yrtph.2009.02.004

Simons FER (2008) Emergency treatment of anaphylaxis. BMJ 336:1141–1142. doi:10.1136/bmj.39547.452153.80

Singh H, Raghava GP (2001) ProPred: prediction of HLA-DR binding sites. Bioinformatics 17(12):1236–1237

Singh AK, Mehta AK, Sridhara S, Gaur SN, Singh BP, Sarma PU, Arora N (2006) Allergenicity assessment of transgenic mustard (*Brassica juncea*) expressing bacterial codA gene. Allergy 61(4):491–497

Sonika R, Anil J (2013) Molecular modelling: a new scaffold for drug design. Int J Pharm Pharm Sci 5(1):5–8

Stadler MB, Stadler B (2003) Allergenicity prediction by protein sequence. FASEB J 17(9):1141–1143

Viquez OM, Konan KN, Dodo HW (2003) Structure and organization of the genomic clone of a major peanut allergen gene. Ara h 1. Mol Immunol 40(9):565–571

Vrtala S (2008) From allergen genes to new forms of allergy diagnosis and treatment. Allergy 63(3):299–309. doi:10.1111/j.1398-9995.2007.01609.x

Wichers HJ, De Beijer T, Savelkoul HF, Van Amerongen A (2004) The major peanut allergen Ara h 1 and its cleaved-off N-terminal peptide; possible implications for peanut allergen detection. J Agric Food Chem 52 (15):4903–4907

Woods RK, Stoney RM, Raven J, Walters EH, Abramson M, Thien FC (2002) Reported adverse food reactions overestimate true food allergy in the community. Eur J Clin Nutr 56(1):31–36

Yang CY, Wu JD, Wu CH (2000) Sequence analysis of the first complete cDNA clone encoding an American cockroach Per a 1 allergen. Biochim Biophys Acta 1517(1):153–158

Zhang ZH, Koh JL, Zhang GL, Choo KH, Tammi MT, Tong JC (2007) AllerTool: a web server for predicting allergenicity and allergic cross-reactivity in proteins. Bioinformatics. 23(4):504–506

Zhang ZH, Tan SC, Koh JL, Falus A, Brusic V (2006) ALLERDB database and integrated bioinformatic tools for assessment of allergenicity and allergic cross-reactivity. Cell Immunol 244(2):90–96

Plant MicroRNAs: An Overview

Kompelli Saikumar and Viswanathaswamy Dinesh Kumar

Abstract

MicroRNAs (miRNAs) are an abundant class of endogenous, noncoding ribonucleic acids (RNAs). They negatively regulate the expression of a wide range of genes, at the posttranscriptional level, that are involved in plant developmental pathways. MiRNAs are present in both plants and animals, but they differ in their biogenesis, mode of action, and evolution. Biogenesis and function of miRNAs require special class of conserved RNAi machinery core proteins. This chapter discusses the methods for the identification of miRNAs using deep sequencing, direct cloning, and computational approaches. Tools available for identifying plant miRNA genes and their targets are listed and discussed along with a brief overview of current knowledge about miRNA clusters, mirtrons, miRNA promoters, and designing and applications of an miRNA.

Keywords

MicroRNA • Conservation • *Arabidopsis thaliana* • Deep sequencing • Direct cloning • Artificial microRNA • Promoter • MicroRNA identification

1 Introduction

MicroRNAs are a small class of RNA molecules that play important role in development by regulating gene expression by binding to target mRNA with high complementarity (Bartel 2004). The length of plant miRNA ranges from 19 to 24 nucleotides. MiRNAs were first identified in the nematode *Caenorhabditis elegans* in 1993 (Lee et al. 1993). A total of 21,264 mature miRNAs are reported in 193 species including plants, animals, fungi, and viruses and have been deposited in the miRNA database (miRBase 19.0; http://www.mirbase.org/).

The numbers of plant miRNAs are increasing as a result of identification of new or conserved miRNAs through computational (bioinformatics) and/or experimental methods such as deep sequencing (Liang et al. 2010; Xin et al. 2010;

K. Saikumar (✉) • V.D. Kumar
Directorate of Oilseeds Research, Rajendra Nagar, Hyderabad 500 030, India
e-mail: Saikumar_kompelli@yahoo.co.in; dineshkumarv@yahoo.com

K.K. P.B. et al. (eds.), *Agricultural Bioinformatics*,
DOI 10.1007/978-81-322-1880-7_8, © Springer India 2014

Li et al. 2011b; Yang et al. 2011b; Wang et al. 2011), direct cloning (Wang et al. 2007; Sanan-Mishra et al. 2009), and other methods. Many miRNA families are evolutionarily conserved and plant miRNAs are highly conserved compared to animal miRNAs. Several investigations have shown that some miRNAs are highly conserved throughout evolution and can be found in mosses to higher plants while a significant number of non-conserved miRNAs also have been identified as species-specific ones (Jones-Rhoades et al. 2006; Sanan-Mishra et al. 2009; Xin et al. 2010; Wang et al. 2011). The expression of miRNAs is known to be altered in specific environmental conditions. A large proportion of the non-conserved miRNAs is active in tissue-specific or developmental-specific conditions. It has been suggested that some of the non-conserved miRNAs arose in recent evolutionary past and got integrated into plant-specific regulatory networks due to the advantageous influence these non-conserved miRNAs might have in regulation of tissue-specific pathways and functions (Cuperus et al. 2011). In this review, we provide a brief overview of the plant microRNAs with an emphasis on the methodologies being followed in their identification in plant systems.

1.1 Nomenclature of miRNAs and Their Genes

The mature ~22 nucleotide miRNAs from plants are referred by the alphabets miR followed by a number (Reinhart et al. 2002). The miRBase database uses abbreviated three or four letter prefixes to designate the species, such that identifier takes the form Ath-miR156 (in *Arabidopsis thaliana*), Zma-miR160 (in *Zea mays*), and Osa-miR171 (in *Oryza sativa*). Their corresponding genes are italicized and the alphabets are denoted in the uppercase: miR156a refers to the mature miRNA and *MIR156a* refers to the gene that code for the miRNA (Reinhart et al. 2002; Griffiths-Jones 2004). Mature miRNAs of animal origin are detonated by hyphen-separated alphabet, miR, and the number in the order of its discovery – like miR-6. If more than one locus code for the same miRNA, the names are numerically suffixed, as in miR-6-1 and miR-6-2 (Griffiths-Jones et al. 2006).

2 Biogenesis of Plant miRNA

MiRNA genes represent about 1–2 % of eukaryotic genome and constitute an important class of fine-tuning regulators that are involved in several physiological and cellular processes (Sanan-Mishra and Mukherjee 2007; Zeng et al. 2010; Axtell et al. 2011; Li et al. 2011a). The majority of plant miRNA genes are located in the intergenic regions (Lindow and Krogh 2005; Adai et al. 2005; Zhang et al. 2009c; Li et al. 2010b). Majority of the miRNA genes are transcribed by RNA polymerase II in a similar way to protein encoding genes (Lee et al. 2004; Jones-Rhoades et al. 2006), while some miRNAs are known to be transcribed by RNA polymerase III (Faller and Guo 2008). RNA polymerase II-transcribed pri-miRNAs in both plants and animals are capped and polyadenylated (Lee et al. 2004). The genomic terrain that constitutes the microRNA genes is called miR loci. Transcripts of miR loci result in the synthesis of intramolecular, long dsRNA molecules with imperfect fold-back structures called primary miRNA (pri-miRNA). A long primary transcript which may be polycistronic is processed in the nucleus, yielding one or more pre-miRNAs. A class of proteins called Dicer-like 1 (DCL1) proteins in concert with HYPONASTIC LEAVES 1 (HYL 1), a nuclear dsRNA-binding protein, and RNA-binding protein DAWDLE (DDL) process pri-miRNA to form precursor miRNA (pre-miRNA) that is further processed by DCL1, HYL1, and SERRATE (SE), a zinc-finger protein to release approximately 21–25 base-pair (bp) long miRNA-miRNA* (* represents the antisense strand) imperfect duplexes, which in turn is stabilized via methylation (of terminal sugar at 3$'$ end of both the strands) by methyltransferase called HUA ENHANCER 1 (HEN 1), a small RNA methylase. Newly formed miRNA-miRNA* duplexes are degraded by a class of exoribonucleases called SMALL RNA DEGRADING NUCLEASE (SDN) (Ramachandran and Chen 2008). Methylation of miRNA-miRNA* duplex prevents uridylation and subsequent degradation. Methylation step also helps in protecting miRNA-miRNA* duplex from nucleases and/or other types of terminal modifications (Yang et al. 2006; Chen 2005). All these events occur in nucleus where a number of methylated miRNA-miRNA* imperfect duplexes are produced (depending on the length and number of imperfect fold-back structures in pri-miRNA) and exported to the cytoplasm via the action of exportin-5 (e.g., HASTY homolog of exportin-5 in *Arabidopsis*) in an ATP-dependent manner. However, the exact function of HASTY is not fully understood. Loss-of-function mutation in HASTY gene only reduces transport of miRNAs from the nucleus to cytoplasm, indicating that miRNA duplex may also be exported by other proteins (Voinnet 2009). Methylated miRNA-miRNA* imperfect duplex transported to the cytoplasm is unwound to form single-stranded miRNA (matured miRNA). This is selectively loaded on to AGO1 containing RISC (RNA-Induced Silencing Complex) by DCL1 and HYL1. Thermodynamic properties play a major role in determining the destiny of each strand in an miRNA-miRNA* duplex (Schwab et al. 2006; Ossowski et al. 2008; Li et al. 2011a, b). In the miRNA-miRNA* duplex, miRNA strand is called guide strand and miRNA* strand is called passenger strand. Guide strand enters RISC and forms miR-RISC (Chen 2005), while the miRNA* strand is degraded by an unknown mechanism. However, some miRNA* may also enter into RISC and regulate expression of target genes (Guo and Lu 2010). Mature miRNA 3′ overhang binds to the PAZ domain of Argonaute (AGO) protein, a component of RISC, and the PIWI domain of AGO displays RnaseH activity involved in effecting target cleavage. RISC is guided by miRNA (possibly by helicase-scanning mechanism) to the target mRNA that comes to lie along the cleft in PIWI domain of AGO of RISC. PIWI domain cleaves the resulting miRNA-mRNA duplex at 10th and 11th nucleotide from 5′ end of the miRNA match forming two mRNA halves (Kidner and Martienssen 2005; Sanan-Mishra and Mukherjee 2007), which are further degraded by exoribonuclease 4 (XRN4) (Kidner and Martienssen 2005). Thus, gene silencing is effected via degradation of target mRNA (Kidner and Martienssen 2005; Sanan-Mishra and Mukherjee 2007; Chapman and Carrington 2007). Many target genes of miRNAs encode transcription factors, each of which further regulate a set of downstream genes (Bartel 2004; Zhang et al. 2006). Thus, biogenesis and mechanism of action of miRNAs in plants require concerted action and physical interaction of several special core RNA silencing machinery proteins such as DCL1, HYL1, HEN1, SE, AGO, and RDR (Kidner and Martienssen 2005; Sanan-Mishra and Mukherjee 2007). DCL and AGO proteins are highly conserved in plants, while RDR protein orthologues are not found in mammals and insects (Margis et al. 2006). The number of DCL, AGO, and RDR genes varies among plants. These proteins are involved not only in biogenesis but also in the identification and slicing of target gene(s).

The well-characterized plant miRNA biogenesis pathway components were identified through various miRNA-deficient mutants and reverse genetic approach. All these genes were found to act independently as well as to interact with each other to accomplish the miRNA biogenesis pathway. DCL1 and HEN1 were among the first genes whose role in miRNA pathway was genetically established (Park et al. 2002; Reinhart et al. 2002), and the mutant plants showed several developmental defects reflecting the critical role played by miRNAs in plant development. Similarly, loss-of-function mutations in AGO1 cause several development pleiotropic defects as well as defect in miRNA biogenesis pathway, suggesting that this gene may play important discrete role in plant development and small RNA biogenesis (Park et al. 2002; Reinhart et al. 2002; Bartel 2004). Similarly, some evidence suggests that the DDL proteins may have functions other than just miRNA biogenesis because the *ddl* mutant caused stronger abnormalities in plant development than DCL1 mutant (Yu et al. 2008).

It has also been known that microRNAs play diverse and complex roles in plant development, namely, developmental regulation (Kidner and Martienssen 2005; Glazinska et al. 2009; Arenas-Huertero et al. 2009; Li et al. 2011b), flowering time (Glazinska et al. 2009), nodulation (Subramanian et al. 2008), epigenetic modifications, mRNA splicing, biotic and abiotic stress responses (Llave 2004; Zeng et al. 2010; Sanan-Mishra et al. 2009; Zhang et al. 2009a, b, c; Jian et al. 2010; Xin et al. 2010; Li et al. 2011a, b; Khraiwesh et al. 2012; Sunkar et al. 2012), protein synthesis, seed germination (Wang et al. 2011), continuous cropping (Yang et al. 2011b), and endosperm development (Li et al. 2009).

2.1 Differences in Biogenesis and Mechanism of Action of Plant and Animal miRNAs

The miRNAs are evolutionarily ancient and spread throughout the plant and animal lineage. Even though in both plants and animals the miRNAs are generated by cleavage of imperfect stable secondary structure and posttranscriptionally control the gene expression through interactions with their target mRNAs, plant and animal miRNAs exert their control in fundamentally different ways. Differences between biogenesis and mode of action in plant and animal miRNAs are enumerated in Table 1.

3 Methods for Identifying miRNAs

After the first plant miRNA was discovered in *Arabidopsis*, identification and functional validation of plant miRNAs and their target genes has become one of the important research areas in plant biotechnology. Both experimental and computational methods have been widely used to identify miRNAs in plants and animals. Direct cloning and deep sequencing methods are finding wide usage for the identification of animal miRNAs than plant miRNAs because of less conserved nature of animal miRNAs. Evidence shows that many miRNAs are evolutionarily conserved in plants. This conserved nature allows the easy identification of plant miRNAs using computational approaches. Deep sequencing and direct cloning methods are used to identify species-specific miRNAs.

3.1 Direct Cloning Method

Direct cloning method was first used to clone and identify miRNAs in animals. The small RNA cloning procedure is based on adapter ligation. The adapter oligonucleotides are used as primer binding sites for reverse transcription and for identifying the orientation and sequence of the cloned small RNAs. In this method, small RNAs are separated from total RNA and then ligated with adaptors and a small RNA cDNA library is constructed. After sequencing the small RNA library, sequences would be matched against miRBase to identify conserved small RNAs and the remaining sequences are matched against genome sequence for identification of novel small RNAs by the ability of genomic region around the region of homology to form potential stem-loop hairpin secondary structures. Once the precursor sequences meet all other criteria like MFE (minimum free energy), MFEI (MFE Index), and AMFE (adjusted MFE), then small RNAs will be annotated as miRNAs. Initially, majority of plant and animal miRNAs were identified by direct cloning (Wang et al. 2007; Sanan-Mishra et al. 2009), and later other approaches have been adopted.

3.2 Deep Sequencing Method

Deep sequencing technology is an advanced experimental method for identifying and analyzing the expression of miRNAs. Deep sequencing of miRNAs has more advantages than direct cloning technology (Unver et al. 2009). Both Roche 454 and Illumina platforms have been employed to sequence and identify miRNAs, millions of reads are produced from each sample, and it is possible to identify all conserved as well as species-specific miRNAs (Unver et al. 2009). Currently, by using deep sequencing, thousands of miRNAs have been identified from both plants and animals (Liang et al. 2010; Xin et al. 2010; Wang et al. 2011; Li et al. 2011a, b; Yang et al. 2011a, b).

Table 1 Differences between plant and animal microRNAs

Plant miRNA	Animal miRNA
Plant miRNA genes are transcribed by RNA polymerase II (Voinnet 2009)	Mostly animal miRNA genes are transcribed by RNA polymerase II. But subsets of animal miRNAs (C19MC) are transcribed by RNA polymerase III (Faller and Guo 2008)
Plants utilize DCL1 proteins in concert with HYL1 to process pri-miRNA to pre-miRNA that is further processed by DCL1, HYL1, and SERRATE proteins to release miRNA-miRNA* duplex (Voinnet 2009)	Animals use Drosha, an RNaseIII enzyme, which partners with double-stranded RNA-binding domain protein DGCR8 (known as pasha in invertebrates) to process pre-miRNA from pri-miRNA (Faller and Guo 2008; Axtell et al. 2011)
Length of plant pre-miRNA is heterogeneous, ranging from 70 to many hundreds of nucleotides (Axtell et al. 2011)	Length of pre-miRNAs is more consistent in animals, mostly 55–70 nucleotides; Drosophila pre-miRNAs are about 200 nucleotides (Ruby et al. 2007a, b; Axtell et al. 2011)
Plant mature miRNAs are methylated at the 3′end by HEN1 (Yang et al. 2006)	Most animal mature miRNAs are not methylated, but Piwi-interacting RNAs (piRNAs) are methylated by HEN1 (Horwich et al. 2007)
Most plant miRNA encoding region is independent and nonprotein coding. Intronic plant miRNAs are very rare (Zhu et al. 2008; Joshi et al. 2012)	Many animal miRNA encoding regions derive from stand-alone, nonprotein-coding, and approximately 30 % are located on sense strands of introns and few from antisense strand (Bartel 2004; Millar and Waterhouse 2005; Westholm and Lai 2011)
No miRNA production from exons, protein-coding genes, and UTRs (Axtell et al. 2011)	Occasional miRNA production from exons, protein-coding genes, and UTRs (Voinnet 2009; Han et al. 2009a, b; Berezikov et al. 2011)
Occurrence of polycistronic miRNAs are less (Guddeti et al. 2005)	Occurrences of polycistronic miRNAs are high (Voinnet 2009)
61–90 % hairpins of polycistronic miRNAs encode identical miRNAs	Up to 40 % polycistronic miRNAs encode unrelated miRNAs
Plant miRNAs have their own promoters (Voinnet 2009)	Intronic miRNAs utilize the advantage of *cis* regulatory elements that direct the expression of host mRNA (Isik et al. 2010)
Plant miRNAs are produced by nuclear RNaseIII Dicer-like1 (DCL1) enzyme (Margis et al. 2006)	Animal miRNAs are produced by nuclear and cytoplasmic RNaseIII enzymes (Faller and Guo 2008)
In very few cases, miRNA* strand also enters into AGO-RISC complex to cleave target mRNA (Schwab et al. 2006; Ossowski et al. 2008)	Many Drosophila miRNA* strand also preferentially enters into AGORISC complex (Faller and Guo 2008)
The main miRNA effector in plants is AGO1, and to a lesser extent AGO10 and other AGOs; AGO7 carries the exceptional miRNA miR390 (Vaucheret 2006; Axtell et al. 2011)	The main miRNA effector in animals is dAGO1 for Drosophila, ALG1/2 for *C. elegans*, and AGO1 to AGO4 for vertebrates (Liu et al. 2004; Meister et al. 2004; Axtell et al. 2011)
Dicer-independent miRNA pathways are not found in plants (Voinnet 2009)	Dicer-independent miRNA pathways are found in animals (e.g., mirtron pathway) (pre-miRNA ends are defined by splicing instead of Drosha cleavage) (Yang et al. 2010a, b; Axtell et al. 2011)
Formation of mature miRNA is via miRNA-miRNA* duplex intermediate (Voinnet 2009; Zhang et al. 2010)	Mostly formation of mature miRNA is via miRNA-miRNA* duplex intermediate (exception is that miR451 pathway is instructive in that it does not proceed via miRNA-miRNA* intermediate) (Axtell et al. 2011)
AGO-mediated miRNA biogenesis is not found in plants (Axtell et al. 2011)	AGO-mediated miRNA biogenesis is found in animals (Okamura et al. 2008)
Plants have extensive capacity to process long inverted repeat transcript into small RNAs (Dunoyer et al. 2010)	Endogenous processing of long hairpins in mammalian cells appears limited, because extensive dsRNA triggers the antiviral interferon response (Watanabe et al. 2008)
Plant miRNAs often target with perfect/near perfect complementarity (Voinnet 2009)	In animals, target recognition depends mainly on a "seed" sequence (2–8 nucleotide of miRNA) (Brennecke et al. 2005)

(continued)

Table 1 (continued)

Plant miRNA	Animal miRNA
Mostly plant miRNA-binding sites are found in ORFs and less frequently in 5′UTRs, 3′UTRs, and in nonprotein-coding transcripts (Axtell et al. 2011)	Mostly animal miRNA-binding sites are found in 3′UTRs. Binding sites in ORF or 5′UTR appear to be hampered by competition with ribosomes (Lee et al. 2004)
MicroRNA mediated mRNA regulation mostly through slicing of target and less frequently through translation inhibition (Voinnet 2009)	MicroRNA mediated mRNA regulation mostly through translation repression followed by mRNA destabilization and less frequently by slicing of target (Fabian et al. 2010)
Plant miRNA targets have a single target site and are regulated by just one miRNA (Axtell et al. 2011)	Animal miRNA target genes have conserved target sites for different miRNAs (i.e., one target is controlled by many miRNA) (Flynt and Lai 2008)
MicroRNA biogenesis from inverted duplication (MITE) is common in plants (Piriyapongsa and Jordan 2007)	Rare in animals (but common in fly hpRNAs) (Piriyapongsa and Jordan 2008)
MicroRNA biogenesis from initially unstructured sequence is rare in plants (Axtell et al. 2011)	Common in animals (Axtell et al. 2011)
In plants, the pre-miRNAs are much more diverse in length and structure, with the miRNA-miRNA* duplex being at variable positions (Mallory and Vaucheret 2006)	It has been shown that animal pri-miRNAs are processed in a loop-to-base manner to produce a structure containing 21 bp miRNA-miRNA* duplex and a terminal loop at the distal end (Werner et al. 2010)
miRNA-miRNA* duplex is transported from nucleus to cytoplasm with help of exportin-5 protein (Park et al. 2005)	Pre-miRNA itself is exported from nucleus to cytoplasm with help of HASTY protein (Park et al. 2005; Axtell et al. 2011)
Processing of the pre-miRNA transcript to form the miRNA-miRNA* duplex occurs in the nucleus (Axtell et al. 2011)	Processing of the pre-miRNA transcript to form the miRNA-miRNA* duplex occurs in the cytoplasm (Axtell et al. 2011)

High-throughput sequencing technologies have greatly increased the sensitivity of molecular cloning experiments; otherwise, it is still difficult to ensure that all possible tissues, developmental stages, and growth conditions are represented in cloned libraries of small RNAs. Therefore, computational methods would still remain useful in discovery of plant miRNAs.

3.3 Computational Method

While direct cloning and deep sequencing are the most direct ways of discovering miRNAs, bioinformatic approaches have been of useful complementary strategy for identifying miRNAs. Novel miRNAs are discovered through experimental methods, and homologous miRNAs in other databases such as EST/GSS and WGS databases can be identified through computational approaches. Computational approaches are best suited in identifying miRNA families that are difficult to clone due to lower abundance or specific expression, e.g., Ath-miR 395 and 399. Flowchart for identification of miRNAs using different approaches is illustrated in Fig. 1.

However, miRNA sequence similarity cannot guarantee that homologous sequences code for true miRNA due to the fact that miRNAs are only about 19–24 nt in length. To better distinguish true miRNA coding genes from pseudo-genes, computational searches always need to be combined with the major characteristics of miRNAs, particularly stem-loop hairpin structures, MFE, MFEI, and AMFE.

Zhang et al. (2005) developed a methodology analysis to identify miRNAs using EST databases, and the same methodology can be used to identify miRNAs from other genomic sequences, including the GSS database and WGS. Computational approach has been widely used, and many conserved miRNAs have been identified from many plant species. The main limitation of in silico approaches is its inability to identify species-specific and novel miRNAs. However, there are now pipelines (e.g., NOVOMIR, semiRNA) available to identify all potential miRNA genes in the genome sequence.

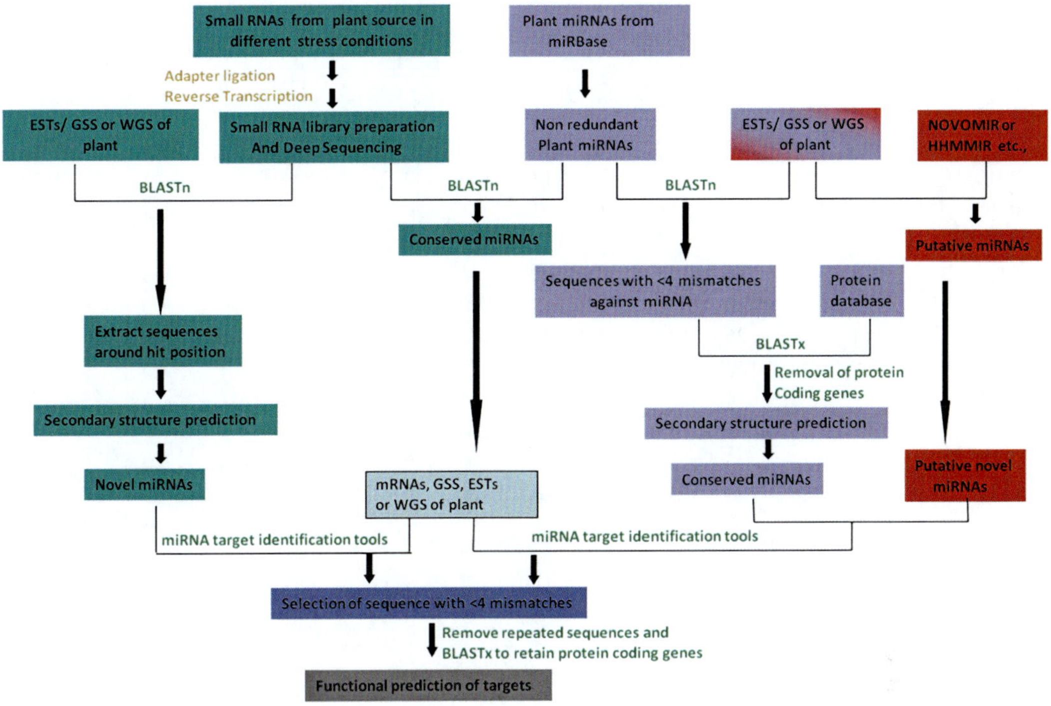

Fig. 1 Flowchart for identification of miRNAs by conventional and computational methods

Computational prediction of miRNA genes is based on the conserved nature of mature miRNA and secondary structure of pre-miRNA. A central step in the prediction of plant miRNA genes is the analysis of secondary structures to identify candidate hairpin precursors. Plant miRNA hairpins are diverse in terms of length and extent of base pairing in the loop region. However, all predicted pre-miRNAs have extensive base pairing in the region containing the miRNAs and miRNA*.

Computational method has been preferred over experimental approaches, such as direct cloning and deep sequencing, because of no requirement of specific software or program skills and all sequences used are readily accessible from the NCBI GenBank Databases. However, this method is limited by the number of sequences available in the database. The majority of plant miRNAs have been discovered in plants with fully sequenced genomes and in plants with a large number of ESTs and GSSs including *Arabidopsis thaliana* (Wang et al. 2004), *Brassica napus* (Wang et al. 2007), *Brassica oleracea*, *Brassica rapa*, *Citrus aestivum* (Song et al. 2009), *Chlamydomonas reinhardtii* (Molnar et al. 2007), Castor bean (Zeng et al. 2010), *Glycine max* (Subramanian et al. 2008; Zhang et al. 2008), *Gossypium herbaceum*, *Gossypium hirsutum* (Ruan et al. 2009), *Gossypium arboretum* (Wang et al. 2012), *Medicago truncatula*, *Nicotiana tabacum* (Frazier et al. 2010), *Oryza sativa* (Sanan-Mishra et al. 2009; Jian et al. 2010; Li et al. 2010b), *Pinus taeda*, *Populus trichocarpa* (Barakat et al. 2007), *Populus euphratica* (Li et al. 2011a), *Phaseolus vulgaris* (Arenas-Huertero et al. 2009), *Solanum lycopersicum*, *Sorghum bicolor*, *Triticum aestivum* (Han et al. 2009b; Xin et al. 2010), and *Zea mays* (Zhang et al. 2009b). The number of highly conserved miRNA families identified in plant genomes is listed in Table 2, and information regarding the currently available plant miRNAs and their distribution in the publicly available miRNA database miRBase 19.0 (August 2012) is listed in Table 3.

The availability of a large number of sequences from many plant species in public

Table 2 Number of conserved miRNA families in plants

Family	Ath	Aly	Osa	Zma	Sbi	Mtr	Gma	Ptr	Ppa	Vvi	Mdo	Rco	Nta	Bdi	Cme
miR156	10	16	19	23	9	10	**28**	12	3	9	**31**	**8**	10	4	10
miR157	4	8	0	0	0	0	0	0	0	0	0	0	0	0	0
miR159	3	6	7	22	2	2	10	6	0	3	3	1	1	1	2
miR319	3	8	4	8	2	2	14	9	7	5	3	4	2	2	4
miR160	3	6	12	13	6	6	7	11	9	5	5	3	4	5	4
miR162	2	4	2	2	1	1	3	3	0	1	2	1	2	1	1
miR164	3	6	6	16	5	4	11	6	0	4	6	4	3	5	4
miR165	2	4	0	0	0	0	0	0	0	0	0	0	0	0	0
miR166	7	14	**26**	26	11	8	26	17	**13**	8	9	6	8	7	9
miR167	4	8	16	20	9	3	10	11	1	5	10	3	5	4	6
miR168	2	4	3	4	1	4	2	4	0	1	2	1	5	1	1
miR169	**15**	**28**	20	**32**	**19**	20	24	**36**	0	**25**	6	3	**19**	11	**20**
miR170	1	2	0	0	0	0	0	0	0	0	0	0	0	0	0
miR171	3	6	14	28	11	9	26	17	2	9	15	7	3	4	9
miR172	6	10	5	8	6	4	15	12	0	4	15	1	10	3	6
miR390	2	4	2	4	1	1	9	5	4	1	6	2	3	1	4
miR393	5	4	3	6	2	3	11	6	0	2	6	1	0	2	3
miR394	2	4	1	4	2	0	9	4	0	3	2	0	1	1	2
miR395	6	15	25	32	12	18	13	11	1	14	9	5	3	**13**	6
miR396	2	4	14	13	5	4	14	9	0	4	7	1	3	5	5
miR397	2	4	4	4	2	0	3	3	0	1	2	1	1	1	1
miR398	3	6	10	4	1	4	3	4	0	3	3	2	1	2	2
miR399	6	19	11	20	11	18	8	12	0	9	10	6	7	2	7
miR403	1	2	0	0	0	0	2	5	0	6	2	2	0	0	0
miR408	1	2	2	3	1	2	7	2	2	1	4	1	1	1	1
miR827	1	2	3	2	0	0	0	1	0	0	1	0	1	1	0
miR828	1	2	0	0	0	0	2	3	0	2	2	0	0	0	1
miR845	2	4	0	0	0	0	0	0	0	5	0	0	0	0	1
miR2111	4	4	0	0	0	**23**	6	2	0	2	2	0	0	0	2
miR473	0	0	0	0	0	0	0	3	0	0	0	0	0	0	0
miR477	0	0	0	0	0	0	0	6	10	1	1	0	2	0	2
miR482	0	0	0	2	0	1	9	5	0	1	5	0	5	0	0
miR529	0	0	2	2	1	21	0	0	7	0	0	0	0	1	0
miR530	0	0	2	0	0	1	5	2	0	0	0	0	3	0	2
miR535	0	0	2	0	0	0	0	0	4	3	4	1	0	0	0
miR536	0	0	0	0	0	0	2	0	6	0	0	0	0	0	0
miR479	0	0	0	0	0	0	0	1	0	1	0	0	2	0	0

Number of loci belonging to conserved miRNA families in plants (Ath. *Arabidopsis thaliana*, Aly. *Arabidopsis lyrata*, Osa. *Oryza sativa*, Zma. *Zea mays*, Sbi. *Sorghum bicolor*, Mtr. *Medicago truncatula*, Gma. *Glycine max*, Ptr. *Populus trichocarpa*, Ppa. *Physcomitrella patens*, Vvi. *Vitis vinifera*, Mdo. *Malus domestica*, Rco. *Ricinus communis*, Nta. *Nicotiana tabacum*, Bdi. *Brachypodium distachyon*, Cme *Cucumis melo*), as listed in miRBase 19.0. miRNA family comprising the highest number of conserved miRNA in each plant is highlighted in *bold*

databases has made in silico analysis an approach of choice to several workers. Therefore, a simple outline of the methodology being followed for in silico identification of miRNAs is provided here.

To identify conserved miRNAs, mature nonredundant miRNA sequences are subjected to a BLASTn search against the available sequences of the organism of interest. To improve the BLASTn search, the Blast parameters settings

Table 3 Distribution of plant microRNAs

Plant species	No. of miRNAs
Medicago truncatula	719
Oryza sativa	708
Glycine max	555
Arabidopsis lyrata	375
Populus trichocarpa	369
Arabidopsis thaliana	338
Zea mays	321
Physcomitrella patens	280
Sorghum bicolor	242
Malus domestica	207
Vitis vinifera	186
Nicotiana tabacum	165
Brachypodium distachyon	136
Cucumis melo	120
Brassica napus	92
Chlamydomonas reinhardtii	85
Theobroma cacao	82
Hordeum vulgare	69
Citrus sinensis	64
Selaginella moellendorffii	64
Ricinus communis	63
Cynara cardunculus	57
Ectocarpus siliculosus	52
Aquilegia caerulea	45
Solanum lycopersicum	44
Brassica rapa	43
Triticum aestivum	42
Picea abies	41
Gossypium hirsutum	39
Pinus taeda	38
Arachis hypogaea	32
Pinus densata	31
Hevea brasiliensis	28
Saccharum ssp.	19
Salvia sclarea	18
Vigna unguiculata	18
Amphimedon queenslandica	16
Helianthus tuberosus	16
Saccharum officinarum	16
Festuca arundinacea	15
Digitalis purpurea	13
Glycine soja	13
Rehmannia glutinosa	13
Solanum tuberosum	11
Manihot esculenta	10
Phaseolus vulgaris	10
Helianthus annuus	9
Acacia auriculiformis	7
Brassica oleracea	7
Elaeis guineensis	7
Citrus trifoliata	6
Citrus clementine	5
Populus euphratica	5
Bruguiera cylindrica	4
Bruguiera gymnorrhiza	4
Citrus reticulata	4
Gossypium raimondii	4
Lotus japonicus	4
Acacia mangium	3
Helianthus argophyllus	3
Helianthus ciliaris	3
Helianthus paradoxus	3
Helianthus petiolaris	3
Aegilops tauschii	2
Helianthus exilis	2
Carica papaya	1
Gossypium arboreum	1
Gossypium herbaceum	1
Triticum turgidum	1

are usually adjusted as follows: expect values are set at 1,000 to increase the number of potential hits; the default word-match size between the query and database sequences can be set at 7; and the number of descriptions and alignments are increased to 1,000. All the sequences with no more than four mismatches will be selected for further investigation. The selected sequences are BLASTed against each other to remove repeated sequences. Then the selected sequences are used for BLASTx analysis; the protein-coding sequences are removed and only the nonprotein-coding sequences are retained. Then, the nonprotein-coding sequences are subjected to web-based tools like Mfold (Zuker 2003) to predict the secondary structures as well as MFEs of pre-miRNA sequences. The Mfold default parameters are used to predict the secondary structures of selected sequences. A sequence is considered as potential miRNA candidate when it meets the following criteria. First, the predicted mature miRNA should not have more than four mismatches compared with a known mature miRNA. Second, the selected sequence must fold

Table 4 MiRNA gene prediction tools

Tool name	Link	Platform	Reference	Used for
NOVOMIR	http://www.biophys.uni-duesseldorf.de/novomir/	Perl/Linux	Teune and Steger (2010)	Plants/viral
HHMMiR	http://www.benoslab.pitt.edu/kadriAPBC2009.html	Hidden Markov Models	Kadri et al. (2009)	Plants/animals
TRIPLET-SVM	http://bioinfo.au.tsinghua.edu.cn/software/mirnasvm/	Linux	Xue et al. (2005)	Plants/animals/virus
MiRAlign	http://bioinfo.au.tsinghua.edu.cn/miralign/	Windows	Wang et al. (2005)	Plants/animals
MiRcheck	http://bartellab.wi.mit.edu/software.html	Perl	Jones-Rhoades and Bartel (2004)	Plants
MiRFINDER	http://www.bioinformatics.org/mirfinder/	Windows/Linux	Huang et al. (2007)	Plants
microHARVESTER	http://www-ab.informatik.uni-tuebingen.de/software/microHARVESTER	Linux	Dezulian et al. (2006)	Plants
MiRPara	http://www.whiov.ac.cn/bioinformatics/mirpara	Perl	Wu et al. (2011)	Plants/animals
findMiRNA	http://sundarlab.ucdavis.edu/mirna/	Windows	Adai et al. (2005)	Plants
BayesmiRNAfind	https://bioinfo.wistar.upenn.edu/miRNA/miRNA/login.php	Naive Bayes	Yousef et al. (2006)	Plants/virus/worm/vertebrates
miMatcher	http://wiki.binf.ku.dk/MiRNA/miMatcher2	SVM	Lindow and Krogh (2005)	Plants
deepBase	http://deepbase.sysu.edu.cn/index.php	Windows	Yang et al. (2010a)	Plants/animals
MicroPC	http://www3a.biotec.or.th/micropc/application.html	Windows	Mhuantong and Wichadakul (2009)	Plants
miRDeepFinder	http://www.leonxie.com/DeepFinder.php	Perl	Xie et al. (2012)	Plants
miRDeep-P	http://faculty.virginia.edu/lilab/miRDP/	Perl	Yang and Li (2011)	Plants
PMRD	http://bioinformatics.cau.edu.cn/PMRD/	Windows	Zhang et al. (2010)	Plants

into an appropriate stem-loop hairpin secondary structure. Third, the mature miRNA should be localized in any one arm of the stem-loop structure. Fourth, there should be no loop or break in the miRNA or miRNA* sequences. Fifth, there should not be more than six mismatches between the predicted mature miRNA sequence and its opposite miRNA* sequence in the secondary structure. Sixth, the predicted secondary structure should have high negative MFE and MFEI values. All MFEs are expressed as negative kcal/mol. To avoid the potential effect of nucleotide length on MFEs, adjusted MFE (AMFE) values are calculated to analyze the MFEs of pre-miRNAs. AMFE means the MFE of an RNA sequence with 100 nt in length, which is equal to MFE/(length of a potential pre-miRNA) *100, and MFEI is equal to AMFE/(G + C) percentage (Zhang et al. 2009b). The application of these criteria for inclusion of RNAs as miRNAs reduces the number of RNAs analyzed, and this minimizes the likelihood that non-miRNAs are included in subsequent analyses, and this significantly reduces the total number of predicted false miRNAs. According to previous reports, precursor miRNA sequences have significantly higher MFEI compared to coding or noncoding sequences, and the candidate miRNA sequence is more likely to be miRNAs when the MFEI is greater than 0.85 (Zhang et al. 2008; Song et al. 2009; Frazier et al. 2010). To avoid mistakenly designating other types of RNAs as miRNA candidate, MFEI and AMFE are also considered when predicting secondary structures. Plant miRNA gene prediction tools widely used for computational discovery are listed in Table 4.

4 Identification of miRNA Targets in Plants

MicroRNAs bind perfectly or near perfectly to mRNAs and repress the gene expression or mediate the cleavage of mRNAs and therefore inhibit gene expression. In plants, microRNA-mediated mRNA cleavage is considered as the major mechanism of action of miRNA, but in animals, miRNA binds to target and repress gene expression (Jones-Rhoades and Bartel 2004; Sanan-Mishra and Mukherjee 2007; Li et al. 2011a, b). Identification and validation of miRNA target is a critical step in miRNA research. MicroRNA binds to target based on complementarity. BLASTn searching has been widely used to predict miRNA targets, followed by validation through wet-lab experiments. RACE-PCR is widely employed to find the cleavage site for validating the predicted targets (Wang et al. 2004; Jones-Rhoades and Bartel 2004; Adai et al. 2005; Fahlgren et al. 2007; Zeng et al. 2010; Li et al. 2010a, b, 2011a, b). Real-time PCR and northern blotting have been used to check expression of miRNAs and their targets (Zhang et al. 2008; Arenas-Huertero et al. 2009; Song et al. 2009; Zeng et al. 2010). However, Blastn searching of miRNA targets needs to be modified because of short length of miRNAs, and traditional Blastn searching was developed for use with long sequence alignment. The conservation of a target site in other plant species also should be considered for identifying miRNA targets and removing the false positives. The following criteria would serve to determine complementary sites between miRNAs and their potential mRNA targets: first, not more than four mismatches between miRNAs and their potential mRNA targets are allowed; second, of these four mismatches, no mismatches are in positions 10 and 11 (assumed to be the cleavage site), and more than two continuous mismatches are not allowed at any other positions; and third, no gaps are allowed between the complementary binding site of potential miRNAs and the predicted mRNA targets. The "seed" region, located at miRNA nucleotide 2–8, is the most important sequence for interaction with mRNA targets (Zhang et al. 2009b). Computational miRNA target identification tools have been developed based on these principles. Many computational tools are available for predicting animal miRNA targets, and only a few tools are available for identifying plant miRNA targets (Table 5).

Table 5 Plant miRNA target prediction tools

Tool name	Link	Platform	Reference	Used for
Target-align	http://www.leonxie.com/targetAlign.php	Windows/Linux	Xie and Zhang (2010)	Plant
TAPIR	http://bioinformatics.psb.ugent.be/webtools/tapir/	Windows	Bonnet et al. (2010)	Plants
miRTour	http://bio2server.bioinfo.uni-plovdiv.bg/miRTour/	Windows	Milev et al. (2011)	Plants
psRNATarget	http://plantgrn.noble.org/psRNATarget/	Windows	Dai and Zhao (2011)	Plants
MicroPC	http://www3a.biotec.or.th/micropc/application_target.html	Windows	Mhuantong and Wichadakul (2009)	Plants
UEA sRNA toolkit	http://srna-tools.cmp.uea.ac.uk/plant/cgi-bin/srna-tools.cgi	Windows/Linux/Mac	Moxon et al. (2008)	Plants/animals
Target Finder	http://carringtonlab.org/resources/targetfinder	Windows/Perl	Fahlgren et al. (2007)	Plants
p-TAREF	http://scbb.ihbt.res.in/new/p-taref/index.html	Windows/Linux	Jha and Shankar (2011)	Plants
starBase	http://starbase.sysu.edu.cn/degradomeSeq.php	Windows	Yang et al. (2011a, b)	Plants
imiRTP	http://admis.fudan.edu.cn/projects/imiRTP.htm	Windows	Ding et al. (2012)	Plants
AthaMap	http://www.athamap.de/	Windows	Bulow et al. (2012)	*Arabidopsis*
psRobot	http://omicslab.genetics.ac.cn/psRobot/stemloop_1.php	Windows/Linux	Wu et al. (2012)	Plant

5 MicroRNA Gene Clusters

MicroRNA gene clusters are group of miRNA genes clustered together within close proximity in the genome. Recent studies on miRNAs have shown that production of polycistronic miRNAs or several different mature miRNAs form the same sequence. Initial studies indicated that about 50 % of total miRNA genes throughout the *Drosophila* genome were clustered, whereas only few miRNA genes might be clustered within human genome (Bartel 2004). Occurrence of miRNA clusters is more common in animals, and only a few miRNA clusters have been found in plants, for example, five in soybean (Zhang et al. 2008), seven in maize (Zhang et al. 2009b), and eight clusters in tobacco (Frazier et al. 2010). Occurrence of miRNA clusters in plants has suggested that miRNA clusters might have evolved in lineage-specific manner (Zhang et al. 2008). Many miRNA clusters in animals are predicted to have formed through a complex history of tandem duplication and loss of individual members as well as tandem duplication of entire clusters (Zhang et al. 2009c). MicroRNA clusters in plants also may have been generated by similar mechanism of miRNA duplication. The clustered animal miRNAs have similar gene expression pattern and transcribed as polycistronic transcript. This cluster is processed into mature miRNAs and the mature miRNAs bring about degradation of the target mRNA molecules very effectively. There has been accumulating evidence suggesting that clustered miRNA genes are often, though not always, located in polycistrons and are co-expressed with neighboring miRNAs and host genes. Yu et al. (2006) demonstrated that in the expression profile of 51 identified human microRNA clusters, 39 miRNA clusters showed the consistent expression of miRNAs in a single cluster, while the remaining miRNA clusters indicated the differential expression of members in a single cluster. It seems that the clustered miRNAs are essential in regulating a complex cell signaling network, which is more efficient and complicated than the regulatory pattern mediated by discrete miRNAs. The major difference between plant and animal miRNA clusters is the number of miRNAs in each cluster. Animal miRNA clusters have as many as 40 miRNAs clustered together, whereas plant miRNA clusters are found to have only small number of about 5–10 (Frazier et al. 2010; Zhang et al. 2008, 2009c).

6 Mirtrons

Recent studies from Drosophila, *C. elegans*, mouse, and human have reported a special class of miRNA, called mirtrons, that are generated via a nonclassical miRNA pathway from spliced-out introns in a Drosha-independent manner. The spliced-out introns fold into hairpin stem-loop structure corresponding to the pre-miRNA, which can then be processed by host RNAi machinery proteins into mature miRNAs. Recently, mirtrons have also been reported in plants. Joshi et al. (2012) identified 70 mirtrons in rice and were short-listed to 16 best sequences through a stringent selection filter. The prediction accuracy was subsequently validated by northern analysis and RT-PCR. Mirtronic introns exhibit extensive pairing between $5'$ and $3'$ splice sequences. The slight diversion in structural features of mirtrons has made it possible to differentiate them from other intronic sequences. The mirtrons have conserved canonical splice sites, with "AG" splice acceptors of mirtronic introns typically adopting a 2 nucleotide $3'$ overhang to these hairpins (Ruby et al. 2007a, b). It is difficult to filter mirtronic introns having hairpin structure from a huge repertoire of introns. MirtronPred (http://bioinfo.icgeb.res.in/mirtronPred) computational algorithm was developed to predict mirtrons from a pool of plant introns (Joshi et al. 2012). The G:U wobble pair has negative effect on secondary structure of mirtrons. More than five G:U wobble pairs in the $5'$ seed regions are believed to impair mirtron structure. About 30 % of miRNA genes are located in the introns of protein-coding genes, indicating some linkage between mRNA transcription and miRNA production. The average mirtron yield is substantially fewer in small RNA libraries compared to canonical miRNAs (Ruby et al. 2007a, b; Sibley et al. 2012; Westholm and Lai 2011).

7 MicroRNA Promoters

The precise regulation of gene expression plays an important role in cell development and cellular responses to biotic and abiotic stress. Identification of the factors involved in regulation of gene expression is one of the major challenges in modern biology. The spatiotemporal expression of genes is controlled by DNA elements called *cis*-elements and other trans-factors such as transcription factors (TFs), enhancers, and signaling molecules. TFs recognize short sequence motifs and bind to them in a sequence-specific manner. Almost no information is available regarding the regulation of miRNA genes themselves. Pri-miRNAs are typically transcribed by RNA polymerase II and have promoter elements that are similar to those of protein-coding genes.

Study of miRNA promoters remains one of the main interesting research problems in the study of miRNA biogenesis pathway and mode of action. Recent studies in *Arabidopsis* and *Oryza* by computational approaches resulted in identification of miRNA promoters and their elements using novel promoter prediction method, called *common query voting* (CoVote) (Zhou et al. 2007).

8 Artificial MicroRNAs

Plant endogenous miRNAs have few (zero to five) mismatches to their targets and trigger local transcript cleavage and subsequent degradation. The highest number of targets empirically confirmed for a specific plant miRNA is only 10 (Schwab et al. 2006), contrasting with the large number of targets of the typical animal miRNA. These observations have raised the question whether the differences are only due to intrinsic properties of the miRNA machinery or at least partially caused by selection against plant miRNAs with a large number of targets. Both the possibilities suggest disadvantages of deploying plant miRNA for target-specific gene silencing. In addition, plant siRNA-mediated gene silencing leads to formation of secondary siRNAs, causing silencing of off-targets (Schwab et al. 2006). This suggests the need for adapting artificial miRNA (amiRNA)-mediated gene-silencing approach in order to avoid the possible ill effect of siRNA approach.

Vectors that contain a hairpin precursor are recognized as second-generation RNAi vectors. Their sequence can be modified in such a way that miRNAs of other defined sequence, called artificial miRNAs (amiRNAs), are to be used *in planta*. These vectors can be used as reverse genetic tools to direct gene silencing, also in non-model systems. The unique features of amiRNAs are transient and tissue-specific gene silencing and the simultaneous silencing of several related genes if a conserved region among them is targeted. The web server Web MicroRNA Designer 3 (http://wmd3.weigelworld.org/cgi-bin/webapp.cgi) can be used to design suitable amiRNA sequences for a variety of different plant species for which a whole-genome annotation or significant EST/cDNA sequence information is available as well as the designing of primer sequences needed to modify the pre-miRNA sequence in the vectors (pRS300 containing *Arabidopsis ath-MIR319a* for dicots or pNW55 containing rice *osa-MIR528* for monocots).

The tool "DESIGNER," from the WMD 3 (Web microRNA Designer 3) platform is used for designing of amiRNAs against intended target gene(s). Designed output amiRNAs are ranked, by the tool, into four color categories: (a) green, the best desirable candidate amiRNAs; (b) yellow and (c) orange, the intermediate desirable candidate amiRNAs; and (d) red, the most undesirable candidate amiRNAs. The higher the degree of desirability, the lower is the score of penalty points and also that of off-targets. The designed candidate amiRNAs are selected based on the following criteria:

(a) Uridine at position 1 and, if possible, adenine at position 10, both of which are overrepresented among natural plant miRNAs and highly efficient siRNAs.
(b) Higher AT content at the 5′ end and higher GC content at the 3′ end with overall GC percentage of 30–50% (if high unwinding is affected, low stable-duplex formation is affected).
(c) Thermodynamic instability/differential end display/strand asymmetry at the 5′ end (high AT content for easy unwinding from mRNA target after the target cleavage).
(d) Avoiding or reducing off-targets by introducing or accepting mismatches at particular places.
(e) In a 21 bp amiRNA, one or no mismatch is allowed between positions 2 and 12, none at the cleavage site (positions 10 and 11), and up to four mismatches between positions 13 and 21, and with not more than two consecutive mismatches at the 3′ end. Mismatches at the 3′ end reduce possible off-target effects and potential transitivity effect due to priming and extension by RNA-dependent RNA polymerase (RdRP).
(f) amiRNA – mRNA must have absolute and relative hybridization energy, with 80–95 % of free energy calculated for perfect match, and an absolute value of −35 to −38 kcal per mole as determined by "RNAcofold" and "Mfold" tools.
(g) No perfect complementarity to their intended targets (for transitivity concern).

It is important to optimize the designed amiRNAs for both effectiveness and specificity. The optimization for specificity (i.e., predicting and avoiding off-target) depends on the availability of transcriptome sequence information. WMD can design amiRNAs to silence single genes as well as multiple genes. In comparison with the conventional siRNA-mediated gene-silencing approach, amiRNA-mediated gene silencing has numerous advantages (Table 6.)

The advantages suggest the need for adopting artificial miRNA (amiRNA)-mediated gene-silencing approach in order to avoid the possible off-target effects of siRNA approach. Successful use of amiRNAs for the specific downregulation of genes is shown for dicotyledonous plants *Arabidopsis*, tomato, and tobacco, for monocot rice (Schwab et al. 2006; Ossowski et al. 2008), and for non-seed bryophyte (*Physcomitrella patens*) (Khraiwesh et al. 2008). In *Arabidopsis thaliana,* amiRNAs were used to identify possible phosphatidase components of the mechanism responsible for polar targeting of PIN (pin-formed) auxin transport proteins, which are essential for the auxin signaling. Recently, amiRNAs were used to elucidate the function of genes involved in circadian clock regulation and pollen development in *Arabidopsis* (Kojima et al. 2011).

Phytoene desaturase (Pds), spotted leaf 11 (Spl11), and *elongated uppermost internode 1 (Eui1)* genes were silenced in *Oryza sativa* using amiRNA technology and resulted in albino phenotype (*pds*), spontaneous lesion formation in the absence of pathogens (*spl11*), and elongation of the uppermost internode at heading stage (*eui1*) (Warthmann et al. 2008).

In a few reports, artificial miRNAs (amiRNAs) were compared to those of natural miRNAs, and it has been found that amiRNAs efficiently silence both the single and multiple target genes with little evidence for the formation of secondary siRNAs, unlike in the case of most natural plant miRNAs (Schwab et al. 2006; Ossowski et al. 2008; Khraiwesh et al. 2008). Also it is possible to generate constructs expressing multiple and unrelated amiRNAs due to their small size as well as designing amiRNAs targeting specific alleles or splice forms for a given gene. It has also been suggested that amiRNAs pose fewer biosafety issues when applied to crop improvement than other gene-silencing and/or transgenic approaches (Liu and Chen 2010).

Table 6 Differences between amiRNA and siRNA

Artificial microRNA	Short interfering RNA
Length of amiRNA is 21 nt	Length of siRNA 21–25 nt
Computational designing of amiRNA is required	Not needed
Sequence of amiRNA is known	Sequence of siRNA is not known
Pre-miRNA backbone is required for expression	Not required
It is not an intron-mediated gene silencing	It is an intron-mediated gene silencing (intron hairpin technology)
Length of pre-miRNA is about 160–200 bp	Length of hairpin is about 200–3,000 bp
Imperfect complementarity is observed in hairpin	Perfect complementarity is observed in hairpin
GC composition of hairpin is 30–50 %	GC composition differs from hairpin to hairpin
One miRNA-miRNA* duplex will come from pre-miRNA	Many duplexes arise from dsRNA
Only one small effective amiRNA molecule will arise from one hairpin	Depending on length of hairpin, many siRNA molecules formed
AmiRNA follows 5′ strand/end asymmetry	Does not follow
In miRNA-miRNA* duplex only miRNA will be entered into RISC	Both strands will have equal chance of entering into RISC
1–2 mismatches at 3′ end of amiRNA to target is allowed	Mismatches are not allowed
Partial complementarity is observed between amiRNA and target mRNA	Perfect complementarity observed between siRNA and target mRNA
Mismatches are allowed between miRNA and target mRNA	Not allowed
1–10 targets can be silenced with one amiRNA	Not known
AmiRNA splicing site at the target is known	SiRNA splicing site at the target is not known
Off-targets can be avoided	Not possible
Transitivity/formation of secondary siRNAs is not observed	Observed
RdRP, SDE3 enzyme activity is not required	RdRP, SDE3 enzyme activity is required for formation of secondary siRNA
Heterogeneous in nature (pre-miRNA and mRNA do not share homology to each other)	Homogeneous in nature

9 Role of miRNAs in Stress Responses

Many plant genes respond to biotic and abiotic stresses. Stress response causes plants to over- or under express certain miRNAs or to synthesize new miRNAs to cope with stress. Several stress-regulated miRNAs have been identified in model plants under various biotic and abiotic stress conditions. Most of the stress-regulated miRNAs do not function independently but rather are involved in overlapping regulatory networks (Khraiwesh et al. 2012). Stress-responsive miRNAs mostly target developmental, stress, and defense-related genes indicating that plant miRNAs can be induced by stress and function in critical defense systems for structural and mechanical stability (Khraiwesh et al. 2012).

Plants and animals employ small RNA-mediated gene silencing as an important mechanism for host immunity against bacterial, viral, and fungal pathogens (Padmanabhan et al. 2009; Gibbings and Voinnet 2010; Khraiwesh et al. 2012). Several recent studies suggest that endogenous small RNA-mediated gene silencing may serve as one of the important mechanisms for gene expression reprogramming in plant immune responses. A number of biotic stress-responsive miRNAs produced in plants infected by pathogenic bacteria, viruses, nematodes, and fungi have been reported (Khraiwesh et al. 2012). miR393 is induced by bacterial elicitor flg22 and positively contributes to pathogen-associated molecular pattern (PAMP)-triggered immunity (PTI) by silencing auxin receptors and subsequently suppressing auxin signaling (Navarro et al. 2006). Auxin promotes plant growth, which provides carbon and nitrogen sources for biotrophic pathogens and increases their virulence and susceptibility. In addition, auxin can also promote pathogenesis by suppressing salicylic acid (SA)-mediated defense responses (Grant and Jones 2009). miR160, miR167, miR390, and miR393 were found to regulate the expression of a group of signaling genes that were involved in various steps of auxin signaling and thus contributed to the inhibition of pathogen growth (Zhang et al. 2011). Recent discovery also revealed that miR393-guided posttranscriptional regulation plays a crucial role in the plant defense against pathogens through targeting transport inhibitor response 1 (TIR1), an auxin receptor (Navarro et al. 2006). Transgenic *Arabidopsis* overexpressing miR393a shows enhanced resistance to virulent *P. syringae* pv. tomato (Shukla et al. 2008).

Small RNA profiling analysis using deep sequencing revealed that in addition to miR393, miR167 and miR160 were also induced by a nonpathogenic *Pseudomonas syringae* pv. tomato (Pst) DC3000 strain with a mutated type III secretion system hrcC (Fahlgren et al. 2007). Profiling of AGO1-bound small RNAs after flg22 treatment revealed the role of miR160a, miR398b, and miR773 in callose deposition, indicating their involvement in PAMP-triggered immunity (Li et al. 2010a). miR398 is involved in both biotic and abiotic responses (Sunkar et al. 2006; Jagadeeswaran et al. 2009). It is downregulated by bacterial pathogen Pst DC3000 (avrRpt2) or Pst (avrRpm1). In *Pinus taeda*, the expression of ten tested miRNA (pta-miRNA) families was significantly repressed in the fusiform rust fungus (*Cronartium quercuum*) of infected galled stem compared to healthy stem (Lu et al. 2007). Furthermore, the expression of approximately 82 plant disease-related transcripts was detected to be altered in response to miRNA regulation in pine (Lu et al. 2007). In *Brassica*, miR1885, which was predicted to target TIR-NBS-LRR class genes, was induced in Turnip mosaic virus-infected plants (He et al. 2008). Another report in loblolly pine indicated that expressions of 10 miRNAs were decreased in response to the rust fungus (Lu et al. 2007). Recent reports have also indicated that miRNAs play important role in regulating plant rhizobium interaction during nitrogen (N) fixation and tumor formation by *Agrobacterium* (Wang et al. 2009). miRNAs in plants have been found to respond to abiotic stress such as drought, cold, and heavy metals (Sunkar and Zhu 2004; Khraiwesh et al. 2012). The phytohormone abscisic acid (ABA) is involved in plant responses to environmental stresses. The first indication that miRNAs may be involved in ABA-mediated responses came from observation of ABA hypersensitivity in *Arabidopsis* mutant. ABA or gibberellin treatment controlled the expression of miR159 and regulated floral and organ development, and miR159 was upregulated in ABA-treated *Arabidopsis* seedlings. In contrast, miR398a appeared to be downregulated by ABA (Khraiwesh et al. 2012).

In response to key nutrient deprivation, miR395 and miR399 were induced by sulfate and phosphate starvation, respectively (Chiou et al. 2006). miR395 participated in sulfate assimilation and allocation (Jones-Rhoades and Bartel 2004), and miR399 was crucial in maintaining phosphate homeostasis (Chiou et al. 2006). Studies also indicated that miR168, miR17, and miR396 were responsive to high salinity, drought, and cold stress in *Arabidopsis* (Khraiwesh et al. 2012). Several abiotic stresses, including excess heavy metals, led to oxidative stress by generating ROS in plants, and miRNAs were shown to be important for plant responses to heavy metal stress (Sunkar and Zhu 2004; Ding et al. 2011). Northern blot analysis showed that the expression of miR398 in *Arabidopsis* was downregulated by heavy metals, such as copper (Cu) and iron (Fe), and was important for accumulation of Cu, Zn superoxide dismutase (CSD), a scavenger of superoxide radicals (Sunkar and Zhu 2004); miR393 and miR171 responded to heavy metals in *Brassica napus* (Huang et al. 2010) and *Medicago truncatula* (Zhou et al. 2008). Additionally, Huang et al. (2010) isolated 19 potential novel miRNAs that were responsive to Cd, and miRNA microarrays have been used to profile the expression patterns of annotated miRNAs (miRBase Release 11.0; http://www.mirbase.org/)

in rice roots under Cd stress, leading to the identification of 19 Cd-responsive miRNAs. Among them, miR156, miR162, miR166, miR168, miR171, miR390, miR396, miR444, and miR1318 miRNAs were downregulated, and only miR528 was upregulated under Cd stress. Target gene prediction and metal stress-responsive *cis*-element analysis of the Cd-responsive miRNA promoters provided further evidence for the potential involvement of these miRNAs in response to Cd stress. The data set of the identified Cd-responsive miRNAs is potentially important for additional characterization of the molecular mechanisms underlying Cd tolerance in rice. However, further functional analysis of Cd-responsive miRNAs is required to confirm their function in heavy metal tolerance in plants (Ding et al. 2011). Functional analyses have demonstrated that several plant miRNAs play a vital role in plant resistance to abiotic as well as biotic stresses (Mallory and Vaucheret 2006).

There are several methods available for identifying stress-responsive miRNAs, i.e., northern blot, reverse transcription PCR (RT-PCR), conventional sequencing, and microarray analysis. GC content may also be considered an important parameter for predicting stress-regulated miRNAs in plants because of significant increase in the GC content of stress-responsive miRNAs. Using these technologies it is possible to carry out a large-scale survey of the expression patterns of all the annotated miRNAs in a given plant species (Zhao et al. 2007; Liu et al. 2008). Profiling the microRNAs at any given condition is an important area of research. Identifying and understanding the role of microRNAs implicated in regulating the cellular processes has paved way for using them in manipulating the traits of interest through transgenic approach either by overexpressing or through ectopic expression (Zhou and Luo 2013).

Acknowledgment We would like to thank Dr. Prashanth Suravajhala (founder, Bioclues.org) for guidance, enormous patience, and encouraging me at every stage, till the completion of book chapter. We acknowledge Mr. Velu Mani Selvaraj for contributing toward miRNA gene nomenclature and for critical reading that improved the clarity of this chapter. We thank all our lab-mates and family members for their support in writing this chapter.

References

Adai A, Johnson C, Mlotshwa S, Archer-Evans S, Manocha V, Vance V (2005) Computational prediction of miRNAs in *Arabidopsis thaliana*. Genome Res 15:78–91

Arenas-Huertero C, Perez B, Rabanal F, Blanco-Melo D, De la Rosa C, Estrada-Navarrete G (2009) Conserved and novel miRNAs in the legume *Phaseolus vulgaris* in response to stress. Plant Mol Biol 70:385–401

Axtell MJ, Westholm JO, Lai EC (2011) Vive la difference: biogenesis and evolution of microRNAs in plants and animals. Genome Biol 12:221

Barakat A, Wall PK, Diloreto S, Depamphilis CW, Carlson JE (2007) Conservation and divergence of microRNAs in *Populus*. BMC Genomics 8:481

Bartel DP (2004) MicroRNAs: genomics, biogenesis, mechanism, and function. Cell 116:281–297

Berezikov E, Robine N, Samsonova A, Westholm JO, Naqvi A, Hung JH (2011) Deep annotation of *Drosophila melanogaster* microRNAs yields insights into their processing, modification, and emergence. Genome Res 21:203–215

Bonnet E, He Y, Billiau K, Van de Peer Y (2010) TAPIR, a web server for the prediction of plant microRNA targets, including target mimics. Bioinformatics 26:1566–1568

Brennecke J, Stark A, Russell RB, Cohen SM (2005) Principles of microRNA-target recognition. PLoS Biol 3:e85

Bulow L, Bolivar JC, Ruhe J, Brill Y, Hehl R (2012) 'MicroRNA Targets', a new AthaMap web-tool for genome-wide identification of miRNA targets in *Arabidopsis thaliana*. BioData Min 5:7

Chapman EJ, Carrington JC (2007) Specialization and evolution of endogenous small RNA pathways. Nat Rev Genet 8:884–896

Chen X (2005) MicroRNA biogenesis and function in plants. FEBS Lett 579:5923–5931

Chiou TJ, Aung K, Lin SI, Wu CC, Chiang SF, Su CL (2006) Regulation of phosphate homeostasis by MicroRNA in *Arabidopsis*. Plant Cell 18:412–421

Cuperus JT, Fahlgren N, Carrington JC (2011) Evolution and functional diversification of MIRNA genes. Plant Cell 23:431–442

Dai X, Zhao PX (2011) psRNATarget: a plant small RNA target analysis server. Nucleic Acids Res 39:W155–W159

Dezulian T, Remmert M, Palatnik JF, Weigel D, Huson DH (2006) Identification of plant microRNA homologs. Bioinformatics 22:359–360

Ding Y, Chen Z, Zhu C (2011) Microarray-based analysis of cadmium-responsive microRNAs in rice (*Oryza sativa*). J Exp Bot 62:3563–3573

Ding J, Li D, Ohler U, Guan J, Zhou S (2012) Genome-wide search for miRNA-target interactions in *Arabidopsis thaliana* with an integrated approach. BMC Genomics 13(Suppl 3):S3

Dunoyer P, Brosnan CA, Schott G, Wang Y, Jay F, Alioua A (2010) An endogenous, systemic RNAi pathway in plants. EMBO J 29:1699–1712

Fabian MR, Sonenberg N, Filipowicz W (2010) Regulation of mRNA translation and stability by microRNAs. Annu Rev Biochem 79:351–379

Fahlgren N, Howell MD, Kasschau KD, Chapman EJ, Sullivan CM, Cumbie JS (2007) High-throughput sequencing of *Arabidopsis* microRNAs: evidence for frequent birth and death of MIRNA genes. PLoS One 2:e219

Faller M, Guo F (2008) MicroRNA biogenesis: there's more than one way to skin a cat. Biochim Biophys Acta 1779:663–667

Flynt AS, Lai EC (2008) Biological principles of microRNA-mediated regulation: shared themes amid diversity. Nat Rev Genet 9:831–842

Frazier TP, Xie F, Freistaedter A, Burklew CE, Zhang B (2010) Identification and characterization of microRNAs and their target genes in tobacco (*Nicotiana tabacum*). Planta 232:1289–1308

Gibbings D, Voinnet O (2010) Control of RNA silencing and localization by endolysosomes. Trends Cell Biol 20:491–501

Glazinska P, Zienkiewicz A, Wojciechowski W, Kopcewicz J (2009) The putative miR172 target gene InAPETALA2-like is involved in the photoperiodic flower induction of *Ipomoea nil*. J Plant Physiol 166:1801–1813

Grant MR, Jones JD (2009) Hormone (dis)harmony moulds plant health and disease. Science 324:750–752

Griffiths-Jones S (2004) The microRNA registry. Nucleic Acids Res 32:D109–D111

Griffiths-Jones S, Grocock RJ, van Dongen S, Bateman A, Enright AJ (2006) miRBase: microRNA sequences, targets and gene nomenclature. Nucleic Acids Res 34:D140–D144

Guddeti S, Zhang DC, Li AL, Leseberg CH, Kang H, Li XG, Zhai WX, Johns MA, Mao L (2005) Molecular evolution of the rice miR395 gene family. Cell Res 15:631–638

Guo L, Lu Z (2010) The fate of miRNA* strand through evolutionary analysis: implication for degradation as merely carrier strand or potential regulatory molecule? PLoS One 5:e11387

Han J, Pedersen JS, Kwon SC, Belair CD, Kim YK, Yeom KH (2009a) Posttranscriptional cross regulation between Drosha and DGCR8. Cell 136:75–84

Han Y, Luan F, Zhu H, Shao Y, Chen A, Lu C (2009b) Computational identification of microRNAs and their targets in wheat (*Triticum aestivum* L.). Sci China C Life Sci 52:1091–1100

He XF, Fang YY, Feng L, Guo HS (2008) Characterization of conserved and novel microRNAs and their targets, including a TuMV-induced TIR-NBS-LRR class R gene-derived novel miRNA in *Brassica*. FEBS Lett 582:2445–2452

Horwich MD, Li C, Matranga C, Vagin V, Farley G, Wang P (2007) The Drosophila RNA methyltransferase, DmHen1, modifies germline piRNAs and single-stranded siRNAs in RISC. Curr Biol 17:1265–1272

Huang TH, Fan B, Rothschild MF, Hu ZL, Li K, Zhao SH (2007) MiRFinder: an improved approach and software implementation for genome-wide fast microRNA precursor scans. BMC Bioinformatics 8:341

Huang SQ, Xiang AL, Che LL, Chen S, Li H, Song JB (2010) A set of miRNAs from *Brassica napus* in response to sulphate deficiency and cadmium stress. Plant Biotechnol J 8:887–899

Isik M, Korswagen HC, Berezikov E (2010) Expression patterns of intronic microRNAs in *Caenorhabditis elegans*. Silence 1:5

Jagadeeswaran G, Saini A, Sunkar R (2009) Biotic and abiotic stress down-regulate miR398 expression in *Arabidopsis*. Planta 229:1009–1014

Jha A, Shankar R (2011) Employing machine learning for reliable miRNA target identification in plants. BMC Genomics 12:636

Jian X, Zhang L, Li G, Wang X, Cao X, Fang X (2010) Identification of novel stress-regulated microRNAs from *Oryza sativa* L. Genomics 95:47–55

Jones-Rhoades MW, Bartel DP (2004) Computational identification of plant microRNAs and their targets, including a stress-induced miRNA. Mol Cell 14:787–799

Jones-Rhoades MW, Bartel DP, Bartel B (2006) MicroRNAS and their regulatory roles in plants. Annu Rev Plant Biol 57:19–53

Joshi PK, Gupta D, Nandal UK, Khan Y, Mukherjee SK, Sanan-Mishra N (2012) Identification of mirtrons in rice using MirtronPred: a tool for predicting plant mirtrons. Genomics 99:370–375

Kadri S, Hinman V, Benos PV (2009) HHMMiR: efficient de novo prediction of microRNAs using hierarchical hidden Markov models. BMC Bioinformatics 10 (Suppl 1):S35

Khraiwesh B, Ossowski S, Weigel D, Reski R, Frank W (2008) Specific gene silencing by artificial MicroRNAs in *Physcomitrella patens*: an alternative to targeted gene knockouts. Plant Physiol 148:684–693

Khraiwesh B, Zhu JK, Zhu J (2012) Role of miRNAs and siRNAs in biotic and abiotic stress responses of plants. Biochim Biophys Acta 1819:137–148

Kidner CA, Martienssen RA (2005) The developmental role of microRNA in plants. Curr Opin Plant Biol 8:38–44

Kojima S, Shingle DL, Green CB (2011) Post-transcriptional control of circadian rhythms. J Cell Sci 124:311–320

Lee RC, Feinbaum RL, Ambros V (1993) The *C. elegans* heterochronic gene lin-4 encodes small RNAs with antisense complementarity to lin-14. Cell 75:843–854

Lee Y, Kim M, Han J, Yeom KH, Lee S, Baek SH (2004) MicroRNA genes are transcribed by RNA polymerase II. EMBO J 23:4051–4060

Li D, Zheng Y, Wan L, Zhu X, Wang Z (2009) Differentially expressed microRNAs during solid endosperm development in coconut (*Cocos nucifera* L.). Sci Hortic 122:666–669

Li Y, Zhang Q, Zhang J, Wu L, Qi Y, Zhou JM (2010a) Identification of microRNAs involved in pathogen-associated molecular pattern-triggered plant innate immunity. Plant Physiol 152:2222–2231

Li YF, Zheng Y, Addo-Quaye C, Zhang L, Saini A, Jagadeeswaran G (2010b) Transcriptome-wide identification of microRNA targets in rice. Plant J 62:742–759

Li B, Qin Y, Duan H, Yin W, Xia X (2011a) Genome-wide characterization of new and drought stress responsive microRNAs in *Populus euphratica*. J Exp Bot 62:3765–3779

Li H, Dong Y, Yin H, Wang N, Yang J, Liu X (2011b) Characterization of the stress associated microRNAs in *Glycine max* by deep sequencing. BMC Plant Biol 11:170

Liang C, Zhang X, Zou J, Xu D, Su F, Ye N (2010) Identification of miRNA from *Porphyra yezoensis* by high-throughput sequencing and bioinformatics analysis. PLoS One 5:e10698

Lindow M, Krogh A (2005) Computational evidence for hundreds of non-conserved plant microRNAs. BMC Genomics 6:119

Liu Q, Chen YQ (2010) A new mechanism in plant engineering: the potential roles of microRNAs in molecular breeding for crop improvement. Biotechnol Adv 28:301–307

Liu J, Carmell MA, Rivas FV, Marsden CG, Thomson JM, Song JJ (2004) Argonaute2 is the catalytic engine of mammalian RNAi. Science 305:1437–1441

Liu HH, Tian X, Li YJ, Wu CA, Zheng CC (2008) Microarray-based analysis of stress-regulated microRNAs in *Arabidopsis thaliana*. RNA 14:836–843

Llave C (2004) MicroRNAs: more than a role in plant development? Mol Plant Pathol 5:361–366

Lu S, Sun YH, Amerson H, Chiang VL (2007) MicroRNAs in loblolly pine (*Pinus taeda* L.) and their association with fusiform rust gall development. Plant J 51:1077–1098

Mallory AC, Vaucheret H (2006) Functions of microRNAs and related small RNAs in plants. Nat Genet 38(Suppl):S31–S36

Margis R, Fusaro AF, Smith NA, Curtin SJ, Watson JM, Finnegan EJ (2006) The evolution and diversification of Dicers in plants. FEBS Lett 580:2442–2450

Meister G, Landthaler M, Patkaniowska A, Dorsett Y, Teng G, Tuschl T (2004) Human Argonaute2 mediates RNA cleavage targeted by miRNAs and siRNAs. Mol Cell 15:185–197

Mhuantong W, Wichadakul D (2009) MicroPC (microPC): a comprehensive resource for predicting and comparing plant microRNAs. BMC Genomics 10:366

Milev I, Yahubyan G, Minkov I, Baev V (2011) MiRTour: plant miRNA and target prediction tool. Bioinformation 6:248–249

Millar AA, Waterhouse PM (2005) Plant and animal microRNAs: similarities and differences. Funct Integr Genomics 5:129–135

Molnar A, Schwach F, Studholme DJ, Thuenemann EC, Baulcombe DC (2007) MiRNAs control gene expression in the single-cell alga *Chlamydomonas reinhardtii*. Nature 447:1126–1129

Moxon S, Schwach F, Dalmay T, Maclean D, Studholme DJ, Moulton V (2008) A toolkit for analysing large-scale plant small RNA datasets. Bioinformatics 24:2252–2253

Navarro L, Dunoyer P, Jay F, Arnold B, Dharmasiri N, Estelle M (2006) A plant miRNA contributes to antibacterial resistance by repressing auxin signaling. Science 312:436–439

Okamura K, Chung WJ, Ruby JG, Guo H, Bartel DP, Lai EC (2008) The *Drosophila* hairpin RNA pathway generates endogenous short interfering RNAs. Nature 453:803–806

Ossowski S, Schwab R, Weigel D (2008) Gene silencing in plants using artificial microRNAs and other small RNAs. Plant J 53:674–690

Padmanabhan C, Zhang X, Jin H (2009) Host small RNAs are big contributors to plant innate immunity. Curr Opin Plant Biol 12:465–472

Park W, Li J, Song R, Messing J, Chen X (2002) CARPEL FACTORY, a Dicer homolog, and HEN1, a novel protein, act in microRNA metabolism in *Arabidopsis thaliana*. Curr Biol 12:1484–1495

Park MY, Wu G, Gonzalez-Sulser A, Vaucheret H, Poethig RS (2005) Nuclear processing and export of microRNAs in *Arabidopsis*. Proc Natl Acad Sci U S A 102:3691–3696

Piriyapongsa J, Jordan IK (2007) A family of human microRNA genes from miniature inverted-repeat transposable elements. PLoS One 2:e203

Piriyapongsa J, Jordan IK (2008) Dual coding of siRNAs and miRNAs by plant transposable elements. RNA 14:814–821

Ramachandran V, Chen X (2008) Degradation of microRNAs by a family of exoribonucleases in *Arabidopsis*. Science 321:1490–1492

Reinhart BJ, Weinstein EG, Rhoades MW, Bartel B, Bartel DP (2002) MicroRNAs in plants. Genes Dev 16:1616–1626

Ruan MB, Zhao YT, Meng ZH, Wang XJ, Yang WC (2009) Conserved miRNA analysis in *Gossypium hirsutum* through small RNA sequencing. Genomics 94:263–268

Ruby JG, Jan CH, Bartel DP (2007a) Intronic microRNA precursors that bypass Drosha processing. Nature 448:83–86

Ruby JG, Stark A, Johnston WK, Kellis M, Bartel DP, Lai EC (2007b) Evolution, biogenesis, expression, and target predictions of a substantially expanded set of Drosophila microRNAs. Genome Res 17:1850–1864

Sanan-Mishra N, Mukherjee SK (2007) A peep into the plant miRNA world. Open Plant Sci J 2007(1):1–9

Sanan-Mishra N, Kumar V, Sopory SK, Mukherjee SK (2009) Cloning and validation of novel miRNA from basmati rice indicates cross talk between abiotic and biotic stresses. Mol Genet Genomics 282:463–474

Schwab R, Ossowski S, Riester M, Warthmann N, Weigel D (2006) Highly specific gene silencing by artificial microRNAs in *Arabidopsis*. Plant Cell 18:1121–1133

Shukla LI, Chinnusamy V, Sunkar R (2008) The role of microRNAs and other endogenous small RNAs in plant stress responses. Biochim Biophys Acta 1779:743–748

Sibley CR, Seow Y, Saayman S, Dijkstra KK, El Andaloussi S, Weinberg MS (2012) The biogenesis and characterization of mammalian microRNAs of mirtron origin. Nucleic Acids Res 40:438–448

Song C, Fang J, Li X, Liu H, Thomas Chao C (2009) Identification and characterization of 27 conserved microRNAs in *citrus*. Planta 230:671–685

Subramanian S, Fu Y, Sunkar R, Barbazuk WB, Zhu JK, Yu O (2008) Novel and nodulation-regulated microRNAs in soybean roots. BMC Genomics 9:160

Sunkar R, Zhu JK (2004) Novel and stress-regulated microRNAs and other small RNAs from *Arabidopsis*. Plant Cell 16:2001–2019

Sunkar R, Kapoor A, Zhu JK (2006) Posttranscriptional induction of two Cu/Zn superoxide dismutase genes in *Arabidopsis* is mediated by down regulation of miR398 and important for oxidative stress tolerance. Plant Cell 18:2051–2065

Sunkar R, Li YF, Jagadeeswaran G (2012) Functions of microRNAs in plant stress responses. Trends Plant Sci 17:196–203

Teune JH, Steger G (2010) NOVOMIR: De novo prediction of microRNA-coding regions in a single plant-genome. J Nucleic Acids 2010

Unver T, Namuth-Covert DM, Budak H (2009) Review of current methodological approaches for characterizing microRNAs in plants. Int J Plant Genomics 2009:262463

Vaucheret H (2006) Post-transcriptional small RNA pathways in plants: mechanisms and regulations. Genes Dev 20:759–771

Voinnet O (2009) Origin, biogenesis, and activity of plant microRNAs. Cell 136:669–687

Wang XJ, Reyes JL, Chua NH, Gaasterland T (2004) Prediction and identification of *Arabidopsis thaliana* microRNAs and their mRNA targets. Genome Biol 5:R65

Wang X, Zhang J, Li F, Gu J, He T, Zhang X (2005) MicroRNA identification based on sequence and structure alignment. Bioinformatics 21:3610–3614

Wang L, Wang MB, Tu JX, Helliwell CA, Waterhouse PM, Dennis ES (2007) Cloning and characterization of microRNAs from *Brassica napus*. FEBS Lett 581:3848–3856

Wang Y, Li P, Cao X, Wang X, Zhang A, Li X (2009) Identification and expression analysis of miRNAs from nitrogen-fixing soybean nodules. Biochem Biophys Res Commun 378:799–803

Wang L, Liu H, Li D, Chen H (2011) Identification and characterization of maize microRNAs involved in the very early stage of seed germination. BMC Genomics 12:154

Wang M, Wang Q, Wang B (2012) Identification and characterization of microRNAs in Asiatic cotton (*Gossypium arboreum* L.). PLoS One 7:e33696

Warthmann N, Chen H, Ossowski S, Weigel D, Hervé P (2008) Highly specific gene silencing by artificial miRNAs in rice. PLoS One 3:e1829

Watanabe T, Totoki Y, Toyoda A, Kaneda M, Kuramochi-Miyagawa S, Obata Y (2008) Endogenous siRNAs from naturally formed dsRNAs regulate transcripts in mouse oocytes. Nature 453:539–543

Werner S, Wollmann H, Schneeberger K, Weigel D (2010) Structure determinants for accurate processing of miR172a in *Arabidopsis thaliana*. Curr Biol 20:42–48

Westholm JO, Lai EC (2011) Mirtrons: microRNA biogenesis via splicing. Biochimie 93:1897–1904

Wu Y, Wei B, Liu H, Li T, Rayner S (2011) MiRPara: a SVM-based software tool for prediction of most probable microRNA coding regions in genome scale sequences. BMC Bioinformatics 12:107

Wu HJ, Ma YK, Chen T, Wang M, Wang XJ (2012) PsRobot: a web-based plant small RNA meta-analysis toolbox. Nucleic Acids Res 40:W22–W28

Xie F, Zhang B (2010) Target-align: a tool for plant microRNA target identification. Bioinformatics 26:3002–3003

Xie F, Xiao P, Chen D, Xu L, Zhang B (2012) miRDeepFinder: a miRNA analysis tool for deep sequencing of plant small RNAs. Plant Mol Biol 80:75–84

Xin M, Wang Y, Yao Y, Xie C, Peng H, Ni Z (2010) Diverse set of microRNAs are responsive to powdery mildew infection and heat stress in wheat (*Triticum aestivum* L.). BMC Plant Biol 10:123

Xue C, Li F, He T, Liu GP, Li Y, Zhang X (2005) Classification of real and pseudo microRNA precursors using local structure-sequence features and support vector machine. BMC Bioinformatics 6:310

Yang X, Li L (2011) MiRDeep-P: a computational tool for analyzing the microRNA transcriptome in plants. Bioinformatics 27:2614–2615

Yang Z, Ebright YW, Yu B, Chen X (2006) HEN1 recognizes 21–24 nt small RNA duplexes and deposits a methyl group onto the 2′ OH of the 3′ terminal nucleotide. Nucleic Acids Res 34:667–675

Yang JH, Shao P, Zhou H, Chen YQ, Qu LH (2010a) DeepBase: a database for deeply annotating and mining deep sequencing data. Nucleic Acids Res 38: D123–D130

Yang JS, Maurin T, Robine N, Rasmussen KD, Jeffrey KL, Chandwani R (2010b) Conserved vertebrate mir-451 provides a platform for Dicer-independent, Ago2-mediated microRNA biogenesis. Proc Natl Acad Sci U S A 107:15163–15168

Yang JH, Li JH, Shao P, Zhou H, Chen YQ, Qu LH (2011a) StarBase: a database for exploring microRNA-mRNA interaction maps from Argonaute CLIP-Seq and Degradome-Seq data. Nucleic Acids Res 39:D202–D209

Yang Y, Chen X, Chen J, Xu H, Li J, Zhang Z (2011b) Differential miRNA expression in *Rehmannia glutinosa* plants subjected to continuous cropping. BMC Plant Biol 11:53

Yousef M, Nebozhyn M, Shatkay H, Kanterakis S, Showe LC, Showe MK (2006) Combining multi-species genomic data for microRNA identification using a Naive Bayes classifier. Bioinformatics 22:1325–1334

Yu J, Wang F, Yang GH, Wang FL, Ma YN, Du ZW (2006) Human microRNA clusters: genomic organization and expression profile in leukemia cell lines. Biochem Biophys Res Commun 349:59–68

Yu B, Bi L, Zheng B, Ji L, Chevalier D, Agarwal M (2008) The FHA domain proteins DAWDLE in *Arabidopsis* and SNIP1 in humans act in small RNA biogenesis. Proc Natl Acad Sci U S A 105:10073–10078

Zeng C, Wang W, Zheng Y, Chen X, Bo W, Song S (2010) Conservation and divergence of microRNAs and their functions in *Euphorbiaceous* plants. Nucleic Acids Res 38:981–995

Zhang BH, Pan XP, Wang QL, Cobb GP, Anderson TA (2005) Identification and characterization of new plant microRNAs using EST analysis. Cell Res 15:336–360

Zhang B, Pan X, Cannon CH, Cobb GP, Anderson TA (2006) Conservation and divergence of plant microRNA genes. Plant J 46:243–259

Zhang B, Pan X, Stellwag EJ (2008) Identification of soybean microRNAs and their targets. Planta 229:161–182

Zhang J, Xu Y, Huan Q, Chong K (2009a) Deep sequencing of *Brachypodium* small RNAs at the global genome level identifies microRNAs involved in cold stress response. BMC Genomics 10:449

Zhang L, Chia JM, Kumari S, Stein JC, Liu Z, Narechania A (2009b) A genome-wide characterization of microRNA genes in maize. PLoS Genet 5:e1000716

Zhang Y, Zhang R, Su B (2009c) Diversity and evolution of MicroRNA gene clusters. Sci China C Life Sci 52:261–266

Zhang W, Gao S, Zhou X, Xia J, Chellappan P, Zhang X (2010) Multiple distinct small RNAs originate from the same microRNA precursors. Genome Biol 11:R81

Zhang W, Gao S, Zhou X, Chellappan P, Chen Z, Zhang X (2011) Bacteria-responsive microRNAs regulate plant innate immunity by modulating plant hormone networks. Plant Mol Biol 75:93–105

Zhao B, Liang R, Ge L, Li W, Xiao H, Lin H (2007) Identification of drought-induced microRNAs in rice. Biochem Biophys Res Commun 354:585–590

Zhou M, Luo H (2013) MicroRNA-mediated gene regulation: potential applications for plant genetic engineering. Plant Mol Biol 83:59–75

Zhou X, Ruan J, Wang G, Zhang W (2007) Characterization and identification of microRNA core promoters in four model species. PLoS Comput Biol 3:e37

Zhou ZS, Huang SQ, Yang ZM (2008) Bioinformatic identification and expression analysis of new microRNAs from *Medicago truncatula*. Biochem Biophys Res Commun 374:538–542

Zhu QH, Spriggs A, Matthew L, Fan L, Kennedy G, Gubler F (2008) A diverse set of microRNAs and microRNA-like small RNAs in developing rice grains. Genome Res 18:1456–1465

Zuker M (2003) Mfold web server for nucleic acid folding and hybridization prediction. Nucleic Acids Res 31:3406–3415

ESTs in Plants: Where Are We Heading?

Sameera Panchangam, Nalini Mallikarjuna, and Prashanth Suravajhala

Abstract

Expressed sequence tags (ESTs) are the most important resources for transcriptome exploration. Next-generation sequencing technologies have been generating gigabytes of genetic codes representing genes, partial and whole genomes most of which are EST datasets. Niche of EST in plants for breeding, regulation of gene expression through miRNA studies, and their application for adapting to climatic changes are discussed. Some of the recent tools for analysis of EST exclusive to plants are listed out. Systems biology though in its infancy in plants has influenced EST mapping for unraveling gene regulatory circuits, which is illustrated with a few significant examples. This review throws a glance at the evolving role of ESTs in plants.

Keywords

Expressed Sequence Tags (ESTs) • Plant ESTs • EST analysis pipelines • miRNA • Systems biology

1 Introduction

Bioinformatics has provided us an impetus to learn systems biology. The bioinformatics tools have not only allowed us to understand what systems biology could make use of but also on how it dissects the behavior of complex biological organization and processes in terms of molecular constituents. It involves the study of all genes expressed as messenger RNAs and characterization of the proteins and metabolites under different conditions (Kirschner 2005). Significant advancement in high-throughput (HT) technologies such as microarrays, automated sequencing, and mass spectrometry has generated huge amount of data which can be optimized by various computational tools for accelerated process of discovery. Access to a number of next-generation sequencing (NGS) technologies such as Roche/454, Illumina, and ABI SOLiD has drastically reduced the cost and time of sequencing and increased the length of sequence reads. These NGS technologies are being utilized

S. Panchangam (✉) • N. Mallikarjuna
Department of Cell Biology, International Crop Research, Institute for Semi-Arid Tropics, Patancheru 502319, AP, India
e-mail: sameera.panchangam@gmail.com

P. Suravajhala
Bioclues Organization, IKP Knowledge Park, Picket, Secunderabad 500 009, Andhra Pradesh, India

K.K. P.B. et al. (eds.), *Agricultural Bioinformatics*,
DOI 10.1007/978-81-322-1880-7_9, © Springer India 2014

for de novo sequencing, genome re-sequencing, whole genome, and transcriptome analysis (Morozova and Marra 2008). Despite these advantages and availability of whole genome sequences of more than 180 organisms (http://www.genomenewsnetwork.org/; http://www.ebi.ac.uk/genomes/), the plethora of datasets constituting umpteen genomes is not fully understood. Therefore, it is believed that expressed sequences tags (EST) especially from unsequenced genomes will continue to play an important role in post genome sequencing and will apply NGS technologies in transcriptome sequencing. "Poor man's genome" as they are known, ESTs are short (200–800 nucleotide bases in length), unedited, randomly selected, single-pass sequence reads derived from cDNA libraries (Adams et al. 1991; Nagaraj et al. 2006). Since the use of ESTs as the primary source of human gene discovery in 1991, there has been manifold growth in the generation and accumulation of EST data for a range of organisms from bacteria to vertebrates (Lee and Shin 2009). In combination with NGS, ESTs have proven to be an extremely valuable resource for high-throughput gene discovery, identification of novel genes, splice variants, gene location, and intron–exon boundaries within genomic sequence assemblies. They are a cost-effective alternative to whole genome sequencing (WGS), for annotation of genes and development of molecular markers in organisms with large genome size and in species which lack draft genome sequences (Dias et al. 2000).

2 Identifying Niche of ESTs for Desired Traits in Plants

Plant breeders constantly strive to develop improved varieties of crops for desirable traits through conventional breeding techniques which are laborious and time-consuming as careful phenotypic and genotypic selection is needed. Most of the traits of interest in plant breeding such as high yield, height, drought resistance, disease resistance in many species, etc., are quantitative, also called polygenic, continuous, multifactorial, or complex traits, which further complicate the breeding program (Semagn et al. 2010). However, advances in genomics and DNA marker technology have helped to develop molecular markers, which are now widely used to track loci and genome regions in several crop-breeding programs. With this molecular markers tightly linked with a large number of agronomic and disease resistance, traits have become available in major crop species (Jain et al. 2002; Gupta and Varshney 2004). Some sequence tagged sites (STS) are also enriched and have potentially been used as markers for PCR (polymerase chain reaction). Most of these markers developed in the past were related to genomic DNA (gDNA) and therefore could belong to either the transcribed region or the non-transcribed region of the genome. These markers were termed as random DNA markers (RDMs) (Andersen and Lübberstedt 2003). As a result, a large number of genes have been identified in the recent past through "wet lab" as well as in silico studies, and a wealth of sequence data have been accumulated in public databases (e.g., http://www.ncbi.nlm.nih.gov; http://www.ebi.ac.uk) in the form of BAC (bacterial artificial chromosome) clones, ESTs, full-length cDNA clones, and genes. The availability of enormous amount of sequence data from complete or partial genes has made it possible to develop molecular markers directly from the parts of genes (Varshney et al. 2007). Genic molecular markers (GMMs) that developed from coding sequences like ESTs or fully characterized genes frequently have been assigned known functions. EST-based markers such as SSRs (simple sequence repeats), RFLPs (restriction fragment length polymorphisms), AFLPs (amplified fragment length polymorphisms), and SNPs (single nucleotide polymorphisms) and novel markers such as expressed sequence tag polymorphisms (ESTPs), conserved orthologous set (COS) markers, etc., have been developed for many crop species (Gupta and Rustgi 2004). Orphan crops like peanut, sorghum and millets, groundnut, cowpea, common bean, chickpea, pigeon pea, cassava, yam, and sweet potato (Varshney et al. 2012) and many other important horticultural and forest species with large and complex genomes whose whole genome sequences are not yet available greatly benefit from the EST data.

For example, genes encoding key enzymes for fatty acid and seed storage protein biosynthesis, bacterial wilt disease, and novel genes discovered in peanut were derived from ESTs belonging to different tissues, different growth stages, and under different abiotic and biotic stresses (Feng et al. 2012).

More recently, microRNAs (miRNA) have received a lot of attention due to their role in regulation of gene expression which finds applications in functional genomics and study of various pathways in organisms. In plants, miRNAs are involved in diverse aspects of growth and development such as leaf morphology and polarity, root formation, transition from embryogenic to vegetative phase, flowering time, floral organ identity, and reproduction (Mallory and Vaucheret 2006; Sun 2012). They are also found to be involved in defense mechanisms, hormone signaling, and abiotic and biotic stress responses (Lu et al. 2008). 21,264 entries representing hairpin precursor miRNAs, expressing 25,141 mature miRNA products, in 193 (>170 plants) species are available (www.mirbase.org/). It is generally accepted that plant miRNAs have extensive complementarity to their targets, and their prediction usually relies on the use of empirical parameters deduced from known miRNA–target interactions. The biogenesis of miRNAs suggests that it is possible to find new miRNAs by homology searching of known miRNAs in ESTs, especially in plants whose whole genome sequence data is unavailable (Sunkar and Jagadeeswaran 2008). Since ESTs represent transcribed sequences, their analyses provide direct evidence for miRNA expression through simple tools for comparative genomics which in turn helps in identification of conserved miRNAs (Zhang et al. 2005). Both experimental methods and computational approaches have been adopted to identify miRNAs in plants, and the latter has been identified as the simplest and most effective method (Sun 2012). Several groups have attempted to identify novel miRNAs and decode their interaction with protein coding transcripts by examining ESTs (Nasaruddin et al. 2007; Das and Mondal 2010; Boopathi and Pathmanaban 2012; Muvva et al. 2012). Despite the tremendous applications of miRNA in plant biotechnology and the growing interest, our knowledge about the regulatory mechanisms and functions of miRNAs remains very limited (Liu et al. 2012). The limited number of experimentally validated miRNA targets, the spatio-temporal specific regulation of miRNA, and the lack of graphical-user interface models without the need for programming skills are major constraints. However, user-friendly software packages, which enable computational identification of miRNA and its target (C-mii), functional annotation of miRNAs (miRFANS), transcription factor–miRNA regulation (TransmiR), PMRD, etc., are now publicly available which are exclusive to plants (Liu et al. 2012; Numnark et al. 2012).

"Climate change," "sustainable agriculture," and "Ecogenomics" are some of the paradigms that have influenced researches of late. Genomics and bioinformatics have great potential in addressing various topics in these areas through approaches such as association mapping, genome scans, transcript profiling, and gene regulatory networks, thus leading to an understanding of the genetic architecture of climate change adaptation (Franks and Hoffman 2012). Gene transcription profiling, in particular, is one important step toward identifying those genes and metabolic pathways that underlie ecologically important traits, and ESTs can bridge genomics and molecular ecology because they can provide a means of accessing the gene space of almost any organism (Bouck and Vision 2007). EST libraries are a cost-effective tool to characterize genes important under particular conditions, as well as the starting point for the development of molecular genetic markers, such as gene-linked microsatellites and single nucleotide polymorphisms (SNP). In marine species, gene-linked microsatellites (EST-SSR = simple sequence repeats) were successfully identified, for example, in the ecologically important sea grass *Zostera marina* (eelgrass) to elucidate the molecular genetic basis of adaptation to environmental extremes. Approximately one-third of the eelgrass genes were characteristic for the stress response of the terrestrial plant model *Arabidopsis thaliana* (Reusch et al. 2008). Similarly, EST-based SSR markers for breeding of drought-resistant durum wheat in Mediterranean dry lands (Habash et al. 2009), over 400 markers for various traits in important tropical fruits like mango and banana (Arias et al. 2012), and linkage mapping studies and identification of markers for beech bark disease resistance in American beech (Mason et al. 2013) are some recent examples of the potential application of ESTs in varied species for adaptation to climate change.

3 ESTs in Plants: Various Pipelines for EST Analysis

The number of EST entries in GenBank dbEST is 74,186,692 as on January 1, 2013 (http://www.ncbi.nlm.nih.gov/dbEST/dbEST_summary.html). Handling the huge and ever accumulating data efficiently is an important and daunting task (Pertea et al. 2003). Since ESTs are single-pass reads and represent only a small portion of the mRNA, they are prone to errors and inherent deficiencies. Problems such as low-quality regions within the sequence, redundancy, differentially expressed genes in the host, contaminants like vectors, linkers, chimeric sequences, and natural sequence variations need to be dealt with, before further analysis. Several tools have been developed for each of the steps involved in EST analysis in the past few years (Hotz-Wagenblatt et al. 2003; Mao et al. 2003; Kumar et al. 2004; Conesa et al. 2005). A generic protocol of the different steps in the analysis of EST datasets and a list of various tools has been dealt with in considerable detail by Nagaraj et al. (2006). Some of the steps require the use of intensive computing power and an in-depth knowledge of bioinformatics which is not available to small research groups without access to bioinformatics personnel and advanced computer systems. As rightly pointed out by many researches, an ideal EST analysis tool should possess a few characteristics such as (1) to be fully automated in a pipeline covering all the steps from the input chromatogram files to a clean, annotated web-searchable EST database; (2) to be highly modular and adaptable; (3) to be able to run in parallel in a personal computer (PC) cluster, thus benefiting from the multiprocessing capabilities of these systems; (4) to use third-party freely available programs, in order to ease the incorporation of the improvements made by others programmers; (5) to include a highly configurable and extensible user-friendly interface to perform data mining by combining any search criteria, fitting the final user needs; and (6) to be based on an open-source license to allow a continuous development by a community of users and programmers, as well as its customization for the needs of different projects (Forment et al. 2008). As new tools are being constantly developed and the existing ones being updated to meet the requirements, a few of the most recent tools are listed here (Table 1).

4 Systems Biology and Impact on EST Mining

Structural genomics and, more recently, functional genomics have become the base of sustainable agriculture, forestry, industry, and environment (Campbell et al. 2003; Diouf 2003; Mazur et al. 1999; Somerville and Somerville 1999; Walbot 1999). Much of the efforts were directed toward the identification of markers for agronomic traits and physical and nutritional traits, genes encoding biosynthetic enzymes and production of secondary and intermediary metabolites, and understanding of the biochemical pathways in crop and some forage plants (Girke et al. 2003; Sweetlove et al. 2003; Varshney et al. 2007). Systems biology has created sweeping changes in our approach to genomics and plant biology. The focus now is on the molecular, cellular, and organismic changes in plants such as totipotency (dedifferentiation and regeneration ability), apomixis (vegetative seed production), embryogenesis (somatic, zygotic, and microspore), induction of haploids, heterosis or hybrid vigor, flower development, symbiotic nitrogen fixation, etc. For example, transcriptomic, proteomic, and metabolomic studies have led to a deeper understanding of microspore embryogenesis in barley (*Hordeum vulgare* L.), rapeseed (*Brassica napus* L.), tobacco (*Nicotiana* spp.), wheat (*Triticum aestivum* L.), and maize (*Zea mays*), which are now considered model species to study the mechanisms of stress-induced androgenesis (Maraschin et al. 2005). Analysis of

Table 1 EST analysis tools developed after the year 2006

Name	Description	Category	Reference
EST2uni	Processing, clustering, annotation	F/D	Forment et al. (2008)
ESTPiper	Sequencing, assembly, annotation, probe design	F/W/D	Tang et al. (2009)
ESTPass	Processing, annotation		Lee et al. (2007)
ESMP	EST-SSRs pipeline	F/W	Sarmah et al. (2012)
ParPEST	Parallel computing	RA	D'Agostino et al. (2005)
PESTAS	Processing, assembly, annotation	RA/W	Nam et al. (2009)
SCRAF	Sort and assemble 454-EST sequences	F/W	Barker et al. (2009)
OREST	Analysis, annotation	F/W	Waegele et al. (2008)
ConiferEST	Conifer EST mining, processing, annotation	F/W	
KAIKObase	Silkworm database	F/W	Shimomura et al. (2009)
OrchidBase	Processing, clustering, annotation	F/W/D	Tsai et al. (2013)
GarlicEST	Mining, annotation, expression profiling	F/W	Kim et al. (2009)
TomatoEST	Tomato functional genomics data	F/W	Agostino et al. (2007)
MELOGEN	Melon EST database	RA	González-Ibeas et al. (2007)
bEST-DRRD	Barley ESTs involved in DNA repair and replication	F/W	Gruszka et al. (2012)
MoccaDB	Orthologous markers in Rubiaceae	F/W	Plechakova et al. (2009)

F free, *W* web based, *D* downloadable, *RA* restricted access

20,000 ESTs from fresh and cultured microspores of barley revealed clusters of differentially expressed genes and identification of 16 genes which could serve as markers for induction of androgenesis and progression of microspore embryogenesis (Malik et al. 2007). Strategies with fluorescent-labeled probes for in situ hybridization and immunofluorescence have provided unique images of the spatial and temporal pattern of the expression of genes and proteins and of the subcellular rearrangements that accompany microspore embryogenesis (Testillano and Risueño 2009). Another key trait that has defied scientific unraveling is the phenomenon of heterosis (Bircher et al. 2003). A systems biological approach to define how plant genomes interact to create phenotype is needed to arrive at a final resolution of this phenomenon.

Metabolic engineering and synthetic biology are an integral part of systems biology. From an engineering perspective, synthetic biology insists on standardized parts (e.g., genes, proteins, circuits) that can be assembled using bioinformatics and simulation tools to build functionality (Osbourn et al. 2012). Though they are still at infancy in plant research, the impact of systems biology on plants is ever increasing and well documented (Fernie 2012). Traditionally for gene detection, the two main approaches are EST mapping and computational gene prediction combined with homology-based search methods (Wortman et al. 2003). Cometh systems biology, the combination of two or more approaches, has helped in improved annotation of the genome and identification of novel genes and proteins (Allmer et al. 2006). These technologies provide validation

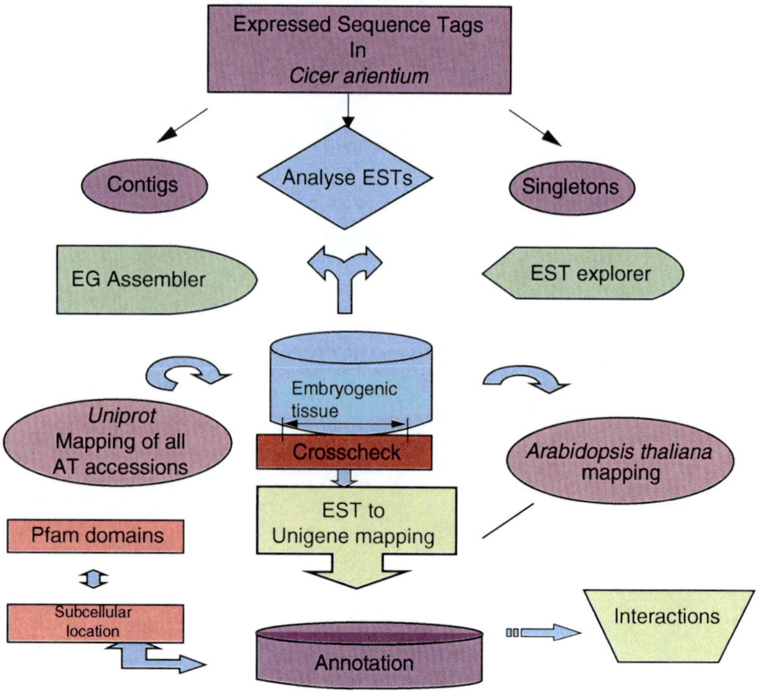

Fig. 1 Steps involved in analysis of Cicer ESTs

of the in silico gene models and enable fast and comprehensive analysis of the molecular plant phenotype (Naumann et al. 2007; Weckwerth 2008) as well as providing complementary means for probing the completeness of genome annotations. A case in example is integrated analysis of the molecular repertoire of *Chlamydomonas reinhardtii,* wherein bioinformatics annotation methods combined with GCxGC/MS-based metabolomics and LC/MS-based shotgun proteomics profiling technologies have been applied to characterize abundant proteins and metabolites, resulting in the detection of 1,069 proteins and 159 metabolites. By integrating genomic annotation information with experimentally identified metabolites and proteins, a draft metabolic network for *Chlamydomonas* was constructed which also provides entry points for further targeted gene discovery or biochemical pathway research (May et al. 2008). Metabolomics integrated with transcriptomic and proteomic studies have led to the identification of key steps involved in response to nitrogen deficiency in maize (Amiour et al. 2012). Yet another example of the application of EST analysis for discerning organization of cells besides predicting biological functions and providing insight into a variety of biochemical processes is the construction of protein interaction networks (PIN) (Guan and Kiss 2008). Despite the availability of advanced methods connecting orthology mapping and comparative approaches for predicting PIN, annotation of those proteins like "predicted" or "similar to" or "hypothetical" poses many challenges. To tackle this, a six-point classification system to validate protein interactions based on diverse features was proposed by Suravajhala and Sundararajan (2012). Using the six-point classification system, the genes related to embryogenesis and apomixis in chickpea were predicted based on the model apomictic plants Poa, Pennisetum, and apomeiotic mutant *Arabidopsis thaliana* (Panchangam et al. 2012). Here, EST analysis pipeline employed for annotation of proteins related to embryogenesis in chickpea is represented as a flowchart (Fig. 1).

Systems biology approach is not limited to crop plants and breeding but is also finding its way into unraveling different metabolic pathways in fruits, vegetables, and aromatic plants. A combined metabolomic, proteomic, and transcriptomic approach was employed to investigate fruit development in tomato which led to identification of a novel gene regulatory mechanism for ethylene biosynthesis during the post climacteric ripening of the fruit (Van de Poel et al. 2012). A similar study was carried out in apple to obtain proteome information on fruit ripening in response ethylene treatment (Zheng et al. 2013). A database of molecular networks occurring in grapevine was built based on EST datasets, leading to 39,423 unique potential genes and proteins. Among them, 7,265 genes have been assigned to 107 pathways, including 86 metabolic pathways, 3 transporter pathways, 9 genetic information processing pathways, and 9 signal pathways focused mainly on phytohormone signaling (Grimplet et al. 2008). Metabolic pathways occurring in many medicinal and aromatic plants have been reviewed by Khanuja et al. (2012).

5 Conclusion and Future Directions

EST analysis holds an important spot in plant breeding by not only aiding the development of molecular markers for traits and annotation of genes but also providing insights into key developmental processes, regulation of gene expression, and to reveal the complete proteomic repertoire of an organism (Nagaraj et al. 2007). Although EST databases are no substitute for whole genome scaffolds, they certainly played a key role in pre-genome sequencing era and will continue to be promising resources for various in vitro and in silico experiments (Feng et al. 2012). The ability to generate large amount of data has become quick and cheap due to NGS technologies and has transformed various areas of biology which were previously unattainable, particularly for non-model systems that lack extensive genomic resources. Next-generation sequencing has great potential for accurate transcriptome characterization because of the large amount of data obtained at considerably lower costs compared to traditional methods, and with the decreasing costs transcriptome sequencing will be dramatically improved in the near future. EST sequencing along with NGS technologies is revolutionizing applications that revolve around gene expression. With deeper sequencing (e.g., 6–20 plates), researchers attain a level of transcriptome that has never been possible before due to the higher cost of earlier technologies. Not only will these studies sequence more than 90 % of the transcriptome, the coverage per gene will approach traditional sequencing. This should allow researchers to use these genes to identify pathways and determine tissue-specific expression for lowly expressed genes and will be critical for genome annotation (Kerr Wall et al. 2009). In retrospection, ESTs thus do not lose to whole genome sequencing, but coupled with NGS technologies and simulation/computational tools, they have revolutionary applications for both sequenced and unsequenced genomes. The large-scale development of tools for analysis of genes, transcripts, and proteins has generated vast data which holds great promise for revealing novel plant biology. The focus now is a systems perspective with the cumulative "omics" approach (e.g., genomics, epigenomics, transcriptomics, proteomics, metabolomics, interactomics, ionomics, phenomics, etc.) (Liberman et al. 2012). The way to sustainable agriculture in the very near future is to move from genetic manipulation of parts of genomes to more engineering-based approach, combining traditional plant breeding techniques with systems biology and predictive science.

References

Adams MD, Kelley JM, Gocayne JD, Dubnick M, Polymeropoulos MH, Xiao H, Merril CR, Wu A, Olde B, Moreno R, Kerlavage AR, Mccombie WR, Venter JC (1991) Complementary DNA sequencing: expressed sequence tags and the human genome project. Science 252:1651–1656

Allmer J, Naumann B, Markert C, Zhang M, Hippler M (2006) Mass spectrometric genomic data mining: novel insights into bioenergetic pathways in *Chlamydomonas reinhardtii*. Proteomics 6:6207–6220

Amiour N, Imbaud S, Clément G, Agier N, Zivy M, Valot B, Balliau T, Armengaud P, Quilleré I, Cañas R, Tercet-Laforgue T, Hirel B (2012) The use of metabolomics integrated with transcriptomic and proteomic studies for identifying key steps involved in the control of nitrogen metabolism in crops such as maize. J Exp Bot 63(14):5017–5033

Andersen JR, Lübberstedt T (2003) Functional markers in plants. Trends Plant Sci 8:554–559

Arias RS, Borrone JW, Tondo CL, Kuhn DN, Irish BM, Schnell RJ (2012) Genomics of tropical fruit crops. In: Schnell RJ, Priyadarshan PM (eds) Genomics of tree crops. Springer, Dordrecht, p 209

Barker MS, Dlugosch KM, Reddy ACC, Amyotte SN, Rieseberg LH (2009) SCARF: maximizing next-generation EST assemblies for evolutionary and population genomic analyses. Bioinformatics 25 (4):535–536

Bircher JA, Auger DL, Riddle NC (2003) In search of the molecular basis of heterosis. Plant Cell 15:2236–2239

Boopathi N, Pathmanaban R (2012) Additional insights into the adaptation of cotton plants under abiotic stresses by in silico analysis of conserved miRNAs in cotton expressed sequence tag database (dbEST). Afr J Biotechnol 11(76):14054–14063

Bouck A, Vision T (2007) The molecular ecologists' guide to expressed sequence tags. Mol Ecol 16 (5):907–924

Campbell MM, Brunner AM, Jones HM, Strauss SH (2003) Forestry's fertile crescent: the application of biotechnology to forest trees. Plant Biotechnol J 1:141–154

Conesa A, Gotz S, Garcia-Gomez JM et al (2005) Blast2GO: a universal tool for annotation, visualization and analysis in functional genomics research. Bioinformatics 21:3674–3676

D'Agostino N, Aversano M, Chiusano ML (2005) ParPEST: a pipeline for EST data analysis based on parallel computing. BMC Bioinform 6(4):9

D'Agostino N, Aversano M, Chiusano ML (2007) Tomato EST database: *in silico* exploitation of EST data to explore expression patterns in tomato species. Nucleic Acids Res 35:901–905

Das A, Mondal TK (2010) Computational identification of conserved microRNAs and their targets in Tea (*Camellia sinensis*). Am J Plant Sci 1:77–86

Dias NE, Correa RG, Verjovski-Almeida S, Briones MR, Nagai MA, Wilson DS, Zago MA, Bordin S, Costa FF, Goldman GH et al (2000) Shotgun sequencing of the human transcriptome with ORF expressed sequence tags. Proc Natl Acad Sci U S A 97:3491–3496

Diouf D (2003) Genetic transformation of trees. Afr J Biotechnol 2:328–333

Feng S, Wang X, Zhang X, Dang PM, Holbrook CC, Culbreath AK, Wu Y, Guo B (2012) Peanut (*Arachis hypogaea*) expressed sequence tag project: progress and application. Comp Funct Genomic 2012:373768

Fernie AR (2012) Grand challenges in plant systems biology: closing the circle(s). Front Plant Sci 3:35

Forment J, Gilabert F, Robles A, Conejero V, Nuez F, Blanca JM (2008) EST2uni: an open, parallel tool for automated EST analysis and database creation, with data mining web interface and microarray expression data integration. BMC Bioinform 9:5

Franks SJ, Hoffman AA (2012) Genetics of climate change adaptation. Annu Rev Genet 12(46):185–208

Girke T, Ozkan M, Carter M, Raikhel NV (2003) Towards a modelling infrastructure for studying plant cells. Plant Physiol 132:410–414

González-Ibeas D, Blanca J, Roig C, González-To M, Picó B, Truniger V, Gómez P, Deleu W, Caño-Delgado A, Arús P, Nuez F, García-Mas J, Puigdomènech P, Aranda MA (2007) MELOGEN: an EST database for melon functional genomics. BMC Genomics 8:306–312

Grimplet J, Dickerson JA, Victor KJ, Cramer GR, Fennell AY (2008) Systems biology of the Grapevine. In Proceedings of the 2nd annual national viticulture research conference, University of California, Davis, 9–11 July

Gruszka D, Marzec M, Szarejko I (2012) The barley EST DNA Replication and Repair Database (bEST-DRRD) as a tool for the identification of the genes involved in DNA replication and repair. BMC Plant Biol 12:88–94

Guan H, Kiss-Toth E (2008) Advanced technologies for studies on protein interactomes. Adv Biochem Eng Biotechnol 110:1

Gupta PK, Rustgi S (2004) Molecular markers from the transcribed/expressed region of the genome in higher plants. Funct Integr Genomic 4:139–162

Gupta PK, Varshney RK (2004) Cereal genomics: an overview. In: Gupta PK, Varshney RK (eds) Cereal genomics. Kluwer Academic, Dordrecht, p 639

Habash DZ, Kehel Z, Nachit M (2009) Genomic approaches for designing durum wheat ready for climate change with a focus on drought. J Exp Bot 60 (10):2805–2815. doi:10.1093/jxb/erp211

Hotz-Wagenblatt A, Hankeln T, Ernst P et al (2003) ESTAnnotator: a tool for high throughput EST annotation. Nucleic Acids Res 31:3716–3719

Jain SM, Brar DS, Ahloowalia BS (2002) Molecular techniques in crop improvement. Kluwer Academic, Dordrecht

Kerr Wall P, Leebens-Mack J, Chanderbali AS, Barakat A, Wolcott E, Liang H, Landherr L, Tomsho LP, Hu Y, Carlson JE, Ma H, Schuster SC, Soltis DE, Soltis PS, Altman N, dePamphilis CW (2009) Comparison of next generation sequencing technologies for transcriptome characterization. BMC Genomics 10:347

Khanuja SPS, Jhang T, Shasany AK (2012) Medicinal and aromatic plants: a case example of evolving secondary metabolome and biochemical pathway diversity. In: Sharma VP (ed) Nature at work: ongoing saga of evolution. Springer, Dordrecht, p 355

Kim D-W, Jung T-S, Nam S-H, Kwon H-R, Kim A, Chae S-H, Choi S-H, Kim D-W, Kim RN, Park H-S (2009) GarlicESTdb: an online database and mining tool for garlic EST sequences. BMC Plant Biol 9:61–67

Kirschner MW (2005) The meaning of systems biology. Cell 121:503–504

Kumar CG, LeDuc R, Gong G et al (2004) ESTIMA, a tool for EST management in a multi-project environment. BMC Bioinform 5:176

Lee B, Shin G (2009) CleanEST: a database of cleansed EST libraries. Nucleic Acids Res 37:686–689

Lee B, Hong T, Byun SJ, Woo T, Choi YJ (2007) ESTpass: a web-based server for processing and annotating expressed sequence tag (EST) sequences. Nucleic Acids Res 35:159–162

Liberman LM, Sozzani R, Benfey PN (2012) Integrative systems biology: an attempt to describe a simple weed. Curr Opin Plant Biol 15(2):162–167

Liu H, Jin T, Liao R, Wan L, Xu B, Zhou S, Guan J (2012) miRFANs: an integrated database for *Arabidopsis thaliana* microRNA function annotations. BMC Plant Biol 12:68

Lu Y, Gan Q, Chi X, Qin S (2008) Roles of microRNA in plant defense and virus offense interaction. Plant Cell Rep 27(10):1571–1579

Malik MR, Wang F, Dirpaul JM, Zhou N, Polowick PL, Ferrie AMR, Krochko JE (2007) Transcript profiling and identification of molecular markers for early microspore embryogenesis in *Brassica napus*. Plant Physiol 144:134–154

Mallory CA, Vaucheret H (2006) Functions of microRNAs and related small RNAs in plants. Nat Genet 38:31–36

Mao C, Cushman JC, May GD, Weller JW (2003) ESTAP – an automated system for the analysis of EST data. Bioinformatics 19:1720–1722

Maraschin SF, De Priester W, Spaink HP, Wang M (2005) Androgenic switch: an example of plant embryogenesis from the male gametophyte perspective. J Exp Bot 417:1711–1726

Mason ME, Koch JL, Krasowski M, Loo J (2013) Comparisons of protein profiles of beech bark disease resistant and susceptible American beech (*Fagus grandifolia*). Proteome Sci 11:2

May P, Wienkoop S, Kempa S, Usadel B, Christian N, Rupprecht J, Weiss J, Recuenco-Munoz L, Ebenhöh O, Weckwerth W, Walther D (2008) Metabolomics- and proteomics-assisted genome annotation and analysis of the draft metabolic network of *Chlamydomonas reinhardtii*. Genetics 179:157–166

Mazur B, Krebbers E, Tingey S (1999) Gene discovery and product development for grain quality traits. Science 285:372–375

Morozova O, Marra MA (2008) Applications of next-generation sequencing technologies in functional genomics. Genomics 92(5):255–264

Muvva C, Tewari L, Aruna K, Ranjit P, Zahoorullah SMD, Matheen KAMD, Veeramachaneni H (2012) In silico identification of miRNAs and their targets from the expressed sequence tags of *Raphanus sativus*. Bioinformation 8:2

Nagaraj SH, Gasser RB, Ranganathan S (2006) A hitchhiker's guide to expressed sequence tag (EST) analysis. Brief Bioinform 8(1):6–21

Nagaraj SH, Deshpande N, Gasser RB, Ranganathan S (2007) ESTExplorer: an expressed sequence tag (EST) assembly and annotation platform. Nucleic Acids Res 35:143–147

Nam S-H, Kim D-W, Jung T-S, Choi Y-S, Kim D-W, Choi H-S, Choi S-H, Park H-S (2009) PESTAS: a web server for EST analysis and sequence mining. Bioinformatics 25(14):1846–1848

Nasaruddin MN, Harikrishna K, Othman YR, Hoon SL, Harikrishna AJ (2007) Computational prediction of microRNAs from Oil Palm (*Elaeis guineensis* Jacq.) expressed sequence tags. Asian Pac J Mol Biol Biotechnol 15(3):107–113

Naumann B, Busch A, Allmer J, Ostendorf E, Zeller M et al (2007) Comparative quantitative proteomics to investigate the remodelling of bioenergetic pathways under iron deficiency in *Chlamydomonas reinhardtii*. Proteomics 7:3964–3979

Numnark S, Mhuantong W, Ingsriswang S, Wichadakul D (2012) C-mii: a tool for plant miRNA and target identification. BMC Genomics 13:7–16

Osbourn AE, O'Maille PE, Rosser SJ, Lindsey K (2012) Synthetic biology. New Phytol 196(3):671–677

Panchangam S, Mallikarjuna N, Suravajhala P (2012) Apomixis in Chickpea: biology and bioinformatics. Poster session presented at VI international conference on legume genetics and genomics, Hyderabad, India

Pertea G, Huang X, Liang F, Antonescu V, Sultana R, Karamycheva S, Lee Y, White J, Cheung F et al (2003) TIGR Gene Indices Clustering Tools (TGICL): a software system for fast clustering of large EST datasets. Bioinformatics 19:651–652

Plechakova O, Tranchant-Dubreuil C, Benedet F, Couderc M, Tinaut A, Viader V, De Block P, Hamon P, Campa C, de Kochko A, Hamon S, Poncet V (2009) MoccaDB – an integrative database for functional, comparative and diversity studies in the Rubiaceae family. BMC Plant Biol 9:123–129

Reusch TBH, Reusch AS, Preuss C, Weiner J, Wissler L, Beck A, Klages S, Kube M, Reinhardt R, Bornberg-Bauer E (2008) Comparative analysis of expressed sequence tag (EST) libraries in the seagrass *zostera marina* subjected to temperature stress. Mar Biotechnol 10:297–309

Sarmah R, Sahu J, Dehury B, Sarma K, Sahoo S, Sahu M, Barooah M, Sen P, Modi MK (2012) ESMP: a high-throughput computational pipeline for mining SSR markers from ESTs. Bioinformation 8:4

Semagn K, Bjornstad A, Xu Y (2010) The genetic dissection of quantitative traits in crops. Electron J Biotechnol 13:5

Shimomura M, Minami H, Suetsugu Y, Ohyanagi H, Satoh C, Antonio B, Nagamura Y, Kadono-Okuda K, Kajiwara H, Sezutsu H, Nagaraju J, Goldsmith MR, Xia Q, Yamamoto K, Mita K (2009) KAIKObase: an integrated silkworm genome database and data mining tool. BMC Genomics 10:486–493

Somerville C, Sommerville S (1999) Plant functional genomics. Science 285:380–383

Sun G (2012) MicroRNAs and their diverse functions in plants. Plant Mol Biol 80(1):17–36. doi:10.1007/s11103-011-9817-6

Sunkar R, Jagadeeswaran G (2008) *In silico* identification of conserved microRNAs in large number of diverse plant species. BMC Plant Biol 8(37)

Suravajhala P, Sundararajan VS (2012) A classification schema to validate protein interactors. Bioinformation 8(1):34–39

Sweetlove LJ, Last RL, Fernie AR (2003) Predictive metabolic engineering: a goal for systems biology. Plant Physiol 132:420–425

Tang Z, Choi J-H, Hemmerich C, Sarangi A, Colbourne JK, Dong Q (2009) ESTPiper – a web-based analysis pipeline for expressed sequence tags. BMC Genomics 10:174

Testillano PS, Risueño MC (2009) Tracking gene and protein expression during microspore embryogenesis by confocal laser scanning microscopy. In: Touraev A (ed) Advances in haploid production in higher plants. Springer, Dordrecht, p 339

Tsai WC, Fu CH, Hsiao YY, Huang YM, Chen LJ, Wang M, Liu ZJ, Chen HH (2013) OrchidBase 2.0: comprehensive collection of Orchidaceae floral transcriptomes. Plant Cell Physiol 54(2):7

Van de Poel B, Bulens I, Markoula A, Hertog M, Dreesen R, Wirtz M, Vandoninck S, Oppermann Y, Keulemans J, Hell R, Waelkens E, De Proft MP, Sauter M, Nicolai BM, Geeraerd AH (2012) Targeted systems biology profiling of tomato fruit reveals coordination of the yang cycle and a distinct regulation of ethylene biosynthesis during postclimacteric ripening. Plant Physiol 160:1498–1514

Varshney RK, Mahendar T, Aggarwal RK, Börner A (2007) Genic molecular markers in plants: development and applications. In: Varshney RK, Tuberosa R (eds) Genomics-assisted crop improvement: genomics approaches and platforms, vol 1. Springer, Dordrecht, pp 13–29

Varshney RK, Ribaut J-M, Buckler ES, Tuberosa R, Rafalski JA, Langridge P (2012) Can genomics boost the productivity of orphan crops? Nat Biotechnol 30:1172–1176. doi:10.1038/nbt.2440

Waegele B, Schmidt T, Mewes HW, Ruepp A (2008) OREST: the online resource for EST analysis. Nucleic Acids Res 1(36(Web Server issue)):W140–W144

Walbot V (1999) Genes, genomes, genomics. What can plant biologists expect from the 1998 National Science Foundation plant genome research program? Plant Physiol 119:1151–1155

Weckwerth W (2008) Integration of metabolomics and proteomics in molecular plant physiology: coping with the complexity by data-dimensionality reduction. Physiol Plant 132:176–189

Wortman JR, Haas BJ, Hannick LI, Smith RK Jr, Maiti R, Ronning CM, Chan AP, Yu C, Ayele M, Whitelaw CA, White OR, Town CD (2003) Annotation of the arabidopsis genome. Plant Physiol 132(2):461–468

Zhang BH, Pan XP, Wang QL, Cobb GP, Anderson TA (2005) Identification and characterization of new plant microRNAs using EST analysis. Cell Res 15:336–360

Zheng Q, Song J, Campbell-Palmer L, Thompson K, Li L, Walker B, Cui Y, Li X (2013) A proteomic investigation of apple fruit during ripening and in response to ethylene treatment. J Proteomics. http://dx.doi.org/10.1016/j.jprot.2013.02.006

Bioinformatics Strategies Associated with Important Ethnic Medicinal Plants

Priyanka James, S. Silpa, and Raghunath Keshavachandran

Abstract

Plants have been used as a source of medicine since historic times, and herbal drugs play an important role in the treatment of various ailments. Several commercially important modern medicines are of plant-based origin. One of the main purposes to investigate medicinal plants is to understand the main extracts from plants like alkaloids, flavonoids, etc., which can be used as therapeutics for human diseases. Bioinformatics play a crucial role in the analysis and interpretation of high-throughput data generated using molecular biology-based techniques. Bioinformatics approaches leverage plant-based knowledge discovery by offering new tools for the identification of genes and pathways involved in the production of secondary metabolites and also help to identify therapeutically important active compounds. Here we review bioinformatics strategies associated with important ethnic medicinal plants.

Keywords

Medicinal plants • Secondary metabolites • Database • MySql • PhP

Abbreviations

AP-1	Activator protein 1
BAX	BCL2-associated X protein
Bcl-2	B-cell lymphoma 2
cAMP	Cyclic adenosine monophosphate
c-Fos	Cellular oncogene Fos
CREB	cAMP response element-binding protein
COX2	Cyclo oxygenase2
CYP3A4	Cytochrome P450, family 3, subfamily A, polypeptide 4
Egf	Epidermal growth factor
EST	Expressed sequence tags
HTML	HyperText Markup Language
HTTP	Hypertext Transfer Protocol
IKK	I kappa B kinase
IUPAC	International Union of Pure and Applied Chemistry
MCF-7	Michigan Cancer Foundation-7 (Breast cancer cell line)
MDR	Multi drug resistance
NF-KB	Nuclear transcription factor kappa B

P. James (✉) • S. Silpa • R. Keshavachandran
Bioinformatics Centre, Kerala Agricultural University
Vellanikkara, Thrissur, Kerala 680656, India
e-mail: priyankajames2007@gmail.com

K.K. P.B. et al. (eds.), *Agricultural Bioinformatics*,
DOI 10.1007/978-81-322-1880-7_10, © Springer India 2014

NIK	NF-κB-inducing kinase
ODC	Ornithine decarboxylase
PhP	Hypertext preprocessor
RDBMS	Relational Database Management System
SQL	Structured Query Language
TNF-α	Tumour necrosis factor alpha
TPA	Tissue plasminogen activator

1 Introduction

Medicinal plants are nature's gift to human beings that help to attain a disease-free healthy life. Most of the population of the world still rely on herbs for their medicine. The inherent disadvantage of toxicity of synthetic drugs, especially when used for a long period, has led to the search for easily available safe remedies with less harmful effects (Balunas and Kinghorn 2005). In medicinal plants we look for therapeutically useful chemicals which are generally termed as secondary metabolites, produced in response to stress which protects them from organisms, diseases or the environment. Plants manufacture many different types of secondary metabolites which have been subsequently exploited by humans due to their profound physiological effect on the mammalian system and thus are known as active principles of the plant (Briskin 2000). Secondary metabolites that are useful in medicine are mostly polyphenols, alkaloids, glycosides, terpenes, flavonoids, coumarins, etc. Crude medicines of plant origin are used in the Indian system of medicine or "Ayurveda", and purified products are used in allopathic medicines. The identification of biologically active compounds in herbs is an essential requirement for the determination of plant-based drugs. This review discusses the significance of data management associated with medicinal plant research with focus on medicinal plants, bioactive compounds and their medicinal effects using bioinformatics strategies.

2 Bioinformatics Approaches to Medicinal Plant Research

Bioinformatics offers essential techniques for the analysis and interpretation of high-throughput data generated using molecular biology-based techniques. Huge volume of data generated using such techniques have made bioinformatics crucial for the analysis, interpretation and integration of data to infer knowledge from the whole system point of view (Sharma and Sarkar 2012). The availability of completely sequenced genomes offers opportunity for comparative analysis at genomic and proteomic levels. EST sequence data of plants can be exploited for the prediction of exons and identification of networks involved in secondary metabolite production (Yu et al. 2004; Singh et al. 2011). A comparative genomics study in plants using model crops revealed the evolutionary conservation of genes (Mahalakshmi and Ortiz 2001; Matthews 2003), and this information can be used for the improvement of other food crops (Paterson et al. 2005). Advancement in molecular biotechnology opened up the possibility for massive profiling experiments at different biological levels such as genomics, proteomics, metabolomics and transcriptomics. Metabolomics is extremely important for medicinally relevant plant species because it gives a description on biological pathways, and one of the main objectives of studying medicinal plants is to understand the secondary metabolites or natural products that can serve as health-care products or lead compounds for new drug development (Edwards and Batley 2004).

3 Maintenance of Knowledge Bases for Medicinal Plants

Interdisciplinary approaches based on molecular biology and biochemistry have led to an explosive growth in biological data which in turn requires the incorporation of computers for research process. One of the main applications of bioinformatics is the creation and maintenance of scientific databases. The volume of information related to medicinal plants and phytochemicals accumulated over the years is scattered and unstructured. This requires a comprehensive platform for the integration of knowledge related to medicinal plants (Sharma and Sarkar 2012). Here, we provide detailed information on active compounds, relevant pathway and therapeutic applications of few ethnic medicinal plants. Data on important native medicinal plants collected from the literature are given in Table 1.

Table 1 List of medicinal plants

Scientific name	Active constituents	Therapeutic applications
Aegle marmelos (Maity et al. 2009)	Skimmianine, lupeol, eugenol	Anticancer, antidiabetic, antimalarial, antibacterial, hepatoprotective
Acorus calamus (Bisht et al. 2011)	Anthraquinones, terpenoids	Antifungal, anticancer
Adhatoda zeylanica (Ahmad et al. 2009)	Vasicine, vasicinone, vasicinol, vasicinine and vasicoline	Hypoglycaemic, antibacterial, antiulcer, hepatoprotective
Andrographis paniculata (Jarukamjorn and Nemoto 2008)	Andrographolide, isoandrographolide, andrographin	Anticancer, antimalarial, hypoglycaemic
Azadirachta indica (Pankaj et al. 2011)	Nimbidin, nimbin, nimbolide, azadirachtin	Anti-inflammatory, spermicidal, antibacterial, anticancer, can be used as pesticide
Bacopa monnieri (Sudharani et al. 2011)	Brahmine, herpestine	Antioxidant, hepatoprotective, anticancer
Boerhavia diffusa (Mahesh et al. 2012)	Boeravinone A-F, ursolic acid	Immunomodulatory, antidiabetic, antistress, anticancer
Carica papaya (Krishna et al. 2008)	Chymopapain, papain, carpaine, carpasemine	Antimicrobial, antifungal, antimalarial, hepatoprotective
Centella asiatica (Gohil et al. 2010)	Asiaticosides, brahmoside, brahminoside	Wound healing, antidepressant, cognitive, antioxidant, anti-inflammatory
Cyclea peltata (Patel et al. 2010)	Tetrandrine	Anticancer
Eclipta prostrata (Jadhav et al. 2009)	Dasyscyphin C, eclalbatin, wedelolactone, ecliptalbine, verazine	Antiviral, anticancer, antibacterial, lipid lowering
Hemidesmus indicus (Austin 2008)	Hemidesmol, lupeol, indicine, hemidine	Anticancer, antivenom, antidiarrhoeal, anti-inflammatory
Moringa oleifera (Fahey and Sc 2005)	4-(4′-O-acetyl-a-L-rhamnopyranosyloxy) benzyl isothiocyanate, niazimicin, pterygospermin	Hypotensive, anticancer, antibacterial
Nelumbo nucifera (Mukherjee et al. 1996)	Nuciferine, asimilobin, nimbolide	Antidiabetic, antidiarrhoeal, cardiac tonic
Ocimum sanctum (Singh et al. 2012; Vishwabhan et al. 2011)	Eugenol, ursolic acid, carvacrol, linalool, apigenin	Antidiabetic, anticancer, antibacterial, antiviral, immunomodulatory
Phyllanthus emblica (Khan 2009)	Ellagic acid, chebulinic acid, emblicanins, cyclophosphamide	Anticancer, antidiabetic, antioxidant, immunomodulatory, gastroprotective, antiulcer
Rauvolfia serpentina (Dey and De 2010)	Reserpine, rauwolfine, rauhimbin	Hypotensive, useful against schizophrenia and epilepsy
Saraca asoca (Pradhan et al. 2009)	Epicatechin, quercetin, apigenin	Anticancer, antimenorrhagic, antioxytocic
Santalum album (Scartezzini and Speroni 2000)	Santalol, beta santalenes, nuciferol	Antipyretic, antioxidant
Tinospora cordifolia (Sankhala et al. 2012)	Berberine, palmatine, choline	Anticancer, hypoglycaemic, immuno modulatory, anti-inflammatory
Vetiveria zizanioides (Chou et al. 2012)	Beta vetivone, alpha vetivone, beta vatirenene	Antimicrobial, antioxidant, useful against rheumatism
Vitex negundo (Zargar et al. 2011)	Epicatechin, quercetin, catechin, myricetin, kaempferol	Anti-inflammatory, antioxidant activity

Table 1 contains the botanical names of the medicinal plants in alphabetical order along with active components and biological activity.

4 Ethnic Herbs with Medicinal Values

4.1 *Aloe barbadensis*

Aloe barbadensis has a vast traditional role in indigenous systems of medicine like Ayurveda, Siddha, Unani and Homoeopathy. The plant contains flavonoids, terpenoids, lectins, anthraquinones, polysaccharides, sterols, etc. *Aloe vera* contains aloe emodin, an anthraquinone, which has the ability to suppress or inhibit the growth of several malignant cancer cells including human lung carcinoma (Lee 2001), human tongue cancer cells (Chen et al. 2010), leukaemia cell lines (Chen et al. 2004) and hepatoma cell lines (Kuo et al. 2002), and it is believed to have antineoplastic properties (Thomson PDR 2004). Aloe emodin induces apoptosis in human tongue squamous cancer cells accompanied by mitochondrial dysfunction, cytochrome C release, activation of caspase-3 and cell-cycle arrest involving increase in p53 expression level and upregulation of p21 (Chiu et al. 2009). Acemannan, one of the polysaccharides found in *Aloe vera*, has several important therapeutic properties, including acceleration of wound healing, inhibition of inflammation and increase in the white blood cells or macrophages and T cells and antiviral effects (Nandal and Bhardwaj 2012). Acemannan in the presence of interferon gamma induces apoptosis in mouse monocytic-macrophage cell line RAW 264.7, by the inhibition of bcl-2 (B-cell lymphoma 2) expression (Ramamoorthy and Tizard 1998). *Aloe vera* juice and phytosterols derived from *Aloe vera* gel have antihyperglycaemic effect and would be useful for the treatment of type 2 diabetes mellitus.

4.2 *Andrographis paniculata*

Andrographis paniculata (Acanthaceae) commonly known as "kalmegh" is a traditional vital medicinal herb of Ayurveda, and it is considered as a popular remedy for the treatment of various disorders. Herbal extract of the plant has immunostimulant (Puri et al. 1993), anti-inflammatory (Shen et al. 2002), antiviral (Chang et al. 1991), anti-HIV (Calabrese et al. 2000), antithrombotic (Zhao and Fang 1991), anticancer (Kumar et al. 2004; Rajagopal et al. 2003; Matsuda et al. 1994) and antiplatelet aggregation (Amroyan et al. 1999) properties. It is also used for myocardial ischaemia (Guo et al. 1995) and respiratory tract infections (Coon and Ernst 2004). One of the major constituents of *A. paniculata* is andrographolide, a diterpenoid, which shows cytotoxic activity against human epidermoid carcinoma and lymphocytic leukaemia cells. Andrographolide inhibits human breast cancer cell MCF-7 by induction of cell-cycle inhibitor protein p27 and also decreased expression of cyclin-dependent kinase (Satyanarayana et al. 2004). Andrographolide possessed inhibitory effect of DNA topoisomerase II. Andrographolide induces apoptosis in TD-47 human breast cancer cell line by the increased expression of p53, BAX and caspase-3 and decreased expression of bcl-2 (Sukardiman et al. 2007).

4.3 *Curcuma longa*

Curcumin, yellow pigment that is present in the rhizome of turmeric *(Curcuma longa* L.) and related species, has been used in Ayurvedic medicine for centuries, as it is non-toxic and has a variety of therapeutic properties including antioxidant, analgesic, anti-inflammatory and antiseptic activities. Curcumin is one of the most extensively investigated phytochemicals, with regard to chemopreventive potential via its effect on a variety of biological pathways involved in mutagenesis, oncogene expression, cell-cycle regulation, apoptosis, tumorigenesis and metastasis (Wilken et al. 2011). TNF-α-induced cyclooxygenase-2 (COX2) gene transcription and NF-KB activation by the NIK/IKK signalling complex, probably at the level of IKKα/β, were inhibited by curcumin in human colon epithelial cells (Plummer et al. 1999). Curcumin prevents the 12-otetradecanoylphorbol-13-acetate (TPA)-induced activation of both nuclear factor KB (NF-KB) and activator protein-1 (AP-1).

The inhibition of NF-KB was by blocking the degradation of IKBα and by subsequent nuclear translocation of the p65 subunit of NF-KB (Han et al. 2002). Curcumin downregulates cyclin D1 expression, and this occurs at the transcriptional and post-transcriptional level (Bharti et al. 2003; Mukhopadhyay et al. 2001).

4.4 *Piper nigrum*

Black pepper, the king of spices, is one of the important spices originated in the tropical evergreen forests of the Western Ghats of India. Piperine, an amide isolated from *Piper* species (Piperaceae), is responsible for the pungency of black pepper. It has also been used in some forms of traditional medicine and has many pharmacological actions such as antidepressant-like activity (Li et al. 2007; Wattanathorn et al. 2008), antioxidant activity (Vijayakumar et al. 2004), inhibition of platelet aggregation (Park et al. 2007), anti-inflammatory activity (Kumar et al. 2007), antitumor activity (Manoharan et al. 2009; Wongpa et al. 2007) and antifungal activity (Ahmad et al. 2012), and it is known to have insecticidal activity against mosquitoes and flies. Piperine inhibits breast stem cell self-renewal mediated through inhibition of Wnt signalling (Kakarala et al. 2011). Piperine induced apoptosis and increased the percentage of cells in G2/M phase in 4 T1 murine breast cancer model (Lai et al. 2012). Piperine is a potent inhibitor of transcription factors such as nuclear factor-kappaB (NF-KB), activated transcription factor (ATF-2), c-Fos and cAMP response element-binding protein (CREB) (Pradeep and Kuttan 2004). Piperine, the major constituent of pepper, inhibits the functions of human P-glycoprotein and CYP3A4 which affects drug metabolism and can reverse multidrug resistance (MDR) by multiple mechanisms in tumour cells (Bhardwaj et al. 2002; Li et al. 2011).

4.5 *Zingiber officinale* Rosc

Ginger (*Zingiber officinale* Rosc), valued as a spice, is known for various medicinal properties in almost all systems of medicine and is used to cure a variety of diseases. Varieties of bioactive compounds including zingiberol, zingiberone, zingiberene, pungent and non-pungent components such as shogaol, gingerol and zingerone have been isolated from the plant and were analysed pharmacologically (Wang et al. 2012). Rhizome of ginger has been recommended for use in arthritis, rheumatism, sprains, muscular aches, pains, sore throats, cramps, constipation, indigestion, vomiting, hypertension, dementia, fever, infectious diseases and helminthiasis (Ali et al. 2008). Gingerol, a phenolic substance that is responsible for the spicy taste of ginger (*Zingiber officinale* Roscoe), has been reported to inhibit tumour promotion and PMA-induced ornithine decarboxylase (ODC) activity and Tnf-α production in mouse skin. 6-Gingerol has been found to inhibit epidermal growth factor (Egf) of mouse, epidermal JB6 cells, and reduced the activation of activator protein-1 (AP-1) which plays a critical role in neoplastic transformation (Bode et al. 2001). 6-Gingerol is capable of inducing apoptotic cell death in p53-mutant cancer cells by arresting mutant p53-expressing cells at G1 phase (Park et al. 2006).

5 Future Prospects and Challenges

Herbal drugs and isolated natural products play a predominant role in the pharmaceutical industry which increased the demand of the plant-based compounds for medicinal purpose. Bioinformatics facilitate medicinal plant research by offering essential methods and tools for rapid analysis and interpretation of data obtained by a wide variety of approaches. The key challenges to bioinformatics rely on current flood of raw data, increasing complexity and diversity of data, the variability of conditions and quality of experimental data. The exponential growth in biological information and the incorporation of that raw information into highly integrated databases on the World Wide Web present several challenges in data storage and retrieval. Multidisciplinary approaches can analyse heterogeneous information related to plants by integrating computational and experimental components for identifying therapeutically important secondary metabolites that may enable plant-based drug discovery. We are planning to host the full compendium of database for medicinal plants, acronymed *MedBase* at http://www.kaubic.in/databases.html

6 MedBase Creation

6.1 Manual Curation of Medicinal Plants

Ethnic medicinal herbs with relevant therapeutic applications were collected and recorded through extensive manual curation of scientific literatures and databases. Information retrieved from literature includes scientific name, common name, classification hierarchy, parts used, phytocompound, therapeutic applications and geographic location of the plant. Chemical properties like IUPAC name, molecular weight, molecular formula, number of hydrogen bond donors and acceptors, structure, etc., were assigned to phytocompounds via compound database search.

6.2 Database Architecture and Implementation

MySQL 5.0 (http://www.mysql.com/), an object-relational database management system (RDBMS), was used at the back end to store collected data as tables and perform SQL (Structured Query Language) queries that provide speed and flexibility in data retrieval. Hypertext Preprocessor (PHP) programming language along with HTML and JavaScript was used as the front end in order to provide dynamism to the Web interface. MedBase would be deployed on Apache HTTP server that runs on a server managed by the Windows operating system.

References

Ahmad S, Garg M, Ali M et al (2009) A phytopharmacological overview on *Adhatoda zeylanica* Medic. syn. *A. vasica* (Linn.) Nees. Nat Prod Radiance 8:549–554

Ahmad N, Fazal H, Abbasi BH, Farooq S, Ali M, Khan MA (2012) Biological role of *Piper nigrum* L. (black pepper): a review. Asian Pac J Trop Biomed 2: S1945–S1953

Ali BH, Blunden G, Tanira MO, Nemmar A (2008) Some phytochemical, pharmacological and toxicological properties of ginger (*Zingiber officinale* Roscoe): a review of recent research. Food Chem Toxicol 46 (2):409–420

Amroyan E, Gabrielian E, Panossian A, Wikman G, Wagner H (1999) Inhibitory effect of andrographolide from *Andrographis paniculata* on PAF-induced platelet aggregation. Phytomedicine 6:27–31

Austin A (2008) A review on Indian Sarsaparilla, *Hemidesmus indicus*. J Biol Sci 8:1–12

Balunas MJ, Kinghorn AD (2005) Drug discovery from medicinal plants. Life Sci 78:431–441

Bhardwaj RK, Glaeser H, Becquemont L, Klotz U, Gupta SK, Fromm MF (2002) Piperine, a major constituent of black pepper, inhibits human P-glycoprotein and CYP3A4. J Pharmacol Exp Ther 302(2):645–650

Bharti AC, Donato N, Singh S, Aggarwal BB (2003) Curcumin (diferuloylmethane) down-regulates the constitutive activation of nuclear factor-kappa B and IkappaBalpha kinase in human multiple myeloma cells, leading to suppression of proliferation and induction of apoptosis. Blood 101:1053–1062

Bisht VK, Negi JS, Bhandari AK, Sundriyal RC (2011) Anti-cancerous plants of Uttarakhand Himalaya: a review. Int J Cancer Res 7:192–208

Bode AM, Ma WY, Surh YJ, Dong Z (2001) Inhibition of epidermal growth factor-induced cell transformation and activator protein 1 activation by [6]-gingerol. Cancer Res 61(3):850–853

Briskin DP (2000) Update on phytomedicines medicinal plants and phytomedicines. Linking plant biochemistry and physiology to human health. Plant Physiol 124:507–514

Calabrese C, Berman SH, Babish JG, Ma X, Shinto L, Dorr M, Wells K, Wenner CA, Standish LJ (2000) A phase I trial of andrographolide in HIV positive patients and normal volunteers. Phytother Res 14:333–338

Chang RS, Ding L, Chen GQ, Pan QC, Zhao ZL, Smith KM (1991) Dehydroandrographolide succinic acid monoester as an inhibitor against the human immunodeficiency virus. Proc Soc Exp Biol Med 97:59–66

Chen HC, Hsieh WT, Chang WC, Chung JG (2004) Aloe-emodin induced in vitro G2/M arrest of cell cycle in human promyelocytic leukemia HL-60 cells. Food Chem Toxicol 42:1251–1257

Chen YY, Chiang SY, Lin JG (2010) Emodin, aloe-emodin and rhein induced DNA damage and inhibited DNA repair gene expression in SCC-4 human tongue cancer cells. Anticancer Res 30:945–952

Chiu TH, Lai WW, Hsia TC et al (2009) Aloe-emodin induces cell death through S-phase arrest and caspase-dependent pathways in human tongue squamous cancer SCC-4 cells. Anticancer Res 29:4503–4511

Chou S-T, Lai C-P, Lin C-C, Shih Y (2012) Study of the chemical composition, antioxidant activity and anti-inflammatory activity of essential oil from *Vetiveria zizanioides*. Food Chem 134:262–268

Coon JT, Ernst E (2004) *Andrographis paniculata* in the treatment of upper respiratory tract infections: a systematic review of safety and efficacy. Planta Med 70:293–298

Dey A, De JN (2010) *Rauvolfia serpentine* (L). Benth. exKurz – a review. Asian J Plant Sci 9:285–298

Edwards D, Batley J (2004) Plant bioinformatics: from genome to phenome. Trends Biotechnol 22:232–237

Fahey JW, Sc D (2005) *Moringa oleifera* : a review of the medical evidence for its nutritional, therapeutic, and prophylactic properties. Part 1. Tree Life J 1:5

Gohil KJ, Patel JA, Gajjar AK (2010) Pharmacological review on *Centella asiatica*: a potential herbal cure-all. Indian J Pharm Sci 72:546–556

Guo ZL, Zhao HY, Zheng XH (1995) An experimental study of the mechanism of *Andrographis paniculata* Nees (APN) in alleviating the Ca (2+)-overloading in the process of myocardial ischemic reperfusion. J Tongji Med Univ 15:205–208

Han S-S, Keum Y-S, Seo H-J, Surh Y-J (2002) Curcumin suppresses activation of NF-kappaB and AP-1 induced by phorbol ester in cultured human promyelocytic leukemia cells. J Biochem Mol Biol 35:337–342

Jadhav VM, Thorat RM, Kadam VJ, Salaskar KP (2009) Chemical composition, pharmacological activities of *Eclipta alba*. J Pharm Res 2:18–20

Jarukamjorn K, Nemoto N (2008) Pharmacological aspects of *Andrographis paniculata* on health and its major diterpenoid constituent andrographolide. J Health Sci 54:370–381

Kakarala M, Brenner DE, Korkaya H, Cheng C, Tazi K, Ginestier C, Liu S, Dontu G, Wicha MS (2011) Targeting breast stem cells with the cancer preventive compounds curcumin and piperine. Breast Cancer Res Treat 122(3):777–785

Khan KH (2009) Roles of *Emblica officinalis* in medicine – a review. Bot Res Int 2:218–228

Krishna KL, Paridhavi M, Patel JA (2008) Review on nutritional, medicinal and pharmacological properties of Papaya (*Carica papaya* Linn.). Nat Prod Radiance 7:364–373

Kumar RA, Sridevi K, Kumar NV, Nanduri S, Rajagopal S (2004) Anticancer and immunostimulatory compounds from *Andrographis paniculata*. J Ethnopharmacol 92:291–295

Kumar S, Singhal V, Roshan R, Sharma A, Rembhotkar GW, Ghosh B (2007) Piperine inhibits TNF-alpha induced adhesion of neutrophils to endothelial monolayer through suppression of NF-kappaB and IkappaB kinase activation. Eur J Pharmacol 575(1–3):177–186

Kuo PL, Lin TC, Lin CC (2002) The antiproliferative activity of aloe-emodin is through p53-dependent and p21-dependent apoptotic pathway in human hepatoma cell lines. Life Sci 71:1879–1892

Lai LH, Fu QH, Liu Y, Jiang K, Guo QM, Chen QY, Yan B, Wang QQ, Shen JG (2012) Piperine suppresses tumor growth and metastasis in vitro and in vivo in a 4T1murine breast cancer model. Acta Pharmacol Sin 33(4):523–530

Lee HZ (2001) Protein kinase C involvement in aloe-emodin and emodin-induced apoptosis in lung carcinoma cell. Br J Pharmacol 134:1093–1103

Li S, Wang C, Wang M, Li W, Matsumoto K, Tang Y (2007) Antidepressant like effects of piperine in chronic mild stress treated mice and its possible mechanisms. Life Sci 80:1373–1381

Li S, Lei Y, Jia Y, Li N, Wink M, Ma Y (2011) Piperine, a piperidine alkaloid from *Piper nigrum* re-sensitizes P-gp, MRP1 and BCRP dependent multidrug resistant cancer cells. Phytomedicine 19(1):83–87

Mahalakshmi V, Ortiz R (2001) Plant genomics and agriculture: from model organisms to crops, the role of data mining for gene discovery. Electron J Biotechnol 4(3):169–178

Mahesh AR, Kumar H, Mk R, Devkar RA (2012) Detail study on *Boerhaavia diffusa* plant for its medicinal importance – a review. Res J Pharm Sci 1:28–36

Maity P, Hansda D, Bandyopadhyay U, Mishra DK (2009) Biological activities of crude extracts and chemical constituents of Bael, *Aegle marmelos* (L.) Corr. Indian J Exp Biol 47(11):849–861

Manoharan S, Balakrishnan S, Menon VP, Alias LM, Reena AR (2009) Chemopreventive efficacy of curcumin and piperine during 7,12 dimethylbenz[a] anthracene-induced hamster buccal pouch carcinogenesis. Singapore Med J 50(2):139–146

Matsuda T, Kuroyanagi M, Sugiyama S, Umehara K, Ueno A, Nishi K (1994) Cell differentiation-inducing diterpenes from *Andrographis paniculata* Nees. Chem Pharm Bull 42:1216–1225

Matthews DE (2003) Grain genes, the genome database for small-grain crops. Nucleic Acids Res 31:183–186

Mukherjee PK, Balasubramanian R, Saha K et al (1996) A review on *Nelumbo nucifera* Gaertn. Anc Sci Life 15:268–276

Mukhopadhyay A, Bueso-Ramos C, Chatterjee D, Pantazis P, Aggarwal BB (2001) Curcumin down regulates cell survival mechanisms in human prostate cancer cell lines. Oncogene 20:7597–7609

Nandal U, Bhardwaj RL (2012) *Aloe vera* for human nutrition, health and cosmetic use – a review. Int Res J Plant Sci 3:38–46

Pankaj S, Lokeshwar T, Mukesh B, Vishnu B (2011) Review on neem (*Azadirachta indica*): thousand problems one solution. Int Res J Pharm 2:97–102

Park YJ, Wen J, Bang S, Park SW, Song SY (2006) [6]-Gingerol induces cell cycle arrest and cell death of mutant p53-expressing pancreatic cancer cells. Yonsei Med J 47(5):688–697

Park BS, Son DJ, Park YH, Kim TW, Lee SE (2007) Antiplatelet effects of acid amides isolated from the fruits of *Piper longum* L. Phytomedicine 14(12):853–855

Patel B, Das S, Prakash R, Yasir M (2010) Natural bioactive compound with anticancer potential. Int J Adv Pharm Sci 1:32–41

Paterson AH, Freeling M, Sasaki T (2005) Grains of knowledge: genomics of model cereals. Genome Res 15:1643–1650

Plummer SM, Holloway KA, Manson MM, Munks RJ, Kaptein A, Farrow S, Howells L (1999) Inhibition of cyclo-oxygenase 2 expression in colon cells by the chemopreventive agent curcumin involves inhibition of NF-kappaB activation via the NIK/IKK signalling complex. Oncogene 18:6013–6020

Pradeep CR, Kuttan G (2004) Piperine is a potent inhibitor of nuclear factor-kappaB (NF-kappaB), c-Fos, CREB, ATF-2 and proinflammatory cytokine gene expression in B16F-10 melanoma cells. Int Immunopharmacol 4(14):1795–1803

Pradhan P, Joseph L, Gupta V et al (2009) *Saraca asoca* (Ashoka): a review. J Chem Pharm Res 1:62–71

Puri A, Saxena R, Saxena RP, Saxena KC, Srivastava VTJ (1993) Immunostimulant agents from *Andrographis paniculata*. J Nat Prod 56:995–999

Rajagopal S, Kumar RA, Deevi DS, Satyanarayana C, Rajagopalan R (2003) Andrographolide, a potential cancer therapeutic agent isolated from *Andrographis paniculata*. J Exp Ther Oncol 3:147–158

Ramamoorthy L, Tizard IR (1998) Induction of apoptosis in a macrophage cell line RAW 264.7 by acemannan, a beta-(1,4)-acetylated mannan. Mol Pharmacol 53:415–421

Sankhala LN, Saini RK, Saini BS (2012) A review on chemical and biological properties of *Tinospora cordifolia*. Int J Med Aroma Plant 2:340–344

Satyanarayana C, Deevi DS, Rajagopalan R, Srinivas N, Rajagopal S (2004) DRF 3188 a novel semi-synthetic analog of andrographolide: cellular response to MCF 7 breast cancer cells. BMC Cancer 4:26

Scartezzini P, Speroni E (2000) Review on some plants of Indian traditional medicine with antioxidant activity. J Ethnopharmacol 71:23–43

Sharma V, Sarkar IN (2012) Bioinformatics opportunities for identification and study of medicinal plants. Brief Bioinform 14(2):238–250. doi:10.1093/bib/bbs021

Shen YC, Chen CF, Chiou WF (2002) Andrographolide prevents oxygen radical production by human neutrophils: possible mechanism(s) involved in its anti-inflammatory effect. Br J Pharmacol 135:399–406

Singh VK, Singh AK, Chand R, Kushwaha C (2011) Role of bioinformatics in agriculture and sustainable development. Int J Bioinform Res 3:221–226

Singh N, Verma P, Pandey BR, Bhalla M (2012) Review article therapeutic potential of *Ocimum sanctum* in prevention and treatment of cancer and exposure to radiation: an overview. Int J Pharm Sci Drug Res 4:97–104

Sudharani D, Krishna KL, Deval K, Safia AKP (2011) Pharmacological profiles of *Bacopa monnieri*: a review. Int J Pharm 1:15–23

Sukardiman H, Widyawaruyanti A, Sismindari, Zaini NC (2007) Apoptosis inducing effect of andrographolide on TD-47 human breast cancer cell line. Afr J Tradit Comp Altern Med 4:345–351

Thomson PDR (2004) PDR for herbal medicines, 3rd edn. PDR, Montvale

Vijayakumar RS, Surya D, Nalini N (2004) Antioxidant efficacy of black pepper (*Piper nigrum* L.) and piperine in rats with high fat diet induced oxidative stress. Redox Rep 9(2):105–110

Vishwabhan S, Birendra VK, Vishal S (2011) A review on ethnomedical uses of *Ocimum sanctum* (Tulsi). Int Res J Pharm 2:1–3

Wang W, Zhang L, Li N, Zu Y (2012) Chemical composition and in vitro antioxidant, cytotoxicity activities of *Zingiber officinale* Roscoe essential oil. Afr J Biochem Res 6:75–80

Wattanathorn J, Chonpathompikunlert P, Muchimapura S, Priprem A, Tankamnerdthai O (2008) Piperine, the potential functional food for mood and cognitive disorders. Food Chem Toxicol 46(9):3106–3110

Wilken R, Veena MS, Wang MB, Srivatsan ES (2011) Curcumin: a review of anti-cancer properties and therapeutic activity in head and neck squamous cell carcinoma. Mol Cancer 10:12

Wongpa S, Himakoun L, Soontornchai S, Temcharoen P (2007) Antimutagenic effects of piperine on cyclophosphamide-induced chromosome aberrations in rat bone marrow cells. Asian Pac J Cancer Prev 8 (4):623–627

Yu US, Lee SH, Kim YJ, Kim S (2004) Bioinformatics in the post-genome era. J Biochem Mol Biol 37:75–82

Zargar M, Azizah AH, Roheeyati AM et al (2011) Bioactive compounds and antioxidant activity of different extracts from *Vitex negundo* leaf. J Med Plant Res 5:2525–2532

Zhao HY, Fang WY (1991) Antithrombotic effects of *Andrographis paniculata* Nees in preventing myocardial infarction. Chin Med J 104:770–775

Mining Knowledge from Omics Data

Katsumi Sakata, Takuji Nakamura, and Setsuko Komatsu

Abstract

To extract knowledge from complex data in multiply layered biological information, a multiple omics-based approach would be a powerful method. In this chapter, we introduce a representative approach to integrate multiple omics data, and discuss topics that relate to mining knowledge from the omics data. First, we introduce an approach to map multiple omics data on a metabolic network and obtaining a panoramic view across "omes" (e.g. genomes, transcriptomes and metabolomes). As an example, we detail a study in which an integrated analysis was conducted for early stage soybean, which suggested that flooding stress caused differential expression between different biological layers. Second, we describe statistical tests for evaluating the significance of observations. In that section, we explain the meaning of the *p*-value and introduce applications of statistical testing to expression analyses for proteins in the early seedling stage of soybean. Finally, we focus on a metric, a measure of distance, which indicates the similarity between data objects. The metrics generally influence results of clustering, a popular method in profile analyses of omics data. In that section, we present a metric that is robust against measurement noise, and compare performance of the metric with other metrics.

Keywords

Omics • *p*-value • Metric • Clustering • Soybean

K. Sakata (✉)
Department of Life Science and Informatics,
Maebashi Institute of Technology,
Maebashi 371-0816, Japan
e-mail: ksakata@maebashi-it.ac.jp

T. Nakamura
National Hokkaido Agricultural Research Center,
National Agriculture and Food Research Organization
(NARO), Sapporo 062-8555, Japan

S. Komatsu
National Institute of Crop Science, National Agriculture
and Food Research Organization (NARO),
Tsukuba 305-8518, Japan

K.K. P.B. et al. (eds.), *Agricultural Bioinformatics*,
DOI 10.1007/978-81-322-1880-7_11, © Springer India 2014

Abbreviations

2-DE	Two-dimensional polyacrylamide gel electrophoresis
BLAST	Basic Local Alignment Search Tool
CE/MS	Capillary electrophoresis-mass spectrometry
EST	Expressed sequence tag
TCA cycle	Tricarboxylic acid cycle
UPGMA	Unweighted pair group method with arithmetic mean

1 Panoramic View Across "Omes" by Mapping Data on a Metabolic Network

The integration of omics data would be useful to extract knowledge concerning biological functions from complex data across multiple "omes". Metabolism is a scientific field with a rich history, and our biochemical knowledge grew throughout the early twentieth century, as represented by the prolific works of Hans Krebs (reviewed by Kornberg 2000). Accordingly, metabolic networks were clarified early and consolidated as public databases, such as KEGG (Kanehisa et al. 2008). Metabolomic networks are more stable, i.e., they rarely change, compared with other networks, e.g., transcriptome networks such as the recently published large scale human regulatory network (Gerstein et al. 2012). The potency of the metabolic networks is also demonstrated by their ability to integrate information from genome to phenome. Metabolomic approaches may have an intermediary bridge-building role (Fiehn et al. 2001), offering not only comprehensive analysis of a large number of metabolites, but also targeting several compounds simultaneously (Nakamura et al. 2010). In the present chapter, we introduce and discuss some recent studies integrating omics data on a metabolic-network-based scheme.

An integrated analysis was conducted for early stage soybean (*G. max* cultivar Enrei) (Sakata et al. 2009). The samples were collected from the radicle, hypocotyls, hypocotyl and roots from 1 to 6 days after sowing. In the analysis, 106 mRNAs, 51 proteins and 89 metabolites were observed to vary over time under flooding stress, which was applied to the samples 2 days after sowing. In the transcriptome analysis, data from high coverage gene expression profiling analysis (Fukumura et al. 2003) were filtered based on the fold-change of peak intensities between normal and flooding treatment. One hundred and six peaks out of the 29,388 detected were obtained that met the criteria of showing at least a 25-fold change after 12 h treatment and a ten-fold change during treatment. The sequences of the 106 mRNAs were subjected to Blastn homology searches (Altschul et al. 1990) against the EST sequence database for *G. max* (DDBJ, http://www.ddbj.nig.ac.jp) and the soybean genome sequence database (phytozome, http://www.phytozome.net/soybean.php). Matching cDNA sequences were subjected to Blastx homology searches against the GenBank protein database and the Soybean Uni-Gene Database (Komatsu et al. 2009). For the proteome data, quantitative analysis identified 51 proteins, detected by two-dimensional polyacrylamide gel electrophoresis (2-DE) and identified by mass spectrometry (MS), with at least a 1.5-fold change in abundance during flooding treatment (Komatsu et al. 2010). In the metabolome analysis, 89 metabolites were identified using capillary electrophoresis-mass spectrometry (CE/MS), which showed changes in abundance during flooding treatment, and were mapped on and around tricarboxylic acid (TCA) cycle (Nakamura et al. 2012) (Fig. 1a).

The mRNAs, proteins and metabolites were categorized and tabulated by their function, as shown in the multiple omics table (Fig. 1b). Significant relationships across the mRNAs, proteins and metabolites were observed. The relationships between mRNAs and proteins were determined by a homology search using the sequence. Metabolites and the others were related by referring to information on the metabolic network associated with each mRNA or protein from the integrated database: IntEnz (Fleischmann et al. 2004) and KEGG (Kanehisa et al. 2008).

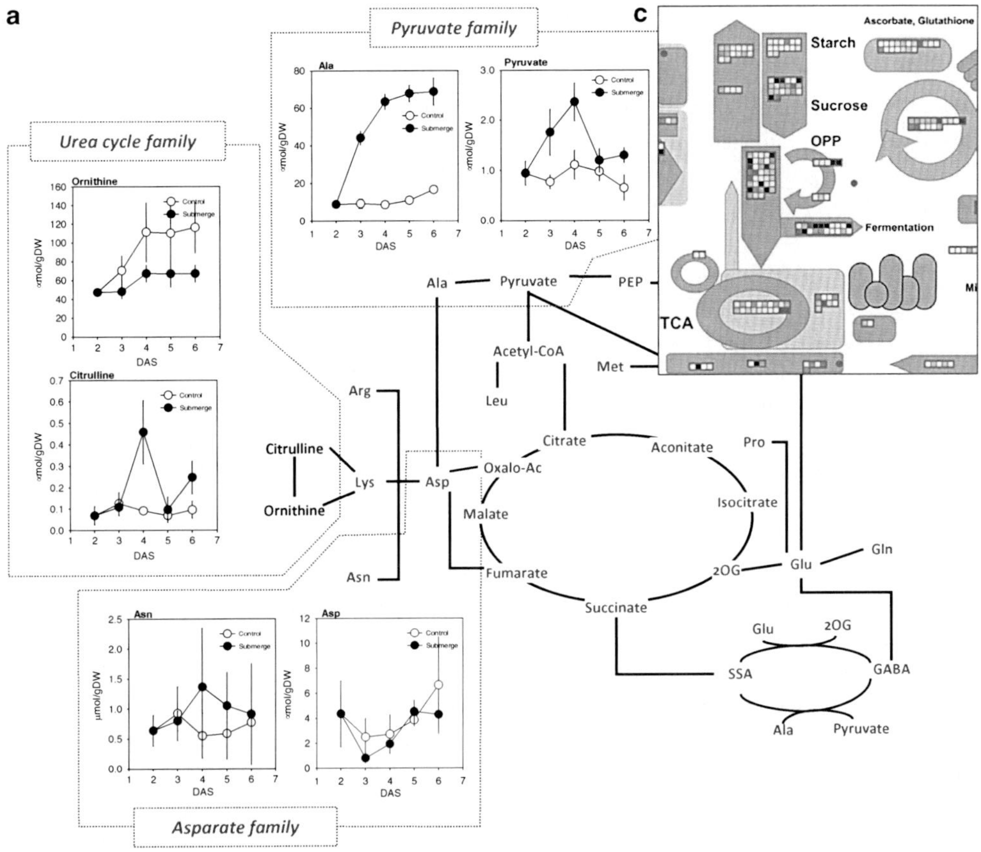

b

Metabolome (89)				Proteome (51)					Transcrip	
Family	Metabolite	Profile(Control / Treatment)		Description	No.	AC	Profile(Control / Treatment)		No.	Description
(SORT)	(SORT)	(SORT)	(SORT)	(SORT)	(SORT)	(SORT)	(SORT)	(SORT)	(SORT)	(SORT)
Metabolism										
Glycolysis	G1P	Down	Down3	UDP-glucose pyrophosphorylase [Amorpha fruticosa]	248	Gm0040z0014	Down2.5	(-)3		
Glycolysis	FBP	(-)5	Up	fructose-bisphosphate aldolase, cytoplasmic isozyme	383	Gm0036x00298	Down3	Down2.25		
Glycolysis	GA3P	(-)3	(-)3	fructose-bisphosphate aldolase, cytoplasmic isozyme	383	Gm0036x00298	Down3	Down2.25		
Glycolysis	3PGA	(-)3	Up3	cytosolic phosphoglycerate kinase [Pisumsativum]	402	Gm0054x00118	(-)2.25	Down2.25		
Glycolysis	2,3-DPG	Down	(-)4	cytosolic phosphoglycerate kinase [Pisumsativum]	402	Gm0054x00118	(-)2.25	Down2.25		
Purine	ADP	Down3	(-)3	nucleoside diphosphate kinase I	792	Gm0072x00084	(-)2.25	(-)2.25		
Purine	GDP	Down3	Down3	nucleoside diphosphate kinase I	792	Gm0072x00084	(-)2.25	(-)2.25		
Purine	ATP	Down5	(-)3	nucleoside diphosphate kinase I	792	Gm0072x00084	(-)2.25	(-)2.25		
Purine	GTP	(-)5	Down3	nucleoside diphosphate kinase I	792	Gm0072x00084	(-)2.25	(-)2.25		
Pyridine	CDP	Down	Down3	nucleoside diphosphate kinase I	792	Gm0072x00084	(-)2.25	(-)2.25		
Pyridine	UDP	Down	Down3	nucleoside diphosphate kinase I	792	Gm0072x00084	(-)2.25	(-)2.25		
Pyridine	CTP	Down5	(-)3	nucleoside diphosphate kinase I	792	Gm0072x00084	(-)2.25	(-)2.25		
Pyridine	UTP	Down3	Down3	nucleoside diphosphate kinase I	792	Gm0072x00084	(-)2.25	(-)2.25		
Oxidative metabolism	NADP	(-)3	(-)4						2 [GRAPH]	CYP96A cytochrome P450 [Arabidopsis thaliana]
Oxidative metabolism	NADPH	Down5	(-)4						2 [GRAPH]	CYP96A cytochrome P450 [Arabidopsis thaliana]
Shikimate pathway	Cinnamate	Down3	(-)3						2 [GRAPH]	CYP96A cytochrome P450 [Arabidopsis thaliana]

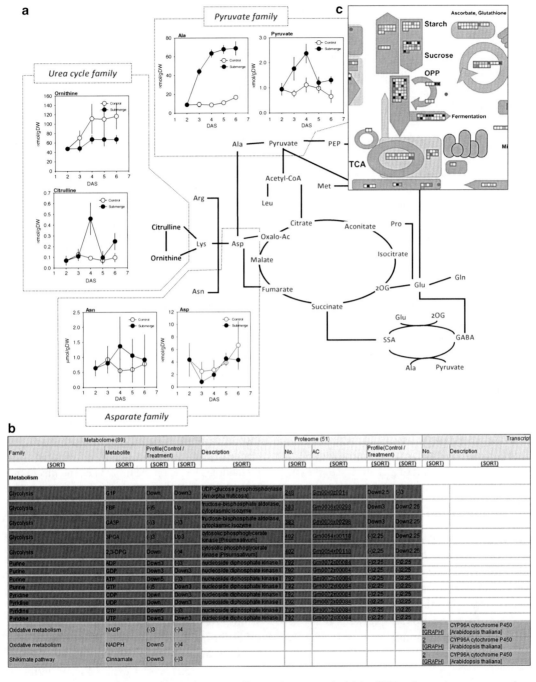

Fig. 1 A representative omics analysis in the seedling stage of soybean. (**a**) Metabolome network including temporal profiles of metabolites observed to vary over time under flooding stress compared with the control condition. (**b**) Omics table indicating significant relationships across the mRNAs, proteins and metabolites by the cells of the same color. (**c**) mRNAs observed to vary over time under flooding stress could also be mapped on the metabolome network. The diagram (Nanjo et al. 2011) shows the fold change of expression level compared with the control condition using MAPMAN software (Thimm et al. 2004)

A tagging method was developed to characterize profiles across multiple "omes" (Sakata et al. 2009), which will be useful for retrieving data based on the profiles. In the study, temporal profiles of mRNA, proteins and metabolites were analyzed. The overall profile (up- or down-regulated), and the time when the first extremum was observed during the experimental period, were investigated (Fig. 2a). The assigned tags are displayed in the "Profile" column of the omics table (Fig. 1b). Stored data in the table may be easily retrieved using a text search based on the tags. The first part of the tag indicates an overall profile during the experimental period: "Up" indicates that the final value was 1.5-fold larger and "Down" indicates it was 1.5-fold smaller than the initial value; "(−)" indicates that the fold change between the initial and final value was smaller than 1.5. The second part indicates the time when the first extremum (maximum or minimum) was observed during the experimental period. For example, if the tag is "Up3", the final value was greater than 1.5-fold higher than the initial value and the first extremum point was observed 3 days after seedling emergence. The tag is simple, and it contains information about the time of the first extremum. Information about the first extremum is useful for understanding the start of a change in expression. A general view of the profiles of mRNAs, proteins and metabolites in control (normal growth) and treated (growth under flooding) seedlings was investigated using the tag (Fig. 2b). Differences between profiles in multiple "omes" were visualized based on the profile tags, such that 78 out of 92 mRNAs that had been classified in the overall profile as "Down" or "(−)" in the control sample were changed to "Up" in the treated sample.

2 Statistical Tests for Validating the Results

Results from a study become convincing when they are evaluated for statistical significance. For the evaluations, we calculate the *p*-value, the probability of obtaining a test statistic at least as extreme as the one that was actually observed, if the null hypothesis is true (Wilcox 1997). When the *p*-value is less than the significance level, which is often 0.05 or 0.01 (http://www.jerrydallal.com/LHSP/p05.htm), we reject the null hypothesis. When the null hypothesis is rejected, the result is said to be statistically significant.

We applied statistical tests to the temporal profile analyses based on the temporal profile tags (Sakata et al. 2009). This represented a good example for understanding the statistical tests. In Fig. 2b, we classified the mRNAs, proteins and metabolites based on the overall temporal profile. Out of 92 mRNAs, 78 that had been classified as "Down" or "(−)" in the control sample were changed to "Up" in the treated sample. A test for homogeneity of variance was conducted, which showed that the variances differed between the mRNAs and proteins ($p = 0.01$), and between the mRNAs and metabolites ($p = 0.01$). The *p*-values are usually indicated as shown above, as a probability inside of a parenthesis. In the study, the validated results suggested that flooding stress caused more differential expression in the transcriptome than in the proteome or the metabolome of soybean. In Fig. 2c, we classified mRNAs, proteins and metabolites based on the time when the first extremum (maximum or minimum) was observed. For the mRNAs, the histograms were significantly different between the control and treated samples (upper panel of Fig. 2c). The number of mRNAs without an extremum decreased from 80 in the control sample to 21 in the treated sample. A test of homogeneity of variance showed that the variances differed between the control and treated sample for the mRNAs ($p = 0.05$), but did not differ for the proteins or metabolites ($p = 0.05$). These results also suggest that flooding stress causes more differential expression, in terms of timing of expression, in the transcriptome than in the proteome or metabolome.

Another example of the statistical tests is the evaluation of the specificity of the expressed proteins in a given organ during the seedling stage of soybean (Ohyanagi et al. 2012). In the

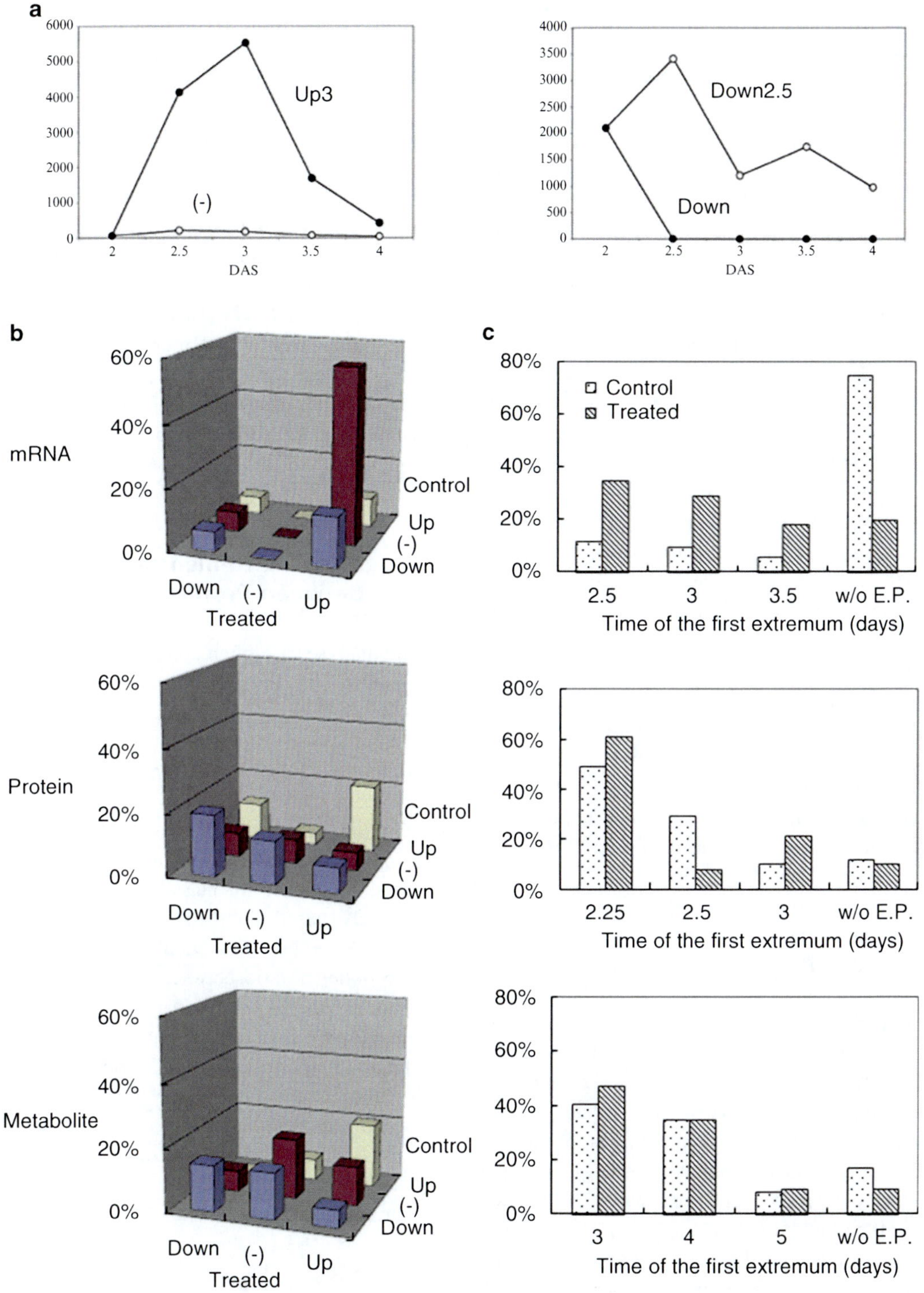

Fig. 2 Assigned temporal profile tags and classification based on the temporal profile tags. (**a**) Examples of the tags. "Up", "Down" and "(−)" indicate the overall profiles, and the number indicates the time when the first extremum (maximum or minimum) was observed during the experimental period. (**b**) Classification based on the overall profile. (**c**) Classification based on the time when the first extremum (maximum or minimum) was observed. "w/o E.P." means "without an extremum point" (Sakata et al. 2009)

study, 3,399 proteins were detected from seven organs, based on 2-DE analyses, and 210 non-redundant proteins were identified through protein sequencing or peptide peak determination by MS. Identified proteins were investigated as representatives of the proteins expressed in the corresponding organs (Fig. 3a).

The 210 proteins identified in the organs were evaluated. The numbers of proteins identified in common between 4, 3, 2 and 1 organs were 2, 7, 18 and 183, respectively. The p-values of the 27 (=2 + 7 + 18) out of 210 proteins that were expressed in two or more organs and 183 that were expressed in only one organ were calculated. In the first (1) calculation, it was assumed the 30,000 proteins (~ the number of genes in a higher plant) were randomly expressed and the probability (p) that a protein expresses in an organ was calculated as $p = 486/30{,}000 = 0.0162$. Here the number of expressed proteins is assumed to equal the average number of detected proteins in an organ, which was 486 (maximum: 847, minimum: 173). The second (2) calculation was the probability (P_k) that a protein expresses in k out of seven organs: P_k = Combination (7,k) $\times p^k \times (1-p)^{7-k}$. The third calculation (3) was the probability ($P_{k>0}$) that a protein expresses in one or more organs: $P_{k>0} = 1-P_0$. Fourth (4), the probability ($P_{(1)}$) that a protein expresses only in one organ among proteins expressing in one or more organs was calculated: $P_{(1)} = P_1/P_{k>0}$. Last, (v) the probability that 183 proteins are expressed only in one organ and that 27 proteins are expressed in more than one organ was calculated: Combination (210,27) $\times P_{(1)}^{183} \times (1-P_{(1)})^{27} = 2.8\text{e}{-6}$. As mentioned above, the small p-value (2.8e–6) suggests that proteins do not express at random, but are specifically expressed in a given organ. The number of identified proteins, 210, was small compared with the number of genes in a higher plant, but the specific expression of the proteins in a given organ was validated.

Furthermore, the above investigations also show that the probability that a protein is randomly expressed in three or more organs is $1-P_0-P_1-P_2 = 1.4\text{e}{-4}$. Nearly 1/800 of the probability that a protein is expressed in one or more organs, $1-P_0 = 0.108$. Thus, the p-values suggest that the proteins identified in three organs, AB046874 (*Glycine max* mRNA for allergen Gly m Bd 28 K partial cds), AF338252 (*Glycine max* BiP-isoform), AF456323 (*Glycine max* cyclophilin), K02646 (Soybean glycinin subunit), P21241 (RuBisCO subunit binding-protein beta subunit), P52572 (probable peroxiredoxin (EC 1.11.1.15)) and S47563 (nucleoside-diphosphate kinase), and identified in four organs, P10743 (Stem 31 kDa glycoprotein precursor) and P31233 (20 kDa chaperonin, chloroplast), are significantly represented in these organs.

3 A Metric to Measure Similarity Between Profiles

Profile analyses are often conducted for clustering, and to characterize fluctuating compounds such as transcripts, proteins and metabolites. In such analyses, the results are susceptible to a metric, a measure of distance indicating the similarity between data objects. Sometimes, we focus on overall profiles rather than the magnitude of each datum. In this section, we would like to present a metric for classifying data objects exclusively based on their overall expression profiles. A conventional metric for such cases was to obtain a ratio between each datum in a data series and a control datum, such as the datum at an initial time point and calculate a Euclidian metric between the series of ratios. This calculation was sensitive to measurement noise in the control datum. An improved metric, d_{nov}, which includes a normalization coefficient, λ, between data series $\mathbf{x}$ and $\mathbf{y}$, was proposed (Mitsui et al. 2008):

$$d_{\text{nov}} = |\log(\mathbf{x}) - \log(\lambda \mathbf{y})|,$$

where $|\log(\mathbf{x})\text{-}\log(\lambda \mathbf{y})| = \min_{\lambda>0} |\log(\mathbf{x})\text{-}\log(\lambda \mathbf{y})|$. In the formulae $\log(\mathbf{x})$ means $[\log(x_1), \log(x_2), \ldots, \log(x_n)]^T$.

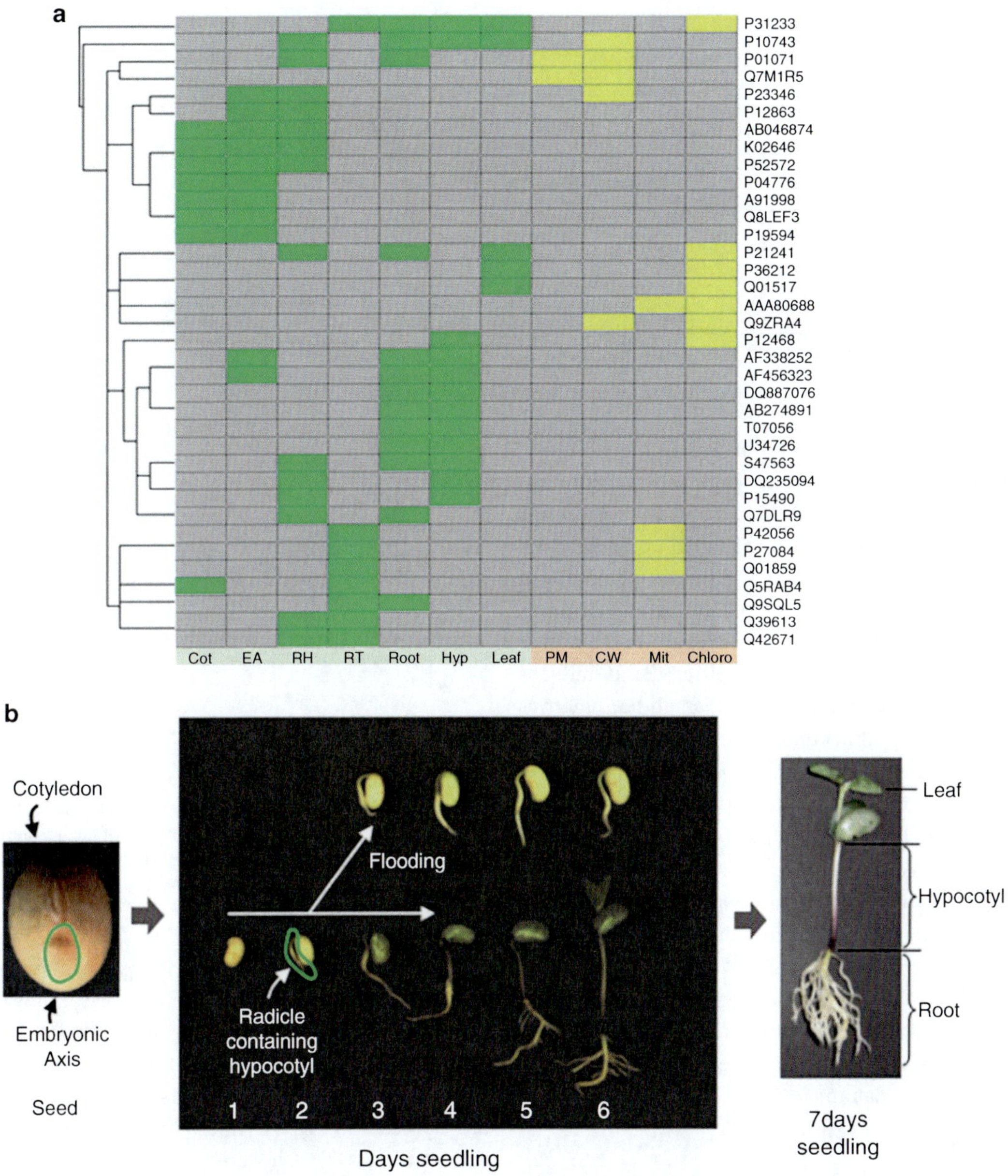

Fig. 3 Clustering results for 36 proteins identified in more than one organ and/or subcellular compartment in the early seedling stage of soybean (Ohyanagi et al. 2012). (**a**) Accession numbers of the proteins are indicated on the right-hand side. *Colored boxes* indicate an identified protein found in the organ (*green*) or a subcellular compartment (*yellow*). Samples from seven organs, *Cot* (cotyledon), *EA* (embryonic axis), *RH* (radical plus hypocotyl), *RT* (root tip), *Root*, *Hyp* (hypocotyls), and Leaf; and four subcellular compartments, *PM* (plasma membrane), *CW* (cell wall), *Mit* (mitochondrion), and *Chloro* (chloroplast), were investigated. Hierarchical clustering was performed using Gene Cluster3.0 (de Hoon et al. 2004) with Euclidean distance and centroid linkage method. The resulting clusters were visualized using JavaTreeview (Saldanha 2004). (**b**) Soybean materials used for the analysis. Cotyledon and embryonic axis were collected from seeds of soybean (*G. max* L.) cv. Enrei. Seeds were germinated for 2 days and then flooded by water. Seedlings were allowed to grow until the sixth day after germination. Leaves, hypocotyls and roots were collected from 7-day-old seedlings (Sakata et al. 2009)

The square of the metric becomes:

$$d_{\mathrm{nov}}{}^2 = \sum_{i=1}^{n}\left((\log\lambda)^2 - 2(\log x_i - \log y_i)\log\lambda + (\log x_i - \log y_i)^2\right)$$
$$= n(\log\lambda)^2 - 2\left(\sum_{i=1}^{n}(\log x_i - \log y_i)\right)\log\lambda + |\log(\mathbf{x}) - \log(\mathbf{y})|^2.$$

Thus we can obtain optimal λ:

$$\lambda = \exp\left(\frac{\sum_{i=1}^{n}(\log x_i - \log y_i)}{n}\right). \qquad (1)$$

The metric has the following features: (1) it automatically normalizes data, and the normalization is based on all the data in the data series (see Eq. 1). It reduces the effect of noise included in the control datum. (2) The metric is applicable to constant data, because it eliminates a denominator of sample standard deviation that was included in the Pearson metric. (3) It reduces the tendency observed in a Euclidian metrics (such as L_2 norm) that the largest-scaled feature dominates the others (Han and Kamber 2000), by using the logarithmic scale. (4) The metric has a mathematical symmetry and triangular inequality.

The improved metric was compared with a conventional metric for the gene expression profiles from stress treated *Arabidopsis thaliana*. The conventional metric used for the comparison was the log-ratio metric, which is a Euclidian metric based on the ratio between a data series and control datum:

$$d_{\mathrm{LR}} = \left|\log\left(\frac{x_2}{x_1}\right) - \log\left(\frac{y_2}{y_1}\right), \ldots, \log\left(\frac{x_n}{x_1}\right) - \log\left(\frac{y_n}{y_1}\right)\right|$$

where x_1 and y_1 are control data. The comparison was performed based on two types of time courses, the step-function-like (sf) and peak-and-flat (pf) time course in the *Arabidopsis thaliana* dataset, TAIR Expression Set:1007966835 (http://www.arabidopsis.org/). Those expression profiles were stated to be representative in previous studies (Pandey et al. 2004; Ronen et al. 2002). Based on the TAIR dataset, a test dataset was generated. In the test dataset, artificial noise was added to baseline time courses by multiplying random numbers. If the magnitude of the noise (mn) is two, the artificial data lies between 1/2 and two-fold of the baseline time course. The maximum mn was set as two; such noise levels are typical in DNA microarray experiments (Tu et al. 2002). Clustering was performed by the unweighted pair group method with arithmetic mean (UPGMA) until the time courses merged into two clusters based on the two metrics.

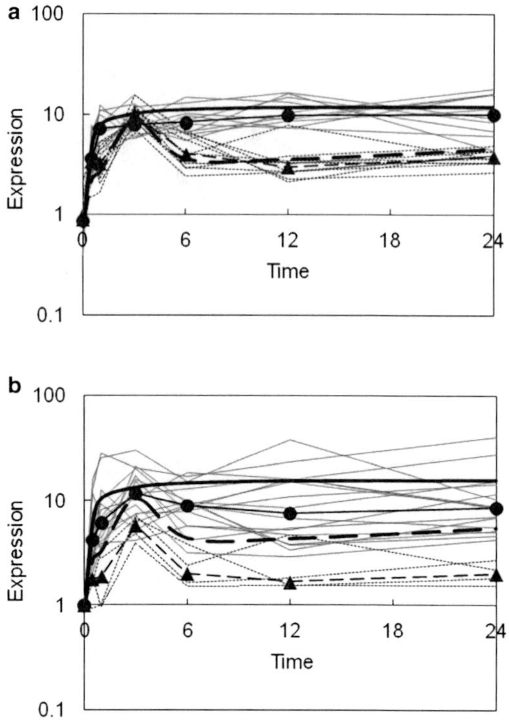

Fig. 4 Clustering results obtained when the noise level (mn) was set at two. The *gray line* and the *dotted line* show members of the two groups after clustering. The smoothened *thick line* and the marked line show the baseline time course and the central value of each group, respectively. (**a**) Clustering result based on the improved metric. (**b**) Result based on the log-ratio metric (Mitsui et al. 2008)

Figure 4 shows the clustering results based on the improved metric (a) and the log-ratio (b). In the case of the log-ratio (Fig. 4b), six pf time courses were wrongly grouped into sf time courses and the central value shifted below the baseline time course. In the case of the improved metric (Fig. 4a), each time course was correctly clustered and the central value mimicked the baseline time course.

The improved metric should also be useful for interaction analyses, based on a mathematical model including a power function, such as the law of mass action:

$$z = \alpha x^a y^b \tag{2}$$

where, x and y denote an expression level of the 1st and 2nd transcripts, respectively, and z denotes a time derivative of the expression level of the 3rd transcript. The exponents, a and b, denote the intensity of interaction by the 1st and 2nd transcripts. The same set of estimates of the exponents ($\hat{a}, \hat{b}$) will be obtained by fitting the model (Eq. 2) to the following two datasets: $\{\mathbf{x}_1, \mathbf{y}_1, \mathbf{z}_1\}$ and $\{\mathbf{x}_2\ (= k_x\ \mathbf{x}_1), \mathbf{y}_2 (= k_y\ \mathbf{y}_1), \mathbf{z}_2\ (= k_z\ \mathbf{z}_1)\}$. Thus, it is desirable that data series $\mathbf{x}_1$ and $\mathbf{x}_2$ are classified in the same group if $\mathbf{x}_2 \approx k_x\ \mathbf{x}_1$. The improved metric actually gives zero as the distance between $\mathbf{x}_1$ and $\mathbf{x}_2$ if, and only if, $\mathbf{x}_2 = k_x\ \mathbf{x}_1$. This property will be appropriate to the clustering. On the other hand, the estimate of exponents ($\hat{a}, \hat{b}$) will be different between the following datasets: $\{\mathbf{x}_1, \mathbf{y}_1, \mathbf{z}_1\}$ and $\{\mathbf{x}_1{}^h\ (h \neq 1), \mathbf{y}_1, \mathbf{z}_1\}$. Thus, it is also desirable that a metric gives $\neq 0$ as the distance between $\mathbf{x}_1$ and $\mathbf{x}_1{}^h$ $(h \neq 1)$. The improved metric indeed gives $\neq 0$ as the distance. The Pearson metric gives zero between $\mathbf{x}_1$ and $\mathbf{x}_1{}^h$ in some cases.

References

Altschul SF, Gish W, Miller W et al (1990) Basic local alignment search tool. J Mol Biol 215:403–410

de Hoon MJL, Imoto S, Nolan J et al (2004) Open source clustering software. Bioinformatics 20:1453–1454

Fiehn O, Kloska S, Altmann T (2001) Integrated studies on plant biology using multiparallel techniques. Curr Opin Biotechnol 12:82–86

Fleischmann A, Darsow M, Degtyarenko K et al (2004) IntEnz, the integrated relational enzyme database. Nucleic Acids Res 32:D434–D437

Fukumura R, Takahashi H, Saito T et al (2003) A sensitive transcriptome analysis method that can detect unknown transcripts. Nucleic Acids Res 31:e94

Gerstein MB, Kundaje A, Hariharan M et al (2012) Architecture of the human regulatory network derived from ENCODE data. Nature 489:91–100

Han J, Kamber M (2000) Data mining: concepts and techniques. Kaufmann, Morgan

Kanehisa M, Araki M, Goto S et al (2008) KEGG for linking genomes to life and the environment. Nucleic Acids Res 36:D480–D484

Komatsu S, Yamamoto R, Nanjo Y et al (2009) A comprehensive analysis of the soybean genes and proteins expressed under flooding stress using transcriptome and proteome techniques. J Proteome Res 8:4766–4778

Komatsu S, Sugimoto T, Hoshino T et al (2010) Identification of flooding stress responsible cascades in root and hypocotyls of soybean using proteome analysis. Amino Acids 38:729–738

Kornberg H (2000) Krebs and his trinity of cycles. Nat Rev Mol Cell Biol 1:225–228. doi:10.1038/35043073. PMID 11252898

Mitsui S, Sakata K, Nobori H et al (2008) A novel metric embedding optimal normalization mechanism for clustering of series data. IEICE Trans Inf Syst E91-D:2369–2371

Nakamura T, Okazaki K, Benkeblia N et al (2010) Metabolomics approach in soybean. In: Bilyeu K, Ratnaparkhe MB, Kole C (eds) Genetics, genomics and breeding of soybean, 1st edn, Genetics, genomics, and breeding of crop plants. Science Publishers, Enfield

Nakamura T, Yamamoto R, Hiraga S et al (2012) Evaluation of metabolite alteration under flooding stress in soybeans. Jpn Agric Res Q 46:237–248

Nanjo Y, Maruyama K, Yasue H et al (2011) Transcriptional responses to flooding stress in roots including hypocotyl of soybean seedlings. Plant Mol Biol 77:129–144

Ohyanagi H, Sakata K, Komatsu S (2012) Soybean proteome database 2012: update on the comprehensive data repository for soybean proteomics. Front Plant Sci 3:110. doi:10.3389/fpls.2012.00110

Pandey GK, Cheong YH, Kim K-N et al (2004) The calcium sensor calcineurin B-like 9 modulates abscisic acid sensitivity and biosynthesis in Arabidopsis. Plant Cell 16:1912–1924

Ronen M, Rosenberg R, Shraiman BI et al (2002) Assigning numbers to the arrows: parameterizing a gene regulation network by using accurate expression kinetics. Proc Natl Acad Sci U S A 99:10555–10560

Sakata K, Ohyanagi H, Nobori H et al (2009) Soybean proteome database: a data resource for plant differential omics. J Proteome Res 8:3539–3548

Saldanha AJ (2004) JavaTreeview–extensible visualization of microarray data. Bioinformatics 20:3246–3248

Thimm O, Bläsing O, Gibon Y et al (2004) MAPMAN: a user-driven tool to display genomics data sets onto diagrams of metabolic pathways and other biological processes. Plant J 37:914–939

Tu Y, Stolovitzky G, Klein U (2002) Quantitative noise analysis for gene expression microarray experiments. Proc Natl Acad Sci U S A 99:14031–14036

Wilcox RR (1997) Introduction to robust estimation and hypothesis testing. Academic, San Diego

Cloud Computing in Agriculture

L.N. Chavali

Abstract

The cloud computing environment is an offshoot of the traditional common computing environment with a mission. Cloud computing is driven by factors such as aging of the current IT infrastructure and changes in IT landscape. Refreshing the infrastructure is restricted due to limited capital investments. Virtualisation and cloud computing have transformed how the IT services are delivered at a lower cost by migrating the services to cloud computing. IT-enabled agricultural farming, such as precision farming, is information intensive, which can impact the rural economy. Therefore, the future growth of agriculture depends how the new technologies such as cloud computing are adopted with a focus on farmer needs. The use of appropriate technologies should help a farmer in terms of accessibility and affordability. Cloud computing in agriculture provides an enabling environment for innovation and services with a flexible regulatory environment.

Keywords

Cloud computing • SaaS • PaaS • IaaS • Virtualization • ICT • Multi-tenant • Public cloud • Private cloud

1 Introduction

The communities in villages have been passing on for ages the best practices in the form of knowledge of farming to each successive generation. Agriculture is traditionally practised by families by passing on their knowledge as inherited from their forefathers. Agricultural productivity is low and unreliable in poor rural areas, leading to food insecurity. Information exchange can play an important role and can help reduce the poverty. Therefore, use of information and communication technology (ICT) can act as a positive force and can promise agricultural growth, poverty reduction and sustainable resource utilisation. The emerging trend in sustainable development is the use of 'informatics' as the appropriate tool.

With the knowledge of agriculture being very ancient, leveraging IT to this knowledge domain is a big step forward. Intuition and experience is the basis for farming in most villages, even today. Agriculture has undergone several fundamental

L.N. Chavali (✉)
Plot #358, Gautham Nagar, Malkajgiri, Hyderabad 500047, India
e-mail: lnchavali@yhaoo.com

K.K. P.B. et al. (eds.), *Agricultural Bioinformatics*,
DOI 10.1007/978-81-322-1880-7_12, © Springer India 2014

changes during the twentieth century, including extensive dependence on farm machinery, intensive fertiliser and agrochemical management, crop breeding, high-yielding hybrid varieties and genetic manipulation. Agricultural development requires resources, infrastructure, technology and institutions. Tacit knowledge is gained from experiences and practice, whereas explicit knowledge is based more on theory and rationality. The information system of the future must have both forms of knowledge and must encourage the conversion of knowledge between the forms as a continuous process.

The global trends in the use of technology are:

(a) Specialisation and growth in sustainable agriculture, organic farming, niche farming and direct marketing
(b) Genetic engineering to lower costs, improve productions and promote environmentally friendly practices
(c) Increased use of technology, such as computers and Global Positioning Systems (GPS), in production and marketing of agricultural commodities

With cloud computing burgeoning, one can expect this to support the above-mentioned global trends and be very suitable for converting tacit knowledge into explicit knowledge. The cloud services can be based on plan-do-check-act life cycle processes (Hori et al. 2010) of farming. Data visualisation, data mining and knowledge management are the fabrics of cloud services in agriculture. The field of agriculture is not just a production activity, but it encompasses other activities such as communication between people, sales and logistics. Mobile telephones are dream-come-true to villagers. The farmers in villages try to replace ICT with mobile phones. The mobile phone is no longer used only for audio communication, but the villagers of today have the complete knowledge of mobile operations and have understood its value and benefits in terms of integrated functionality. Therefore, M-agriculture or mobile applications for agriculture play an important role in improving the efficiency and productivity. Bioinformatics is the application of information technology to manage biological data that helps in decoding plant genomes. During the last two decades, colossal data has been generated in biological science: firstly, with the onset of sequencing the genomes of model organisms and, secondly, rapid application of high-throughput experimental techniques in laboratory research. Biological research that earlier used to start in laboratories, fields and plant clinics now commences at the computational level using computers for analysis of the data, experiment planning and hypothesis development. Application of various bioinformatics tools in biological research enables storage, retrieval, analysis, annotation and visualisation of results and promotes better understanding of biological system in fullness. This will help in plant health-care-based disease diagnosis to improve the quality of plants. Microorganisms play a crucial role in agriculture, and bioinformatics provides the genomic information of these organisms.

The tools of bioinformatics play a significant role in providing the information about the genes present in the genomes of these species. The sequencing of the genomes of plants and animals should have enormous benefits for the agricultural community. Bioinformatics tools can be used to search for the genes within these genomes and to elucidate their functions. This specific genetic knowledge could then be used to produce stronger, more drought, disease and insect resistant crops and improve the quality of livestock making them healthier with more disease resistance and production.

(a) The integration of bioinformatics will influence plant science and lead to crop improvements in the following areas: identification of important genes through genomics, expression analysis and functional genomics in conjunction with the design and construction of transgenic plants, thus allowing new target genes to be identified which will improve quantitative and qualitative traits in commercially important crops.
(b) The design of agrochemicals based on analysis of the components of signal perception and transduction pathways. Selecting targets to identify potential compounds using chemoinformatics tools that can be used as herbicides, pesticides or insecticides.
(c) The utilisation of plant genetic resources to preserve genetic diversity in agricultural species. The need for taxonomic data goes far

beyond the field of classical taxonomy, and a catalogue of all species, with phenotypic and genotypic attributes, is required. The core taxonomic effort gives stability to the work of regulatory, management and conservation bodies.

(d) Efficient utilisation of biological repositories of clones, cell lines, organisms and seeds. Typically, existing repositories are not linked to each other in the databases. Many commercial databases and repositories are also part of the bioinformatics infrastructure but operate largely outside the present-day cooperative activities.

Bioinformatics is instrumental in (Singh et al. 2011):

1. Improving nutritional quality as scientists have been successful in transferring genes into rice to increase levels of vitamin A, iron and other micronutrients
2. Developing greater tolerance for soil alkalinity for producing crops in poor soil areas
3. Improving other food crops from the information obtained from model crop systems

2 ICT in Agriculture

Agricultural development requires resources, infrastructure, technology and institutions. The marketing in the present days needs information systems as the market serves as drivers for agricultural growth.

The adoption to ICT in agriculture has played an essential role in education, training, e-services and rural development projects. ICT adoption in agriculture at village level through information centres comprising of open-source software will reduce the gap of digital divide in rural population especially in developing countries like India. The main drivers of use of ICT in agriculture (Rudgard et al. 2009) are described below.

2.1 Low-Cost and Pervasive Connectivity

The Internet and mobile technologies have enabled cost reduction and better competition.

2.2 ICT Tools

Laptops, phones, instruments and software are available cheap in today's world because of innovation. The technologies such as short message services (SMS) have been helping farmers in receiving alerts and doing transactions in case of selling in big way.

2.3 Advances in Data Storage and Exchange

The adoption of ICT in agriculture has improved the ability to access and share data remotely. This has provided opportunities to include more stakeholders in agricultural research.

2.4 Open Access

The information that was hardly accessible previously is now publicly accessible because of ICT. The social media such as Facebook has enabled knowledge sharing and collaboration even in the field of agriculture.

Genetic modification, bioinformatics, genome research and information technology play an important role in stimulating growth in agriculture, and fusion of these technologies is the key need of the hour today. The purpose of ICT in agriculture is to bring about changes in the lives of disadvantaged communities. ICT in agriculture will render information that is otherwise not available to rural farmers and provide outcomes which otherwise not possible.

Access to information, knowledge and skill development is very pivotal to improving the livelihood of rural farmers. In order to meet these objectives, it is essential to have an agricultural delivery mechanism using ICT, as it is ideally suited for enhanced cooperation, communication in terms of faster exchange of agricultural information and innovation among growing array of actors in agriculture and to provide services on demand with 'Agriculture Online'.

ICT is rudimentary to the business models of 'brokers'—extension agents, consultants, companies contracting farmers and others who are emerging to broker advice, knowledge, collaboration and interaction among groups and communities throughout the agricultural sector. Empowering millions of farmers in rural and various institutions to create, access, use and collaborate information is only possible with ICT. Furthermore, it also facilitates e-learning, and high-speed and reliable communication networks provide opportunities for farmers to connect with extension workers, agribusiness, researchers, etc.

Policy changes among research institutions, extension agents and governments for mobile phones and Internet usage in agriculture around the world have contributed to the innovation in agriculture with ICT. Because of the change, ICT is enabling agriculture research to be more inclusive and with a focus on developmental goals. The knowledge sharing toolkit (http://www.kstoolkit.org) consists of methods and tools to promote collaboration through each stage of the research project cycle.

ICT enables agricultural online surveys using mobile applications such as iFormBuilder (http://www.iformbuilder.com), and the research data is gathered from mobile devices such as smartphones, mobile phones using short messaging service (SMS) text messages, personal data assistants (PDAs), global positioning system (GPS) units and devices to measure indicators of soil nutrient levels. This data is analysed using a variety of ICT tools. There are various organisations offering GIS data, remote-sensing data, sequence data of crop genomes, etc. which can be analysed using analytical tools. These analytical tools are extortionate and should be made available to developing countries free of cost in order to promote innovation in agricultural.

The importance of emerging tools of information technology, namely, the differential global positioning system, the laser surveying system and handheld computers, is paramount in agricultural research in general and precision agriculture in particular. These tools are successfully employed for developing natural resource inventories and its management, development and updating of cadastral map, preparation of thematic maps under participatory geographical information system, etc.

The software packages used by agricultural producers are data oriented with the most common one being designed for financial accounting. The percentage of farms owning a computer continues to grow. Most commercial farms now own a computer and have access to the Internet, many with high-speed connections. Much of the software for on-farm usage is mostly stand-alone. Most of the packages on farm are related to accounting for keeping financial records, taxation and production management to address the livestock problems. Database systems are available for keeping track of information on fields and subfields, particularly fertilisers and pesticides applied, varieties planted and yields achieved.

There are web-based online nutrient systems available which are an online solution that provides an easy mechanism to recommend the major fertilisers N, P and K for various crops based on attainable nutrients in the soils and for targeted yield of crops.

National Agricultural Technology Project (NATP) entitled 'Integrated National Agricultural Resources Information System' has developed a data mart on field crops. The target users for the decision support system developed under this project are: (1) research managers, (2) research scientists and (3) general users. This is probably the first attempt of data warehousing of agricultural resources in the world. This provides systematic and periodic information to research scientists, planners, decision-makers and developmental agencies in the form of online analytical processing (OLAP) decision support system.

Information systems help in tracking genetic performance, balancing rations, monitoring health problems, facilities scheduling, controlling the housing environment and so forth. It is acknowledged that information systems will enable significant reduction in cost per unit of output (10–15 %) over that of more traditional, smaller farming operations. The information systems can be used to gain a strategic competitive advantage. Because of decline in the growth of the poultry industry, pilot studies on poultry informatics were initiated through collection of first-hand information on the current status of the poultry industry. Intensive work on poultry informatics could be expected to provide impetus to the poultry production activities.

Cloud computing services have immense potential to improve agricultural innovation systems. As resources can be provisioned on demand over Internet, the access to the shared pool of computing resources created an opportunity for data-sharing initiatives that were once prohibitively very expensive. This has led to easing the data collection and aggregation process, which is critical for research, extension and education.

3 Challenges in Agriculture

In developing countries like India and China, agriculture is a major source of employment for roughly half the labour force. Agricultural output is connected directly to gross domestic product and also accounts for growth of exports in countries where it is treated as an industry.

In general, the yield is impacted by area, seed quality, fertiliser quality, credit availability and mechanisation. The impact on yield is a major challenge as it affects the output. If the land/ revenue records are computerised, it would help estimate: (a) irrigated areas, (b) fallow lands and (c) wastelands. The main challenge is to supply water to areas under fallow or the ones covered with shrubs in order to increase the agriculture output. The variety of seeds available is an additional challenge, as it requires varying water and fertilisers. The high yield of hybrid varieties provides an opportunity to release a portion of the irrigated area for industrial purpose. Therefore, the information on seed quality and its availability and organic, inorganic and bio-fertilisers' availability are important for improving the productivity and for decision-making. Timely information on the availability of tractors, spraying machines, crop-cutting choppers, etc. is crucial, as they play an important role in farming.

Few major challenges around the world that agriculture is faced today are given below:

(a) Low yields per hectare, volatility in production and disparities in productivity over regions and crops.
(b) Achieving the accelerated agricultural growth by improving productivity.
(c) Developing sustainable water management strategies for a drought-prone environment.
(d) Combating dry land salinity.
(e) Combating the decision to grow or not to grow genetically modified foods.
(f) Domestically, some commodities are facing increased competition from imports.
(g) Deforestation.
(h) Lack of poor access to proper infrastructure, credit and modern technology.
(i) Lack of efficient and effective supply chain of preharvest and postharvest segments of agricultural operations that can impact farm investment.
(j) Fast-changing cultivated land to multistoried flats, no farmer beneficial schemes or insurance policies.
(k) Lack of proper storage or distribution and of proper price to farmer if there is surge in production.

4 Cloud Computing

4.1 Basics

Cloud computing paradigm is distributed on a broader aspect large scale and driven by economies of scale, in which a pool of abstracted virtualised, dynamically scalable, managed computing power, storage, platforms and services are delivered on demand to external customers over the Internet. Cloud computing is a delivery model for technology-enabled services that drives greater agility, speed and cost savings by providing on-demand access via a network to an elastic pool of shared computing assets (e.g. services, applications, frameworks, platforms, servers, storage and networks) that can be rapidly provisioned and released with minimal service provider interaction and scaled as needed to enable pay per use.

John Foley describes cloud computing as 'on-demand access to virtualized IT resources that are housed outside of your own data center, shared by others, simple to use, paid for via subscription, and accessed over the Web'.

Forrester defines cloud computing as 'A pool of abstracted, highly scalable, and managed compute infrastructure capable of hosting end-customer applications and billed by consumption'.

The National Institute of Standards and Technology (NIST) Cloud Computing Project defines cloud as 'a model for enabling convenient, on-demand network access to a shared pool of configurable computing resources (e.g., networks, servers, storage, applications, and services) that can be rapidly provisioned and released with minimal management effort or service provider interaction'.

Lewis Cunningham defines cloud as 'using the internet to access someone else's software running on someone else's hardware in someone else's data centre'.

Cloud computing is a radically new approach to the delivery of ICT services which promises:

(a) 'Anywhere' access to shared computing resources
(b) 'Freedom' from capital expenditure on back-end computing equipment and software
(c) Ability to provision computing services very quickly and cheaper than traditional models
(d) Ability to pay for such services on some form of metered or per-use basis.

Clouds provide a powerful and often otherwise unattainable IT infrastructure at a modest cost. In addition, they free individuals and small businesses from worries about quick obsolescence and lack of flexibility.

Cloud computing provides the facility to access shared resources and common infrastructure, offering services on demand over the network to perform operations that meet changing business needs. The location of physical resources and devices being accessed is typically not known to the end user. It also provides facilities for users to develop, deploy and manage their applications 'on the cloud', which entails virtualisation of resources that maintains and manages itself.

Cloud computing signifies the slow changes of the consumer and business experience with their local gadget. The data processes and back-end operations of a business are now available in the cloud. Small businesses do not have to spend considerable amounts for local installation. The desktop experience in large and small companies is slowly being replaced with applications that could be launched at anytime and with almost any gadget.

The cloud computing solution is made up of: (1) clients, (2) the data centre and (3) distributed servers. Clients are devices that the end users use to interact with cloud for accessing the information such as PDA, smartphone, iPhone and Web browsers. Data centre is an area or place where the collection of storage devices, servers and other communication equipment are hosted. The servers can be accessed from the Internet or intranet. The growing trend in IT is to virtualise the servers, i.e. one physical server can accommodate multiple virtual servers catering to different functions.

Public bodies such as Department of Agriculture, Food and Marine are co-locating or consolidating their back-end ICT infrastructure or elements of it in other public service data centres to reduce hosting costs and to share the costs of power, cooling and basic management. Cloud computing now gives public bodies the opportunity to consider the consumption of ICT services on metered basis as an alternative to traditional provisioning models.

The distributed servers are geographically separated and placed in different locations. This offers more flexibility in terms of failover or security options. The locations of the servers are not known to cloud subscriber, but the subscriber feels that all servers are placed next to each other. The advantage with the distribution is that more servers can be added in a new location and make them part of the cloud. Expertise, experience and right resources are the key in building cloud computing solutions. Cloud computing plays an important role in providing solutions in the absence of right resources. In the beginning, to build a cloud-based solution may be expensive, but the advantages provided by such solutions are far more than the initial spending.

The differences between conventional computing and cloud computing are described in Table 1.

Most agricultural software as of today uses the windowing environment, which makes it easier for the user to access the information and to move data from one application to another or to link applications.

Table 1 Conventional vs. cloud computing

Sl. no.	Conventional computing	Cloud computing
1.	Provisioned manually	Self-provisioned
2.	Managed by system administrator	Managed by API
3.	Fixed capacity	Elastic capacity
4.	Pay for capacity	Pay for use
5.	Dedicated hardware	Shared hardware
6.	Capital and operational expenses	Operational expenses

Cloud-based solutions or applications enable users to move beyond the desktop experience, and therefore, the effect of such solutions on the client or user end should be considered. For example, the trader at the agricultural market yard may have mails in Google, insurance data in insurance company servers, etc., all available in distributed servers and data centres. The users hold the data in the cloud while expecting it to be maintained. Therefore, the data is distributed over the Internet and is then available on local desktops or local area network databases. The clients in Fig. 1 equate to these users. Employing Web 2.0 programming language techniques is common while launching interactive applications at the user end. Cloud applications use extensively techniques such as AJAX and Ruby on Rails during development. The security remains the major challenge to end users despite having the best user interfaces. Apart from this challenge, the user experience is dependent on how fast the local hardware or gadget processes the data that it receives from the cloud.

Some of the popular SaaS vendors have been mentioned in Table 2 (Kang et al. 2010).

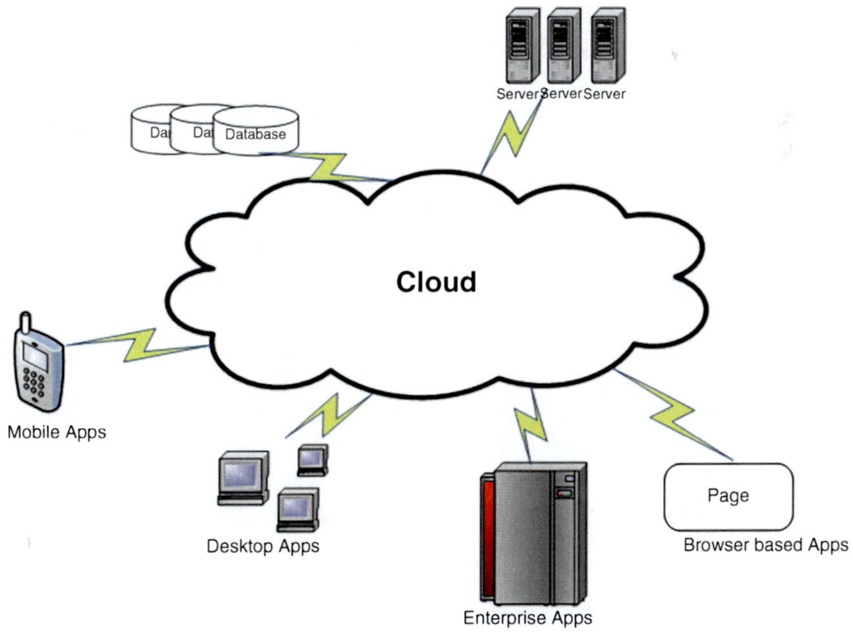

Fig. 1 Conceptual view of cloud computing

Table 2 SaaS vendors in the market

Vendor	Description	Service
Amazon	Computing resources	Infrastructure as a service
Salesforce	Web-based CRM	Platform as a service
Microsoft	Office tools	Software as a service
Google	Web office tools	Software as a service

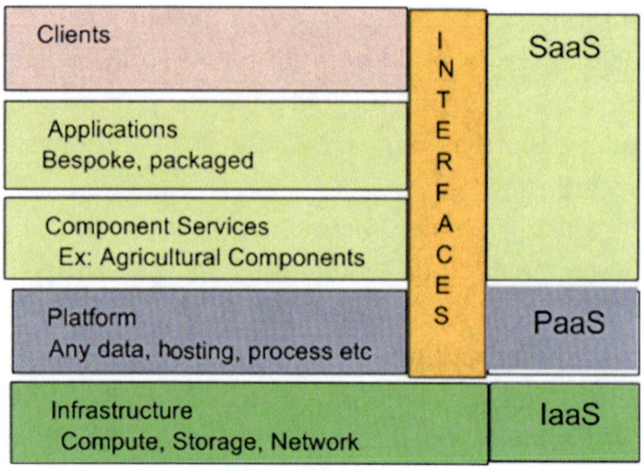

Fig. 2 Layers of cloud services

4.2 Features

Cloud computing brings an array of new features compared to any other computing paradigms. They are briefly described below:

(a) *Scalability and On-Demand Services*—Cloud computing provides resources and services for users on demand, i.e. a consumer can unilaterally provision computing capabilities such as server time and network storage as needed automatically, without requiring human interaction with a service provider. The resources are scalable over several data centres. Cloud environments provide high scalability and provide services to cater to the needs of business for larger audiences.

(b) *Quality of Service (QoS)*—Cloud computing can guarantee QoS for users in terms of hardware or CPU performance, bandwidth and memory capacity. The availability of servers is high and more reliable as the chances of infrastructure failure are minimal.

(c) *User-Centric Interface*—Cloud interfaces are location independent, and they can be accessed by well-established interfaces such as Web services and Web browsers. Also, the cloud can be accessed by mechanisms that promote use by heterogeneous thick client platforms (e.g. mobile phones, laptops and PDAs).

(d) *Autonomous System*—Cloud computing systems are autonomous systems managed transparently to users. The cloud works in 'distributed mode' environment. The software and data inside clouds can be automatically reconfigured and consolidated to a simple platform depending on user's needs. The cloud environment is more agile as it shared resources among users and tasks, i.e. the provider's computing resources are pooled to serve multiple consumers using a multi-tenant model, with different physical and virtual resources dynamically assigned and reassigned according to consumer demand.

(e) *Pricing*—Cloud computing does not require upfront investment. No capital expenditure is required. Users may pay and use or pay for services and capacity as they need them, i.e.by automatically controlling and optimising resource usage by leveraging a metering capability, the resource usage can be monitored, controlled and reported. SLAs between the provider and the user must be defined when offering services in pay-per-use mode for the sake of transparency and quality of the services. This may be based on the complexity of services offered.

(f) *APIs*—Application programming interfaces may be offered to the users so that they can access services on the cloud.

In Fig. 2, client refers to UI channels and all types of human-operated devices such as browser, mobile devices, smartphones, etc. The application and component services refer to bespoke or custom-built applications or components, off-the-shelf solutions such as CRM, billing, etc., as per requirements. The platform is comprised of all the software to support application and service layers. Also, it acts as a bridge between application and IaaS. The infrastructure layer consists of servers (computing), storage and network infrastructure including virtualisation. The external service consumers use the interface exposed by the service to interact with service using API (application programming interface). Service providers are expanding their available offerings to include the entire traditional IT stack, from hardware and platforms to application components, software services, and whole applications, as shown in Fig. 2. The common thread in cloud computing offerings across all levels of the stack is the consumer-provider relationship and a dependence on the network to connect the two parties (Raines 2009).

The cloud infrastructure provides the physical resources which are distributed at multiple sites to support cloud computing. The virtualisation technologies with concepts such as virtual machine (VM) have made the infrastructure layer very efficient by allowing higher utilisation of physical resources. Because of advances in storage virtualisation, it is possible to rent out the storage incrementally over the Internet: network-based large-scale storage on demand is an example. Component as a service has well-defined interfaces for system-to-system integration, as the service components are distributed. Systems integration is increasing both in complexities within organisations and across external organisations. SOA attempts to streamline integration across systems by providing components. Cloud computing and SOA are synonymous, and they share some common characteristics such as service orientation. Cloud computing and SOA can be pursued independently, or concurrently, as complementary activities. From Fig. 2, it is clear that the services in the cloud are provided to end users through three delivery modes, namely, software as a service (SaaS), platform as a service (PaaS) and infrastructure as a service (IaaS) which have been described in detail in the section below.

4.3 Delivery Models

Agricultural research, education, extension and training are the essential four pillars of sustainable Agriculture. Information and communications technologies (ICTs) have tremendous role to play in these four components.

Sustainable development depends on the prudent use of natural resources such as soil, water, livestock, plant genetics, fisheries, forestry, climate, rainfall and topography. The productivity and sustainable development has a relationship as per the new research studies.

The farmers can gather the information at present through various sources such as from other farmers, money lenders, teachers, public phone operator, postman and health workers, government officials, agriculture extensionists, agriculture fairs and agricultural universities and through radios, televisions and newspapers.

However, the information that is available in some such sources is not readily available or accessible to farmers, and the illiteracy of farmers prevents them from gaining any benefit from these sources. The local agricultural centres do not have the updated information on crop varieties, pest control and government schemes and subsidies. Therefore, a service-oriented framework of information technology is essential for sustainable development.

The services based on information technology should work as a catalyst to help farmers who are small, having their lands away from markets and in ecologically fragile areas and who lack credit and tools to enhance their productivity. The services based on information technology can be hosted in a cloud.

4.3.1 Software as a Service (SaaS)

SaaS as a dominant and distribution service model is an underlying technology that supports Web services and service-oriented architecture (SOA). In the SaaS model, users sign up for services hosted by providers on the Internet and use them without the knowledge of its location or implementation. The applications are accessible from various client devices through either a thin client interface, such as a Web browser (e.g. Web-based e-mail), or a programme interface. SaaS applications are designed for end users, delivered over the Web. The services are offered to the community as a service on demand, which will be used by users from NGOs, cooperative societies, distributors and retailers in market yards and farmers. The entities that are utilising these services need not invest upfront in servers or software licences. The major advantage with SaaS is its ease of use and access. However, the end user is not aware of complexities or technologies used to develop the application or

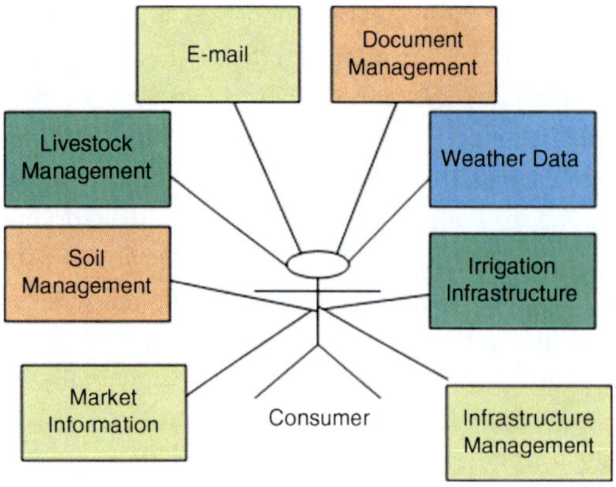

Fig. 3 Agricultural services to SaaS consumer

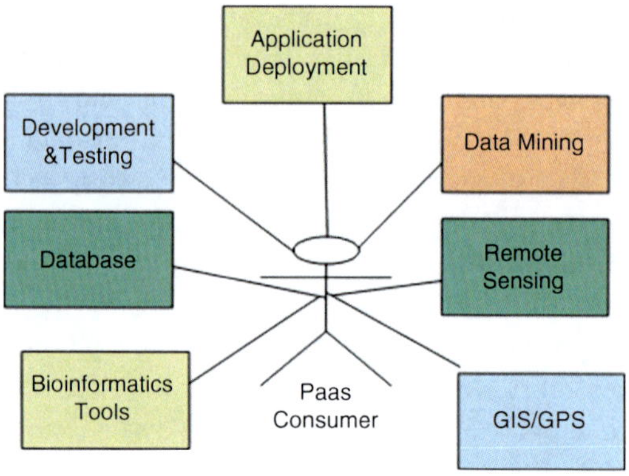

Fig. 4 Agricultural services to PaaS consumer

services. SaaS is a service-oriented framework with high deployment efficiency and a supportable platform as opposed to ASP (application service provider), which is focussed more on architecture-oriented solutions with low deployment efficiency.

The end users do not manage or control the underlying cloud infrastructure including the network, servers, operating systems, storage or even individual application capabilities. The characteristics of SaaS include:

(a) Web access to applications
(b) Software that is managed from a central location
(c) Software delivered in a 'one-to-many' model
(d) Users not required to handle software upgrades and patches
(e) Application programming interfaces (APIs) that allow for integration

Figure 3 presents some example of SaaS services available to cloud consumers in agriculture.

4.3.2 Platform as a Service (PaaS)

PaaS provides an application-centric development environment over IaaS (infrastructure as a service) to build, deploy, deliver and manage their applications, i.e. it provides the capability to the consumer to deploy onto the cloud infrastructure consumer-created or acquired applications created using programming languages, libraries, services and tools supported by the provider. It provides an operating environment for delivering a variety of applications and is essentially an outgrowth of the SaaS application delivery model. PaaS is the set of tools and services designed to make coding and deploying those applications quick and efficient. The applications that are built are run on the provider's infrastructure. Multiple concurrent users utilise the same application. Also, developers can leverage the additional infrastructure facilities like authentication and data access provided by the platform. It is the responsibility of the provider to meet manageability and scalability requirements of the applications. The limitation with this service is the dependency on the provider's cloud infrastructure. The end user does not manage or control the underlying cloud infrastructure including network, servers, operating systems or storage. PaaS is analogous to SaaS except that, rather than being software delivered over the Web, it is a platform for the creation of software, delivered over the Web. PaaS service delivery is employed where multiple developers are working on the same project and also when to leverage the existing assets such as data source or informatics tools while developing the application. The limitations of PaaS may include proprietary tools or languages, vendor lock-in and portability.

Figure 4 presents some example of PaaS services available to cloud consumers in agriculture.

4.3.3 Infrastructure as a Service (IaaS)

It is the delivery of computer infrastructure (typically a platform virtualisation environment) as a service. IaaS provides the entire infrastructure

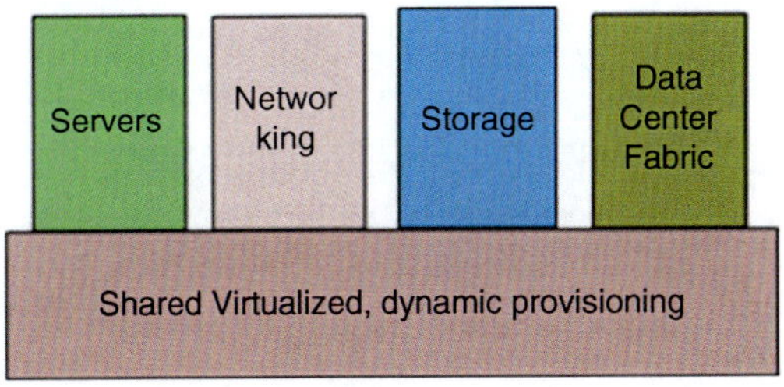

Fig. 5 Infrastructure building blocks

stack in terms of shared resources such as servers, storage systems, networking equipment, and data centre space as shown in Fig. 5.

IaaS differs from SaaS in that instead of software, IaaS delivers hardware. The details of resources or its location are not revealed to the client or customer but allocated on a need basis to manage workloads. Resources are distributed as service with dynamic scaling. The client will typically deploy his/her own software on the infrastructure with full control over server infrastructure, i.e. the user does not manage or control the underlying cloud infrastructure but she/he has control over operating systems, storage and deployed applications and possibly limited control of select networking components. The IaaS comes with its own limitation like higher prices for services, server downtime issues, etc. Other notable models that are built on the basic foundation of IaaS are computing as a service (CaaS) and hardware as a service (HaaS).

Apart from the above service delivery models, a number of variations exist, namely, security as a service (SECaaS), monitoring as a service (MaaS), communication as a service (CaaS), software testing as a service (STaaS), business process as a service (BPaaS), IT as a service (ITaaS), database as a service (DBaaS) and many more other variations being defined on a daily basis.

4.3.4 Comparison

Table 3 briefly explains the key features of each service.

The organisation has complete control over all technology components in self-owned data centres. Table 4 describes the degree of control an organisation exercises over technology components in different service delivery models when compared with self-owned data centre.

4.4 Deployment Models

There are different types of clouds that you can subscribe to depending on your needs. As a home user or small business owner, you will most likely use public cloud services:

1. *Public Cloud*—A public cloud can be accessed by any subscriber with an Internet connection and access to the cloud space. Typically, public clouds are administrated by third parties or vendors over the Internet, and services are offered on pay-per-use basis as resources are dynamically provisioned over the Internet via Web application or Web services. These are also called provider clouds. Security is a significant concern in public clouds.

   The deployments that are ideal for public cloud would also be most appropriate for any public-facing and nonsensitive activity, namely:

   (a) Open data initiatives
   (b) Public information repositories
   (c) Public collaboration or surveying facilities
   (d) Analytics involving nonsensitive or non-confidential data
   (e) The front-end elements of online services or apps that do not store sensitive data
   (f) Simulation testing of the availability, robustness and functionality of online services
   (g) Developing, testing and piloting new applications or solutions where deep integration with back-end data of a sensitive or confidential nature is not required

   Figure 6 presents a simple view of public cloud (Chavali and Sireesh Chandra 2011) and its consumers.

2. *Private Cloud*—A private cloud is established for a specific group or organisation and limits access to just that group with a full control over data security and quality of service. This cloud computing environment resides within the boundaries of an organisation and is used exclusively for the organisation's benefits.

Table 3 Feature comparison of cloud services

Service name	Description	Features	Examples
SaaS	Highly scalable Internet-based applications are hosted on the cloud and offered as services to the end user	1. Pay-per-use software	Google mail, Cisco WebEx office, myspace.com, Yahoo! Maps API, Google Calendar API, salesforce.com AppExchange, etc.
		2. Fast and easy to deploy	
		3. Managed by the vendor	
		4. Short- or long-term use	
		5. Rich Internet applications as websites	
		6. Collaboration and e-mail	
		7. API-specific services for integration	
PaaS	Platforms to build applications using SDLC are provided	1. Build your own cloud service	Microsoft SQL server data services, Google App Engine, Linux, Apache, PHP, restricted J2EE, Ruby, etc.
		2. Scalable test environment	
		3. Platform managed by vendor	
		4. Database	
		5. Message queue	
		6. App server	
IaaS	Storage, database and CPU power are provided on demand	1. Pay-per-hour storage, networking	Amazon EC2—creation of Linux virtual machines on the fly, GoGrid, 3tera, etc.
		2. Massive scalability	
		3. Rapid provisioning	
		4. Distributed servers geographically	
		5. Virtual servers	
		6. Vlan	
		7. Logical disks	

These are also called internal clouds. Private clouds can be built and managed by a company's own IT organisation or by a cloud provider.

3. *Community Cloud*—A community cloud is shared among two or more organisations that have similar cloud requirements such as security requirements, policy and compliance considerations. It may be managed by a third party or by the organisations themselves.
4. *Hybrid Cloud*—A hybrid cloud is essentially a combination of at least two clouds, where the clouds included are a mixture of public, private or community.

There are inherent risks with cloud service delivery and deployment models. SaaS in a public cloud has the highest inherent risk with least control, whereas IaaS in a private cloud has the least risk with maximum control.

4.5 Challenges

There is an inherent risk relationship between service delivery and deployment models. The risk grows with less direct control as we move from private to hybrid to public deployment models, so also in the service delivery model as we move from IaaS to PaaS to SaaS. Some of the typical risks associated with cloud computing are given below:

1. Data privacy
2. Data recovery and availability
3. Provider viability
4. Regulatory and compliance restrictions
5. Performance reliability

Table 4 Level of controls in service delivery model

Sl. no.	Component	SaaS	PaaS	IaaS
1.	Application	X	√	√
2.	Middleware	X	X	√
3.	Operating system	X	X	√
4.	Virtual machine	X	X	X
5.	Server	X	X	X
6.	Storage	X	X	X
7.	Networking	X	X	X

X—vendor manages in cloud
√—client manages in cloud

5 SaaS Levels

With the emergence of SaaS, many ASP players have tried to shift to this new model. To adapt to the concept of SaaS, the maturity model presented in this section may help to define the common key functions of SaaS service and to build a successful SaaS service. In the SaaS model, it has many role players such as developer, vendor, customer, etc. As per the Microsoft maturity model, there are four maturity levels, namely, ad hoc/custom, configurable, multi-tenant and scalable. These four models have been described below.

5.1 Level 1: Ad Hoc/Custom

Whenever a new user is added, a new instance of the software is created. If the user needs something specific, the instance of the software is changed. Each user is essentially running on his/her own 'version' of the software. Multiple instances of the software by different users run without sharing the content, i.e. own database and own schema as shown in Fig. 7.

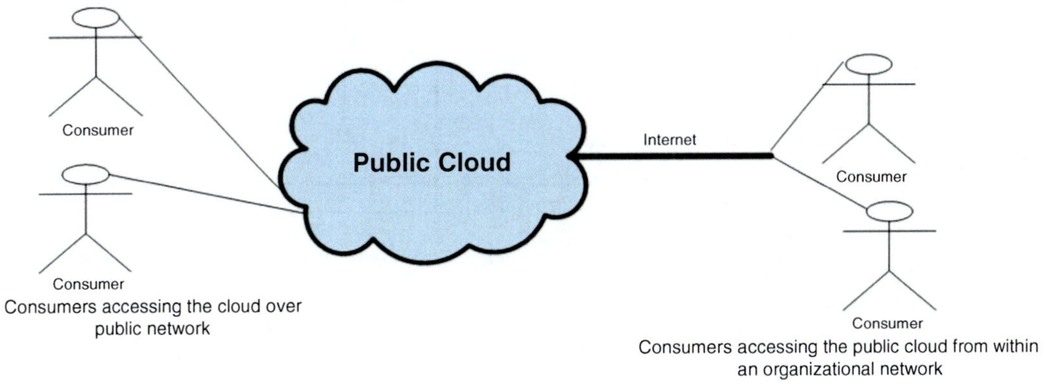

Fig. 6 Public cloud overview

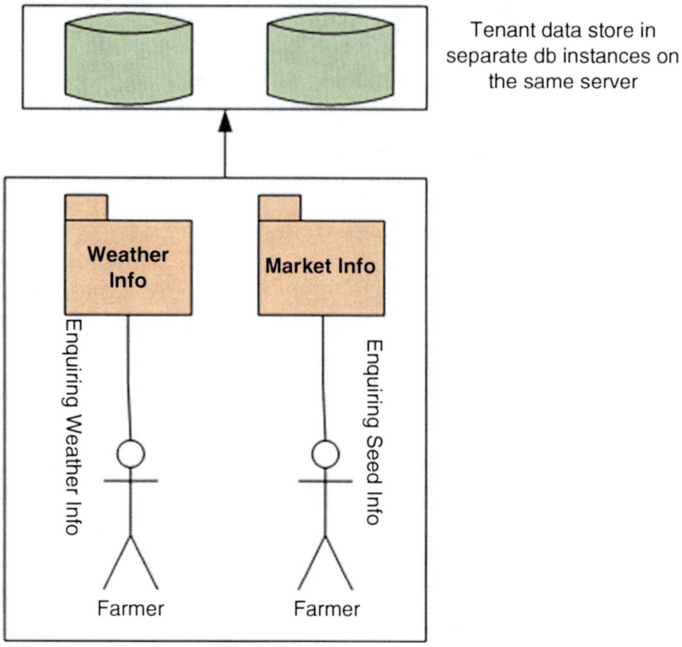

Fig. 7 Weather and market information on the same server; one instance per actor

5.2 Level 2: Configurable per Customer

This level aims at standardisation to provide shared service with the discrete instances of software (or user's application) and configurability. Every user runs on the same *version* of the software, and any customisations are done via configuration. This level supports shared database and dedicated schema, i.e. tenant data is stored in the same database but in respective schema (Fig. 8).

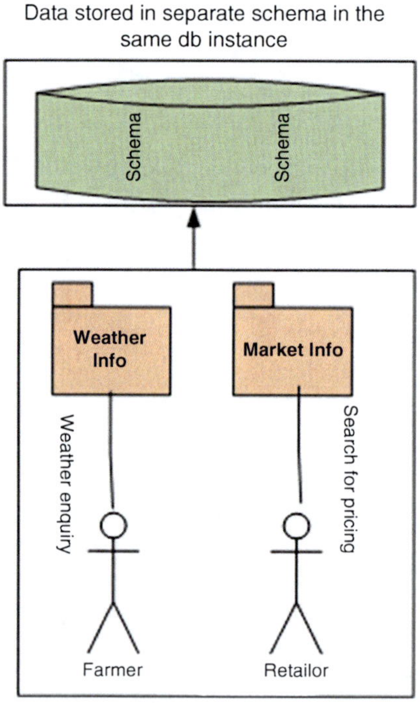

Fig. 8 Multiple data storage on a single server, configurable per actor

5.3 Level 3: Multi-tenant and Configurable

This level supports integration with focus on multi-tenancy (sharing resources across users). All users run on a single version of the software and on one 'instance' as shown in Fig. 9. The database schema and database are shared itself in order to accommodate multi-tenant simultaneously. There is no custom code built in to meet user-specific functions, which are achieved by configuration. This level gets into a limitation while accommodating large number of tenants.

5.4 Level 4: Multi-tenant, Configurable and Scalable

This level is regarded as a level of virtualisation. This level mainly focusses on maximisation of

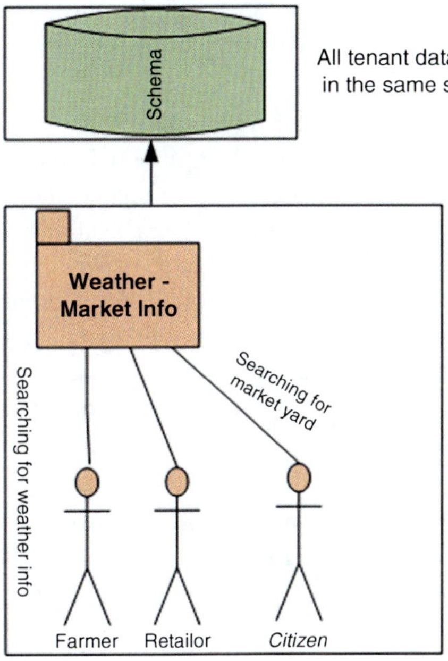

Fig. 9 Sharing of information among different users on a single database; multi-tenant

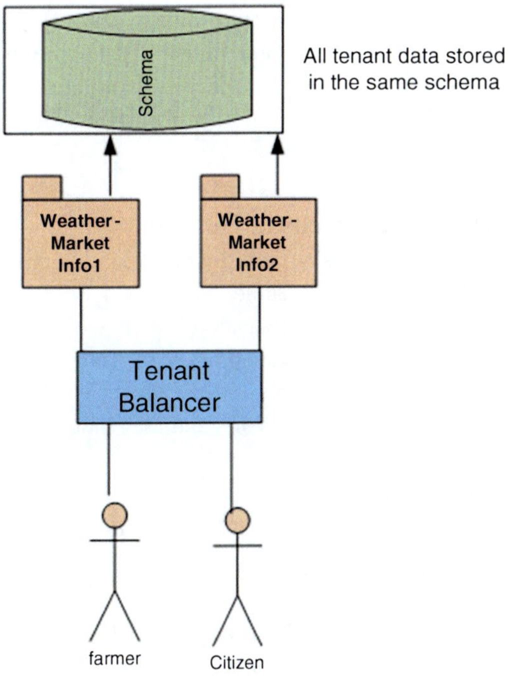

Fig. 10 Data storage in different servers; multi-tenant scalable

practical use of resources via service modulation and encapsulation. The economies decide the extent of scaling of tenants rather than the architecture deciding it, i.e. there is an optimal number of tenants per instance. In Fig. 10, Market Info1 and Market Info2 (latest service) represent two versions of the Market Info having the same database and schema, i.e. different code is run to meet requirements of users.

6 Cloud Services in Agriculture

6.1 SaaS Services

The list of agricultural services are described below (http://dacnet.nic.in/AMMP/AMMO.htm).

6.1.1 Pesticides, Fertilisers and Seeds (PFS)

This service aims at providing information on:

1. Good agricultural practices
2. Current prices and availability in an area closest to the farmer
3. Dealer networks
4. Quality control and assurance mechanism for pesticides (including insecticides), fertilisers and seeds
5. Registration and licensing for manufacturing and marketing of pesticides and fertilisers
6. Process of licence to retail seeds
7. Process registration of seed growers and certification of seeds
8. Publicly display quality testing of the samples drawn for pesticides, fertilisers and seeds
9. Expert advice and grievance management related to pesticides, fertilisers and seeds

6.1.2 Soil Health (SH)

This service aims at providing information on:

1. Soil health conditions
2. Practices suitable to the soil type
3. Balanced use of fertilisers
4. Soil-testing lab results
5. Soil surveys for different crops and different agro climatic zones
6. Expert advice on seeds for the same or alternate crops
7. Grievance management through multiple service delivery channels to the farmers

Farmers will also be provided expert advice depending on expected yield and maturity period after considering soil condition and other agro-climatic parameters.

6.1.3 Crops and Farm Machinery (CFM)

This service aims at providing information on:

1. Best practices for each crop stage for management of plant population
2. Monitoring of crops for pests and diseases
3. Expert advice on different aspects of crop management
4. Grievance management through multiple service delivery channels to the farmers
5. Farm machinery availability, quality and guidance to farmers

6.1.4 Forecasted Weather (FW)

This service aims at providing disaggregated information in each agro-ecological subregion on forecasted weather, agro-met advisory, SMS alerts for weather forecast and crop impact and grievance management through multiple service delivery channels to the farmers.

The main purpose of this service is to provide reliable weather information in real time such as:

1. Satellite imaginary or animations
2. Rainfall bulletins
3. Forecasts based on geographical location
4. Summaries of weather over period of time

6.1.5 Commodity Prices and Arrivals (CPA)

This service aims at providing information on:

1. Prices, including minimum support prices (MSPs)
2. SMS alerts on prices, arrivals and commodity indices
3. SMS alerts for buyers, sellers and transporters
4. Prices, arrivals and commodity indices
5. Crop quality associated with MSP
6. Historic price trends
7. Expert advice on issues related to future prices and arrivals trends
8. Mechanism for grievance management

It will provide an e-platform for interaction and for marketing of agricultural produce.

6.1.6 Electronic Certification for Exports and Imports (EC)

This service aims at providing information on certification procedure, fees, competent authorities, automation of certification process on a workflow basis, SMS-based status alerts and mechanism for grievance management.

6.1.7 Marketing Infrastructure (MI)

This service aims at providing information on:

1. Marketing infrastructure at the regulated market yards, postharvest facilities available at the yards
2. Storage infrastructure like availability, capacity and fees for storages/warehouses of both the private and public sector
3. Information needs of the farmers on credit linkages

6.1.8 Monitoring and Evaluation of Schemes (MES)

This service aims at providing information on schemes and programmes implemented at the state:

1. Physical progress and fund utilisation
2. Automation of issue and submission of utilisation certificate
3. Mechanism for grievance management
4. Search facility to common public and government officials using customisable queries
5. Evaluating and monitoring reports on different schemes

6.1.9 Irrigation Infrastructure (II)

This service aims at providing information on:

1. Water release schedule
2. Best practices on irrigation
3. Ground water, availability and viability of tube wells in an area
4. Water level in reservoirs and area that can be irrigated from it
5. Irrigation equipment
6. Expert advice and mechanism for grievance management

6.1.10 Livestock Management (LM)

This service will provide information on:

1. Livestock management-related activities at the state level
2. Expert advice on livestock during normal and drought circumstances
3. Availability of fodder in the nearest region of the farmer

6.1.11 Market Information (MKI)

This service provides information related to market such as:

1. Market fee and market charges
2. Market functionaries and market laws
3. Information on market committees
4. Prices of agricultural commodities both maximum and minimum
5. Promotional information such as market credit, accepted standards, etc.
6. Trading information on food products

6.2 IaaS Services

The cloud providers are now beginning to segment cloud infrastructure and services in their own data centres exclusively for government use; the probability is that it would not be any more expensive for cloud providers to do similarly in public service-owned facilities as it would negate the necessity for them to invest in or enhance facilities, power, cooling, networking and access control for public service purposes.

The infrastructure in a cloud combines together servers, storage, networking and security at the data centre to manage the infrastructure. The cloud system is built on service automation and converged infrastructure. OEM such IBM, HP, etc. provides various solutions to automate cloud solutions, and a detailed discussion on how to automate infrastructure is beyond the scope of this section. For more information on cloud offerings in infrastructure, refer to OEM websites. However, server automation at data centre using VMware has been illustrated briefly below for the sake of reader's understanding from application service perspective.

One of the primary benefits of server virtualisation with VMware (http://www.vmware.com) is that it allows IT organisations to consolidate the servers. All applications of agricultural services or all the components of an application can be made available in a single server as virtualised services depending the complexity of each component service/application and computing power of the server, which means a single physical server can support multiple virtual machines (VMs). The applications that would normally require a dedicated server can now share a single physical server. The application, namely, Information on Soil Health, has seven service components, and these components are spread on two physical servers. Figure 11 (Chavali and Sireesh Chandra 2011) describes VMs specific to each component service within the data centre. The server virtualisation results in a reduction in the number of servers in a data centre, which leads to significant savings in capital expenditure (CapEx) and operational expenditure (OpEx). Virtual machines are created using software that supports virtualisation like ESX server. A virtual network is created between VMs hosted on the same physical system with no virtual network traffic consuming bandwidth on the physical network. The virtual infrastructure created using VMware contains VMotion (http://www.vmware.com/products/vmotion), Distributed Resource Scheduler (DRS) and high availability (HA). VMotion allows VMs to be seamlessly relocated to different systems while keeping their MAC addresses. There is no downtime required to move VMs within compatible pools of servers. In Fig. 11, the component services have been deployed on Windows and Linux running inside of VMware ESX server as virtual machines. Also, other infrastructure applications running as virtual machines are shown for the sake of understanding as an example.

Server virtualisation (Chavali and Sireesh Chandra 2011) helps:

1. To increase hardware utilisation and reduce hardware requirements with server consolidation (also known as physical-to-virtual or P2V transformation)
2. To reduce required data centre rack space, power cooling, cabling, storage and network components by reducing the sheer number of physical machines
3. To improve application availability and business continuity independent of hardware and operating systems
4. To improve responsiveness to business needs with instant provisioning and dynamic optimisation of application environment

Virtualization has brought about an advantage to make data centre more dynamic, providing performance, flexibility and capacity at a much lower cost, and enabled the automatic and dynamic

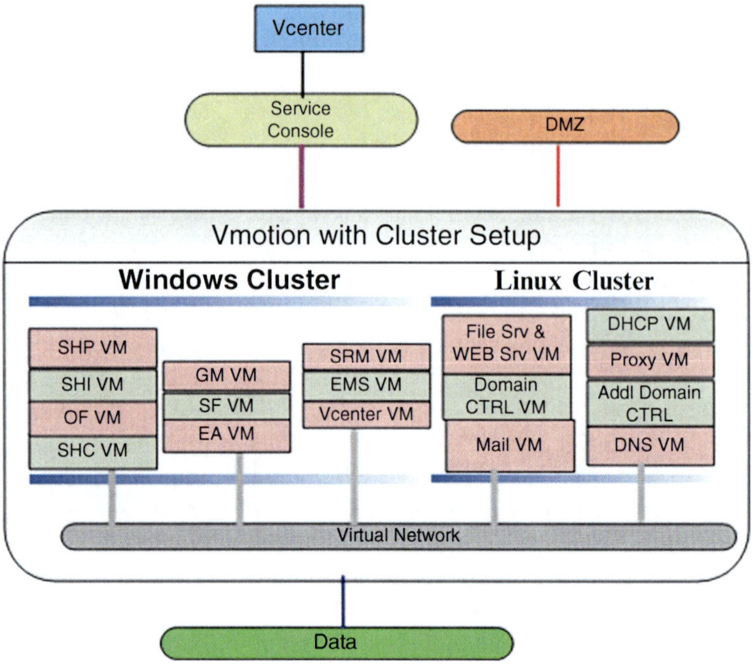

Fig. 11 Virtual network of services

allocation of resources depending on workloads and business requirements. VMware technology can replicate data from primary to secondary with Site Recovery Manager (SRM) thereby creating resiliency from a disaster recovery perspective as well as creating the snapshot of these virtual machines to replicate them to other sites.

The data centre applications like enterprise management software (EMS) tools are used to control and manage the infrastructure to meet on-demand delivery.

7 Cloud Architecture

Figure 12 presents Cloud Computing Reference Architecture (Fang et al. 2011), which is a high-level architecture intended for understanding of requirements, uses, characteristics and standards of cloud computing.

Cloud Computing Actors

As shown in Fig. 12, the NIST cloud computing reference architecture defines five major actors: cloud consumer, cloud provider, cloud carrier, cloud auditor and cloud broker. Each actor is an entity (a person or an organisation) that participates in a transaction or process or performs tasks in cloud computing.

Cloud Consumer

A cloud consumer represents a person or organisation that maintains a business relationship with, and uses the service from, a cloud provider. A cloud consumer browses the service catalogue from a cloud provider, requests the appropriate service, sets up service contracts with the cloud provider and uses the service. The cloud consumer may be billed for the service provisioned and needs to arrange payments accordingly.

Cloud Provider

A cloud provider is the entity (a person or an organisation) responsible for making a service available to interested parties. A cloud provider acquires and manages the computing infrastructure required for providing the services, runs the cloud software that provides the services and makes the arrangements to deliver the cloud services to cloud consumers through network access.

Cloud Broker

As cloud computing evolves, the integration of cloud services can be too complex for cloud consumers to manage. A cloud consumer may request cloud services from a cloud broker,

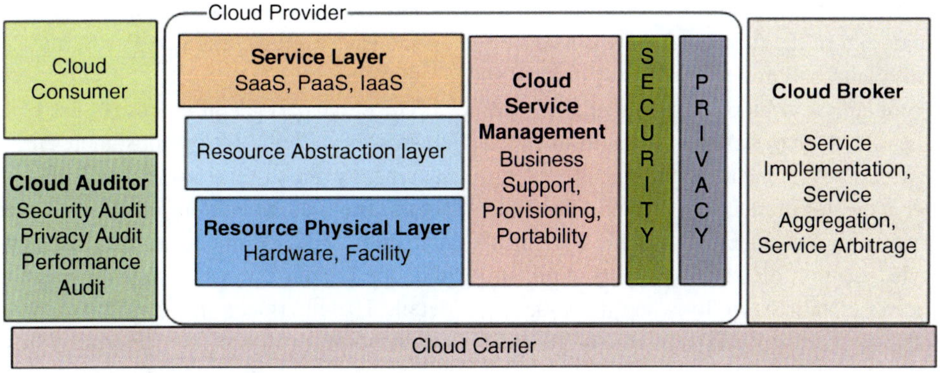

Fig. 12 NIST cloud computing reference architecture

instead of contacting a cloud provider directly. A cloud broker is an entity that manages the use, performance and delivery of cloud services and negotiates relationships between cloud providers and cloud consumers.

Cloud Carrier

A cloud carrier acts as an intermediary that provides connectivity and transport of cloud services between cloud consumers and cloud providers. Cloud carriers provide access to consumers through network, telecommunication and other access devices. For example, cloud consumers can obtain cloud services through network access devices, such as desktop computers, laptops, mobile phones and other mobile Internet devices (MIDs). The distribution of cloud services is normally provided by network and telecommunication carriers or a transport agent, where a transport agent refers to a business organisation that provides physical transport of storage media, such as high-capacity hard drives.

The detailed discussion on the concept of reference architecture is beyond the scope of this section and can be understood further by referring to the National Institute of Standards and Technology (https://www.nist.org).

7.1 Cloud Application Reference Architecture

Cloud architecture is service oriented, i.e. the cloud solution adhering to cloud architecture is a service-oriented architecture (SOA) solution. All cloud services are SOA services; however, all SOA services are not cloud services. Cloud computing does not replace SOA, or the use of distributed software components as an integration technology. The cloud reference architecture is nothing but SOA reference architecture combined with SOA reference architecture service categories and NIST cloud characteristics. The cloud application reference architecture provides reference architecture for use when creating customer-specific architectures and design. In general, there are three views to any cloud solution, namely, business view, function view and technical view. The definition of service type and deployment model starts at business view and continues to be activity in the subsequent functional and technical views.

7.1.1 Business View

The cloud provider should keep in view of the business drivers, the issues and the benefits while transforming the business to a cloud, i.e. that the model to customers and guaranteed continuity of the business on the service platform is the key to transform to cloud. The constraints such as cost reduction, governance, access, etc. play a critical role in solution realisation. While developing the cloud use cases, the following access mechanisms by user should be included as part of use cases:

(a) Private cloud
(b) Enterprise to cloud
(c) End user to cloud
(d) Enterprise to cloud to end user
(e) Enterprise to cloud to enterprise

7.1.2 Functional View

The major aim of the functional view is to understand what the solution will do and can offer. While scoping the solution, it is essential to consider the cloud characteristics such as metered usage, access over Internet, horizontal scaling, multi-tenancy, location independence, etc. apart from evaluating considerations such as architectural style, security, performance, integration, etc. A service is cloud ready only when it can access over Internet on pay-per-use basis with on-demand scaling. In most of the bespoke applications, the primary architectural style is either SOA (service-oriented architecture) or parallel computing, and the functional view of services is represented by using layered model. It also provides a view of the cloud actors from the solution perspective.

7.1.3 Technical View

This view primarily deals with functions, which are related to technical issues of SaaS service such as database management, configurable user interface and business workflow and integration technology. This is expanded to include multi-tenancy (shared database, distributed database schema), configuration (UI, metadata, workflow, etc.), integration (mash-up API, Web service, etc.) and security (authorisation and authentication).

7.2 Agricultural Solution

The solution architecture of agricultural application in cloud is as shown in Fig. 13. The number of portals/websites of agriculture as displayed in Annexure II needs to interact with each other for seamless delivery of information/services to farmers. These agricultural portals interact with one another in cloud by adopting service-oriented architecture (SOA) framework. The interoperability is built on XML (Extensible Markup Language) and Web services standards. The important features of the solution are:

- Cloud application architecture
- Develop business functionality as SaaS (software as a service)
- Provide Web-based interface with service connection
- Extensible to support multiple access devices such as desktop computer, IVRS, mass media, mobiles, etc.

The cloud service provider may act as service creator as well as service consumer, and a consumer may act as service creator. In Fig. 13, all actors such as transition manager, operations manager, security manager, etc. and business and operational processes are not presented in detail. The interfaces in Fig. 13 will have API (application programming interface) for consumer, provider and creator's interaction. The security includes, but not limited to, security policy, vulnerability management, data policy enforcement, etc. which are not covered in detail.

The agricultural applications are divided into core, common and other applications in Fig. 13. The services may be owned and created by different vendors but having a framework for services eradicate duplicate services and ensures better manageability. For example, pesticide registration services would be exposed through the Web services, and the same can be consumed whenever a manufacturing licence is to be issued. The manufacturer applying for a licence would give his/her registration number, and it would be verified against the pesticide registration service.

A cloud integrator is a product or service that helps a business negotiate the complexities of cloud migrations. A cloud integrator service (sometimes referred to as integration as a service) is like a systems integrator (SI) that specialises in cloud computing.

Government, private enterprises, farmers and integrators are the stakeholders in agricultural solution. They access the applications through delivery channels such as XML, SMS, SMTP and Web and receive the information from the services through the delivery channel services. All the content in these applications is not open to public user if it is protected information. The users are allowed to access such protected information through authentication and authorisation process. Even the authorised users are allowed to see the only functionality based on the role associated with login credentials. The content management will manage large number of content objects, and a document and workflow service takes care of processes and business rules for processing data. Identity

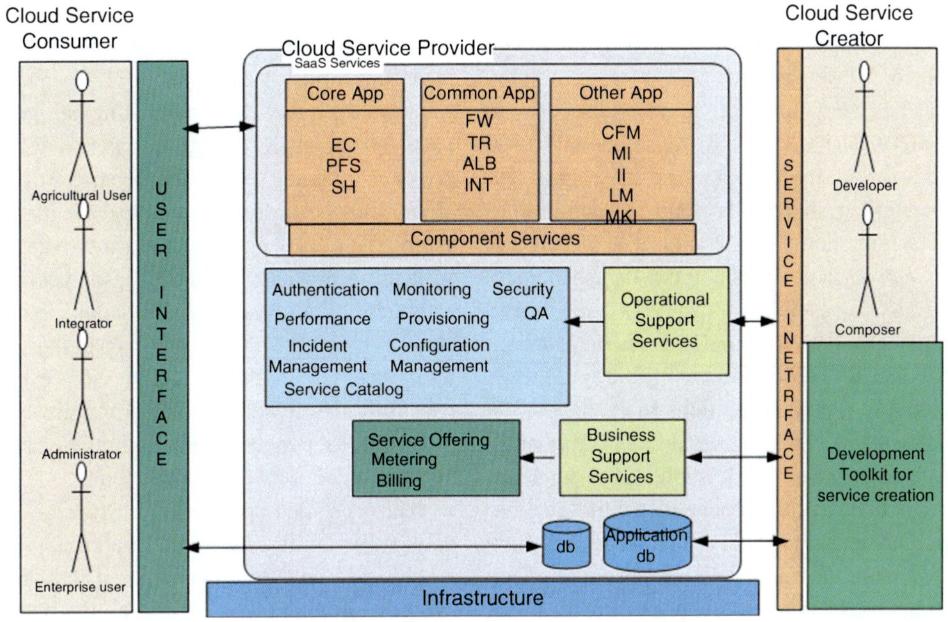

Fig. 13 Service architecture for agricultural services

management is performed through directory services such as LDAP. The service interface layer enables the access to different services in the solution to communication with each other by means of data exchange, message routing, etc.

8 Conclusions

In countries like India, there are many websites and portal providing information and services in agriculture sector. The websites of various departments, viz., agriculture, irrigation, fertilisers, ICAR, etc., do not share Web services among them, thereby leading to visiting the individual website by stakeholder in agriculture sector for tracing the information. As every website is managed independently having their own flavour, it is inconvenient to follow and trace the information. To get over some of these problems, agricultural ministries in some countries have initiated a central agricultural portal consisting of linkages to various other departments, which serves as 'one-stop single-window solution'.

For example, Central Agricultural Portal (http://dacnet.nic.in/) in India provides a platform by offering cluster of services for all stakeholders (farmers, private sector and the government, research scholars) to access information, avail services, collaborate and share knowledge. It would play a critical role in providing single access point to government information and services related to agricultural sector, eliminating the need to navigate multiple websites and applications. It would provide a secure and individualised view of multiple online resources and interactive services.

AGMARNET (http://www.agmarnet.nic.in) is an Agricultural Marketing Information System in India that exchanges data with market yards and provide information to all stakeholders in real-time basis. It is having a database on 300 commodities and about 2,000 varieties and with a nationwide market information network of around 700+ agricultural products wholesale markets (APWM) and 70+ State Agricultural Boards/Directorates. Its aim is to facilitate farmers to bargain better prices for their product. Refer to the Appendix for more information on agricultural portals in India.

China has established rural market information network by connecting cities, provinces, townships, wholesale markets, intermediary agencies and business houses. It has enabled different delivery methods.

The US Department of Agriculture (USDA) is an incredibly complex enterprise, with a distributed workforce, 27 subagencies and a broad mission that touches everything from homeland security to food safety. The USDA has transitioned 120,000 federal workers from on-premise messaging and collaboration to Microsoft's cloud computing solution. The migration to the cloud is part of USDA's (http://www.usda.gov/wps/portal/usda/usdahome) vision to consolidate disparate messaging environments onto a single, unified platform, which will reduce costs, boost workforce productivity and improve communications and collaboration across the agency. When several SaaS components are bundled together, then it is referred as 'aggregation'. The bundles of components are presented to the user where there is little or no choice and allow user to choose the component as per requirement. SaaS aggregation is necessary when there are many suppliers of SaaS, many users and many servers.

An aggregation platform for SaaS is the key for service offerings in cloud. For example, HP AP4SaaS (White 2011) serves as the single point of access for all applications (SaaS and hosted services), delivering a 'one-stop shop' for cloud. The aggregated platform provides a smart information dissemination system and aims to establish a farmer-centric integrated distributed information system for agriculture like Central Agricultural Portal of India.

A global public domain database AGRIS (International System for Agricultural Science and Technology) whose database is maintained by FAO has a varied content provided by more than 150 participating institutions from 65 different countries. It has more than four million structured bibliographical records on agricultural science and technology. The AGRIS search system permits the use of keywords to perform sophisticated searches from the *AGROVOC* thesaurus, specific journal titles or names of countries, institutions and authors.

The aggregate platform is a single virtual platform to aggregate all the available information on agriculture for online access. The relevant links in the portal are managed by the respective agencies or departments. The sensitive information in the service offerings is protected by data encryption, and only authorised users can access such sensitive data.

It has been observed widely that cloud computing is moving from a few early adopters to becoming an increasingly mainstream choice of ICT strategy. It is acknowledged that cloud computing has the potential to fundamentally change the nature of ICT delivery over time and to provide benefits in terms of efficiencies, cost-effectiveness, speed to market, the leveraging of new opportunities, improving mobility and access and deploying resources on core activities. As a consequence, it is anticipated that cloud computing will be a key part of the strategic future of ICT in the public service, eventually becoming the default and primary delivery mode to develop a public service community cloud to negate the necessity for private clouds.

It has been demonstrated that while cloud computing certainly provides opportunities for efficiencies and cost savings, it has not yet evolved to the degree required by the public service in terms of security, reliability, service levels, standards, jurisdictional, legal and contractual arrangements, technical interoperability, licensing, dynamic and real-time availability, availability of requisite skills and commercial models.

Spread of wireless networks, low-cost mobile phones and high reach of Wi-Fi networks in rural areas made it easy to gather information dynamically. The information gathered from mobile phones can be sent to the server in the cloud data centre instantly and thus improving data integration. M-agriculture is playing a crucial role in information gathering and analysis.

Most of the agricultural sites of today are SOA compliant and, in my opinion, are at Level 2 of SaaS, i.e. shared db, dedicated schema and standardised SLA. It has been progressing from standardisation to integration phase where db and schema are shared with multi-tenancy and SLA would be measured (Level 3 of SaaS). The ideal implementation is to have SaaS services with true virtualisation, i.e. virtualisation of data layer with database and schema distributed (Level 4 of SaaS).

Agriculture being service in public, there is a growing trend to make the data, which is collected or developed by different public institutions, publicly available so that its value can be realised. Open data policy will enable sharing and access to nonsensitive data generated by different government institution. *'A dataset is said to be open if anyone is free to use, reuse, and redistribute it – Open Data shall be machine readable and it should also be easily accessible'*. The data portal of India (http://data.gov.in) provides collated access to data sets published by different departments in open format and has agriculture community (http://data.gov.in/community/agriculture-community) which can use the public data for developing new components and applications.

Crops will react differently to different soil, water and heat conditions. Crop data and soil data may be in terabytes. The strategy of putting this data in a traditional data warehouse may be expensive because of the kind of data quantities being produced today. Big data is basically about finding ways of storing and analysing massive amounts of data. Big data solutions take large amounts of structured and unstructured data and run algorithms against data sets to refine data. The open data policies throw opportunities for big data solutions which were never thought before in agriculture.

Data is no longer simply stored in a relational table only because of the emergence of non-relational data structures most of which are open source. The cloud delivery model without the data needed for computation is a problem despite having faster provisioning of infrastructure. The public cloud is being made data ready by feeding the data in trickles. It is easy to burst into the public cloud in agriculture with the availability of voluminous data at present in public domain.

Annexure I

Definitions and Abbreviations

ACL	Access control list
AGMARKNET	Agricultural Marketing Information System Network
AgRIS	Agricultural Resources Information System
AGRIS	International System for Agricultural Science and Technology
AJAX	Asynchronous JavaScript and XML
API	Application programming interface
ASP	Application service provider
CMS	Content management system
CRM	Customer relationship management
DB	Database
DRS	Distributed Resource Scheduler
EMS	Enterprise management software
G2B	Government to Business
G2C	Government to Consumer
G2G	Government to Government
GDP	Gross domestic product
GIS	Geographic information system
GPS	Global positioning system
HA	High availability
HTTPS	Secure Hyper Text Transfer Protocol
IaaS	Infrastructure as a service
ICAR	Indian Council of Agricultural Research
ICT	Information and Communications Technology
IT	Information technology
LDAP	Lightweight Directory Access Protocol
MAC	Media access control
MID	Mobile internet device
MSP	Minimum support price
NGO	Non-governmental organisation
NIST	Nation Institute of Standards and Technology
OEM	Original equipment manufacturer
OLAP	Online analytical processing
PaaS	Platform as a service
PDA	Personal data assistants
QoS	Quality of service
SaaS	Software as a service
SI	System integrator
SLA	Service-level agreement
SMS	Short message service
SMTP	Simple Mail Transfer Protocol
SOA	Service-oriented architecture
SRM	Site Recovery Manager
SSL	Secured sockets layer
VM	Virtual machine
XML	Extensible Markup Language

Annexure II

List of Agricultural Websites

Sl. no.	Website	Description	URL
1.	AGMARKNET	Primarily the farming community	http://agmarknet.nic.in
2.	Department of Agriculture and Cooperation	Central/state agriculture departments	http://agricoop.nic.in http://dacnet.nic.in/farmer/new/home-new.html
3.	Agriculture Census website	Ministry of Agriculture and its users	http://agcensus.nic.in
4.	DACNET	Ministry of Agriculture	http://dacnet.nic.in
5.	SEEDNet	Central/state governments, National Seed Corporation, State Seed Corporation, seed labs, universities, KVKs, farmers	http://seednet.gov.in
6.	Farm mechanisation	Central/state agriculture departments	http://farmmach.gov.in
7.	Rashtriya Krishi Vikas Yojana	Ministry of Agriculture and its users	http://rkvy.nic.in
8.	Agriculture extension	Ministry of Agriculture and extension functionaries	http://vistar.nic.in
9.	Department of Agricultural Research and Education	Department of Agricultural Research and Education	http://dare.gov.in
10.	Directorate of Plant Protection Quarantine and Storage		http://ppqs.gov.in
11.	Indian Council of Agricultural Research	ICAR Library	http://icarlibrary.nic.in
12.	Water resources	Common citizen, state government, organisations under MoWR, research institutes	http://mowr.gov.in
13.	Central Insecticide Board and Registration Committee	Pesticides industry	http://cibrc.nic.in
14.	Plant quarantine	Importer, exporter	http://Plantquarantineindia.nic.in
15.	Macro Management of Agriculture Scheme	It is in public domain—anyone can visit and benefit	http://dacnet.nic.in/macronew
16.	National Horticulture Mission	Farmers	http://nhm.nic.in
17.	National Bamboo Mission	Farmers	http://nbm.nic.in
18.	Directorate of Economics and Statistics	Citizen	http://dacnet.nic.in/eands
19.	Retail Prices Information System	Citizen	http://dacnet.nic.in/rpms
20.	Land Use Statistics Information System	Citizen	http://dacnet.nic.in/lus
21.	Food processing industries	Stakeholders of food processing sector including entrepreneurs, industry, exporters, policymakers, government	http://mofpi.nic.in
22.	Indian Institute of Crop Processing Technology	Stakeholders of agriculture sector including entrepreneurs, industry, exporters, policymakers, government, R&D institutions and farmers	http://www.iicpt.edu.in
23.	Department of Animal Husbandry, Dairying and Fisheries		http://dahd.nic.in http://dms.nic.in

(continued)

Sl. no.	Website	Description	URL
24.	National Meat and Poultry Processing Board (NMPPB)	Stakeholders of meat and poultry sector including entrepreneurs, industry, exporters, policymakers and government	http://nmppb.gov.in
25.	AgRIS	State/district-level agriculture and allied sectors' departments	http://agris.nic.in
26.	Pest Disease Monitoring System		http://dacnet.nic.in/pdmis
27.	ProFarmer	Weekly publications from Australia	www.profarmer.co.au
28.	Weather zone	Weather resource system information of Australia	www.weatherzone.co.au
30.	Agricultural portal	Australian portal services	www.agriculture.gov.au
31.	Online food market	E-platform for food industry in China	www.21food.cn

References

Agricultural Mission Mode Project, Ministry of Agriculture, Govt of India. http://dacnet.nic.in/AMMP/AMMO.htm

Chavali LN, Sireesh Chandra V (2011) Information technology consolidation with virtualization in a contract research organization. Curr Trends Bioinform Pharm 5(3):1344–1352

Fang Liu, Jin Tong, Jian Mao, Bohn R, Messina J, Badger L, Leaf D (2011) NIST cloud computing reference architecture. Special publication 500–292, National Institute of Standards and Technology, Gaithersburg MD 20899

Hori M, Kawashima E, Yamazaki T (2010) Application of cloud computing to agriculture and prospects in other fields. Fujitsu Sci Tech J 46(4):446–454

Kang S et al (2010) A general maturity model and reference architecture for SaaS service. In: DASFAA 2010, Part II, LNCS 5982. Springer, Berlin/Heidelberg, pp 337–346

Raines G (2009) Cloud computing and SOA. In System engineers at Mitre service oriented architecture (SOA) series

Rudgard S et al (2009) Enhancing productivity on the farm. In ICT as enablers of agricultural innovative systems. The World Bank, Washington, DC. http://www.ictinagriculture.org/sourcebook/ict-agriculture-sourcebook

Singh VK, Singh AK, Chand R, Kushwaha C (2011) Role of bioinformatics in agriculture and sustainable development. Int J Bioinform Res. http://www.bioinfo.in/contents.php?id=21

White Paper (2011) Understanding the HP cloud system reference architecture. http://www.hp.com/go/cloudsystem

Bioinformatic Tools in the Analysis of Determinants of Pathogenicity and Ecology of Entomopathogenic Fungi Used as Microbial Insecticides in Crop Protection

Uma Devi Koduru, Sandhya Galidevara, Annette Reineke, and Akbar Ali Khan Pathan

Abstract

Insect pathogenic fungi have a huge potential as microbial components of biopesticides which serve as benign components in plant protection. The infection cycle of these fungi is well known. Realising their potential and scope to improve their utility in phytomedicine, extensive work on the molecular biology of pathogenesis has been done in the past decade. Wet bench techniques like gene isolation, cloning and characterisation and gene knockout experiments to transcriptomics techniques like cDNA-AFLP, microarray, qPCR, cDNA, EST and SSH library construction, as well as whole genome sequencing and analysis of data with a suite of bioinformatic tools and pipelines integrated with several biological databases, were done to understand the process/processes involved at each stage of the infection cycle of the insect pathogenic fungi. These are in particular adherence of spores to the insect cuticle, factors that aid in coping with the physical stress conditions in the surrounding environment, formation of an infection peg, penetrance into the insect, factors that abet in overcoming insect defence systems and growth in the insect, production of toxic secondary metabolites that lead to insect death and surfacing out from the insect cadaver as well as sporulating to iterate the infection cycle on yet another insect. The picture that emerged is detailed in this chapter. The genes/proteins involved and the analyses that aided in their identification are described. Environmental genomics through multitag 454 pyrosequencing of rRNA sequence reads in deciphering the effect of the

U.D. Koduru (✉) • S. Galidevara
Department of Botany, Andhra University,
Visakhapatnam 530003, India
e-mail: umadevikoduru@gmail.com

A. Reineke
Institute of Phytomedicine, Hochschule Geisenheim
University, Von-Lade-Str., 1, D-65366 Geisenheim,
Germany

A.A.K. Pathan
Department of Biochemistry, College of Science, King
Saud University, Riyadh, Kingdom of Saudi Arabia

K.K. P.B. et al. (eds.), *Agricultural Bioinformatics*,
DOI 10.1007/978-81-322-1880-7_13, © Springer India 2014

inundative application of an entomopathogenic fungus on the native soil fungal diversity is described. The chapter highlights the bioinformatics-bolstered investigation of the factors that influence the affectivity of insect pathogenic fungi as microbial biopesticides.

Keywords
Entomopathogenic fungi • Pathogenicity genes • Fungal diversity • Pyrosequencing • In silico tools and pipelines • Databases

1 Introduction

Ever since the hunter and gatherer man swapped spears for spades, he began his contest with insects for the plant produce. The conflicting goals of reduced pesticide usage and maintaining adequate agricultural production have provided strong impetus for the development of cost-effective alternatives to conventional chemical pesticides (Lacey and Goettel 1995). Pest control today is emerging slowly from an era of dependence on factory-produced toxic, broad spectrum chemical pesticides under a paradigm of "Integrated Pest Management" (IPM) in which biological control is the mainstay. Applied in well-conceived IPM systems, biopesticides have the potential to moderate the usage of chemical pesticides thereby reducing the selection pressure on insect pests. Consequently, the affectivity time span of the novel and effective chemical pesticides can be prolonged, and the highly efficient food production systems that they support can become more long lasting (Wraight et al. 2001). Biocontrol is a component in the IPM of all crop plants. Despite the renaissance of biocontrol from the late 1970s, entomopathogens still remain one of the greatest untapped resources for insect pest management (Goettel et al. 2001). The most popular microbial bioinsecticides are based on bacteria – *Bacillus thuringiensis* (Bt) being the most popular.

Fungal-based microbial biopesticides have a huge potential (Butt et al. 2001). Fungal entomopathogens are being used, though modestly, for management of insect pests of field crops and glass house plants, orchards and ornamentals, range turf and lawn, stored products and forestry and for abatement of pest and vector insects of veterinary and medical importance (Lacey et al. 2001). In comparison to the currently popular bacterial-based biopesticides, fungal-based products score on many counts making them more favourable organisms for microbial biopesticides: (1) Since the mode of entry of a fungus being mainly through the insect cuticle, even insects which do not munch on plant parts such as sucking insects (e.g. aphids, mites, mealy bugs, white flies, etc.) can be targeted unlike with bacterial-based insecticides. Fungi thus act as contact insecticides. They can also infect soil-inhabiting insects and cryptic species like the stem borers (due to endophytic habit) and bark beetles. (2) Death of the insect due to fungal infection occurs due to a combination of effects and events – therefore, the risk of development of insect resistance is decimal unlike with Bt-based biopesticides. (3) Fungi have the ability to recycle in the pest population due to the expression of mycosis (tuft of conidiphores with copious amount of conidia) on insect cadavers thus offering a long-term check. Fungal infections of insects take the form of a rapid epidemic (termed epizootics) rather than a chronic endemic type. Fungal-based biopesticides thus lend to both inoculative (applied at very low concentration and with an autonomous population increase) and augmentative (low concentration but the environment is modified to favour their development) applications (Cook 2000). (4) Entomopathogenic fungi have long persistence in the soil and therefore a long-term suppression activity (Khan et al. 2012).

Mycoinsecticides are yet to make a dent in the biopesticide market considering the impetus given to biological control in IPM. Biopesticides gained the notoriety of inconsistent performance due to their dependence on environmental conditions for survival and activity. Unlike the instant acting chemicals, they are slow in action. The marketability of the mycopesticides is believed to be directly related to speed of kill (St. Leger and Wang 2009). Of the ~750 known species of entomogenous fungi, only 6 have thus far been registered for use as mycopesticides (Butt et al. 2001). These are *Beauveria bassiana*, *Metarhizium* spp., *Verticillium* (*Lecanicillium*) *lecanii*, *Nomuraea rileyi*, *Paecilomyces (Isaria) fumosorosea* and *Beauveria brongniartii*. Among them, *B. bassiana* and *M. anisopliae* are the most popular. Several products with different trade names based on these two fungal species are in the market (De Faria and Wraight 2007; Jackson et al. 2010). Both these fungi have a very large host range of ~700 insect species for *B. bassiana* and ~200 for *M. anisopliae* (Humber 1991; Butt and Goettel 2000). *B. bassiana* is a generalist fungus with no host preference (Wraight et al. 2003; Rehner and Buckley 2005; Uma Devi et al. 2008) and thus can be used as a broad spectrum insecticide.

Entomopathogenic fungal genera have also been reported to afford protection of crops from pathogenic fungi, lead an endophytic existence and endow beneficial effects on the plant by their existence in the rhizosphere (Vega et al. 2009).

Extensive work has been done in understanding the genes involved in entomopathogenic fungal pathogenicity and virulence, tolerance to stress, their persistence in the soil and impact on soil microbes. Bioinformatics served as a handmaiden for the molecular biology techniques employed in these studies. Beginning with the study of one/few genes in a go, aided by simple sequence similarity studies with the most popular bioinformatic tools such as BLAST (Altschul et al. 1990), the investigations evolved into high-throughput studies in the post-genomic era. During the last decade, studies on fungi–host interactions and comparative transcriptomics especially those related to different strains/species of entomopathogenic fungi have progressed dramatically. This is due to the advent of next-generation nucleic acid sequencing (pyrosequencing) and advances in bioinformatics. Collectively, these advancements allow de novo assembly of high-quality eukaryotic genomes and high-throughput protein identification and functional assignment. Until 2010, pathogenicity genes were characterised based on transcriptomic profiling through cDNA-AFLP technique, microarray, qPCR, EST and cDNA and SSH libraries from entomopathogenic fungi cultured in vitro on cuticle extracts and in vivo in insects (Freimoser et al. 2003; Wang and St. Leger 2005; Wang et al. 2005; Freimoser et al. 2005; Khan et al. 2007). Gene knockout experiments were also conducted in some instances to validate the role of genes predicted to have a role in virulence/pathogenicity. The whole genome sequencing has been completed for three entomopathogenic fungal genera – *Metarhizium anisopliae*, *M. acridium* (Gao et al. 2011); *Cordyceps militaris* (Zheng et al. 2011); and *Beauveria bassiana* (Xiao et al. 2012). Pathogenicity genes have been identified in these genomes through mining with bioinformatic tools, and in some instances, their role was validated. The pathogen–host interactions database (PHI-base) that catalogues experimentally verified pathogenicity, virulence and effector genes from fungal, Oomycete and bacterial pathogens, which infect animal, plant, fungal and insect hosts (Baldwin et al. 2006), has been useful in computational identification of pathogenicity genes through comparative genomics. Bioinformatic analysis has been facilitated through several tools and pipelines integrated to different databases (Table 1). These studies have given an extensive, though not a complete, picture of the genes involved in the pathogenicity of entomopathogenic fungi (Table 1).

Molecular and physiological understanding of the infection process of the fungus and identification of the genes involved increases our ability to introduce a potential strain to manage an outbreak of an insect pest. These studies can foster development of virulent fungal bioinsecticide agents tolerant to abiotic stress and also in identifying genes suitable for development of transgenic crops as was done with *cry* genes of *Bacillus thuringiensis*.

When a product is registered as a biopesticide, considerations are given to its safety to humans and beneficial animals. Information on its effect on soil microbes and the period of its persistence can also be important matters in considering

Table 1 Genes/proteins implicated in the infection process of entomopathogenic fungi and abiotic stress tolerance: molecular biology techniques, bioinformatic tools, pipelines and databases employed for analysis

<table>
<tr><td colspan="2">Infection cycle and genes involved</td></tr>
<tr><td>Stages of infection</td><td>Genes/proteins implicated</td></tr>
<tr><td>Pre-penetration (on the insect cuticle)
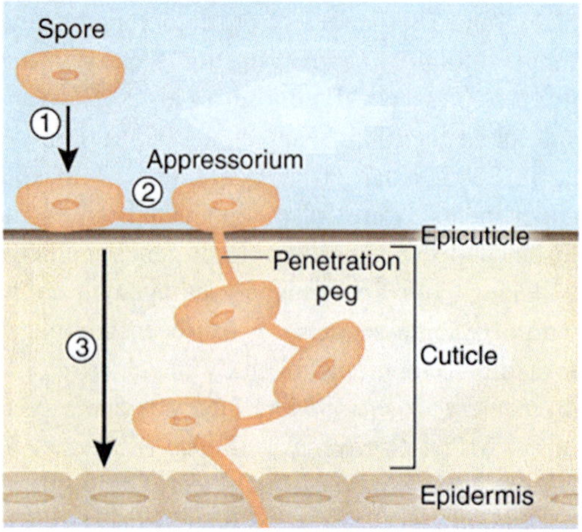
</td><td>Mitogen-activated protein kinase (Bbmapk1), high osmolarity glycerol response kinase (Bbhog1), hydrophobin encoding gene (ssg), cell wall protein (Mad1, Mad2), carboxylate transporter gene (BbJEN1), lipases, MrCYP52 gene, dioxygenase gene, glutathione S transferase, cytochrome P450 (CYP52), proteases (subtilisin-like Pr1, Pr2, Pr1J, Pr1A), trypsin-like protease, carboxypeptidase, metalloproteases like, methionine aminopeptidase, serine protease (bassiasinI), chitosanase, chitinases (BbchitI, VlchitI, CHITI)</td></tr>
<tr><td>Post-penetration (in the insect body cavity – hemocoel with body fluid – the hemolymph)
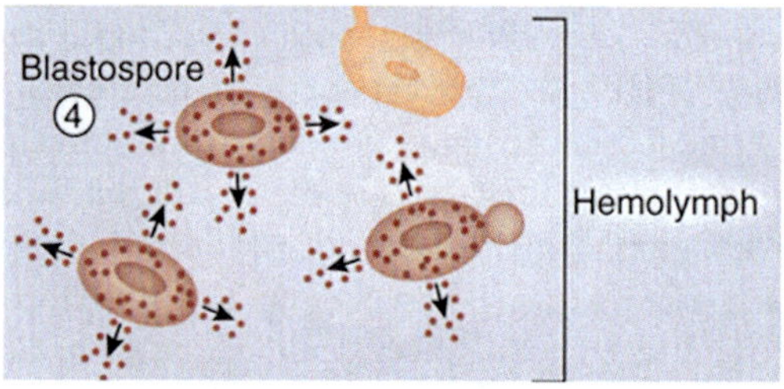
</td><td>Acid phosphatase, choline dehydrogenase (osmosensor), collagenous protein (MCL1), cytochrome P450s, small secreted cysteine-rich proteins (SSCP), G-protein-coupled receptor (GPCR), secondary metabolites (polyketides, PK), non-ribosomally synthesised peptides (NRSP) like destruxin, depsipeptide, beauvericin, israolides and mitochondrial ATPase inhibitory peptides, multimeric enzymes like terpenoid synthase /cyclase</td></tr>
<tr><td>After insect death – emergence of mycelia from the cadaver and sporulation
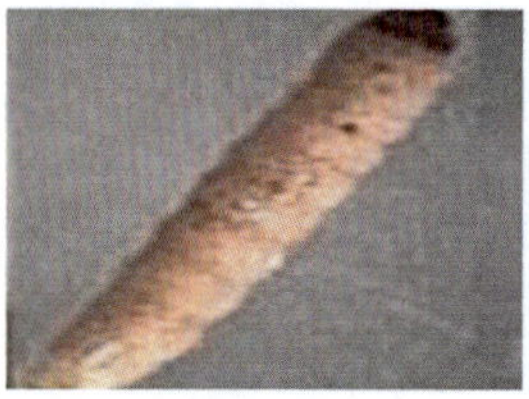
Mummified insect cadaver
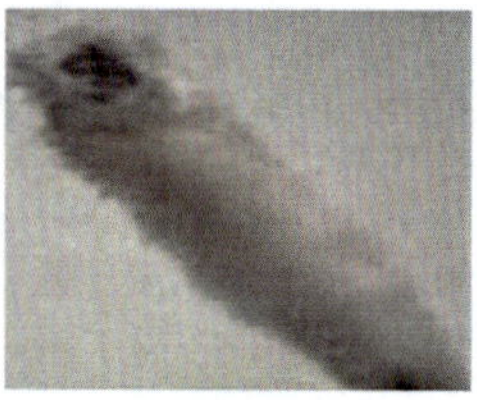
Mycotic cadaver with sporulating fungus</td><td>Subtilisin-like protease (Pr1), hydrophobin gene (ssga), mitogen-activated kinase (BbMPK)</td></tr>
<tr><td colspan="2">Abiotic factors influencing affectivity and genes involved</td></tr>
<tr><td>UV rays in sunlight, heat stress, fungicide</td><td>Protein kinase A (MapKA1), osmosensor (MOS1), laccase, superoxide dismutase (Bb-SOD2), β tubulin, adenylate cyclase</td></tr>
<tr><td colspan="2">Molecular biology techniques</td></tr>
<tr><td>DNA/genome sequencing</td><td>Sanger, 454, Ion torrent, Illumina and PacBio sequencing methods</td></tr>
<tr><td>Transcriptome profiling</td><td>cDNA-AFLP, microarray, 454 pyrotranscriptome sequencing, generation of cDNA, EST and SSH libraries, qPCR, gene knockout</td></tr>
</table>

(continued)

Table 1 (continued)

In silico tools and pipelines	
Tools	antiSMASH, BLAST, Blast2GO, PSI-BLAST, CAFÉ, CLUSTAL-W, Bioedit, interproscan, PareEva, PEDANT, POGO, ProFASTA, PSIPRED, sMEGA, NRPSP predictor, SignalP 3.0, SMURF, TargetP1.1, RIPCAL
Pipelines	FastGroupII, RDP, DOTUR, QIIME, SeqTrim, SCATA, WATERS, CANGS, PANGEA, PyroNoise, MEGAN, CLOTU
Databases	
Primary databases	Gene bank, DDBJ, European nucleotide archive
Genome databases	CAMERA, PATRIC, SEED
Protein sequence databases	Uniport, PIR, Swiss-Prot, PEDANT, PROSITE, Pfam
Proteomic database	PRIDE, MitoMiner
Protein structure databases	Protein databank, SCOP, CATH
Protein model databases	SWISS-MODEL, Protein Model Portal
RNA databases	Rfam, tmRDB, SRPDB, RDB, SILVA
Specialised databases	Repeat masker library, Tandem repetitive binder, Rep base, pathogen–host interaction (PHI) database, Dr. Nelson's P450 database, KinBase, GPCRDB sequences , PEDANT, Library of catalytic and carbohydrate-binding module enzymes, MEROPS

First two images were adapted from Thomas and Read (2007).Third image is from our laboratory

the utility of the biocontrol agent. Culture-independent techniques of soil DNA extraction and 454 pyrosequencing of specific genes coupled with bioinformatic analysis using suitable databases for, e.g. the rRNA gene and EF1-α gene databases (Mahé et al. 2012; Quast et al. 2013; Hirsch et al. 2013) were used in these assessments.

An account of these informatics-bolstered investigations of molecular data on the pathogenicity, affectivity and persistence of entomopathogenic fungi is detailed below.

All the entomogenous fungi registered as biopesticides are Ascomycetes being predominantly mitosporic. They all a have a similar infection cycle described below.

2 Infection Process of an Entomopathogenic Fungus

The infection pathway consists of the following steps: (1) attachment of spore (conidiurm) to the insect cuticle, (2) germination of the spore, (3) penetration through the cuticle, (4) overcoming the host response and immune defence reactions, (5) proliferation within the host and secretion of secondary metabolites leading to death of the insect, (6) saprophytic outgrowth from the dead insect and production of new spores (Boucias et al. 1988) (Fig. 1). Under permissible temperature and moisture in the immediate vicinity, the conidia that adhere (facilitated by the chemical composition and physical architecture of the spore wall) to the insect cuticle germinate. The germ tube develops a terminal swelling that differentiates into an infection structure called appressorium which subsequently produces an infection peg that penetrates the insect cuticle. Both mechanical (turgor pressure in the appressorium) and physiological responses like secretion of several enzymes – lipases, proteases, chitinases, etc. (which degrade chemical components of insect cuticle) – mediate penetration of the fungus into insect body. Lipids – the main nutrient reserves of the fungal spores – are transported as lipid bodies to the appressorium and degraded to glycerol which increases the hydrostatic pressure and provides driving force for mechanical penetration (Wang and St. Leger

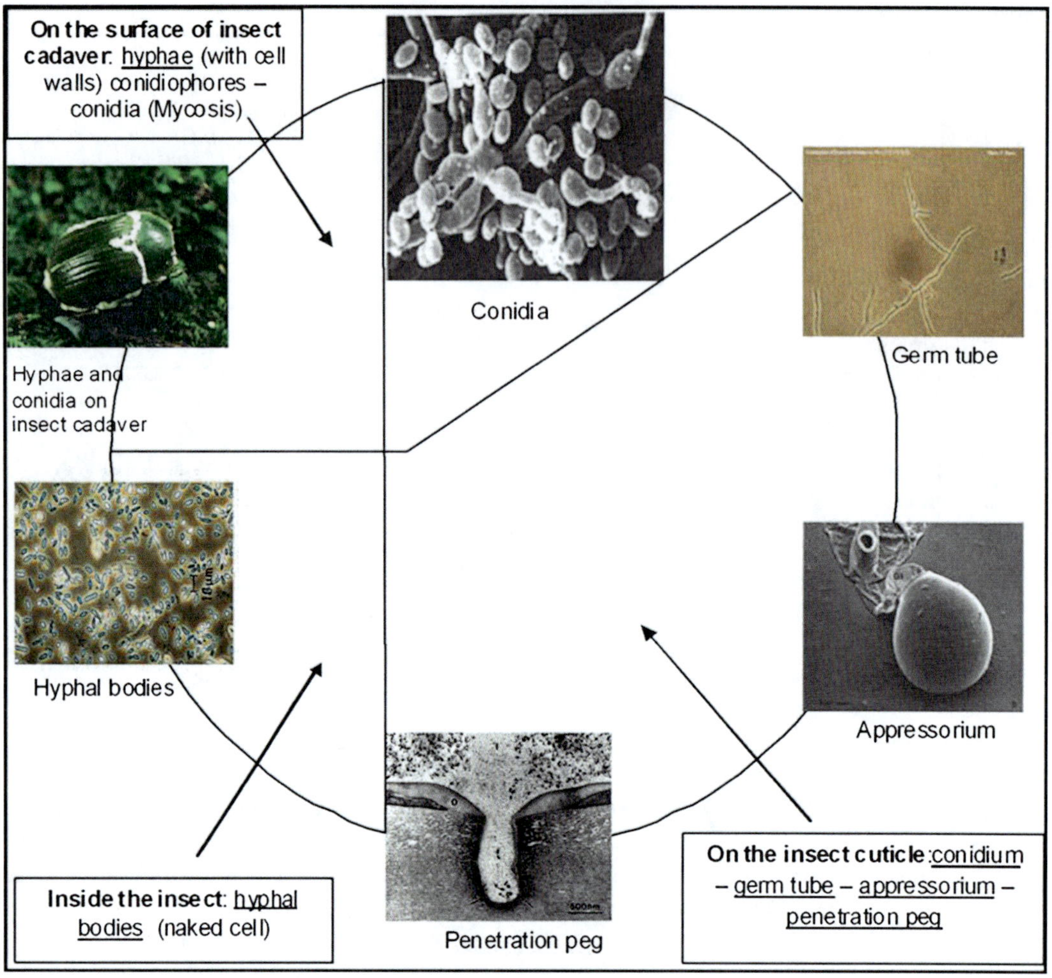

Fig. 1 The pathogenic life cycle of an insect pathogenic fungus

2007a; Fang et al. 2009). After successful penetration, within the insect body cavity (hemocoel), the fungus produces hyphal bodies – yeast-like cells with no cell wall (to evade the immune response of the insect), which are distributed passively through circulating hemolymph (blood + body fluid of the insect) enabling the fungus to invade other tissues of the insect. The hyphal bodies multiply extensively through budding depleting nutrients in the hemolymph and the fat body. Then they produce toxic secondary metabolites. The insect finally succumbs and dies. The exoskeleton of infected insects is preserved intact while the fungus grows extensively within the insect body. The dead insects are mummified. With the end of pathogenic phase, the fungus begins its saprophytic phase – it grows through the cuticle of the insect emerging out and profusely produces spores on the cadaver. Thus, the life cycle of entomopathogenic fungi consists of two phases: a normal mycelial growth phase on the outside surface (cuticle) of the insect body and a yeast-like budding phase in its body cavity (hemocoel). Every step in the life cycle of fungus is gene controlled.

During the early infection stage, pathogenic fungi encounter various adverse factors from the host insect, such as antifungal substances on the cuticle, oxidative stress during infection, osmotic stress inside the host body and behavioural changes such as locust fever (Liu et al. 2012). After insect colonisation it is hypothesised that the fungus evades, overwhelms or subverts the host's immune system by changing morphology, producing toxins, or by unknown but presumably biochemical mechanisms (Charnley 1989).

The ability of the fungus to cause infection is also influenced by the physical environmental conditions like UV rays in the sunlight, temperature (mainly heat shock), humidity and chemicals used in plant protection and fertilisation. The dark green pigments in the conidia of *Metarhizium anisopliae* (Roberts and St. Leger 2004) and impermeable cell membrane are believed to confer abiotic stress tolerance. Molecular approaches coupled with analysis using bioinformatic tools revealed the role of protein kinase A (MapKA1) (Fang et al. 2009), osmosensor (MOS1) (Wang et al. 2008), laccase (Fang et al. 2010), superoxide dismutase (Bb-SOD2) (Xue et al. 2010), mutation in β tubulin genes (Butterrs et al. 2003) and adenylate cyclase (Liu et al. 2012) in tolerance to abiotic stress, adverse conditions from the insect host and, in some instances, virulence of insect pathogenic fungus. The identification of these genes and localisation of proteins in certain occasions was possible through similarity analysis with BLASTP (Altschul et al. 1990) and MitoProtII (Claros and Vincens 1996). MitoProtII aids in identifying the N-terminal end of the protein with a mitochondrial targeting sequence and its cleavage site (http://ihg.gsf.de/ihg/mitoprot.html). Most of the eukaryotic superoxide dismutases are known to be localised in the mitochondria. It was identified with MitoProtII that Bb-SOD2 in *B. bassiana* was rather present in the cytosol (Xue et al. 2010). Sequence analysis of key mutational site in β tubulin gene of *B.bassiana* showed that β tubulin mediates resistance towards the fungicide Bavistin (carbendazim) (Butterrs et al. 2003).

3 Pathogenicity Genes

At each stage of infection process, the fungus must be able to detect and respond to a number of cues. So genes involved in signal transduction, regulation of gene expression and the actual determinants in pathogenicity like enzymes which are responsible for substrate breakdown/ utilisation and synthesis of secondary metabolites like toxins are all expressed during the infection process. All the three classes of genes are important in pathogenicity.

Most work on identification of pathogenicity genes has been done on the two commercially important fungi –*B. bassiana* and *Metarhizium* spp. Differential gene expression studies through microarray and cDNA-AFLP identified specific responses of these fungi to different insect cuticles which demonstrated their ability to rapidly adjust their gene expression patterns to adapt to insect cuticle (Freimoser et al. 2005; Khan 2006). So far 13,975 and 10,658 ESTs were reported in *B. bassiana* and *M. anisopliae*, respectively (http://www.ncbi.nlm.nih.gov/nucest/). Almost 50 % of the arrayed ESTs were found to be upregulated in cuticle-containing media and have similarities with sequences in PHI (pathogen–host interaction database) (http://www.phi-base.org). The function of a considerable proportion of these ESTs is yet to be discovered. Homology modelling studies and interproscan analysis and other bioinformatic tools and databases (Table 1) can be handy in predicting the structure and thus function of the products from these ESTs. The genes assumed to be involved in different stages of infection of entomopathogenic fungi were listed (Uma Devi et al. 2012). They were identified by various bioinformatic tools and databases (Table 1). Sequence comparisons and conserved motifs suggest that about 60 % of the ESTs of *M. anisopliae* expressed during growth on cuticle encode secreted enzymes and toxins.

3.1 Genes Involved in Attachment to the Insect, Infection Peg Formation and Penetration

Adhesin genes (MAD1 and MAD2) which help during the attachment of the spore (Wang and St. Leger 2007a, b), a cell protective coat protein that helps the pathogen to elude the host immunity recognition battery (Wang and St. Leger 2006) and a perilipin-like protein that regulates appressorium turgor pressure and differentiation (Wang and St. Leger 2007b) were identified through transcriptomic analysis of *Metarhizium* spp. A MAP (mitogen-activated protein) kinase gene, Bb*mpk1*, was identified in *B. bassiana* which was found essential for spore adhesion, formation of appressoria and ability to cross the insect cuticle both during initial entry for infection and exit after the death of the insect (Zhang et al. 2010). The MAP signalling pathways play an important role in perceiving and transducing extracellular signals to regulate gene expression. The *Bbmpk1* regulated genes involved in the formation of appressorium were identified through bioinformatic analysis of SSH (suppressive subtractive hybridisation) library (Zhang et al. 2010). The *Bbmpk1* gene has been implicated in regulating vesicular traffic, lipid metabolism, microtubule dynamics and development for differentiation of appressoria (Li et al. 2004). Spore adhesion and appressorium formation was also found to be affected by an HOG (high osmolarity glycerol response) kinase gene – *Bbhog1* – in *B. bassiana* (Zhang et al. 2009). Both *Bbmpk1* and *Bbhog1* genes were found to affect the transcript levels of hydrophobin encoding genes (Zhang et al. 2010). Hydrophobins are responsible for the hydrophobicity of conidia and implicated in their adhesion to insect and formation of appressoria (Holder et al. 2007).

The insect cuticle is constituted by proteins and chitins and is covered by an epicuticle – a hydrocarbon-rich waxy layer. While the protein and chitin components of insect cuticles are similar in all insects, the epicuticular components are highly heterogenous even within the same insect genus (Wang and St Leger 2005). Insect pathogenic fungi penetrate these two layers of insect cuticle by solubilising the chemical components and utilising them for nutrition. For this purpose they secrete large numbers of degradative enzymes.

A gene that mediates degradation of hydrocarbons in the epicuticle has been identified in *Metarhizium robertsii* (Lin et al. 2011). The gene MrCYP52 belongs to cytochrome P450 monooxygenase 52 family (Lin et al. 2011). The MrCYP52 gene was found to be involved in hydrolysis of alkanes in the epicuticle of grasshoppers which provides nutrition for germination and formation of infection structures (Lin et al. 2011). In *M. anisopliae* var. *acridum* cultured on the locust cuticle, a dioxygenase gene and three glutathione S transferase genes that act on toxic substrates known to be present in the epicuticle of locusts were found upregulated (Wang and St. Leger 2005).

The genes encoding the enzymes that degrade the components of the insect cuticle and their precise role as deciphered from investigations are described below.

3.1.1 Proteases

Distinct extracellular serine proteases such as subtilisin-like and trypsin-like proteases, metalloproteases and several families of exo-acting peptidases are implicated in host cuticle degradation (St. Leger et al. 1997; Bagga et al. 2004). Eight subtilisin genes were expressed on the cuticle medium in *M. anisopliae* (Freimoser et al. 2003; Bagga et al. 2004). The most important among them were identified as subtilisin-like *Pr1* and *Pr1J* (Freimoser et al. 2005). The *Pr1A* gene in *M. anisopliae* is the most studied fungal subtilisin protease and the only one generally accepted to have a role in insect cuticle breach (Shah et al. 2005). It is the predominant protein produced during insect cuticle degradation, and its transcripts are 10 times more abundant than *Pr1J*, the second most expressed protease (Bagga et al. 2004). The *Pr1A* gene from M. *anisopliae* was cloned and sequenced (Zhang et al. 2008). The promoter of the *Pr1* gene has putative binding sites for regulatory proteins similar to carbon catabolite repressor (CREA) (Screen et al. 1997), nitrogen metabolite regulator (AREA) and the eukaryotic cAMP response element binding protein (CREB) (Clarkson and Charnley 1996). It is activated by low nitrogen levels and switched off once the hyphae reach the nitrogen-rich hemolymph. The *Pr1* gene was found to be again upregulated when the fungus emerges out of the dead insect (Small and Bidochka 2005). The *Pr1* and *Pr2* genes were also highly expressed in *B. bassiana* on cuticle media (Khan et al. 2007; Donatti 2008; Dias et al. 2008). In *B. bassiana*, *Pr1H* was found to be differentially expressed being expressed on cuticles of *Spodoptera litura* larvae and *Epilachna vigintioctopunctata* but not on cuticles of *Aphis craccivora* and *Periplaneta americana* (Khan et al. 2007). An EST homologous to membrane-bound methionine aminopeptidase (MAP) enzyme of metalloprotease family was detected in *B. bassiana* cultured on *Aphis craccivora* but not on cuticles of *Spodoptera litura* larvae, *Epilachna vigintioctopunctata* and *Periplaneta americana* (Khan et al. 2007). A serine protease bassiasin I was reported in B. *bassiana* (Kim et al. 1999). After penetrating the insect cuticle and adapting to the hemocoel, the fungus turns off the serine protease genes to avoid activating the prophenoloxidase system (insect defence mechanism) (Wang et al. 2005).

In post-genome sequencing, it has been discovered that the genomes of entomopathogenic fungi code for many more proteolytic enzymes than do noninsect pathogenic fungi. The proteases were found to belong to different families: aspartic, cysteine, glutamine, metallo, serine and threonine peptidases. A total of 429 proteases in *B. bassiana*, 431 in *Metarhizium robertsii* and 360 in *M. acridium* were identified through whole genome blast analysis with MEROPS (the database of proteolytic enzymes, their substrates and inhibitors) (Rawlings et al. 2012). *B. bassiana* was found to code for significantly more trypsins (23 *vs* an average of two in plant pathogens), subtilisins (43 *vs* 17) and carboxypeptidases (52 *vs* 32) (Xiao et al. 2012). The dramatic expansion of proteases is suggestive of its adaptation to a broad host range (Xiao et al. 2012). However, *B. bassiana* and the plant pathogens were found to have similar numbers of aspartic, cysteine, threonine and metallo peptidases (Xiao et al. 2012).

3.1.2 Chitinases

Chitinases have been reported to be pathogenicity determinants in entomopathogenic fungi (Charnley and St. Leger 1991; St. Leger et al. 1995, 1996). The key role played by chitinases and lipases in the process of infection has been demonstrated in *Nomuraea rileyi* (El-Sayed et al. 1989, 1993). Overproduction of an endochitinase gene *Bbchit* was demonstrated to significantly enhance the virulence of *B. bassiana* (Fang et al. 2005). An endochitinase gene *Vlchitl* from *Verticillium lecanii* has been cloned and characterised for use in strain improvement (Zhu et al. 2008). In *M. anisopliae*, a chitinase gene was found to be expressed within 1 h after culture on cuticle-containing medium (Freimoser et al. 2005; Baratto et al. 2006). Some controversial experimental results have also been reported with regard to the potential role of chitinases in infection process. A chitinase deficient mutant of *Verticillium lecanii* was still capable of infecting aphids (Jackson et al. 1985). Several classes of chitinases with different functional roles have been discovered in entomopathogenic fungi (Fang et al. 2005). The diverse nature of chitinase function is due to differences in the chitin binding domains and signal peptide. Bioinformatics tools like homology modelling based on SWISS-MODEL programme (3D structure) and PSIPRED (structure prediction), Signal IP3.0 (signal peptide prediction), conserved domain recognition in the chitinase sequences using the NCBI Conserved Domain search software (http://www.ncbi.nlm.nih.gov/Structure/cdd/cdd.shtml) and splice site analysis through FSPLICE offer an important avenue to facilitate understanding the role of residues within the chitinase insertion domain in chitinase function during pathogenesis (Fang et al. 2005).

Phylogenetic analyses of the A, B and C subgroups of chitinases revealed that several gene duplication events have occurred since *B. bassiana*/*Cordyceps militaris*, *Metarhizium* spp. and *Trichoderma* spp. diverged from a common ancestor, suggesting their abundance in each clade is due to convergent evolution (Xiao et al. 2012).

3.1.3 Lipases

The role of individual lipases in pathogenicity in *Metarhizium* has not been demonstrated, although a lipase activity inhibitor was found to block infection processes in *M. anisopliae* (Gao et al. 2011).

3.2 Genes Expressed After Entry into the Insect (in the Hemocoel)

The fungus after breaching the insect cuticle enters the body cavity (hemocoel) and begins to multiply. At this stage, it produces several enzymes and toxins. In *M. anisopliae*, acid phosphatase was found to play a key role in the utilisation of organic phosphate in the insect hemolymph (Xia et al. 2001). The insect hemolymph is solute rich with a high osmotic pressure. An osmosensor was detected in *M. anisopliae* which might mediate cellular responses to the prevailing high osmotic pressure of the hemolymph in the hemocoel of the insect (Charnley 2003; Wang et al. 2008). Osmolyte synthesising enzyme like choline dehydrogenase is believed to provide osmotic balance with the environment and protect enzymes and other cellular components from high osmotic pressure (Pollard and Wyn Jones 1979). A collagenous protein (MCL1) was found in *M. anisopliae* cultured in hemolymph (Wang and St. Leger 2006). The collagenous domain of this protein is believed to serve as an anti-adhesive protective coat to mask antigenic structural components of the fungal cell wall and prevent phagocytosis and encapsulation by host hemocytes (Wang and St. Leger 2006).

It is possible to mine for genes coding for new immunoprotective agents in the genomes of entomopathogenic fungi which have been sequenced with VaxiJen server (http://www.ddg-pharmfac.net/vaxijen/VaxiJen/VaxiJen.html; Doytchinova and Flower 2008).

3.2.1 Toxic Secondary Metabolites

Many toxins have been reported in *B. bassiana* and *M. anisopliae*. They belong to three principal classes: polyketides (PK), non-ribosomally synthesised peptides (NRSP) and alkaloids (Gibson et al. 2007). They are all secondary metabolites. The genes responsible for secondary metabolite biosynthesis, export and transcriptional regulation are often found as contiguous gene clusters in the genomes of entomopathogenic fungi. Web-based software SMURF (www.jcvi.org/smurf/) and antiSMASH (http://antismash.secondarymetabolites.org/) was employed to systematically predict clustered secondary metabolite genes in genomic context and domain content (Khaldi et al. 2010; Medema et al. 2011). Three of the putative biosynthesis clusters were found highly conserved in the insect pathogenic fungi investigated to date but are absent in other fungi (Xiao et al. 2012). Genome survey of *B. bassiana* found that there are 45 non-ribosomal peptide synthetase (NRPS), polyketide synthase (PKS) and terpenoidsynthase/cyclase core genes; this gene number is lesser than those recognised in *Metarhizium* spp. (Xiao et al. 2012). Polyketide synthesis is orchestrated by polyketide synthases (PKSs) which are multimeric enzymes that function much like fatty acid synthases (Gibson et al. 2007). It is predicted that most of the entomopathogenic fungi have multiple PKS genes (Gibson et al. 2007). Four PKS genes were identified in *Verticillium coccosporum* (Bingle et al. 1999), seven in *Tolypocladium inflatum* (Gibson et al. 2007). *Hirsutella thompsonii* was found to produce phomalactone, a tetraketide with insecticidal properties (Gibson et al. 2007).

Several kinds of non-ribosomally synthesised peptides (NRSP) have been identified in entomopathogenic fungi. *Tolypocladium* spp. were found to produce at least four types of biologically active NRSPs: the potent mitochondrial ATPase inhibitory efrapeptins (Gupta et al. 1992), the immunosuppressive cyclosporins (Billich and Zocher 1987; Bisset 1983), the antibacterial amino isobutyric acid (Aib)-containing cicadapeptins (Krasnoff et al. 2004) and the amino acid-derived diketopiperazines (Chu et al. 1993). *Paecilomyces* spp. was found to produce leucinostatins, another class of mitochondrial ATPase inhibitory peptides (Bisset 1983).

Another class of NRSPs is low molecular weight cyclic depsipeptides termed destruxins. Among them, the low molecular weight compounds beauvericin, enniatins, isarolides and bassianolide have been demonstrated to be insecticidal (Gupta et al. 1994; Castlebury et al. 1999). Beauvericin, an ionophoric cyclic peptide of six amino acids synthesised by the multifunctional enzyme beauvericin synthetase, was found toxic to brine shrimp and mosquito larvae (Hamill et al. 1969). At least 26 different destruxins have been identified in *M. anisopliae*. Destruxins A, B and E were predominant in *M. anisopliae* (Gupta et al. 1989; Amiri-Besheli et al. 2000). Destruxins A and E are more toxic than the others (Dumas et al. 1994). These toxins have often been implicated as the cause of death of insects infected with *M. anisopliae* (Butt et al. 1994; Vestergaard et al. 1995). Genome mining of *Beauveria bassiana* identified 11 NRPS and three NRPS_PKS hybrid gene clusters. NRPS clusters unique to *Beauveria* genome, namely, NRPS BbBEAS and NRPS BBBSLS, were reported to be involved in the biosynthesis of insecticidal toxins beauvericin and bassianolide, respectively (XU et al. 2008; XU et al. 2009; Xiao et al. 2012). Genes involved in the biosynthesis of cyclic peptides like beauvericin, bassianolide and tenellin have been functionally verified using transcriptome and biochemical analysis (Molnar et al. 2010; Xiao et al. 2012). A direct correlation between destruxin production and pathogenicity was not always possible as strains which produced low levels of these toxins were also found virulent (Amiri-Besheli et al. 2000). The destruxins effect calcium channels in the insect host and cause paralysis of the insect (Samuels et al. 1988), or they may reduce the immunological competence of the insect (Cerenius et al. 1990).

Bacterial-like toxins have also been reported from entomopathogenic fungi. *Beauveria bassiana* has 13 heat-labile enterotoxins and six in *Metarhizium* (Xiao et al. 2012). The presence of bacterial toxins in *B. bassiana* similar to Bt Cry-like delta endotoxins suggests that insect pathogens may use the bacterial toxin–antitoxin system to control cell stasis or death. However, the bacterial-like toxins were expressed at very low levels in *B. bassiana* grown under three different environments – on locust hind wings, in cotton bollworm blood and in corn root exudates.

In *B. bassiana*, an EST homologous to P450 monooxygenase was detected when cultured on four different insect cuticles but not when cultured on synthetic medium (Khan et al. 2007). Cytochrome P450s (CYPs) are reported to be involved in biosynthesis of pathogenesis-related secondary metabolites besides participating in many essential cellular processes and detoxification and degradation of xenobiotics (Nelson 1999; Guengerich 2001; Ortiz de Montellano 2005). A detailed structural analysis of P450 gene family deciphered its involvement in oxidation of diverse lipid constituents of insect cuticle and in toxin production (Pedrini et al. 2010). Proteomic analysis tool expasy and splice site analysis through FSPLICE (http://www.softberry.com/berry.phtml.) revealed the diverse nature of P450 gene family with respect to introns in *B. bassiana* (Pedrini et al. 2010). Molecular characterisation and sequence examination through in silico analysis revealed a total of eight cytochrome P450 genes corresponding to different families in *B. bassiana* which may display overlapping substrate specificities (Pedrini et al. 2010). The huge number of cytochrome P450 genes (CYP) (~ 111 in *Metarhizium* spp. and 83 in *B. bassiana*) is an indication of the competence of insect pathogen fungi for detoxification and biosynthesis of pathogenesis-related secondary metabolites (Pedrini et al. 2010). Furthermore, these genes would be potential candidates for over-expression in *B. bassiana* to increase its targeting and virulence towards insect pests (Pedrini et al. 2010). Transgenic approaches coupled with functional analysis based on homology modelling may allow exploring the biotechnological application of CYPs of entomopathogenic fungi.

Small secreted cysteine-rich proteins (SSCPs) are implicated in competence for endophytism and in pathogenicity to insects. In *B. bassiana*, an interproscan analysis revealed that some of the SSCPs had homologues in the PHI database of verified virulence determinants, e.g. five putative cutinases and five trypsins (Xiao et al. 2012). Six Bb SSCPs were identified as concanavalin A-like lectins and potentially could function in interactions with both insects and plants (Rappleye and Goldman 2008). *Beauveria bassiana* has four genes encoding proteins with eight cysteine-containing extracellular membrane (CFEM) domains resembling pathogenicity determinants in plant pathogens (Kulkarni et al. 2003). These findings suggest that insect pathogenic fungi perhaps share common mechanisms for endophytic establishment in the plant and insect cuticle penetration.

The role of G protein alpha subunits in pathogenicity has been extensively studied in *Metarhizium*, and many were found essential because they transduce extracellular signals leading to infection-specific development (Solomon et al. 2004). The genes MAA_03488 and MAC_04984 in *Metarhizium* spp. were found highly expressed during infection of either cockroach or locust cuticles (Kamper et al. 2006; Lafon et al. 2006; Hane et al. 2007; Gao et al. 2011).

4 Assessment of the Persistence and Effect of the Entomopathogenic Fungus on Soil Fungal Community Through Metagenomic Approach

For registration purposes of fungal-based biocontrol agents, any risks concerning the persistence of the applied fungal inoculum have to be evaluated in order to assess the organism's potential to spread and to become established in the environment (Scheepmaker and Butt 2010). In addition, registration authorities in the European Union require information on long-term nontarget effects such as potential competitive displacement of soil microorganisms as well as information on the natural background level of a particular entomopathogenic fungus. In inundative applications of entomopathogenic fungal-based biopesticides, a demonstration is required that they have no undesirable effect on the natural soil microbe biodiversity.

A culture-dependent method of retrieving fungal cultures from soil samples (Hagn et al. 2003) or insect bait methods were earlier used for evaluation of the persistence of entomopathogenic fungi. Later genetic markers such as simple sequence repeats (SSR, so-called microsatellites; Enkerli et al. 2001) were used for identification of the applied fungal strain. Pipelines like SSR locator (Carlos et al. 2008) and msatcommander (Faircloth 2008) can be used for genome wide location of microsatellites to develop primers for different SSR markers.

A culture-independent metagenomic approach would be handy in the assessment of the applied entomopathogenic fungus on soil microbial diversity. One such study has recently been done using multitag 454 sequencing of fungal ITS sequences amplified from soil DNA samples collected in a crop (chilli) field at different time intervals after treatment with *B. bassiana* (Hirsch et al. 2013).

454 DNA sequencing (so-called on the name of the sequencing company 454 Life Sciences, Branford, USA) is a next-generation sequencing technology also technically called pyrosequencing. It is based on sequencing-by-synthesis principle and built on a 4-enzyme real-time monitoring of DNA synthesis by bioluminescence. It is a now popular technique for high-throughput approach. Added to the technical robustness of its chemistry, this method generates a large number of reads per run thus giving much greater coverage of metagenomic sequencing. The basic approach is to identify microbes in a complex community by exploiting universal and conserved targets, such as rRNA genes. By amplifying selected target regions within 18/16S rRNA genes (for eukaryotic/prokaryotic), microbes can be identified by the effective combination of conserved primer-binding sites and intervening variable sequences that facilitate genus and species identification (Buée et al. 2009; Wu et al. 2010). To evaluate the effect of an applied microbial biopesticide on the biodiversity of soil microbes over a period of time, a multitag system for marking samples collected at different time intervals can be used and DNA from all samples can be pooled for 454 sequencing. This multitag 454 sequencing of 16/18S rRNA gene regions spanning conserved and variable regions of DNA isolated from soil samples thus offers a cheap method of sequencing large sample sets comprising several collections over a period of time.

The primary challenge for metagenomic studies is at the analytical end of how to obtain accurate microbial identification for hundreds or thousands of species in a reasonable time. This involves signature sequence matching at 95–97 % similarity in pairwise sequence comparisons. Numerous bioinformatic tools are necessary to tackle the vast amounts of microbiological sequence diversity. The present bioinformatics throughput is too slow and not sufficiently automated for large-scale projects analyses. Clearly, sufficient computational power is necessary, although distributed computing networks and robust server technology may eventually meet current metagenomics data analysis demands in research settings (Petrosino et al. 2009). Beginning with sequence collection and verification, algorithms must be in place to trim sequences and to vet the quality of individual reads (Huse et al. 2007). Another problem is that the PCR may generate sequence chimeras during the amplification process. Chimera-checking software has been developed so that amplicons can be vetted for the presence of "sequence hybrids" in software environments such as Greengenes (De Santis et al. 2006) and RDP (Cole et al. 2007) and with algorithms such as Bellerophon (Ashelford et al. 2005) or Pintail (Huber et al. 2004). Once high-quality sequences have been obtained from mixed species communities, the next challenge is to accurately identify many microbes in parallel. The trimmed sequences can be aligned with generic sequence aligners such as NAST (De Santis et al. 2006) or MUSCLE (Edgar 2004), MAFT (Katoh et al. 2009) and STAP (Wu et al. 2008). Green genes, RDP, ARB and DOTUR are integrated with sequence alignment editor – MUSCLE and WATERS has STAP as the sequence aligner. Generic sequence aligners do not take into consideration the predicted secondary structures of the rRNA sequences while aligning them. The secondary structure of rRNA is an important feature to consider in during alignment because it increases the likelihood that the alignment conserves positional homology between sequences. More recent aligners like RDP and SILVA that take into account 16S rRNA secondary structure models (Cole et al. 2007) are associated with 16s rRNA gene databases and reference MSAs. The use of MSA leads to an overestimate of genetic distances and microbial diversities, and computational complexities are also involved in its implementation (Sun et al. 2009). Pipelines like DOTUR, MOTHUR, FastGroupII and RDP-pyro assign sequences to OTUs based on MSA. Because of requirement of enormous computer memory (computational complexities), these pipelines are not suitable for analysing large metagenomic dataset (Sun et al. 2009) . To overcome these problems it was proposed to calculate genetic distance based on pairwise alignment using Needleman–Wunsch algorithm (Needleman and Wunsch 1970) or Blast algorithm. Recent web pipelines like CLOTU and ESPRIT utilise pairwise alignment before clustering into operational taxonomic units (OTUs). The OTU assigning software/pipelines perform meaningful pairwise alignments and for computation of genetic distances. Assignment of sequences to OTUs is done based on any of the three clustering programmes – nearest neighbour (single linkage cluster analysis), furthest neighbour (complete linkage) and average neighbour. Taxonomic annotations of OTUs are done by database searches using BLASTn against either user-defined databases or NCBI nr database.

The following are existing bioinformatic tools available that include various options for processing and clustering 454 reads: FastGroupII (Yu et al. 2006), RDP (Cole et al. 2007), DOTUR (Schloss and Handelsman 2005), SeqTrim (Falgueras et al. 2010), QIIME (Caporaso et al. 2010), SCATA (http://scata.mykopat.slu.se/), WATERS (Hartman et al. 2010), CANGS (Pandey et al. 2010), PANGEA (Giongo et al. 2010) PyroNoise (Quince et al., 2009), MEGAN (Huson et al. 2011) and CLOTU (Kumar et al. 2011). Each tool has its own algorithm for alignment and OTU assignment. For taxonomic annotation, CLOTU and MEGAN have integrated BLASTn option. In other pipelines – QIIME, PANGEA, CANGS, WATERS and DOTUR – the users need to set up the database and BLAST programme on their local computer for assigning taxonomic affiliation to the 454 reads. In WATERS, it is possible to generate phylogenetic trees of the OTUs based on neighbour joining algorithm-based programmes like FastTree or QuickTree or maximum likelihood algorithm-based programme RAxML. C LOTU is a web-based service platform running on a high performance computing environment, while QIIME, PANGEA, C ANGS and WATERS must be installed locally, making subsequent analysis of extensive datasets time consuming.

In addition to accurate microbial identification, indices and algorithms have been developed to assess microbial diversity in the sample. One of the first questions a microbiologist may likely have about their community of interest is how many kinds of organisms do I have here? And how different is one sample from another? To answer these questions, richness estimation and diversity index are computed. Richness estimation is determined by ACE (Chao and Lee 1992), Chao1 (Chao 1984) or rarefaction curves (Hurlbert 1971) and diversity index by Shannon and Simpson diversity indices. Multivariate analysis is used for comparison of diversity patterns of multiple microbial communities and evaluating the significance of the observed differences between them. Nonmetric multidimensional scaling (NMDS) is used for the former and ANOSIM for the latter. Statistical programmes are available for computation. For diversity analyses, EstimateS (Clowee 2006) or SONS (Schloss and Handelsman 2006) are used. To measure community ordination, statistical tools like PC-ORD (McCune and Mefford 2011), Vegan (Oksanen et al. 2007) and PAST (Hammer et al. 2001) are used. Phylogenetic analysis of the community is done with UniFrac (Lozupone and Knight 2005) or Phylocom (Webb et al. 2008). In pyrosequencing pipelines DOTUR, ESPIRT and WATERS, statistical computation of richness estimates can also be carried out alongside the assignment of OTUs. In WATERS additionally, software for statistical assessment of differences in community structure between multiple samples is possible with the integrated tool – UniFrac. In addition to UniFrac, there are several statistical tools like Tree Climber (Schloss and Handelsman 2006), LIBSHUFF (Singleton et al. 2001), SONS (Shared OTUs and (N) Similarity) and analysis of molecular variance (AMOVA) for comparison of multiple community structures. They use either distance matrix or phylogenetic tree as input file.

Multitag 454 pyrosequencing of fungal ITS-1 sequences has been used for characterising the fungal community structure in an agricultural field in India and for assessing both the fate and potential effects of an artificially applied *B. bassiana* strain on diversity of soil fungal communities (Hirsch et al. 2013). A snapshot of the species identified is represented in Fig. 2. No adverse effect on soil fungal diversity was observed due to application of *B. bassiana* (Hirsch et al. 2013).

5 Summary

Entomopathogenic fungi are used as microbial biopesticides in plant protection. The mechanisms of their infection of the insect host leading to its death and the genes/enzymes that facilitate the process have been worked out. Molecular biology techniques coupled with bioinformatic analysis have unravelled the battery of genes involved in insect pathogenicity in these fungi. Research on specific genes and later high-throughput approaches has revealed the role of proteases, chitinases and lipases in the intrusion of the fungus through the insect cuticle. The genes which facilitate the pre-penetration process of spore germination, infection peg formation, signal transduction and those involved post-penetration into the insect body that include genes that evade insect immune responses, cope with the osmotic conditions in the insect body fluid, produce toxins and finally facilitate the emergence of the fungus from the insect cadaver have been identified. Bioinformatic tools and pipelines and databases have been crucial in these analyses. The ecological impact of inundative application of fungal-based biopesticides on the native soil fungal community has been assessed through metagenomics approach. This analysis requires next-generation sequencing (454 pyrosequencing) and an arsenal of

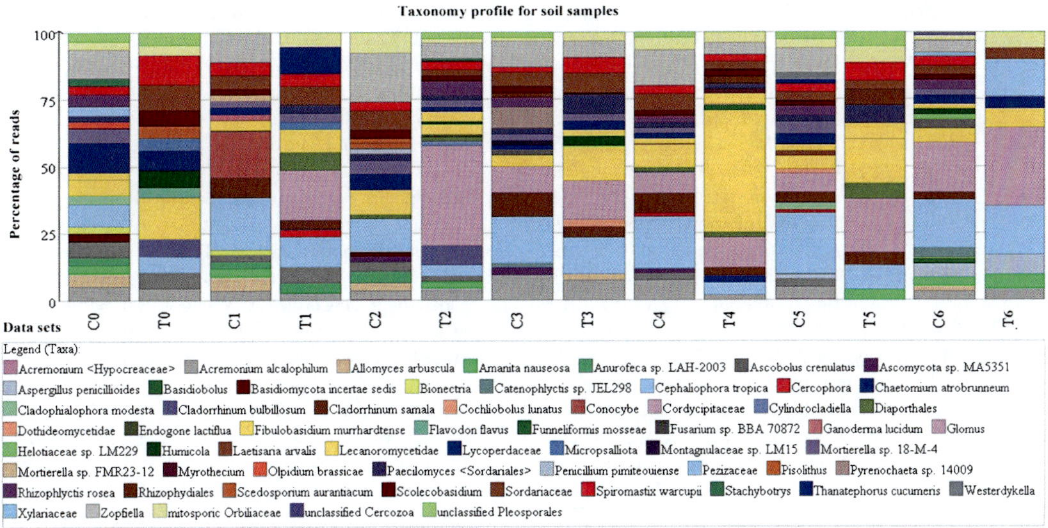

Fig. 2 A snapshot of analysis of fungal diversity in the soil samples collected from a chilli field in India treated with the entomopathogenic fungus *Beauveria bassiana*. The C and T samples are from control (non-treated) and treated plots, respectively – the numbers represent the weeks after treatment. Analysis (with MEGAN (Huson et al. 2011)) of fungal ITS reads generated from soil DNA through multitag 454 pyrosequencing shows the relative abundance of different fungal taxa identified in each sample. The results show that the fungal biodiversity is not affected due to application of *B. bassiana* (results from our laboratory)

bioinformatic tools and statistical assessments for which several options are available.

Acknowledgements The authors are grateful to DST-DFG and DST-DAAD PPP programmes for financial support in carrying out collaborative research on entomopathogenic fungi. Mrs. Sandhya is thankful to CSIR, New Delhi, for a senior research fellowship.

References

Altschul SF, Gish W, Miller W et al (1990) Basic local alignment search tool. J Mol Biol 215:403–410

Amiri-Besheli B, Khambay S, Cameron S et al (2000) Inter-and intra-specific variation in destruxin production by insect pathogenic *Metarhizium* spp. and its significance to pathogenesis. Mycol Res 104:447–4452

Ashelford KE, Chuzhanova NA, Fry JC et al (2005) At least 1 in 20 16S rRNA sequence records currently held in public repositories is estimated to contain substantial anomalies. Appl Environ Microbiol 71:7724–7736

Bagga S, Hu G, Screen SE, Leger RJ (2004) Reconstruction the diversification of subtilisins in the pathogenic fungus *Metarhizium anisopliae*. Gene 324:159–169

Baldwin TK, Winnenburg R, Urban M et al (2006) The pathogen-host interactions database (PHI-base) provides insights into generic and novel themes of pathogenicity. Mol Plant Microbe Interact 19:1451–1462

Baratto CM, Dutra V, Boldo JT et al (2006) Isolation characterization and transcriptional analysis of the chitinase chi2 gene (DQ011663) from the biocontrol fungus *Metarhizium anisopliae* var *anisopliae*. Curr Microbiol 53:217–221

Billich A, Zocher RJ (1987) Enzymatic synthesis of cyclosporin A. J Biol Chem 262:17258–17259

Bingle LEH, Simpson TJ, Lazarus CM (1999) Ketosynthase domain probes identify two subclasses of fungal polyketide synthase genes. Fungal Gen Biol 26:209–223

Bisset J (1983) Notes on Tolypocladium and related genera. Can J Bot 61:1311–1329

Boucias DG, Pendland JC, Latge JP (1988) Nonspecific factors involved in attachment of entomopathogenic deuteromycetes to host insect cuticle. Appl Environ Microbiol 54:1795–1805

Buée M, Reich M, Murat C et al (2009) 454 pyrosequencing analyses of forest soils reveal an unexpectedly high fungal diversity. New Phytol 184:449–456

Butt TM, Goettel MS (2000) Bioassays of entomogenous fungi. In: Navon A, Ascher KRS (eds) Bioassays of entomopathogenic microbes and nematodes. CABI Publishing, Wallingford, pp 141–195

Butt TM, Ibrahim L, Ball BY et al (1994) Pathogenicity of the entomogenous fungi *Metarhizium anisopliae* and *Beauveria bassiana* against crucifer pests and the honey bee. Biocontrol Sci Technol 4:207–214

Butt TM, Jackson C, Magan N (2001) Production, stabilization and formulation of fungal biocontrol agents. In: Butt TM, Jackson C, Magan N (eds) Fungi as biocontrol agents: progress, problems and potential. CABI Publishing, Wallingford, pp 1–8

Butterrs JA, Devi KU, Mohan MC et al (2003) Screening for tolerance to Bavistin, a Benzimidazole fungicide containing methyl benzimidazol-2-yl carbamate (MBC) among strains of the entomopathogenic fungus *Beauveria bassiana* (Balsamo) Vuillemin: sequence analysis of the Beta – tubulin gene to identify mutations conferring tolerance. Mycol Res 107:260–266

Caporaso JG, Kuczynski J, Stombaugh J et al (2010) QIIME allows analysis of high-throughput community sequencing data. Nat Methods 7:335–336

Carlos LM, Palmieri DA, Souza VQ et al (2008) SSR locator: tool for simple sequence repeat discovery integrated with primer design and PCR simulation. Int J Plant Genomics 4:363–374

Castlebury LA, Sutherland JB, Tanne LA et al (1999) Use of a bioassay to evaluate the toxicity of beauvericin to bacteria. World J Microbiol Biotechnol 15:131–133

Cerenius L, Thornqvist PO, Vey A et al (1990) The effect of the fungal toxin destruxin E on isolated crayfish hemocytes. J Insect Physiol 36:785–789

Chao A (1984) Non-parametric estimation of the number of classes in a population. Scand J Stat 11:265–270

Chao A, Lee SM (1992) Estimating the number of classes via sample coverage. J Am Stat Assoc 87:210–217

Charnley AK (1989) Mechanisms of fungal pathogenesis in insects. In: Whipps JM, Lumsden RD (eds) Biotechnology of fungi for improving plant growth. Oxford University Press, London, pp 86–125

Charnley AK (2003) Fungal pathogens of insects: cuticle degrading enzymes and toxins. Adv Bot Res 40:241–321

Charnley AK, Leger RJ (1991) The role of cuticle-degrading enzymes in fungal pathogenesis in insects. In: Cole RT, Hoch HE (eds) Fungal spore disease initiation in plants and animals. Plenum Press, New York/London, pp 267–287

Chu M, Mierzwa R, Truumees I et al (1993) 2 novel diketopiperazines isolated from the fungus *Tolypocladium* sp. Tetrahedron Lett 34:7537–7540

Clarkson JM, Charnley AK (1996) New insights into mechanisms homology modeling and protein engineering strategy of subtilases of fungal pathogenesis in insects. Trends Microbio 4:197–203

Claros MG, Vincens P (1996) Computational method to predict mitochondrial proteins and their target sequences. Eur J Bio 241:779–786

Clowee RK (2006) Estimates: statistical estimation of species richness and shared species from samples.version8, user guide and application published at http://purl.oclc.org/estimates

Cole JR, Chai B, Farris RJ et al (2007) The ribosomal database project (RDP-II): introducing my RDP space and quality controlled public data. Nucleic Acids Res 35:D169–D172

Cook RJ (2000) Advances in plant health management in the 20th century. Annu Rev Phytopathol 38:95–116

De Faria MR, Wraight SP (2007) Mycoinsecticides and mycoacaricides: a comprehensive list with worldwide coverage and international classification of formulation types. Biol Control 43:237–256

De Santis TZ, Hugenholtz P, Larsen N et al (2006) Greengenes, a chimera checked 16S rRNA gene database and workbench compatible with ARB. Appl Environ Microbiol 72:5069–5072

Dias BA, Neves PMOJ, Furlaneto Maia L et al (2008) Cuticle-degrading proteases produced by the entomopathogenic fungus *Beuaveria bassiana* in the presence of coffee berry borer cuticle. Braz J Microbiol 39:301–306

Donatti AC (2008) Production and regulation of cuticle degrading proteases from *Beauveria bassiana* in the presence of *Rhammatocerus schistocercoides* cuticle. Curr Microbiol 56:256–260

Doytchinova IA, Flower DR (2008) Bioinformatic approach for identifying parasite and fungal candidate subunit vaccines. Open Vaccine J 1:22–26

Dumas C, Robert P, Pais M et al (1994) Insecticidal and cytotoxic effects of natural and hemi synthetic destruxins. Comp Biochem Physiol 108:195–203

Edgar RC (2004) MUSCLE: a multiple sequence alignment method with reduced time and space complexity. BMC Bioinform 5:113

El-Sayed GN, Coudron TA, Ignoffo CM (1989) Chitinolytic activity and virulence associated with native and mutant isolates of an entomopathogenic fungus *Nomuraea rileyi*. J Invertebr Pathol 54:394–403

El-Sayed GN, Ignof LTD et al (1993) Cuticular and non-cuticular substrate influence on expression of cuticle degrading enzymes non-cuticular substrate influence on expression of cuticle degrading enzymes. Mycopathologia 122:79–87

Enkerli J, Widmer F, Gessler C et al (2001) Strain-specific microsatellite markers in the entomopathogenic fungus *Beauveria brongniartii*. Mycol Res 105:1079–1087

Faircloth BC (2008) MSATCOMMANDER: detection of microsatellite repeat arrays and automated, locus-specific primer design. Mol Ecol Resour 8:92–94

Falgueras J, Lara AJ, Fernandez-Pozo N et al (2010) SeqTrim: a high-throughput pipeline for pre-processing any type of sequence read. BMC Bioinformatics 11:38

Fang W, Leng B, Xiao Y et al (2005) Cloning of *Beauveria bassiana* chitinase gene Bbchitl and its application to improve fungal strain virulence. Appl Environ Microbiol 71:363–370

Fang W, Pava-Ripoll M, Wang SB et al (2009) Protein kinase A regulates production of virulence determinants by the entomopathogenic fungus *Metarhizium anisopliae*. Fungal Genet Biol 46:277–285

Fang W, Fernandes EKK, Roberts DW et al (2010) A laccase exclusively expressed by *Metarhizium anisopliae* during isotropic growth is involved in pigmentation, tolerance to abiotic stresses and virulence. Fungal Genet Biol 42:602–607

Freimoser FM, Screen S, Bagga S et al (2003) Expressed sequence tag (EST) analysis of two subspecies of *Metarhizium anisopliae* reveals a plethora of secreted proteins with potential activity in insect hosts. Microbiology 149:239–247

Freimoser EM, Hu G, Leger RJ (2005) Variation in gene expression patterns as the insect pathogen *Metarhizium anisopliae* adapts to different host cuticles or nutrient deprivation *in vitro*. Microbiology 151:361–371

Gao Q, Jin K, Ying SH et al (2011) Genome sequencing and comparative transcriptomics of the model entomopathogenic fungi *Metarhizium anisopliae* and *M. acridum*. PLoS Genet 7:1–18

Gibson DM, Krasnoff SB, Churchill ACL (2007) Searching for polyketides in insect pathogenic fungi. In: Rimando AM, Baerson SR (eds) Polyketides: biosynthesis, biological activity, and genetic engineering, vol 955. American Chemical Society, Washington, DC, pp 48–67

Giongo A, Crabb DB, Davis-Richardson AG et al (2010) PANGEA: pipeline for analysis of next generation amplicons. ISME J 4:854–861

Goettel MS, Hajek AE, Siegel JP et al (2001) Safety of fungal biocontrol agents. In: Butt TM, Jackson C, Magan N (eds) Fungi as biocontrol agents: progress, problems and potential. CABI Publishing, Wallingford, pp 347–376

Guengerich FP (2001) Common and uncommon cytochrome P450 reactions related to metabolism and chemical toxicity. Chem Res Toxicol 14:611–650

Gupta S, Roberts DW, Renwick JAA (1989) Preparative isolation of destruxins from *Metarhizium anisopliae* by high performance liquid chromatography. J Liq Chromatogr 12:383–395

Gupta S, Krasnoff SB, Roberts DW et al (1992) Structure of efrapeptins from the fungus *Tolypocladium niveum*: peptide inhibitors of mitochondrial ATPase. J Org Chem 57:2306–2313

Gupta S, Montillot C, Hwang YS (1994) Isolation of novel beauvericin analogues from the fungus *Beauveria bassiana*. J Nat Prod 58:733–738

Hagn A, Pritsch K, Ludwig W et al (2003) Fungal diversity in agricultural soil under different farming management systems, with special reference to biocontrol strains of *Trichoderma* spp. Biol Fert Soils 38:236–244

Hamill RL, Higgens CE, Boaz HE et al (1969) The structure of beauvericin, a new depsipeptide antibiotic toxic to *Artemia salina*. Tetrahedron Lett 49:4255–4258

Hammer Ø, Harper DAT, Ryan PD (2001) PAST: paleontological statistics software package for education and data analysis. Palaeontol Electron 4:9

Hane JK, Lowe RG, Solomon PS et al (2007) Dothideomycete plant interactions illuminated by genome sequencing and EST analysis of the wheat pathogen *Stagonospora nodorum*. Plant Cell 19:3347–3368

Hartman AL, Riddle S, McPhillips T et al (2010) Introducing W.A.T.E.R.S: a workflow for the alignment, taxonomy, and ecology of ribosomal sequences. BMC Bioinformatics 11:317

Hirsch J, Galidevara S, Strohmeier S et al (2013) Effects on diversity of soil fungal community and fate of an artificially applied *Beauveria bassiana* strain assessed through 454 pyrosequencing. Microb Ecol. doi:10.1007/s00248-013-0249-5

Holder DJ, Kirkland BH, Lewis MW et al (2007) Surface characteristics of the entomopathogenic fungus *Beauveria* (Cordyceps) *bassiana*. Microbiology 153:3448–3457

Huber T, Faulkner G, Hugenholtz P (2004) Bellerophon: a programme to detect chimeric sequences in multiple sequence alignments. Bioinformatics 20:2317–2319

Humber RA (1991) Fungal pathogens of aphids. In: Peters DC, Webster JA, Chlouber CS (eds) Proceedings on aphid plant interactions: populations to molecules still water. Okla State University, pp 45–56

Hurlbert SH (1971) The non-concept of species diversity: a critique and alternative parameters. Ecology 52:577–586

Huse SM, Huber JA, Morrison HG et al (2007) Accuracy and quality of massively parallel DNA pyrosequencing. Genome Biol 8:R143–R143.9

Huson DH, Mitra S, Ruscheweyh HJ et al (2011) Integrative analysis of environmental sequences using MEGAN4. Genome Res 21:1552–1560

Jackson CW, Heale JB, Hall RA (1985) Traits associated with virulence to the aphid *Macrosiphoniella sanborni* in eighteen isolates of *Verticillium lecanii*. Ann Appl Biol 106:39–48

Jackson MA, Dunlap CA, Jaronski ST (2010) Ecological considerations in producing and formulating fungal entomopathogens for use in insect biocontrol. BioControl 55:129–145

Kamper J, Kahmann R, Bölker M et al (2006) Insights from the genome of the biotrophic fungal plant pathogen *Ustilago maydis*. Nature 444:97–101

Katoh K, Asimenos G, Toh H (2009) Multiple alignment of DNA sequences with MAFFT. Methods Mol Biol 537:39–64

Khaldi N, Seifuddin FT, Turner G et al (2010) SMURF: genomic mapping of fungal secondary metabolite clusters. Fungal Genet Biol 47:736–741

Khan AAP (2006) A comparative transcriptomic analysis of the generalist entomopathogenic fungus *Beauveria bassiana* (Balsamo) Vuillemin grown on different insect cuticles and synthetic medium through cDNA-AFLP display. Dissertation, Andhra University, Visakhapatnam, India

Khan AAP, Uma Devi K, Vogel H et al (2007) Analysis of differential gene expression in the generalist entomopathogenic fungus *Beauveria bassiana* (Bals.) Vuillemin grown on different insect cuticular extracts and synthetic medium through cDNA-AFLPs. Fungal Genet Biol 44:1231–1241

Khan S, Guo L, Maimaiti Y et al (2012) Entomopathogenic fungi as microbial biocontrol agent. Mol Plant Breed 3:63–79

Kim HK, Hoe HS, Suh D et al (1999) Gene structure and expression of the gene from *Beauveria bassiana* encoding bassiasinI, an insect cuticle-degrading serine protease. Biotechnol Lett 21:777–783

Krasnoff SB, Reátegui RF, Wagenaar MM et al (2004) Cicadapeptins I and II: new Aib-containing peptides from the entomopathogenic fungus *Cordyceps heteropoda*. J Nat Prod 68:50–55

Kulkarni RD, Kelkar HS, Dean RA (2003) An eight-cysteine-containing CFEM domain unique to a group of fungal membrane proteins. Trends Biochem Sci 28:118–121

Kumar S, Carlsen T, Mevik BH et al (2011) CLOTU: an online application for processing and clustering of 454 amplicon reads into OTUs followed by taxonomic annotation. BMC Bioinformatics 12:1–9

Lacey LA, Goettel M (1995) Current development in microbial control of insect pests and prospects for the early 21st century. Entomophaga 40:3–27

Lacey LA, Frutos R, Kaya HK et al (2001) Insect pathogens as biological control agents: Do they have a future? Biol Control 21:230–248

Lafon A, Han KH, Seo JA et al (2006) G-protein and cAMP mediated signaling in aspergilli: a genomic perspective. Fungal Genet Biol 43:490–502

Li Y, Kelly WG, Logsdon JM Jr et al (2004) Functional genomic analysis of the ADP-ribosylation factor family of GTPases: phylogeny among diverse eukaryotes and function in C. elegans. FASEB J 18:1834–1850

Lin L, Fang W, Liao X et al (2011) The MrCYP52 Cytochrome P450 Monooxygenase Gene of Metarhizium robertsii Is Important for Utilizing Insect Epicuticular Hydrocarbons. PLoS ONE 6:e28984

Liu S, Peng G, Xia Y (2012) The adenylate cyclase gene MaAC is required for virulence and multi-stress tolerance of *Metarhizium acridum*. BMC Microbiol 12:163

Lozupone C, Knight R (2005) UniFrac: a new phylogenetic method for comparing microbial communities. Appl Environ Microbiol 71:8228–8235

Mahé S, Duhamel M, Le Calvez TL (2012) PHYMYCO-DB: a curated database for analyses of fungal diversity and evolution. PLoS ONE 7:e431117

McCune B, Mefford MJ (2011) PC-ord. Multivariate analysis of ecological data, version 6.MjM Software, Gleneden Beach, Oregon, U.S.A

Medema MH, Blin K, Cimermancic P et al (2011) antiSMASH rapid identification, annotation and analysis of secondary metabolite biosynthesis gene clusters in bacterial and fungal genome sequences. Nucleic Acids Res 39:W339–W346

Molnar I, Gibson DM, Krasnoff SB (2010) Secondary metabolites from entomopathogenic Hypocrealean fungi. Nat Prod Rep 27:1241–1275

Needleman SB, Wunsch CD (1970) A general method applicable to the search for similarities in the amino acid sequence of two proteins. J Mol Biol 48:443–453

Nelson DR (1999) Cytochrome P450 and the individuality of species. Arch Biochem Biophys 369:1–10

Oksanen J, Kindt R, Legendre P et al (2007) vegan: Community Ecology Package. R package version 1.8-8. Online at: http://r-forge.r-project.org/projects/vegan

Ortiz de Montellano PR (2005) Cytochrome P450: structure, mechanism, and biochemistry, 3rd edn. Kluwer Academic/Plenum Press, New York, p 689

Pandey RV, Nolte V, Schlotterer C (2010) CANGS: a user-friendly utility for processing and analyzing 454 GS-FLX data in biodiversity studies. BMC Res Notes 3:3

Pedrini N, Zhang S, Juarez MP et al (2010) Molecular characterization and expression analysis of a suite of cytochrome P450 enzymes implicated in insect hydrocarbon degradation in the entomopathogenic fungus *Beauveria bassiana*. Microbiology 156:2549–2557

Petrosino JF, Highlander S, Luna RA et al (2009) Metagenomic pyrosequencing and microbial identification. Clin Chem 55:856–866

Pollard A, Wyn Jones KG (1979) Enzyme activities in concentrate solutions of glycine-betaine and other solutes. Planta 144:291–298

Quast C, Pruesse E, Yilmaz P et al (2013) The SILVA ribosomal RNA gene database project: improved data processing and web-based tools. Nucleic Acids Res 41:D590–D596

Quince C, Lanzen A, Curtis TP et al (2009) Accurate determination of microbial diversity from 454 pyrosequencing data. Nat Methods 6:639–641

Rappleye CA, Goldman WE (2008) Fungal stealth technology. Trends Immunol 29:18–24

Rawlings ND, Barrett AJ, Bateman A (2012) MEROPS: the database of proteolytic enzymes, their substrates and inhibitors. Nucleic Acids Res 40:D343–D350

Rehner SA, Buckley E (2005) A Beauveria phylogeny inferred from nuclear 1 ITS and EF1-α sequences: evidence for cryptic diversification and links to Cordyceps teleomorphs. Mycologia 97:84–98

Roberts DW, St Leger RJ (2004) Metarhizium spp., cosmopolitan insect pathogenic fungi: mycological aspects. Adv Appl Microbiol 54:1–70

Samuels RI, Charnley AK, Reynolds SE (1988) The role of destruxins in the pathogenicity of 3strains of *Metarhizium anisopliae* for the tobacco hornworm *Manduca sexta*. Mycopathologia 104:51–58

Scheepmaker JWA, Butt TM (2010) Natural and released inoculum levels of entomopathogenic fungal biocontrol agents in soil in relation to risk assessment and in accordance with EU regulations. Biocontrol Sci Technol 20:503–552

Schloss PD, Handelsman J (2005) Introducing DOTUR, a computer programme for defining operational taxonomic units and estimating species richness. Appl Environ Microbiol 71:1501–1506

Schloss PD, Handelsman J (2006) Introducing Tree Climber, a test to compare microbial community structures. Appl Environ Microbiol 72:2379–2384

Screen S, Bailey A, Charnley K et al (1997) Carbon regulation of the cuticle-degrading enzyme PR1 from

Metarhizium anisopliae may involve a trans-acting DNA-binding protein CRR1, a functional equivalent of the *Aspergillus nidulans* CREA protein. Curr Genet 31:511–518

Shah FA, Wang CS, Butt TM (2005) Nutrition influences growth and virulence of the insect-pathogenic fungus *Metarhizium anisopliae*. FEMS Microbiol Lett 251:259–266

Singleton DR, Furlong MA, Rathbun SL et al (2001) Quantitative comparisons of 16S rRNA gene sequence libraries from environmental samples. Appl Environ Microbiol 67:4374–4376

Small CLN, Bidochka MJ (2005) Up-regulation of Pr1, a subtilisin-like protease, during conidiation in the insect pathogen *Metarhizium anisopliae*. Mycol Res 109:307–313

Solomon PS, Tan KC, Sanchez P et al (2004) The disruption of a G alpha subunit sheds new light on the pathogenicity of *Stagonospora nodorum* on wheat. Mol Plant Microbe Interact 17:456–466

St Leger RJ, Wang C (2009) Entomopathogenic fungi and the genomic era. In: Stock SP, Vandenberg J, Glazer I, Boemare N (eds) Insect pathogens: molecular approaches and techniques. CABI Publishing, Wallingford, pp 366–400

St Leger RJ, Joshi L, Bidochka MJ et al (1995) Protein synthesis in *Metarhizium anisopliae* growing on host cuticle. Mycol Res 99:1034–1040

St Leger RJ, Joshi L, Bidochka MJ et al (1996) Characterization and ultra structural localization of chitinases' from *Metarhizium anisopliae*, *Metarhizium. flavoviride* and *Beauveria bassiana* during fungal invasion of host (*Manduca sexta*) cuticle. Appl Environ Microbiol 62:907–912

St Leger RJ, Joshi L, Roberts DW (1997) Adaptation of proteases and carbohydrates of saprophytic, phytopathogenic and entomopathogenic fungi to the requirements of their ecological niches. Microbiology 143:1983–1992

Sun Y, Cai Y, Liu L et al (2009) ESPRIT: estimating species richness using large collections of 16S rRNA pyrosequences. Nucleic Acids Res 37:e76

Thomas MB, Read AF (2007) Fungal bioinsecticide with a sting. Nat Biotechnol 25:1367–1368

Uma Devi K, Padmavathi J, Uma Maheswara Rao C et al (2008) A study of host specificity in the entomopathogenic fungus Beauveria bassiana (Hypocreales, Clavicipitaceae). Biocontrol Sci Technol 18:975–989

Uma Devi K, Reineke G, Sandhya G et al (2012) Pathogenicity genes in entomopathogenic fungi used as biopesticides. In: Gupta VK, Ayyachamy M (eds) Biotechnology of fungal genes. Science Publishers, Enfield, pp 343–367

Vega FE, GoettelMS BM et al (2009) Fungal entomopathogens: new insights on their ecology. Fungal Ecol 2:149–159

Vestergaard S, Gillespie AT, Butt TM et al (1995) Pathogenicity of the hyphomycete fungi *Verticillium lecanii* and *Metarhizium anisopliae* to the western flower thrips, *Frankliniella occidentalis*. Biocontrol Sci Technol 5:185–192

Wang C, St Leger RJ (2005) Developmental and transcriptional responses to host and non host cuticles by the specific locus pathogen *Metarhizium anisopliae* var.*acridum*. Eukaryot Cell 4:937–947

Wang C, St Leger RJ (2006) A collagenous protective coat enables *Metarhizium anisopliae* to evade insect immune responses. Proc Natl Acad Sci 103:6647–6652

Wang C, St Leger RJ (2007a) The *Metarhizium anisopliae* perilipin homologMPL1regulates lipid metabolism, appressorial turgor pressure, and virulence. J Biol Chem 282:21110–21115

Wang C, St Leger RJ (2007b) The MAD1 adhesin of *Metarhizium anisopliae* links adhesion with blastospore production and virulence to insects and the MAD2 adhesin enables attachment to plants. Eukaryot Cell 6:808–816

Wang CS, Hu G, St Leger RJ (2005) Differential gene expression by *Metarhizium anisopliae* growing in root exudate and host (*Manduca sexta*) cuticle or hemolymph reveals mechanisms of physiological adaptation. Fungal Genet Biol 42:704–718

Wang CS, Duan ZB, Leger RJ (2008) MOS1osmosensor of *Metarhizium anisopliae* is required for adaptation to insect host hemolymph. Eukaryot Cell 7:302–309

Webb CO, Ackerly DD, Kembel SW (2008) Phylocom: software for the analysis of phylogenetic community structure and trait evolution Bioinformatics 24:2098–2100

Wraight SP, Jackson MA, de Kock SL (2001) Production, stabilization and formulation of fungal biocontrol agents. In: Butt TM, Jackson C, Magan N (eds) Fungi as biocontrol agents: progress, problems and potential. CABI Publishing, Wallingford, pp 253–287

Wraight SP, Ramos ME, Williams JE et al (2003) Comparative virulence and host specificity of *Beauveria bassiana* isolates assayed against lepidopteran pests of vegetable crops. J Invertebr Pathol 103:186–199

Wu D, Hartman A, Ward N et al (2008) An automated phylogenetic tree-based small subunit rRNA taxonomy and alignment pipeline (STAP). PLoS ONE 3(7):e2566

Wu GD, Lewis JD, Hoffmann C et al (2010) Sampling and pyrosequencing methods for characterizing bacterial communities in the human gut using 16S sequence tags. BMC Microbiol 30:206

Xia Y, Clarkson JM, Charnley AK (2001) Acid phosphatases of *Metarhizium anisopliae* during infection of the tobacco hornworm *Manduca sexta*. Arch Microbiol 176:427–434

Xiao G, Ying SH, Zheng P et al (2012) Genomic perspectives on the evolution of fungal entomopathogenicity in Beauveria bassiana. Sci Rep 2:483

Xu Y, Orozco R, Wijeratne EM et al (2008) Biosynthesis of the cyclooligomer depsipeptide beauvericin, a virulence factor of the entomopathogenic fungus *Beauveria bassiana*. Chem Biol 15:898–907

Xu Y, Orozco R, Kithsiri Wijeratne EM et al (2009) Biosynthesis of the cyclooligomer depsipeptide bassianolide, an insecticidal virulence factor of *Beauveria bassiana*. Fungal Genet Biol 46:53–364

Xue QX, Wang J, Huang BF et al (2010) A new manganese superoxide dismutase identified from *Beauveria bassiana* enhances virulence and stress tolerance when over expressed in the fungal pathogen. Appl Microbiol Biotechnol 86:1543–1553

Yu Y, Breitbart M, McNairnie P et al (2006) FastGroupII: a web-based bioinformatics platform for analyses of large 16S rDNA libraries. BMC Bioinformatics 7:57

Zhang W, Yueqing C, Yuxian X (2008) Cloning of the subtilisin Pr1A gene from a strain of locust specific fungus *Metarhizium anisopliae* and functional expression of the protein in *Pichia pastoris*. World J Microbiol Biotechnol 24:2481–2488

Zhang Y, Zhao J, Fang W et al (2009) Mitogen-activated protein kinase hog1 in the entomopathogenic fungus *Beauveria bassiana* regulates environmental stress responses and virulence to insects. Appl Environ Microbiol 75:3787–3795

Zhang Y, Zhang J, Jiang X et al (2010) Requirement of a mitogen-activated protein kinase for appressorium formation and penetration of insect cuticle by the entomopathogenic fungus *Beauveria bassiana*. Appl Environ Microbiol 76:2262–2270

Zheng P, Xia Y, Xiao G et al (2011) Genome sequence of the insect pathogenic fungus *Cordyceps militaris*, a valued traditional Chinese medicine. Genome Biol 12:R116

Zhu Y, Pan J, Qiu J et al (2008) Isolation and characterization of a chitinase gene from entomopathogenic fungus *Verticillium lecanii*. Appl Environ Microbiol 39:314–332

Exploring the Genomes of Symbiotic Diazotrophs with Relevance to Biological Nitrogen Fixation

Subarna Thakur, Asim K. Bothra, and Arnab Sen

Abstract

Nitrogen fixation is an important process for the conversion of atmospheric nitrogen to its biologically available form, cutting across various ecosystems. Biological nitrogen fixation (BNF) is accomplished by a group of microorganisms, known as diazotrophs through the catalytic action of nitrogenase enzyme system. N-fixing microbes exhibit a wide diversity in terms of their habitat. Some are free-living, whereas others form symbiotic associations with higher plants. Symbiotic nitrogen fixation is particularly relevant in context of sustainable agriculture. Recent advances in next-generation sequencing technology have ensured the availability of thousands of complete genomes including those of symbiotic diazotrophs. Genomic approaches together with bioinformatics tools are now being used to define and understand the complex molecular relationships that underpin symbiotic nitrogen fixation. Accessibility of new computational tools for genomic and proteomic analysis has accelerated nitrogen fixation research predominantly in the areas of comparative genomics, protein chemistry, and phylogenetic analysis of nitrogen fixation genes. Alternative phylogenetic approaches and protein structure-based studies may prove to be quite prolific to divulge the unfamiliar aspects of symbiotic nitrogen fixation.

Keywords

Diazotrophy • Symbiosis • N_2 fixation • Bioinformatics • Phylogeny

1 Introduction

Nitrogen is an essential nutrient for all living organisms. It is an important component of many organic molecules such as DNA, RNA, and proteins, the building blocks of life. Molecular nitrogen or dinitrogen (N_2) makes up nearly

S. Thakur • A. Sen (✉)
NBU Bioinformatics Facility, Department of Botany, University of North Bengal, Siliguri 734013, India
e-mail: senarnab_nbu@hotmail.com

A.K. Bothra
Bioinformatics Chemoinformatics Laboratory, Department of Chemistry, Raiganj College, Raiganj, India

K.K. P.B. et al. (eds.), *Agricultural Bioinformatics*,
DOI 10.1007/978-81-322-1880-7_14, © Springer India 2014

four-fifths of the atmosphere but is metabolically unavailable directly to higher plants or animals. It is available to some species of microorganism through biological nitrogen fixation (BNF) in which atmospheric nitrogen is converted to ammonia by the enzyme nitrogenase (Postgate 1998). The ammonia is then transferred to the higher plant to meet its nutritional needs for the synthesis of proteins, enzymes, nucleic acids, chlorophyll, etc., and subsequently it enters the food chain. Thus, all eukaryotes (including higher plants and animals) naturally depend on the BNF activity of the N-fixing microbes for their N supply. Microorganisms that fix nitrogen are called diazotrophs. According to current knowledge, only prokaryotes (members of the domains Archaea and Bacteria) are capable of performing BNF (Klipp 2004).

The ability to fix nitrogen is widely, though paraphyletically, distributed across both the bacterial and archaeal domains (Raymond et al. 2004). There are two types of diazotrophic prokaryotes: those that are free-living (e.g., *Azotobacter, Clostridium, Klebsiella*, etc.) and those that form symbiotic relationships (e.g., *Rhizobium*, *Bradyrhizobium, Frankia*, etc.). The free-living diazotrophs require a chemical energy source if they are nonphotosynthetic, whereas the photosynthetic diazotrophs, such as the cyanobacteria, utilize light energy (Leigh 2002). Some diazotrophs called rhizobia enters into symbiotic relationship with legumes like clover and soyabean. The symbiosis between legumes and the nitrogen-fixing rhizobia occurs within nodules mainly on the root and in a few cases on the stem (Burns and Hardy 1975). A similar symbiosis occurs between a number of woody plant species and the diazotrophic actinomycete *Frankia* (Pedrosa et al. 2000). These symbiotic associations are the greatest contributions of fixed nitrogen to agricultural systems.

The exploitation of biological nitrogen fixation for agricultural benefits has long been sought after. Biological nitrogen fixation provides a means to meet the needs of a growing population with a nutritious, environmentally friendly, sustainable food supply. This makes the need for BNF research very compelling in the current scenario. In the last two decades, many exciting happenings in nitrogen fixation took place, genomes have been sequenced, the "omics" approaches have been applied to both symbionts, and new genetically modified crops are becoming commonplace in agriculture. Biochemical research into the workings of nitrogen fixation is generally focused on the enzyme complex called nitrogenase. Other than its usual function, this system has emerged as a model for more general biochemical processes, such as signal transduction, protein–protein interaction, inter- and intramolecular electron transfer, complex metal cluster involvement in enzymatic catalysis, etc. (Peters et al. 1995). It is in this perspective that we have thought of reviewing the present trends in nitrogen fixation research in context to agriculture with special emphasis on genomics, proteomics, and bioinformatics.

2 Various Aspects of Biological Nitrogen Fixation

2.1 Biological Nitrogen Fixation and Sustainable Agriculture

Natural reserves of soil nitrogen are normally low, so commercially prepared N fertilizers must be added to increase plant growth and vigor. Chemical fertilizers had a substantial impact on food production in the recent past and are today an indispensable part of modern agricultural practices. But for the farmers of developing countries, N fertilizers are neither affordable nor widely available. Moreover, the harmful effects on the environment of heavy use of N fertilizer are becoming more evident day by day. Further, the fossil fuels which are used in the production of N fertilizer are becoming scarcer and more expensive. At the same time, the demand for food is going up as populations increase. Therefore, there is a great need to search for all possible avenues. The process of biological nitrogen fixation offers economically attractive and ecologically sound means of reducing external nitrogen input and improving the quality and quantity of internal resources. Biological nitrogen fixation is the reduction of atmospheric N_2 gas to biologically available ammonium, mediated by prokaryotic organisms in symbiotic relationships, associative relationships, and under free-living conditions (Postgate 1998). The fixed nitrogen that is provided by biological nitrogen fixation is less prone to leaching and volatilization, and therefore the biological process contributes an important and sustainable input into agriculture. Nitrogen input through BNF can help maintain soil N reserves as well as substitute for N fertilizer to attain large crop yields (Peoples and Craswell 1992). An understanding of the factors controlling BNF systems in the field is vital for the support and successful adoption in large scale in an agricultural context.

Wani et al. (1995) highlighted the importance of biological nitrogen fixation of legumes in sustainable agriculture in semiarid tropical region. Legumes, one of the most important plant families in agriculture, are often involved in a remarkable symbiosis with nitrogen-fixing rhizobia. Legumes are often considered to be the major nitrogen-fixing systems, as they may derive up to 90 % of their nitrogen from N_2. The quantity of atmospheric N fixed through forage legume biological N fixation can range as high as 200 kg/ha per year (Peoples et al. 1995). The symbiotic association of actinorhizal species helps in improving soil fertility in disturbed sites such as eroded areas, sand dunes, moraines, etc. Actinorhizal plant nitrogen fixation rates are comparable to those found in legumes (Torrey and Tjepkema 1979; Dawson 1983). Nitrogen-fixing *Azolla*–Cyanobacteria symbiosis has been widely used to enrich rice paddies with organic nitrogen in Southeast Asian countries like China, Vietnam, and Southeast Asia (Watanabe and Liu 1992). The rice paddies of Asia, which feed over half of the world's population, depend upon cyanobacterial N_2 fixation (Irisarri et al. 2001).

2.2 Physiological and Phylogenetic Diversity of Diazotrophs

Farmers have known, probably since the time of the Egyptians, that legumes such as pea, lentil, and clover are important for soil fertility. The practices of crop rotation, intercropping, and green manuring were extensively described by then Romans, but it was not until the nineteenth century that an explanation for the success of the legumes in restoring soil fecundity was uncovered. The discovery of nitrogen fixation was attributed to the German scientists Hellriegel and Wilfarth, who in 1886 reported that legumes bearing root nodules could use gaseous nitrogen. Shortly afterward, in 1888, Beijerinck, a Dutch microbiologist, succeeded in isolating a bacterial strain from root nodules. This isolate happened to be a *Rhizobium leguminosarum* strain (Franche et al. 2009). Beijerinck (in 1901) and Lipman (in 1903) were responsible for the isolation of *Azotobacter* spp., while Winodgradsky (in 1901) isolated the first strain of *Clostridium pasteurianum* (Stewart 1969). The discovery of nitrogen fixation in blue-green algae was established much later (Stewart 1969). The identification of nitrogen-fixing microbe from root nodules of nonleguminous plants like *Alder* generated considerable controversy for a while. It was Brunchorst who named the microbe *Frankia subtilis* (Pawlowski 2009). Hiltner (1898) recognized the nodule inhabitant as an actinomycete, Gram-positive bacteria closely related to *Streptomyces*. Pommer (1959) was probably the first person to obtain an isolate, but it did not reinfect its host plant. For a long time, diazotrophy in the actinomycetes was thought to be limited to the genus *Frankia*, but through the years several other actinomycetes have been shown to have nif genes (Gtari et al. 2012). Over the years there have been continual discoveries of new diazotrophs, revealing that this function is performed by a very diverse group of prokaryotes. In the last decades, the use of molecular technologies for the direct detection of the genes of biological nitrogen fixation has shown that the capacity for diazotrophy is even more widespread than previously expected.

Although nitrogen fixation is not found in eukaryotes, it is widely distributed among the Bacteria and the Archaea, revealing considerable biodiversity among diazotrophic organisms. The ability to fix nitrogen is found in most bacterial phylogenetic groups, including green sulfur bacteria, Firmibacteria, actinomycetes, cyanobacteria, and all subdivisions of the Proteobacteria. In Archaea, nitrogen fixation is mainly restricted to methanogens. The ability to fix nitrogen is compatible with a wide range of physiologies including aerobic (e.g., *Azotobacter*), facultatively anaerobic (e.g., *Klebsiella*), or anaerobic (e.g., *Clostridium*) heterotrophs; anoxygenic (e.g., *Rhodobacter*) or oxygenic (e.g., *Anabaena*) phototrophs; and chemolithotrophs (e.g., *Alcaligenes, Thiobacillus, Methanosarcina*) (Young 1992). Diazotrophs show considerable diversity in terms of habitats. They are found as free-living in soils and water, associative symbioses with grasses, actinorhizal associations with woody plants, and cyanobacterial symbioses with various plants. The most widely known and discussed feature of diazotrophs is their symbiotic association with a number of leguminous plants collectively referred to as rhizobia. The rhizobia are Gram negative and belong to the large and important Proteobacteria division and include the genera like *Agrobacterium, Allorhizobium, Azorhizobium, Bradyrhizobium, Mesorhizobium, Rhizobium, Sinorhizobium, Devosia, Methylobacterium,* and *Ochrobactrum* (Franche et al. 2009). These soil bacteria are able to invade legume roots in nitrogen-limiting environments, leading to the formation of a highly specialized organ, the root nodule. These specialized root structures offer an ecological niche for the microbe to fix nitrogen (Mylona et al. 1995). Symbiotic association is not limited to the legumes but to a number of nonlegumes. The most significant among them are the actinorhizal plants–*Frankia* association. The genus *Frankia* consists of filamentous actinomycetes forming symbiotic associations with a number of woody dicot plants like *Casuarina, Hippophae, Alnus, Myrica*, etc., belonging to different families (Benson and Silvester 1993). *Frankia* compartmentalizes nitrogenase within the vesicle structures, which are surrounded by an envelope containing a high content of bacteriohopane lipids and function to protect the enzyme from oxygen inactivation (Berry et al. 1993; Huss-Danell 1997). Over the years diazotrophy has been reported from other actinomycetes as well such as *Mycobacterium flavum, Corynebacterium autotrophicum, Arthrobacter* sp., *Agromyces, etc.* (Gtari et al. 2012). The findings of several authors (Von Bulow and Dobereiner 1975; Dobereiner 1976; Baldani and Baldani 2005) revealed existing associations of tropical grasses with nitrogen-fixing bacteria that, which under favorable conditions, may be contributing significantly to the N economy of these plants. The bacteria belong to the genus *Azospirillum* and are the most promising microorganisms that colonize roots of economically important grasses and cereals (Leigh 2002).

Cyanobacteria have long been known to fix nitrogen. Both heterocystous (like *Anabaena, Nostoc*, etc.) and nonheterocystous cyanobacteria (like *Trichodesmium, Plectonema*, etc.) are capable of diazotrophy (Schlegel and Zaborosch 2003). They are the only organisms that are capable of both O_2-evolving photosynthesis and nitrogen fixation (Klipp 2004). Therefore, face the unique problem of balancing two essential, but incompatible, cellular processes: oxygenic photosynthesis and O_2-sensitive N_2 fixation. In some filamentous cyanobacteria, nitrogen fixation occurs in specialized, terminally differentiated cells called heterocysts that protect the nitrogenase complex from O_2 damage by increasing respiration, terminating photosystem II activity, and forming multilayered cellular membranes that reduce oxygen diffusion, thus creating a microaerobic environment (Adams 2000). However, in members like *Lyngbya, Plectonema*, etc., where heterocyst is absent, nitrogen fixation occurs in internally organized cells (Schlegel and Zaborosch 2003). Another important aspect of cyanobacteria is their association with higher plants. The *Anabaena–Azolla* association (Bohlool et al. 1992) and *Nostoc–Gunnera* association (Mylona et al. 1995) can fix a substantial amount of nitrogen. Cycads in association with cyanobacterial species can also fix nitrogen (Rai et al. 2002).

2.3 Nitrogenase Complex: Enzymatic Machinery

The biochemical machinery required for biological nitrogen fixation is provided by the nitrogenase enzyme system (Eady and Postgate 1974; Hoffman et al. 2009). Nitrogenase is a two-protein component system that catalyzes the reduction of dinitrogen to ammonia coupled to the hydrolysis of ATP (Rees and Howard 2000). The most extensively studied form of nitrogenase is the molybdenum-containing system that consists of two component metalloproteins, the molybdenum–iron (MoFe) protein and the iron (Fe) protein. The smaller component of nitrogenase is the Fe protein, which acts as a redox-active agent and transfers electrons to the MoFe protein for the reduction of substrates from available electron donor in the system (Rees et al. 2005). It has two identical subunits. The Fe protein contains one iron sulfur cluster [4Fe-4S], which bridges the two subunits. The Fe protein has one MgATP-binding site in each subunit that binds to two MgATP molecules. Binding of MgATP to the Fe protein induces conformational changes followed by hydrolysis of MgATP, which facilitate the electron transfer from the Fe protein to the MoFe protein (Rees et al. 2005). Although this transfer of electrons is the main function of the Fe protein, it has some other functions. The Fe protein is needed for initial biosynthesis of the MoFe cofactor. Following the biosynthesis of MoFe cofactor, the insertion of the preformed MoFe cofactor into the MoFe protein requires the Fe protein (Burgess and Lowe 1996). The larger component of nitrogenase is the MoFe protein, which is a $\alpha_2\beta_2$-tetramer, containing two $\alpha\beta$-dimer subunits. Each dimer contains one MoFe cofactor and one P-cluster [8Fe-7S]. The MoFe cofactor is located in the active site of the protein where the reduction of substrates occurs. The main role of the P-cluster is electron transfer by accepting an electron from the Fe protein and donating it to the MoFe cofactor. Each cluster contains eight metals and associated sulfurs that are arranged distinctively. The $\alpha\beta$-dimeric units communicate and contact each other through their subunits (Burgess and Lowe 1996). The P-cluster bridges between each α- and β-subunit, while the MoFe cofactor is placed on α-subunits. In addition to this molybdenum-containing nitrogenase, alternative nitrogenases also exist that are homologous to this system, but with the molybdenum almost certainly substituted by vanadium or iron (Eady 1996). The vanadium-nitrogenase system has two components. It has an Fe protein which is the same as other nitrogenase systems, and the second component is a vanadium–iron (VFe)-containing protein which is different compared to two other systems. This type of nitrogenase has been detected in *A. vinelandii* and *A. chroococcum* (Robson et al. 1986). The third type of nitrogenase, iron only, contains an iron (Fe) protein and another protein, which is very similar to MoFe protein and VFe protein, while it has only Fe as its cofactor. This type of protein has also been detected in *A. vinelandii* nitrogenase (Eady 1996).

Studies by various authors (Thorneley and Lowe 1985; Burgess and Lowe 1996) revealed that the basic mechanism of nitrogenase involves the following: (1) complex formation between the reduced Fe protein with two bound ATP and the MoFe protein, (2) electron transfer between the two proteins coupled to the hydrolysis of ATP, (3) dissociation of the Fe protein accompanied by re-reduction (via ferredoxins or flavodoxins) and exchange of ATP for ADP, and (4) repetition of this cycle until sufficient numbers of electrons and protons have been accumulated so that available substrates can be reduced. In addition to dinitrogen reduction, nitrogenase has been found to catalyze the reduction of protons to dihydrogen, as well as nonphysiological substrates such as acetylene.

2.4 Genetics and Genomics of Biological Nitrogen Fixation

The biochemical complexity of nitrogen fixation is reflected in the genetic organization and in the regulation of expression of the components required for the catalytic activity. Various techniques like mutations, deletion mapping, cloning vectors, etc., have facilitated the identification of genes associated with nitrogen fixation.

The organization and regulation of the genes were revealed in the early 1980s. The organism that appears to have the simplest organization of nitrogen fixation-specific (*nif*) genes, and which is the one best studied at the molecular genetics level, is the facultative anaerobe, *Klebsiella pneumoniae*. Arnold et al. (1988) reported the first ever detailed organization of *nif* genes from this organism. A 24 kb base pair DNA region contains the entire *K. pneumonia nif* cluster, which includes 20 genes. *nif*HDK are the three structural genes encoding for the three subunits of Mo nitrogenase. In most nitrogen-fixing prokaryotes, these three genes form one transcriptional unit, with a promoter in front of the *nif*H gene. A number of studies (Dixon et al. 1980; Paul and Merrick 1989; Rubio and Ludden 2005, 2008) have established that the maturation of apo-Fe protein (NifH) requires the products of *nif*H, *nif*M, *nif*U, and *nif*S, while that of apo-MoFe protein requires at least six genes *nif*E, *nif*N, *nif*V, *nif*H, *nif*Q, and *nif*B which are required for the biosynthesis of FeMoco. There is considerable homology between *nif*DK and *nif*EN, and it has been speculated that the *nif*EN products might form a scaffold for FeMoco biosynthesis that later shifts FeMoco to the *nif*DK complex (Brigle et al. 1987). Imperial and his coworkers (1984) established that the *nif*Q gene product might be involved in the formation of a molybdenum–sulfur precursor to FeMoco. Mutations in *nif*B result in the formation of an immature MoFe protein that lacks FeMo cofactor. It can be activated in vitro by adding FeMo cofactor that has been isolated from wild-type MoFe protein (Roberts et al. 1978). Mutations in the *nif*V gene result in the formation of a nitrogenase with a bound citrate rather than homocitrate. The *nif*V product is homocitrate synthase (Zheng et al. 1997). Thus, on the basis of mutational studies, the function of various other *nif* genes has been confirmed. In contrast to *Klebsiella*, the *nif* organization is a bit complex different in *Azotobacter vinelandii*. In *Azotobacter* the genes coding for the Mo-dependent nitrogenase components (*nif*HDK) and their regulatory and assembly systems are located in two discrete regions (O'Carroll and Dos Santos 2011). The organization of nitrogen-fixing genes along with their genetic regulation in different rhizobia was extensively reviewed by Fischer (1994), and according to him rhizobial *nif* genes are structurally homologous to the 20 *K. pneumoniae nif* genes, and it is inferred that a conserved *nif* gene plays a similar role in rhizobia as in *K. pneumoniae*.

Besides the *nif* genes, the "*fix*"- and "*nod*"-type genes are associated with biological nitrogen fixation and nodule formation in rhizobial species, and many do not have homologues in the free-living diazotroph like *K. pneumonia*. The *fix* genes represent a very heterogeneous class including genes involved in the development and metabolism of bacteroides. Studies by Anthamatten and Hennecke (1991) and Batut et al. (1991) have established that *fix* L, *fix*J, and *fix*K genes encode regulatory proteins. The *fix*ABCX genes code for an electron transport chain to nitrogenase (Fischer 1994). Mutations in any one of the *fix*ABCX genes of *S. meliloti, B. japonicum*, and *A. caulinodans* completely abolish nitrogen fixation. All four *fix*GHIS gene products are predicted to be transmembrane proteins, but further biochemical analysis is required to define their function in rhizobial nitrogen fixation (Fischer 1994). The *fix*NOQP genes encode the membrane-bound cytochrome oxidase that is required for the respiration of the rhizobia in low-oxygen environments (Delgado et al. 1998). Johnston and his coworkers discovered the presence of nodulation genes in a plasmid of *Rhizobium leguminosarum* and mutation of those genes rendered them useless. Later on studies (Schultze and Kondorosi 1998; Perret et al. 2000) ascertained that *nod*, *nol*, and *noe* genes produce nodulation signals. The interplay of different *nod* genes, triggering of the creation of root nodule, signaling cascades, and development of nodule meristem were reported by a number of researchers (Yang et al. 1999; Long 2001; Geurts and Bisseling 2002). In most species, the *nod* ABC genes are part of a single operon. Inactivation of these genes abolishes the ability to elicit any symbiotic reaction in the plant (Long 1989). Over the years other *nod* genes like *nod*D, *nod*EF, *nod*S, *nod*L, and *nod*HPQ have been characterized in many rhizobia. Like the rhizobia, *Azospirillum* includes a megaplasmid and sequences similar to *nod* genes (Elmerich 1984). *Frankia* on the other hand houses a number of *nif* genes, but researchers failed to spot *nod* genes in *Frankia* (Ceremonie et al. 1998).

Understanding of genetic machinery behind biological nitrogen fixation attained new heights with the arrival of complete genome sequences of various diazotrophs. Recent advances in genome sequencing have opened exciting new perspectives in the field of genomics by providing the complete gene inventory of rhizobial microsymbionts. Genomics have enabled thorough analysis of the gene organization of nitrogen-fixing species, the identification of new genes involved in nitrogen fixation, and the identification of new diazotrophic species. *Mesorhizobium loti* strain MAFF303099 (Kaneko et al. 2000) was the first sequence of a symbiotic bacterium, and it was followed by *Sinorhizobium meliloti* (Puhler et al. 2004). The completion of the genomes of *Rhizobium leguminosarum bv viciae* (Young et al. 2006), *Rhizobium etli* (Gonzalez et al. 2006), *Bradyrhizobium* strains, and *Frankia* strains (Normand et al. 2007) and sequences for a number of free-living diazotrophs spanning different habitat and ecological niches bolstered nitrogen fixation. The genome information from all these nitrogen-fixing organisms allows researchers to rapidly apply information obtained from genome sequencing to the developing area of functional genomics, which will provide new insights into the complex molecular relationships that support both symbiotic and nonsymbiotic nitrogen fixation. DNA array technologies are now being used to monitor the expression of a whole genome in a single experiment. The first massive approach to transcriptional analyses of the entire symbiotic replicons was based on a high-resolution transcriptional analysis of the symbiotic plasmid of *Rhizobium* sp. NGR234 (Perret et al. 1999) at the Universite de Geneve, which developed methods to study the regulation of bacterial genes during symbiosis. The transcriptome for *S. meliloti* has been examined under a variety of conditions, including in planta (Ampe et al. 2003; Berges et al. 2003). Functional gene arrays or GeoChips are also being utilized for high-throughput analysis of microbial communities involved in nitrogen fixation. Xie et al. (2011) have utilized GeoChip-based analysis to screen out functional genes associated with N-cycle in extreme environment like acid-mine drainage.

3 The Application of Bioinformatics in BNF Research

As we enter into the post-genomics era, the bioinformatics tools have emerged as important means in research of biological nitrogen fixation. Large-scale genome projects have resulted in the availability of tremendous amount of biological data. This data includes information about genomes which in turn gives the idea about proteins, codon usage, etc. With the current deluge of data, computational methods have become indispensable to biological investigations. The development of bioinformatics and statistical genetics has resulted in the production of a number of tools, which are used to annotate the genome and obtain productive information from them (Hogeweg 2011). Originally developed for the analysis of biological sequences, bioinformatics now encompasses a wide range of subject areas including structural biology, genomics, and gene expression studies.

One of the primary applications of bioinformatics is the organization of the biological data in database that allows researchers to access existing information with ease. Open-access databases like GenBank, EMBL, and DDBJ now house thousands of *nif*H and *nif*D sequences. The numbers of fully sequenced and assembled diazotrophic genomes deposited in the databases have also gone up in the last few years. Simultaneously, new databases exclusively devoted to various aspects of biological nitrogen fixation like NodMutDB (Nodulation Mutant Database) (Mao et al. 2005), RhizoGATE (Becker et al. 2009), RhizoBase (http://genome.kazusa.or.jp/rhizobase/), etc., have also surfaced in recent years. EST programs conducted in the model legume *M. truncatula* have led to the development of databases that allow data mining to identify genes relevant for nitrogen-fixing symbioses, for example, the TIGR *M. truncatula* Gene Index (http://www.tigr.org/tdb/mtgi) (Quackenbush et al. 2000), the *M. truncatula* database MtDB2 (http://www.medicago.org), and the database of the Medicago Genome Initiative (Bell et al. 2001). The data present in the various databases can be analyzed and interpreted in a biologically meaningful manner with the aid of computational tools.

Nowadays, the rapid increase in the number of prokaryotic species with sequenced genomes enables the development of in silico searching tools to identify complex biochemical pathways such as nitrogen fixation. Such assumptions, although very accurate, yield putative results and do not obviate the need for genetic and biochemical confirmation of gene function. Computation prediction tools like BLAST (Basic Local Alignment Search Tool) are being used by researchers for examining the occurrence and distribution of nitrogen fixation genes. The genomes present in the database are being scanned using NifHDK as query sequence (O'Carroll and Dos Santos 2011). Phylogenies for the major *nif* operon genes have been inferred by distance matrix-based methods like neighbor-joining or UPGMA or maximum likelihood-based methods in an attempt to understand the timing and complex genetic events that have marked the history of nitrogen fixation (Raymond et al. 2004). Computational tools are also now routinely employed by researchers (Amadou et al. 2008; Carvalho et al. 2010; Peralta et al. 2011; Black et al. 2012) to compare the entire genomes of diazotrophs, which permits the study of more complex evolutionary events, such as gene duplication, horizontal gene transfer, and the prediction of factors important in bacterial speciation. Comparative genomics of *Frankia* yielded vital information regarding their evolutionary history and linked the inconsistency of genome size with the biogeographic history of the host plants harboring the microbial strains (Normand et al. 2007). Systems biology is another area where computer-based simulation has been used extensively to analyze and visualize the complex connections and circuits of cellular pathways such as nitrogen fixation. Zhao and his colleagues (2012) used several in silico tools for the reconstruction of metabolic network involved in symbiotic nitrogen fixation in *S. meliloti* 1021. It provided a knowledge-based framework for better understanding the symbiotic relationship between rhizobia and legumes. The *nif*H gene is the most widely sequenced marker gene used to identify nitrogen-fixing Bacteria and Archaea. Many PCR primers have been developed to target the *nif*H gene with the purpose of amplifying this gene sequence. Various program tools like Primer designer, PrimerSelect, Primer3, etc., are now available which assist in designing these primers and evaluating the primer through e-PCR (Schuler 1997). Recently Gaby and Buckley (2012) made a thorough in silico evaluation of the various *nif*H primers.

Bioinformatics is also indispensable for the examination of the data obtained in proteome analysis. An excellent resource of Internet-accessible proteome databases is the Expert Protein Analysis System (ExPASy), available online at http://www.expasy.ch/ (Gasteiger et al. 2003). Furthermore software packages have been developed that can take multiple protein-expression profiles and automatically identify quantitative changes of interest. Two-dimensional electrophoresis databases are accessible on the Internet and can be browsed with interactive software and integrated with in-house results. A cluster of Orthologous Groups of proteins (COG) is a new database search and represents an attempt at a phylogenetic classification of proteins from complete genomes (http://www.ncbi.nlm.nih.gov/COG) (Tatusov et al. 2000). It is to serve as a platform for functional annotation of newly sequenced genomes and for studies on genome evolution. In addition, the identification of domains as subsets of proteins has been a very promising approach, implemented by databases such as InterPro (http://www.ebi.ac.uk/interpro/). Proteomic analysis has revealed the direct genome functionality in a number of diazotrophic

Table 1 Some bioinformatics tools used for research in biological nitrogen fixation

Usage	Tools	Description
Primer designing	PrimerSelect, Primer Premier, Array Designer, Primer3, GPRIME, PRIDE, e-PCR, etc.	Analyzes template sequence and design primers for PCR along with their evaluation
Codon usage	CodonW, ACUA, GCUA, CodonExplorer, Codon Plot, CAI calculator, JCAT, CHIPS, etc.	Various codon usage parameters like GC, GC3, NC, Fop, CAI, etc., are estimated
Sequence analysis	BLAST, Artemis, ClustalW, Mumer, FASTA3, CnD, Dambe, DnaSP, etc.	Sequence similarity search, sequence alignment, sequence visualization
Comparative genomics	Mauve, VISTA tools, CMG-Biotools, CGAS, etc.	Comparing and visualizing the similarities and dissimilarities of the genomes
Phylogenetic analysis	MEGA, Phylip, PhyML, PAML, Tree view, fastDNAml, etc.	Tracing the phylogenetic relationship of the genes
Protein sequence analysis	Blocks, MAST, VAST, ProScan, Prosite, ProFunc, NRpred, Bio3D etc.	Identification of motif, domains, and functionally relevant residues from amino acid sequence
Protein structure prediction	Modeler, PredictProtein, SWISS-MODEL, YASARA, HHpred	Prediction of 3D models of proteins based on sequence homology
Metagenomics	MEGAN, MetaCV, GLIMMER, GeneMark, CLaMS, FAMeS, etc.	Composition and operation of complex microbial consortia in environmental samples
Proteomics	Calspec, Aldente, amsrpm Census, Mapper, AgBase, OMSSA, ProteinProspector, etc.	Identify proteins from peptide mass fingerprinting data; quantitative proteomics analysis
Gene expression/ microarray	ArrayExpress, DEGseq, Goober, NBIMiner, MetaDE, etc.	Analysis and visualization of expression data; microarray data analysis

genomes (MacLean et al. 2007). Smit and coworkers (2012) have used various proteomics approaches along with bioinformatics tools for proteomic phenotyping of *Novosphingobium nitrogenifigens*, a free-living diazotroph. Rapid developments of technological expertise in proteomics coupled with the improvement of in silico tools have resulted in a deluge of structural information that guarantees acceleration in nitrogen fixation research.

As we march into the new millennium, practical application of computation tools to decipher meaningful information from available data is inevitable. Some of the important in silico tools used for research in various aspects of biological nitrogen fixation is mentioned in Table 1. Bioinformatics has the potential to elevate the research on biological nitrogen-fixing bacteria and its protein machinery to a next level. The availability of bioinformatics tools has provided an opportunity to focus on the comparative genomics, molecular evolution of the genomes along with conformational and structural details of the proteins involved. Structural studies of proteins will provide a better understanding of the functional evolution of diazotrophy.

3.1 Research Trends in Codon Usage Analysis and Comparative Genomics

In the post-genomics era, the application of bioinformatics tools in comparative genomics has led to the belief that every genome has its own story. Particularly the genetic code and its usage preferences are one of the most interesting aspects of biological science. In the early period, majority of work on codon usage patterns focused upon *E. coli* (Peden 1999). Gradually the bioinformatics analysis of codon usage was applied upon mammalian, bacterial, bacteriophage, viral, and mitochondrial genes. Sharp and Li (1987) were the pioneers in developing the Codon Adaptation Index (CAI) to assess the similarity amid the synonymous codon usage of a gene to that of the reference set. Besides CAI, several indices such as GC content, GC3 content, effective number of codons (Nc) (Wright 1990), relative synonymous codon usage (RSCU) (Sharp et al. 1986), Codon Bias Index (CBI), and Fop (frequency of optimal codons) (Ikemura 1985) are very significant in studies concerning codon usage patterns. Very preliminary work on codon usage of nitrogen-fixing diazotrophs was initiated by Mathur and Tuli (1991). Ramseier and Gottfert (1991) reported differences in codon usage and GC content in *Bradyrhizobium* genes. Moderate codon bias was attributed to translational selection in nitrogen-fixing genes of *Bradyrhizobium japonicum* USDA 110 (Sur et al. 2005). The analysis of synonymous codon usage patterns of three *Frankia* genomes (strains CcI3, ACN14a, and EAN1pec) revealed that codon usage was highly biased, but variations were noticed among the three strains(Sen et al. 2008). Using Codon Adaptation Index (CAI), highly expressed genes in *Frankia* were predicted. Synonymous codon usage analysis in *Azotobacter vinelandii* divulged considerable amount of heterogeneity (Sur et al. 2008). About 503 potentially highly expressed genes were identified, and most of them were linked to metabolic functions of which 10 were associated with the core nitrogen-fixing mechanism. Sen et al. (2012) explored the role of rare TTA codon in the genome of diazotrophic actinomycetes *Frankia*.

Other than codon usage, molecular evolution of genes is another aspect which needs to be investigated. A more reliable index of genetic drift over evolutionary time is the ratio of Ka (nonsynonymous substitutions per site) to Ks (synonymous substitutions per site) for a large set of genes, based on the comparisons of related species. The Ka to Ks ratio, which is almost always less than one, is widely used as an indicator of the extent of purifying selection acting to conserve coding sequences. This parameter has been widely applied in the analysis of adaptive molecular evolution and is regarded as a general method of measuring the rate of sequence evolution in biology. Program packages like PAML (Yang 1997) have been extensively used for the estimation of nucleotide substitution rates based on phylogenetic analysis by maximum likelihood (ML). Ka/Ks parameters have been used to assess the molecular evolution of in plant hemoglobin genes (Guldner et al. 2004), secretory protein genes in *Streptomyces* and yeast (Li et al. 2009b), and in various disease-causing genes. Among diazotrophs, Crossman et al. (2008) measured the rates of synonymous (Ks) and nonsynonymous substitutions (Ka) in orthologous genes of *R. etli* and *R. Leguminosarum*. More recently, synonymous and nonsynonymous substitution rates of orthologs shared by five species of Rhizobiales, three plant symbionts, one plant pathogen, and one animal pathogen have been calculated by Peralta et al. (2011). Apart from the whole genome, molecular evolution of the genes responsible for symbiotic association and nodulation such as nodule-specific genes (Yi 2009) and recently *SymRK* (Mahe et al. 2011) has been specifically analyzed. But still a lot of symbiotic genes from wide range of diazotrophs have still to be analyzed to gather a complete scenario of their evolutionary rate in terms of their sequence features. Accumulations of bacterial whole genome sequences also give the biologists more opportunities to explore and compare the genomes in larger scale. Comparative genomics has given rise to a new concept highlighting the great diversity between closely related strains. A species can be described by its pan-genome, i.e., the sum of a core genome containing genes present in all strains, and a dispensable genome, with genes absent from one or more strains and genes unique to each strain (Medini et al. 2005). Studying the diversity within pan-genomes is of interest for the characterization of the species or genus. Low pan-genome diversity could be reflective of a stable environment, while bacterial species with substantial abilities to adapt to various environments would be expected to have high pan-genome diversity (Snipen and Ussery 2010). In 2005, Tettelin and colleagues introduced the conception of "pan-genome" in *Streptococcus agalactiae* (Tettelin et al. 2005). Soon afterward, pan-genome has been widely used to provide insight into the analysis of the evolution of *S. pneumonia* (Hiller et al. 2007), *H. influenza* (Hogg et al. 2007), *E. coli* (Rasko et al. 2008), and so on. Besides evolution, pan-genome has been widely used to detect strain-specific virulence factors for some pathogens, *L. pneumophila* (D'Auria et al. 2010). Recently symbiotic pan-genome of the nitrogen-fixing bacterium *Sinorhizobium meliloti* has been explored using computational methods,

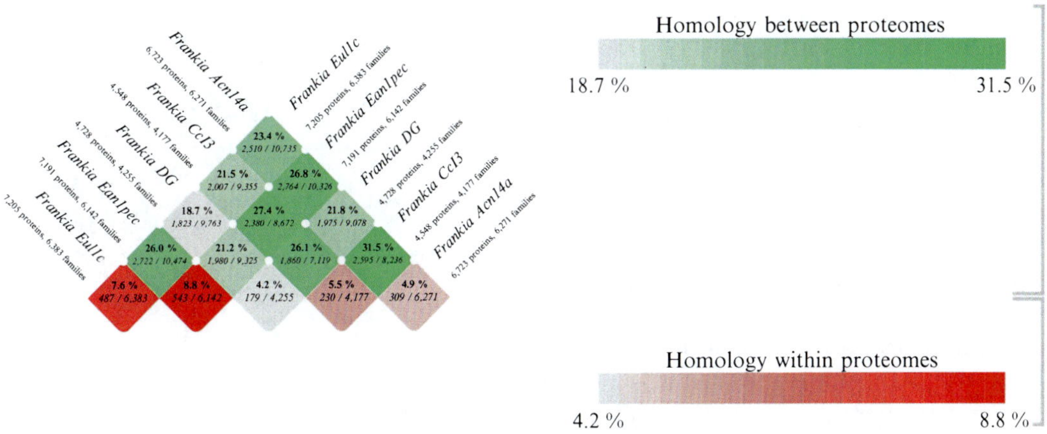

Fig. 1 BLAST matrix for the *Frankia* genomes created by CMG-Biotools platform for comparative genomics. The *darkest green*, indicative of the highest fraction of genes found similar between two genomes

and a set of accessory genetic factors related to the symbiotic process have been defined (Galardini et al. 2011). As complete nucleotide sequences of more chromosome and symbiotic plasmids of nitrogen-fixing organisms become available, we have entered into the phase of comparative genomics. Comparative genomics also enables a much deeper understanding of the origin and evolution of free-living and symbiotic nitrogen fixation. Comparative genomics approach has been utilized by Carvalho et al. (2010) to delineate the evolutionary characterization of diazotrophic and pathogenic bacteria of the order Rhizobiales. Black et al. (2012) have worked upon 14 strains of Rhizobiales to investigate the feasibility of defining a core "symbiome." The authors' group is currently engaged in comparative genomics of nitrogen-fixing actinomycetes, *Frankia*, and members of Rhizobiales using CMG-Biotools – a platform for comparative genomics. The proteomes are compared with BLASTP using the "50/50" rule, i.e., BLASTP hit was considered significant if the alignment produced at least 50 % identity for at least 50 % of the length of the longest gene (either query or subject). The BLAST results are visualized in a BLAST matrix, which summarizes the results of genomic pairwise comparisons. One such BLAST matrix produced for five *Frankia* strains is presented in Fig. 1. The comparison of these whole genomes has revealed valuable information, such as several events of lateral gene transfer, particularly in the symbiotic plasmids and genomic islands that have contributed to a better understanding of the evolution of contrasting symbioses.

3.2 Bioinformatics Approaches for the Characterization of Proteins Related to BNF

Apart from the sequence-based analysis and comparative genomics, the structural biology is one such field which has been hugely benefitted by bioinformatics tools. Structural analyses include protein and nucleic acid structure prediction, comparison, classification, and assessment of structure–function relationship. Often it is seen that structural analysis in turn depends on the results of sequence analysis. For example, protein structure prediction depends on sequence alignment data. Thus the two aspects of bioinformatics analysis are not isolated but often interact to produce integrated results.

Developments in the field of proteomics have resulted in availability of large amount of biological data in the public domain. This data includes amino acid sequences of nitrogenase proteins from a wide range of microbes. However, very little is known about the structure and role of all these proteins. Two technologies, X-ray and NMR, are by far the two most common means used to determine protein structure experimentally. In 1992, Kim and Rees (1992) provided a detailed crystallographic structure of molybdenum–iron protein of the *Azotobacter vinelandii* nitrogenase. The crystal structure of nitrogenase molybdenum–iron protein has also been described from *Clostridium pasteurianum* (Kim et al. 1993). The X-ray crystal structure of *Klebsiella pneumoniae* nitrogenase component 1 (Kp1) has also been determined and refined to a resolution of 1.6 Å (Mayer et al. 1999). The 2.9 Å crystal structure of the NifH protein from *Azotobacter vinelandii* was obtained by Georgiadis et al. (1992). However, tertiary structures of large number of nitrogenase proteins from different diazotrophs particularly those of symbiotic ones have not yet been resolved. The exact mechanism of working of these proteins is also relatively unknown due to the difficulty in obtaining crystals of nitrogen bound to nitrogenase. This is because the resting state of MoFe protein does not bind nitrogen. Moreover, in the recent years, quite a number of discrepancies have also crept out regarding the protein structures resolved by X-ray crystallography leading to retraction of papers (Chang et al. 2006). In this regard, a viable alternative approach is to predict 3D structure of proteins based on homology modeling technique and validate it properly. Homology modeling is a reliable technique that can consistently predict the 3D structure of a protein with precision akin to one obtained at low resolution by experimental means (Marti-Renom et al. 2000). This technique depends upon the alignment of a protein sequence of unknown structure (target) with that of a homologue of known structure (template). This technique is particularly quite important in organisms with slow growth rate which poses difficulties in the purification of subsequent proteins. Browne et al. (1969) published the first report on homology modeling. A model of α-lactalbumin was constructed by taking the coordinates of a hen's egg-white lysozyme and modifying, by hand, those amino acids that did not match the structure. Since the mid-1980s, a large number of homology models of proteins with different folds and functions have been reported in the literature (Johnson et al. 1994; Sali 1995). Homology modeling approaches were first applied for structural analysis of nitrogenase iron protein from *Trichodesmium* sp., a marine filamentous nitrogen-fixing cyanobacteria (Zehr et al. 1997a). Standard homology modeling approaches have also been used to generate reliable models of the nitrogenase Fe protein from thermophilic *Methanobacter thermoautotrophicus* based on the structure of the *Azotobacter vinelandii* nitrogenase Fe protein (Sen and Peters 2006). The authors' group has been involved in the determination of 3D-structure NifH protein from various diazotrophs like *Frankia* (Sen et al. 2010) and *Bradyrhizobium* ORS 278 (Thakur et al. 2012) using homology modeling technique. The model of NifH of *Frankia* (Fig. 2) was based on the template protein which was a nitrogenase iron protein from *Azotobacter vinelandii*. The structure is reliable offering insights into the 3D structural framework as well as structure–function relationship of NifH protein. The models based on homology are quite useful in providing conformational properties and structure–function relationship of these proteins.

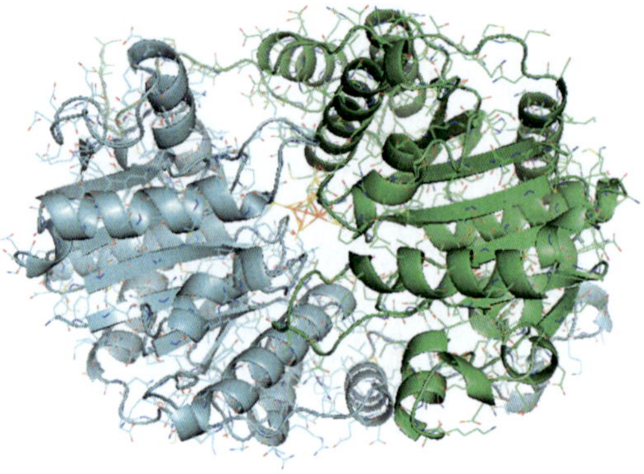

Fig. 2 Three-dimensional model of NifH protein from *Frankia* sp. CcI3 created by homology modeling technique (Sen et al. 2010)

A number of aspects of nitrogenase, particularly structure–function relationships, are interesting areas of fundamental research. The three-dimensional structure of protein like that of nitrogenase is often considered an ideal model system for the study of the complex metal cluster-mediated catalysis, electron transfer, complex metal cluster assembly, protein–protein interactions, and nucleotide-dependent signal transduction. Molecular dynamics simulations offer details about molecular motions as a function of time and are widely used to study protein motions at the atomic level. First protein simulation for 9.2 ps was carried out by McCammon et al. (1977) for bovine pancreatic trypsin inhibitor (BPTI) (McCammon et al. 1977). Case and Karplus work on dynamics of ligand binding to heme protein in 1979 is arguably the first simulation of ligand moving through the protein (Case and Karplus 1979). First application of normal modes to identify low frequency oscillations using the energy minimization of the molecular mechanics force field of protein was described by Brooks and Karplus (1983). This is the basic technique to identify domain-level motions in a protein. First simulation of a protein in explicit waters was done by Levitt and Sharon (1988).

Metalloproteins like nitrogenase are a vast class of biological molecules, which are responsible for many vital functions. Despite the intrinsic difficulties of these systems particularly those related to parameterization of the metal cofactors, they have been the object of several MD simulations. These studies are mainly focused on structural aspects, since the cluster either has a storage role or is involved in an electron-transfer process in these proteins. Among the metalloprotein having FeS cluster cofactor, molecular dynamics simulation has been carried for protein like heme-containing cytochrome P450 (Kuhn et al. 2001), Rubredoxins (Grottesi et al. 2002), 3Fe–4S cluster-containing protein, ferredoxin I (Meuwly and Karplus 2004), adenosine phosphosulfate reductase (dos Santos et al. 2009), and hydA1 hydrogenase (Sundaram et al. 2010). More recently, molecular modeling, dynamics, and docking studies on both *A. vinelandii* and *G. diazotrophicus* FeSII proteins and nitrogenases were carried out by Lery et al. (2010), elucidating molecular aspects of protein–protein interaction. In the MD simulation of metalloproteins, the force field parameters of the metal ion and its ligands need to be defined beforehand taking into account the nature of the metal ion, its coordination number, geometry, oxidation, and spin states and the nature of its ligands. Several sets of parameters have been reported in the literature for the active sites of the most widely studied metalloproteins including the coordination geometries of the metal ligand (Banci and Comba 1997; Norrby and Brandt 2001; Comba and Remenyi 2002). One of the parameters that significantly affect the overall protein structure is the partial charges of the atoms of the metal–ligand moiety. In the bonded model, partial charges are commonly calculated through the RESP (Restrained Electrostatic Surface Potential) methodology (Fox and Kollman 1998) applied to semiempirical or ab initio calculations. The ab initio calculations are mostly performed through density functional theory (DFT) calculations, with the B3LYP functional or Hartree–Fock calculations (Banci 2003). Thus, the development of proper parameters of the metal cofactors needs the amalgamation of quantum calculations in conjunction with classical molecular mechanics calculations. This will enable the description of not only structural features but also of reactivity properties of the metalloproteins.

3.3 Tracing the Evolution of BNF Through Bioinformatics

3.3.1 Classical Approach

Researchers have long sought to answer the question of when nitrogen fixation began and what evolutionary pressures affected it (Postgate and Eady 1988; Berman-Frank et al. 2003). The emergence and evolution of nitrogen fixation ability (diazotrophy) among prokaryotes is complex and has not yet been fully elucidated. The incomplete distribution pattern of this highly conserved enzyme among Bacteria and Archaea has led to the development of conflicting hypotheses on BNF. The first idea theorizes that nitrogen fixation is an ancient function of the last common ancestor of Bacteria–Archaea that was vertically transmitted, but has undergone widespread gene loss among descendants with horizontal transfer in some isolated instances

(Hennecke et al. 1985; Normand and Bousquet 1989; Fani et al. 2000; Berman-Frank et al. 2003). During this postulated time period, reduced nitrogen may have been very abundant and the initial function of nitrogenase was probably very different. One proposed initial function of ancient nitrogenase might be associated with detoxification mechanism for cyanides and other chemicals (Silver and Postgate 1973; Fani et al. 2000). This idea is based on the observation that nitrogenase reduces a number of alternative substrates in addition to N_2, several of which are toxins (e.g., cyanides). The second hypothesis proposes that nitrogen fixation was an anaerobic ability that appeared after the emergence of oxygenic photosynthesis and was subsequently lost in most lineages through horizontal transfer (Postgate 1982; Postgate and Eady 1988). Recently, Hartmann and Barnum (2010) examined Mo-nitrogenase phylogeny and proclaimed a conclusion combining both theories on diazotrophic evolution.

Nitrogenase genes are highly conserved at both the chemical and genetic levels across wide phylogenetic ranges and among closely related organisms. The conservation of nitrogenase genes lends itself for use as a genetic marker for phylogenetic analysis to help answer questions of the evolution of nitrogen fixation and its genes. Raymond et al. (2004) reported that nitrogenase evolved in multiple lineages, and there are evidences of loss, duplications, and horizontal and vertical transfers for the nitrogenase genes and operons during the course of evolution. *nif*D and *nif*K are thought to be the result of an in-tandem gene duplication (Fani et al. 2000; Postgate and Eady 1988), giving the functional components of the enzyme. A second duplication event is thought to have occurred for the *nif*EN genes. Till date most of the studies concerning the evolution of nitrogen fixation have focused on the *nif* genes, primarily the highly conserved *nif*H gene but also the larger but less conserved *nif*D, *nif*K, *nif*E, and *nif*N genes (Normand and Bousquet 1989; Normand et al. 1992; Hirsch et al. 1995; Fani et al. 2000). Sequence alignment-based methods are widely used to study the evolution of relevant *nif* genes. Young (2005) discussed the phylogeny and evolution of nitrogenases in details. According to Young, true NifH proteins can be divided into three types – Type B ("bacterial") is the best represented and includes enzymes from the proteobacteria, cyanobacteria, and firmicutes; Type C ("clostridial") is found in the firmicute bacterium and *Clostridium*, the green sulfur bacterium *Chlorobium*, and also in the archaeon *Methanosarcina*; and Type A is associated with the "alternative" nitrogenases that do not contain molybdenum and is found in both archaea and proteobacteria. There are also a large number of more distant relatives, notable among them light-independent protochlorophyllide (Pchlide). The similarity between these proteins and NifH was analyzed and discussed by Burke et al. (1993), who argued that nitrogen fixation probably originated before photosynthesis, so the photosynthesis enzymes would have been derived from NifH rather than the other way round. The phylogenies of NifDKEN family have also been topic of many research works. Dedysh et al. (2004) utilized the NifD phylogeny to assess the nitrogen fixation capabilities of methanotrophic bacteria. Henson et al. (2004b) reexamined the phylogeny of nitrogen fixation by analyzing only the molybdenum-containing *nif* D gene from a cyanobacteria, proteobacteria, as well as Gram-positive bacteria. The strict requirement of NifH in biological nitrogen fixation and its universal presence in diazotrophs has resulted in this protein serving as a sequence tag or barcode for the identification of nitrogen fixers. Genomic analysis using the sequence of NifH as a query results in BLAST hits that include NifH, VnfH, and AnfH components of the Mo-, V-, and Fe-only nitrogenases, respectively (Raymond et al. 2004). Recently, Dos Santos and his colleagues proposed a new criterion for computational prediction of nitrogen fixation: the presence of a minimum set of six genes coding for structural and biosynthetic components, namely, NifHDK and NifENB (Dos Santos et al. 2012). Latysheva et al. (2012) considered the various *nif* orthologs for performing empirical Bayesian ancestral state reconstructions to investigate the evolution of nitrogen fixation in cyanobacteria.

Over the years, there has been a debate among the workers regarding horizontal gene transfer (HGT) versus vertical descent as the dominant force in the evolution and distribution of N fixation. In the case of an early origin and subsequent vertical descent of the *nif* genes, a comparison of SSU ribosomal phylogeny and the phylogeny of *nif* genes should reveal roughly the same features, assuming that the mutation rates in both genes were similar. In the case of a late development and a mainly horizontal distribution of the genes, the phylogeny of the *nif* genes should deviate significantly from the rRNA-based standard tree. A number of researchers have presented strong evidence that SSU rRNA phylogeny and phylogeny based on the *nif* genes are in general agreement, suggesting that they have evolved in a similar fashion (Hennecke et al. 1985; Young 1992; Zehr et al. 1997b). However, numerous studies have highlighted instances of possible horizontal gene transfer in *nif*D (Parker et al. 2002; Qian et al. 2003; Henson et al. 2004a, b), *nif*H (Normand and Bousquet 1989; Hurek et al. 1997; Cantera et al. 2004; Dedysh et al. 2004), and *nif*K (Kessler et al. 1997) based on incongruence with 16S rRNA trees. Other studies have found support for both vertical descent and horizontal transfer (Hirsch et al. 1995). Haukka et al. (1998) proposed that horizontal gene transfer may have played an increasing role at genus and lower taxonomic levels. This may be especially important in organisms that have *nif* genes located on plasmids (Normand and Bousquet 1989).

3.3.2 Alternative Approaches

For tracing evolution of proteins within a set of divergently evolved proteins, it is useful to construct the phylogenetic trees based on the similarities in the amino acid sequences and the base sequences of the genes. But previous studies seem to suggest that the origin and extant distribution of nitrogen fixation is perplexing from a phylogenetic perspective, largely because of factors that confound molecular phylogeny such as sequence divergence, paralogy, and horizontal gene transfer (Raymond et al. 2004). This leads to the assumption that sequence-based phylogeny is not enough to reveal the complex evolutionary path in BNF. Moreover, many workers (Nadler 1995; Qi et al. 2004; Sims et al. 2009) have also pointed out fallacies in sequence alignment-based methods. Therefore substitute phylogenetic approaches are being sought. Alignment-free condensed matrix method relying on nucleotide triplet is one such alternative approach. The condensed matrix method of studying molecular phylogeny takes into account a set of invariants in a DNA sequence and determines the extent of resemblance among DNA sequences using the invariants (Randic et al. 2001). In the condensed matrix method, all the possible triplets of the *nif* genes were calculated and matrices were formed by using all the possible triplet. Then leading eigenvalues of these matrices were calculated. The eigenvalues were later used for the construction of distance matrices and consequently for tree construction. This approach has been utilized by researchers in phylogenetic analysis of aminoacyl t-RNA synthetase (Mondal et al. 2008), swine flu genomes (Sur et al. 2010), bacterial zeta toxin (Mondal et al. 2011), and nitrogenase proteins (Sur et al. 2010). A cladogram showing the evolution of *nif*H gene in various diazotrophs constructed by the condensed matrix method is presented in Fig. 3. In the phylogram, the placement of *Frankia* ACN14a away from the other actinobacteria and *Synechococcus* sp. JA-3-3Ab being isolated from rest of cyanobacterial strains is apparently quite interesting. Members of various classes of Proteobacteria (alpha, beta, gamma, and delta) are clustered together in the triplet-based phylogenetic tree. Mottled distribution of cyanobacteria is an indication of their polyphyletic origin. Thus, condensed matrix method-based phylogeny is apparently a suitable method for explaining the complex events marking the nitrogen fixation evolution.

Another suitable alternative of protein sequence alignments is the structure-based phylogeny. It is well known that the 3D structures and structural features of homologous proteins are conserved better than their amino acid sequences (Chothia et al. 1986; Hubbard and Blundell 1987). It has been demonstrated

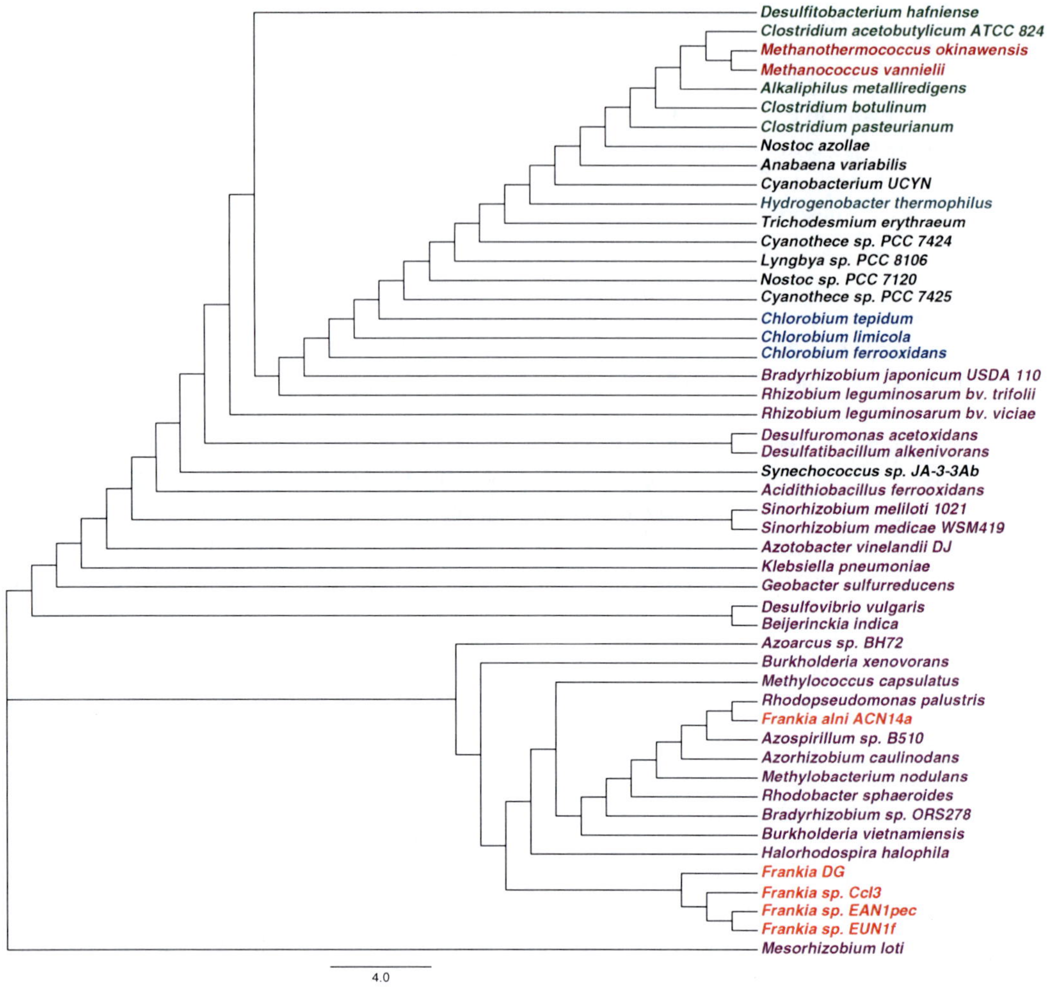

Fig. 3 Phylogenetic tree of nifH gene based on condensed matrix method developed by the author's group. *Colored fonts* are used to indicate different classes of diazotrophs. *Purple* is used for proteobacterial strains; *black* is used for cyanobacterial strains; *blue* for *green* sulfur; *orange* for Actinobacteria; *green* for Firmicutes; *red* for methanogenic; and *gray* for Aquificae

several times that the homologous proteins could diverge beyond recognition at the level of their amino acid sequences but maintain similar structure and function. In several cases of low sequence similarity, proteins retain the fold as well as retain the broad biochemical features and/or functional properties, suggesting an evolutionary connection (Murzin et al. 1995; Russell and Sternberg 1996). Previous studies (Balaji and Srinivasan 2001) have shown that in cases of poor sequence identity, structure-based phylogenies generate better models of evolution of proteins than the traditional sequence-based methods. Hence, it is more appropriate to use similarities in 3D structure of proteins in modeling evolution of distantly related proteins. The construction of phylogenetic trees using 3D structures has been applied for a variety of protein families like short-chain alcohol dehydrogenases (Breitling et al. 2001) and metallo-β-lactamases (Garau et al. 2005). Lately 3D structure-based phylogenetic approach has been utilized for functional characterization of proteins with cupin folds (Agarwal et al. 2009). It was revealed that structure-based clustering of members of cupin superfamily reflects a function-based clustering. Moreover, the comparison of distance matrices utilized in phylogenetic tree construction methods has been considered as an equivalent of comparison of phylogenetic trees based on protein structures (Balaji and Srinivasan 2001; Pazos and Valencia 2001). Therefore, such structure-based approaches can be utilized to assess the phylogenetic relationships of proteins involved in BNF which shares low sequence similarity but high structural resemblance with many proteins with diverse biological functions.

Along with the trajectory of evolution of diazotrophy in various organisms, another feature that needs attention is the functional divergence of the proteins involved in this biological process. Previous workers (Gu 1999; Dermitzakis and Clark 2001; Raes and Van de Peer 2003) have shown that gene duplication events often lead to a shift in protein function from an ancestral role resulting in divergence and as a consequence of which some residues are subjected to altered functional constraints. This implies that evolutionary rates at these sites will vary in different homologous genes of a gene family. Site-specific altered functional constraint (or shifted evolutionary rates) can be detected by comparing the rate correlation between gene clusters, when the phylogeny is given (Gu 1999). This approach has been earlier exploited by researchers to trace the functional divergence in vertebrate hemoglobin (Gribaldo et al. 2003), G-protein alpha subunits (Zheng et al. 2007), OPR gene family in plants (Li et al. 2009a), and anoctamin family of membrane proteins (Milenkovic et al. 2010). However, a broad picture on the functional divergence in the NifH/Bchl protein family is still unavailable.

4 Challenges and Future Prospects

Considerable progress has been made in understanding the machinery of biological nitrogen fixation in the last few decades. The major part of the research has been focused on the structure of nitrogenase, elucidation of the compositions, and functions of all of the *nif*-gene products. In the past, major roadblocks in the BNF research were the struggle associated with the detection of *nif* genes from environmental samples and subsequent crystallization of the nitrogenase enzymes. In the post-genomics era, these hurdles have largely been removed with the advent of metagenomic research and in silico protein modeling techniques. The challenge now is to put all the known information together and, with the combined application of biochemical, genetics, and bioinformatics techniques, to determine how nitrogenase functions at the molecular level. With the rapid increase in the number of complete genomes of varied diazotrophs along with their nitrogen-fixing genes in the public domain, bioinformatics tools have emerged as a potent weapon to tackle the unsolved mysteries of symbiotic and asymbiotic nitrogen fixation. It can be used to extract meaningful interpretation of sequence data. With the advent of new algorithms and computational tools for measuring structural divergence, the problems associated with functional evolution of nitrogenase system can also be tackled in a better way and new glimpses can be gained. Genomic studies aided by the bioinformatics tools offer a global view of the expression, regulation, dynamics, and evolution of the genomes from nitrogen-fixing microbes and have the capability in offering new opportunities to preserve and improve biotic resources.

Acknowledgments AS is grateful to the DBT, Government of India, for providing CREST Award and financial help in setting up Bioinformatics Centre, in the Department of Botany, University of North Bengal. ST acknowledges CSIR for CSIR-SRF Fellowship.

References

Adams D (2000) Heterocyst formation in cyanobacteria. Curr Opin Microbiol 3:618–624

Agarwal G, Rajavel M, Gopal B, Srinivasan N (2009) Structure-based phylogeny as a diagnostic for functional characterization of proteins with a cupin fold. PLoS One 4:e5736

Amadou C et al (2008) Genome sequence of the beta-rhizobium Cupriavidus taiwanensis and comparative genomics of rhizobia. Genome Res 18:1472–1483

Ampe F, Kiss E, Sabourdy F, Batut J (2003) Transcriptome analysis of Sinorhizobium meliloti during symbiosis. Genome Biol 4:R15

Anthamatten D, Hennecke H (1991) The regulatory status of the fixL- and fixJ-like genes in Bradyrhizobium japonicum may be different from that in Rhizobium meliloti. Mol Gen Genet 225:38–48

Arnold W, Rump A, Klipp W, Priefer UB, Puhler A (1988) Nucleotide sequence of a 24,206-base-pair DNA fragment carrying the entire nitrogen fixation gene cluster of Klebsiella pneumoniae. J Mol Biol 203:715–738

Balaji S, Srinivasan N (2001) Use of a database of structural alignments and phylogenetic trees in investigating the relationship between sequence and structural variability among homologous proteins. Protein Eng 14:219–226

Baldani JI, Baldani VL (2005) History on the biological nitrogen fixation research in graminaceous plants: special emphasis on the Brazilian experience. An Acad Bras Cienc 77:549–579

Banci L (2003) Molecular dynamics simulations of metalloproteins. Curr Opin Chem Biol 7:143–149

Banci L, Comba P (1997) Molecular modeling and dynamics of bioinorganic systems. Kluwer Academic, Dordrecht

Batut J, Santero E, Kustu S (1991) In vitro activity of the nitrogen fixation regulatory protein FIXJ from Rhizobium meliloti. J Bacteriol 173:5914–5917

Becker A et al (2009) A portal for rhizobial genomes: RhizoGATE integrates a Sinorhizobium meliloti genome annotation update with postgenome data. J Biotechnol 140:45–50

Bell CJ et al (2001) The Medicago Genome Initiative: a model legume database. Nucleic Acids Res 29:114–117

Benson DR, Silvester WB (1993) Biology of Frankia strains, actinomycete symbionts of actinorhizal plants. Microbiol Rev 57:293–319

Berges H et al (2003) Development of Sinorhizobium meliloti pilot macroarrays for transcriptome analysis. Appl Environ Microbiol 69:1214–1219

Berman-Frank I, Lundgren P, Falkowski P (2003) Nitrogen fixation and photosynthetic oxygen evolution in cyanobacteria. Res Microbiol 154:157–164

Berry AM, Harriott OT, Moreau RA, Osman SF, Benson DR, Jones AD (1993) Hopanoid lipids compose the Frankia vesicle envelope, presumptive barrier of oxygen diffusion to nitrogenase. Proc Natl Acad Sci U S A 90:6091–6094

Black M et al (2012) The genetics of symbiotic nitrogen fixation: comparative genomics of 14 rhizobia strains by resolution of protein clusters. Genes 3:138–166

Bohlool BB, Ladha JK, Garrity DP, George T (1992) Biological nitrogen fixation for sustainable agriculture: a perspective. Plant and Soil 141:1–11

Breitling R, Laubner D, Adamski J (2001) Structure-based phylogenetic analysis of short-chain alcohol dehydrogenases and reclassification of the 17beta-hydroxysteroid dehydrogenase family. Mol Biol Evol 18:2154–2161

Brigle KE, Weiss MC, Newton WE, Dean DR (1987) Products of the iron-molybdenum cofactor-specific biosynthetic genes, *nif*E and *nif*N, are structurally homologous to the products of the nitrogenase molybdenum-iron protein genes, nifD and *nif*K. J Bacteriol 169:1547–1553

Brooks B, Karplus M (1983) Harmonic dynamics of proteins: normal modes and fluctuations in bovine pancreatic trypsin inhibitor. Proc Natl Acad Sci U S A 80:6571–6575

Browne WJ, North AC, Phillips DC, Brew K, Vanaman TC, Hill RL (1969) A possible three-dimensional structure of bovine alpha-lactalbumin based on that of hen's egg-white lysozyme. J Mol Biol 42:65–86

Burgess BK, Lowe DJ (1996) Mechanism of molybdenum nitrogenase. Chem Rev 96:2983–3012

Burke DH, Hearst JE, Sidow A (1993) Early evolution of photosynthesis: clues from nitrogenase and chlorophyll iron proteins. Proc Natl Acad Sci U S A 90:7134–7138

Burns RC, Hardy RW (1975) Nitrogen fixation in bacteria and higher plants. Mol Biol Biochem Biophys 21:1–189

Cantera JJ, Kawasaki H, Seki T (2004) The nitrogen-fixing gene (*nif*H) of Rhodopseudomonas palustris: a case of lateral gene transfer? Microbiology 150:2237–2246

Carvalho FM, Souza RC, Barcellos FG, Hungria M, Vasconcelos AT (2010) Genomic and evolutionary comparisons of diazotrophic and pathogenic bacteria of the order Rhizobiales. BMC Microbiol 10:37

Case DA, Karplus M (1979) Dynamics of ligand binding to heme proteins. J Mol Biol 132:343–368

Ceremonie H, Cournoyer B, Maillet F, Normand P, Fernandez MP (1998) Genetic complementation of rhizobial *nod* mutants with Frankia DNA: artifact or reality? Mol Gen Genet (MGG) 260:115–119

Chang G, Roth CB, Reyes CL, Pornillos O, Chen YJ, Chen AP (2006) Retraction. Science 314:1875

Chothia C et al (1986) The predicted structure of immunoglobulin D1.3 and its comparison with the crystal structure. Science 233:755–758

Comba P, Remenyi R (2002) A new molecular mechanics force field for the oxidized form of blue copper proteins. J Comput Chem 23:697–705

Crossman LC et al (2008) A common genomic framework for a diverse assembly of plasmids in the symbiotic nitrogen fixing bacteria. PLoS One 3:e2567

D'Auria G, Jimenez-Hernandez N, Peris-Bondia F, Moya A, Latorre A (2010) Legionella pneumophila pangenome reveals strain-specific virulence factors. BMC Genomics 11:181

Dawson JO (1983) Dinitrogen fixation in forest ecosystems. Can J Microbiol/(Revue Canadienne de Microbiologie) 29:979–992

Dedysh SN, Ricke P, Liesack W (2004) NifH and NifD phylogenies: an evolutionary basis for understanding nitrogen fixation capabilities of methanotrophic bacteria. Microbiology 150:1301–1313

Delgado MJ, Bedmar EJ, Downie JA (1998) Genes involved in the formation and assembly of rhizobial

cytochromes and their role in symbiotic nitrogen fixation. Adv Microb Physiol 40:191–231
Dermitzakis ET, Clark AG (2001) Differential selection after duplication in mammalian developmental genes. Mol Biol Evol 18:557–562
Dixon R et al (1980) Analysis of regulation of Klebsiella pneumoniae nitrogen fixation (*nif*) gene cluster with gene fusions. Nature 286:128–132
Dobereiner J (1976) Plant genotype effects on nitrogen fixation in grasses. Basic Life Sci 8:325–334
dos Santos ES, Gritta DS, Taft CA, Almeida PF, Ramos-de-Souza E (2009) Molecular dynamics simulation of the adenylylsulphate reductase from hyperthermophilic Archaeoglobus fulgidus. Mol Simulat 36:199–203
Dos Santos PC, Fang Z, Mason SW, Setubal JC, Dixon R (2012) Distribution of nitrogen fixation and nitrogenase-like sequences amongst microbial genomes. BMC Genomics 13:162
Eady RR (1996) Structure-function relationships of alternative nitrogenases. Chem Rev 96:3013–3030
Eady RR, Postgate JR (1974) Nitrogenase. Nature 249:805–810
Elmerich C (1984) Molecular biology and ecology of diazotrophs associated with non-leguminous plants. Nat Biotechnol 2:967–978
Fani R, Gallo R, Lio P (2000) Molecular evolution of nitrogen fixation: the evolutionary history of the *nif*D, *nif*K, *nif*E, and *nif*N genes. J Mol Evol 51:1–11
Fischer HM (1994) Genetic regulation of nitrogen fixation in rhizobia. Microbiol Rev 58:352–386
Fox T, Kollman PA (1998) Application of the RESP methodology in the parametrization of organic solvents. J Phys Chem B 102:8070–8079
Franche C, Lindstrom K, Elmerich C (2009) Nitrogen-fixing bacteria associated with leguminous and non-leguminous plants. Plant and Soil 321:35–59
Gaby JC, Buckley DH (2012) A comprehensive evaluation of PCR primers to amplify the *nif*H gene of nitrogenase. PLoS One 7:e42149
Galardini M et al (2011) Exploring the symbiotic pangenome of the nitrogen-fixing bacterium Sinorhizobium meliloti. BMC Genomics 12:235
Garau G, Di Guilmi AM, Hall BG (2005) Structure-based phylogeny of the metallo-beta-lactamases. Antimicrob Agents Chemother 49:2778–2784
Gasteiger E, Gattiker A, Hoogland C, Ivanyi I, Appel RD, Bairoch A (2003) ExPASy: the proteomics server for in-depth protein knowledge and analysis. Nucleic Acids Res 31:3784–3788
Georgiadis MM, Komiya H, Chakrabarti P, Woo D, Kornuc JJ, Rees DC (1992) Crystallographic structure of the nitrogenase iron protein from Azotobacter vinelandii. Science 257:1653–1659
Geurts R, Bisseling T (2002) Rhizobium Nod factor perception and signalling. Plant Cell Online 14:S239–S249
Gonzalez V et al (2006) The partitioned Rhizobium etli genome: genetic and metabolic redundancy in seven interacting replicons. Proc Natl Acad Sci U S A 103:3834–3839
Gribaldo S, Casane D, Lopez P, Philippe H (2003) Functional divergence prediction from evolutionary analysis: a case study of vertebrate hemoglobin. Mol Biol Evol 20:1754–1759
Grottesi A, Ceruso MA, Colosimo A, Di Nola A (2002) Molecular dynamics study of a hyperthermophilic and a mesophilic rubredoxin. Proteins 46:287–294
Gtari M, Ghodhbane-Gtari F, Nouioui I, Beauchemin N, Tisa LS (2012) Phylogenetic perspectives of nitrogen-fixing actinobacteria. Arch Microbiol 194:3–11
Gu X (1999) Statistical methods for testing functional divergence after gene duplication. Mol Biol Evol 16:1664–1674
Guldner E, Desmarais E, Galtier N, Godelle B (2004) Molecular evolution of plant haemoglobin: two haemoglobin genes in nymphaeaceae Euryale ferox. J Evol Biol 17:48–54
Hartmann LS, Barnum SR (2010) Inferring the evolutionary history of Mo-dependent nitrogen fixation from phylogenetic studies of *nif*K and *nif*DK. J Mol Evol 71:70–85
Haukka K, Lindstrom K, Young JP (1998) Three phylogenetic groups of *nod*A and *nif*H genes in Sinorhizobium and Mesorhizobium isolates from leguminous trees growing in Africa and Latin America. Appl Environ Microbiol 64:419–426
Hennecke H, Kaluza K, Thöny B, Fuhrmann M, Ludwig W, Stackebrandt E (1985) Concurrent evolution of nitrogenase genes and 16S rRNA in Rhizobium species and other nitrogen fixing bacteria. Arch Microbiol 142:342–348
Henson BJ, Hesselbrock SM, Watson LE, Barnum SR (2004a) Molecular phylogeny of the heterocystous cyanobacteria (subsections IV and V) based on *nif*D. Int J Syst Evol Microbiol 54:493–497
Henson BJ, Watson LE, Barnum SR (2004b) The evolutionary history of nitrogen fixation, as assessed by NifD. J Mol Evol 58:390–399
Hiller NL et al (2007) Comparative genomic analyses of seventeen Streptococcus pneumoniae strains: insights into the pneumococcal supragenome. J Bacteriol 189:8186–8195
Hiltner L (1898) Über Entstehung und physiologische Bedeutung der Wurzelknöllchen. Forst Naturwiss Z 7:415–423
Hirsch AM, McKhann HI, Reddy A, Liao J, Fang Y, Marshall CR (1995) Assessing horizontal transfer of *nif*HDK genes in eubacteria: nucleotide sequence of *nif*K from Frankia strain HFPCcI3. Mol Biol Evol 12:16–27
Hoffman BM, Dean DR, Seefeldt LC (2009) Climbing nitrogenase: toward a mechanism of enzymatic nitrogen fixation. Acc Chem Res 42:609–619
Hogeweg P (2011) The roots of bioinformatics in theoretical biology. PLoS Comput Biol 7:e1002021

Hogg JS et al (2007) Characterization and modeling of the Haemophilus influenzae core and supragenomes based on the complete genomic sequences of Rd and 12 clinical nontypeable strains. Genome Biol 8:R103

Hubbard TJ, Blundell TL (1987) Comparison of solvent-inaccessible cores of homologous proteins: definitions useful for protein modelling. Protein Eng 1:159–171

Hurek T, Egener T, Reinhold-Hurek B (1997) Divergence in nitrogenases of Azoarcus spp., Proteobacteria of the beta subclass. J Bacteriol 179:4172–4178

Huss-Danell K (1997) Actinorhizal symbioses and their N2 fixation. New Phytol 136:375–405

Ikemura T (1985) Codon usage and tRNA content in unicellular and multicellular organisms. Mol Biol Evol 2:13–34

Imperial J, Ugalde RA, Shah VK, Brill WJ (1984) Role of the *nif*Q gene product in the incorporation of molybdenum into nitrogenase in Klebsiella pneumoniae. J Bacteriol 158:187–194

Irisarri P, Gonnet S, Monza J (2001) Cyanobacteria in Uruguayan rice fields: diversity, nitrogen fixing ability and tolerance to herbicides and combined nitrogen. J Biotechnol 91:95–103

Johnson MS, Srinivasan N, Sowdhamini R, Blundell TL (1994) Knowledge-based protein modeling. Crit Rev Biochem Mol Biol 29:1–68

Kaneko T et al (2000) Complete genome structure of the nitrogen-fixing symbiotic bacterium Mesorhizobium loti (supplement). DNA Res 7:381–406

Kessler PS, McLarnan J, Leigh JA (1997) Nitrogenase phylogeny and the molybdenum dependence of nitrogen fixation in Methanococcus maripaludis. J Bacteriol 179:541–543

Kim J, Rees DC (1992) Structural models for the metal centers in the nitrogenase molybdenum-iron protein. Science 257:1677–1682

Kim J, Woo D, Rees DC (1993) X-ray crystal structure of the nitrogenase molybdenum-iron protein from Clostridium pasteurianum at 3.0-A resolution. Biochemistry 32:7104–7115

Klipp W (2004) Genetics and regulation of nitrogen fixation in free-living bacteria. Kluwer Academic, Dordrecht/Boston

Kuhn B, Jacobsen W, Christians U, Benet LZ, Kollman PA (2001) Metabolism of sirolimus and its derivative everolimus by cytochrome P450 3A4: insights from docking, molecular dynamics, and quantum chemical calculations. J Med Chem 44:2027–2034

Latysheva N, Junker VL, Palmer WJ, Codd GA, Barker D (2012) The evolution of nitrogen fixation in cyanobacteria. Bioinformatics 28(5):603–606

Leigh GJ (2002) Nitrogen fixation at the millennium. Elsevier, Amsterdam/London

Lery LM, Bitar M, Costa MG, Rossle SC, Bisch PM (2010) Unraveling the molecular mechanisms of nitrogenase conformational protection against oxygen in diazotrophic bacteria. BMC Genomics 11(Suppl 5):S7

Levitt M, Sharon R (1988) Accurate simulation of protein dynamics in solution. Proc Natl Acad Sci U S A 85:7557–7561

Li W, Liu B, Yu L, Feng D, Wang H, Wang J (2009a) Phylogenetic analysis, structural evolution and functional divergence of the 12-oxo-phytodienoate acid reductase gene family in plants. BMC Evol Biol 9:90

Li YD et al (2009b) The rapid evolution of signal peptides is mainly caused by relaxed selection on non-synonymous and synonymous sites. Gene 436:8–11

Long SR (1989) Rhizobium-legume nodulation: life together in the underground. Cell 56:203–214

Long SR (2001) Genes and signals in the Rhizobium-legume symbiosis. Plant Physiol 125:69–72

MacLean AM, Finan TM, Sadowsky MJ (2007) Genomes of the symbiotic nitrogen-fixing of Bacteria legumes. Plant Physiol 144:615–622

Mahe F, Markova D, Pasquet R, Misset MT, Ainouche A (2011) Isolation, phylogeny and evolution of the SymRK gene in the legume genus Lupinus L. Mol Phylogenet Evol 60:49–61

Mao C, Qiu J, Wang C, Charles TC, Sobral BW (2005) NodMutDB: a database for genes and mutants involved in symbiosis. Bioinformatics 21:2927–2929

Marti-Renom MA, Stuart AC, Fiser A, Sanchez R, Melo F, Sali A (2000) Comparative protein structure modeling of genes and genomes. Annu Rev Biophys Biomol Struct 29:291–325

Mathur M, Tuli R (1991) Analysis of codon usage in genes for nitrogen fixation from phylogenetically diverse diazotrophs. J Mol Evol 32:364–373

Mayer SM, Lawson DM, Gormal CA, Roe SM, Smith BE (1999) New insights into structure-function relationships in nitrogenase: a 1.6 A resolution X-ray crystallographic study of Klebsiella pneumoniae MoFe-protein. J Mol Biol 292:871–891

McCammon JA, Gelin BR, Karplus M (1977) Dynamics of folded proteins. Nature 267:585–590

Medini D, Donati C, Tettelin H, Masignani V, Rappuoli R (2005) The microbial pan-genome. Curr Opin Genet Dev 15:589–594

Meuwly M, Karplus M (2004) Theoretical investigations on Azotobacter vinelandii ferredoxin I: effects of electron transfer on protein dynamics. Biophys J 86:1987–2007

Milenkovic VM, Brockmann M, Stohr H, Weber BH, Strauss O (2010) Evolution and functional divergence of the anoctamin family of membrane proteins. BMC Evol Biol 10:319

Mondal UK, Das B, Ghosh TC, Sen A, Bothra AK (2008) Nucleotide triplet based molecular phylogeny of class I and class II aminoacyl t-RNA synthetase in three domain of life process: bacteria, archaea, and eukarya. J Biomol Struct Dyn 26:321–328

Mondal UK, Sen A, Bothra AK (2011) Characterization of pathogenic genes through condensed matrix method, case study through bacterial Zeta toxin. Int J Genet Eng Biotechnol 2:109–114

Murzin AG, Brenner SE, Hubbard T, Chothia C (1995) SCOP: a structural classification of proteins database for the investigation of sequences and structures. J Mol Biol 247:536–540

Mylona P, Pawlowski K, Bisseling T (1995) Symbiotic nitrogen fixation. Plant Cell 7:869–885

Nadler SA (1995) Advantages and disadvantages of molecular phylogenetics: a case study of ascaridoid nematodes. J Nematol 27:423–432

Normand P, Bousquet J (1989) Phylogeny of nitrogenase sequences in Frankia and other nitrogen-fixing microorganisms. J Mol Evol 29:436–447

Normand P, Gouy M, Cournoyer B, Simonet P (1992) Nucleotide sequence of *nif*D from Frankia alni strain ArI3: phylogenetic inferences. Mol Biol Evol 9:495–506

Normand P et al (2007) Genome characteristics of facultatively symbiotic Frankia sp. strains reflect host range and host plant biogeography. Genome Res 17:7–15

Norrby PO, Brandt P (2001) Deriving force field parameters for coordination complexes. Coord Chem Rev 212:79–109

O'Carroll IP, Dos Santos PC (2011) Genomic analysis of nitrogen fixation. Methods Mol Biol 766:49–65

Parker MA, Lafay B, Burdon JJ, van Berkum P (2002) Conflicting phylogeographic patterns in rRNA and *nif*D indicate regionally restricted gene transfer in Bradyrhizobium. Microbiology 148:2557–2565

Paul W, Merrick M (1989) The roles of the *nif*W, *nif*Z and *nif*M genes of Klebsiella pneumoniae in nitrogenase biosynthesis. Eur J Biochem 178:675–682

Pawlowski K (2009) Prokaryotic symbionts in plants. Springer, Berlin/Heidelberg

Pazos F, Valencia A (2001) Similarity of phylogenetic trees as indicator of protein-protein interaction. Protein Eng 14:609–614

Peden JF (1999) Analysis of codon usage, PhD thesis, University of Nottingham

Pedrosa FO, Hungria M, Yates G, Newton WE (eds) (2000) Nitrogen fixation: from molecules to crop productivity. Kluwer Academic Publishers, Dordrecht

Peoples MB, Craswell ET (1992) Biological nitrogen fixation: investments, expectations and actual contributions to agriculture. Plant and Soil 141:13–39

Peoples MB, Herridge DF, Ladha JK (1995) Biological nitrogen fixation: an efficient source of nitrogen for sustainable agricultural production? Plant and Soil 174:3–28

Peralta H, Guerrero G, Aguilar A, Mora J (2011) Sequence variability of Rhizobiales orthologs and relationship with physico-chemical characteristics of proteins. Biol Direct 6:48

Perret X, Freiberg C, Rosenthal A, Broughton WJ, Fellay R (1999) High-resolution transcriptional analysis of the symbiotic plasmid of Rhizobium sp. NGR234. Mol Microbiol 32:415–425

Perret X, Staehelin C, Broughton WJ (2000) Molecular basis of symbiotic promiscuity. Microbiol Mol Biol Rev 64:180–201

Peters JW, Fisher K, Dean DR (1995) Nitrogenase structure and function: a biochemical-genetic perspective. Annu Rev Microbiol 49:335–366

Pommer EH (1959) Ueber die Isolierung des Endophyten aus den Wurzelknollchen Alnus glutinosa Gaertn. und iiber erfolgreiche Re-infektionsversuche. Ber Dtsch Bot Ges 72:138–150

Postgate JR (1982) The fundamentals of nitrogen fixation. Cambridge University Press, Cambridge

Postgate JR (1998) Nitrogen fixation, 3rd edn. Cambridge University Press, Cambridge/New York

Postgate JR, Eady RR (1988) The evolution of biological nitrogen fixation. In: Bother H et al (eds) Nitrogen fixation: hundred years after. Proceedings of the 7th international congress on nitrogen fixation. Gustav Fischer, Stuttgart, pp 31–40

Puhler A, Arlat M, Becker A, Göttfert M, Morrissey JP, O'Gara F (2004) What can bacterial genome research teach us about bacteria-plant interactions? Curr Opin Plant Biol 7:137–147

Qi J, Wang B, Hao BI (2004) Whole proteome prokaryote phylogeny without sequence alignment: a K-string composition approach. J Mol Evol 58:1–11

Qian J, Kwon SW, Parker MA (2003) rRNA and *nif*D phylogeny of Bradyrhizobium from sites across the Pacific Basin. FEMS Microbiol Lett 219:159–165

Quackenbush J, Liang F, Holt I, Pertea G, Upton J (2000) The TIGR gene indices: reconstruction and representation of expressed gene sequences. Nucleic Acids Res 28:141–145

Raes J, Van de Peer Y (2003) Gene duplication, the evolution of novel gene functions, and detecting functional divergence of duplicates in silico. Appl Bioinformatics 2:91–101

Rai AN, Bergman B, Rasmussen U (2002) Cyanobacteria in symbiosis. Kluwer Academic, Dordrecht/Boston

Ramseier TM, Gottfert M (1991) Codon usage and G + C content in Bradyrhizobium japonicum genes are not uniform. Arch Microbiol 156:270–276

Randic M, Guo X, Basak SC (2001) On the characterization of DNA primary sequences by triplet of nucleic acid bases. J Chem Inf Comput Sci 41:619–626

Rasko DA et al (2008) The pangenome structure of Escherichia coli: comparative genomic analysis of E. coli commensal and pathogenic isolates. J Bacteriol 190:6881–6893

Raymond J, Siefert JL, Staples CR, Blankenship RE (2004) The natural history of nitrogen fixation. Mol Biol Evol 21:541–554

Rees DC, Howard JB (2000) Nitrogenase: standing at the crossroads. Curr Opin Chem Biol 4:559–566

Rees DC et al (2005) Structural basis of biological nitrogen fixation. Philos Transact A Math Phys Eng Sci 363:971–984, discussion 1035–1040

Roberts GP, MacNeil T, MacNeil D, Brill WJ (1978) Regulation and characterization of protein products coded by the nif (nitrogen fixation) genes of Klebsiella pneumoniae. J Bacteriol 136:267–279

Robson RL, Eady RR, Richardson TH, Miller RW, Hawkins M, Postgate JR (1986) The alternative nitrogenase of Azotobacter chroococcum is a vanadium enzyme. Nature 322:388–390

Rubio LM, Ludden PW (2005) Maturation of nitrogenase: a biochemical puzzle. J Bacteriol 187:405–414

Rubio LM, Ludden PW (2008) Biosynthesis of the iron-molybdenum cofactor of nitrogenase. Annu Rev Microbiol 62:93–111

Russell RB, Sternberg MJ (1996) A novel binding site in catalase is suggested by structural similarity to the calycin superfamily. Protein Eng 9:107–111

Sali A (1995) Modeling mutations and homologous proteins. Curr Opin Biotechnol 6:437–451

Schlegel HG, Zaborosch C (2003) General microbiology. Cambridge University Press, Cambridge

Schuler GD (1997) Sequence mapping by electronic PCR. Genome Res 7:541–550

Schultze M, Kondorosi A (1998) Regulation of symbiotic root nodule development. Annu Rev Genet 32:33–57

Sen S, Peters JW (2006) The thermal adaptation of the nitrogenase Fe protein from thermophilic Methanobacter thermoautotrophicus. Proteins 62:450–460

Sen A, Sur S, Bothra AK, Benson DR, Normand P, Tisa LS (2008) The implication of life style on codon usage patterns and predicted highly expressed genes for three Frankia genomes. Antonie Van Leeuwenhoek 93:335–346

Sen A, Sur S, Tisa L, Bothra A, Thakur S, Mondal U (2010) Homology modelling of the Frankia nitrogenase iron protein. Symbiosis 50:37–44

Sen A, Thakur S, Bothra AK, Sur S, Tisa LS (2012) Identification of TTA codon containing genes in Frankia and exploration of the role of tRNA in regulating these genes. Arch Microbiol 194:35–45

Sharp PM, Li WH (1987) The rate of synonymous substitution in enterobacterial genes is inversely related to codon usage bias. Mol Biol Evol 4:222–230

Sharp PM, Tuohy TM, Mosurski KR (1986) Codon usage in yeast: cluster analysis clearly differentiates highly and lowly expressed genes. Nucleic Acids Res 14:5125–5143

Silver WS, Postgate JR (1973) Evolution of asymbiotic nitrogen fixation. J Theor Biol 40:1–10

Sims GE, Jun SR, Wu GA, Kim SH (2009) Alignment-free genome comparison with feature frequency profiles (FFP) and optimal resolutions. Proc Natl Acad Sci U S A 106:2677–2682

Smit AM, Strabala TJ, Peng L, Rawson P, Lloyd-Jones G, Jordan TW (2012) Proteomic phenotyping of Novosphingobium nitrogenifigens reveals a robust capacity for simultaneous nitrogen fixation, polyhydroxyalkanoate production, and resistance to reactive oxygen species. Appl Environ Microbiol 78:4802–4815

Snipen L, Ussery DW (2010) Standard operating procedure for computing pangenome trees. Stand Genomic Sci 2:135–141

Stewart WD (1969) Biological and ecological aspects of nitrogen fixation by free-living micro-organisms. Proc R Soc Lond B Biol Sci 172:367–388

Sundaram S, Tripathi A, Gupta V (2010) Structure prediction and molecular simulation of gases diffusion pathways in hydrogenase. Bioinformation 5:177–183

Sur S, Pal A, Bothra AK, Sen A (2005) Moderate codon bias attributed to translational selection in nitrogen fixing genes of Bradyrhizobium japonicum USDA110. Bioinformatics Ind 3:59–64

Sur S, Bhattacharya M, Bothra AK, Tisa LS, Sen A (2008) Bioinformatic analysis of codon usage patterns in a free-living diazotroph, Azotobacter vinelandii. Biotechnology 7:242–249

Sur S, Bothra AK, Ghosh TC, Sen A (2010) Investigation of the molecular evolution of nitrogen fixation using nucleotide triplet based condensed matrix method. Int J Integrative Biol 10:29–65

Tatusov RL, Galperin MY, Natale DA, Koonin EV (2000) The COG database: a tool for genome-scale analysis of protein functions and evolution. Nucleic Acids Res 28:33–36

Tettelin H et al (2005) Genome analysis of multiple pathogenic isolates of Streptococcus agalactiae: implications for the microbial "pan-genome". Proc Natl Acad Sci U S A 102:13950–13955

Thakur S, Bothra AK, Sen A (2012) *In silico* studies of NifH protein structure and its post-translational modification in Bradyrhizobium sp. ORS278. Int J Pharm Bio Sci 3:B22–B32

Thorneley RNF, Lowe DJ (1985) Kinetics and mechanism of the nitrogenase enzyme system. In: Spiro TG (ed) Molybdenum enzymes. Wiley, New York, pp 221–284

Torrey JG, Tjepkema JD (1979) Symbiotic nitrogen fixation in actinomycete-nodulated plants: Preface. Bot Gaz 140(suppl):i–ii

Von Bulow JF, Dobereiner J (1975) Potential for nitrogen fixation in maize genotypes in Brazil. Proc Natl Acad Sci U S A 72:2389–2393

Wani SP, Rupela OP, Lee KK (1995) Sustainable agriculture in the semi-arid tropics through biological nitrogen fixation in grain legumes. Plant and Soil 174:29–49

Watanabe I, Liu CC (1992) Improving nitrogen-fixing systems and integrating them into sustainable rice farming. Plant and Soil 141:57–67

Wright F (1990) The 'effective number of codons' used in a gene. Gene 87:23–29

Xie J et al (2011) GeoChip-based analysis of the functional gene diversity and metabolic potential of microbial communities in acid mine drainage. Appl Environ Microbiol 77:991–999

Yang Z (1997) PAML: a program package for phylogenetic analysis by maximum likelihood. Comput Appl Biosci 13:555–556

Yang GP et al (1999) Structure of the Mesorhizobium huakuii and Rhizobium galegae Nod factors: a cluster of phylogenetically related legumes are nodulated by rhizobia producing Nod factors with alpha, beta-unsaturated N-acyl substitutions. Mol Microbiol 34:227–237

Yi J (2009) The Medicago truncatula genome and analysis of nodule-specific genes, PhD thesis, The University of Oklahoma

Young JPW (1992) Phylogenetic classification of nitrogen-fixing organisms. In: Stacey G, Burris RH, Evans HJ (eds) Biological nitrogen fixation. Chapman & Hall, New York, pp 43–86

Young J (2005) The phylogeny and evolution of nitrogenases. In: Palacios R, Newton WE (eds) Genomes and genomics of nitrogen-fixing organisms. Springer, Dordrecht, pp 221–241

Young JP et al (2006) The genome of Rhizobium leguminosarum has recognizable core and accessory components. Genome Biol 7:R34

Zehr JP, Harris D, Dominic B, Salerno J (1997a) Structural analysis of the Trichodesmium nitrogenase iron protein: implications for aerobic nitrogen fixation activity. FEMS Microbiol Lett 153:303–309

Zehr JP, Mellon MT, Hiorns WD (1997b) Phylogeny of cyanobacterial nifH genes: evolutionary implications and potential applications to natural assemblages. Microbiology 143(Pt 4):1443–1450

Zhao H, Li M, Fang K, Chen W, Wang J (2012) In Silico insights into the symbiotic nitrogen fixation in Sinorhizobium meliloti via metabolic reconstruction. PLoS One 7:e31287

Zheng L, White RH, Dean DR (1997) Purification of the Azotobacter vinelandii nifV-encoded homocitrate synthase. J Bacteriol 179:5963–5966

Zheng Y, Xu D, Gu X (2007) Functional divergence after gene duplication and sequence-structure relationship: a case study of G-protein alpha subunits. J Exp Zool B Mol Dev Evol 308:85–96

Plant-Microbial Interaction: A Dialogue Between Two Dynamic Bioentities

Khyatiben V. Pathak and Sivaramaiah Nallapeta

Abstract

Since the time of evolution, the earth's plant floral community remained associated with ubiquitous population of microbes by a wide array of interactive relationships, ranging from symbiotic to parasitic. The ecology of plant-microbial association has influenced the plant's diversity, metabolism, morphology, productivity, physiology, defence system and tolerance against adversities. Similarly, the microbial population has also been affected in terms of morphology, diversity, community composition, etc. In plant-microbial association, microbes obtain shelter, protection and nutrients from the plants either positively or negatively without affecting the plant's health. In symbiotic plant-microbial interaction, plants provide habitat, nutrients and protection against the adverse environment, while in return, the microbes render several benefits such as protection against pathogens, plant growth promotions, resistance towards abiotic stress, improved nutrient uptake and fitness. The microbial endophytes and epiphytes are generally regarded as plant symbionts. The antagonistic association between these two living systems, in which the plant antagonist kills the microbial pathogens by producing toxic phytochemicals or the microbial parasite adversely affects the plant's fitness by withdrawing essential plant nutrients for their own survival and altering the physiology of the host plant. How do the plants cross talk to the microbes to establish the associations? Various response-related signals drive such cross talks. The omics (genomics, proteomics and metabolomics) approach is being used to unveil the role of complex cryptic signalling process in the plant and microbe interaction. In this chapter, the existing understandings about the plant-microbial interactions and the roles of signalling mechanisms in such interactions have been discussed.

K.V. Pathak (✉)
Bioclues Organisation, IKP Knowledge Park,
Secunderabad 500009, AP, India
e-mail: Khyati835@gmail.com

S. Nallapeta
Bioclues Organization, IKP Knowledge Park, Picket
Secunderabad 500009, AP, India

K.K. P.B. et al. (eds.), *Agricultural Bioinformatics*,
DOI 10.1007/978-81-322-1880-7_15, © Springer India 2014

Keywords
Siderophore • Microbial interaction • Endophyte • Rhizobacteria • Symbiosis

1 Introduction

During the course of evolution, microbes have developed association with the plant environment by establishing various interactions. Almost all the plants existing on earth harbour a single microbe or a variety of microbes (Lindow and Brandl 2003; Rosenblueth and Martínez-Romero 2006; Saharan and Nehra 2011). The microbes colonize in the plants for shelter, nutrient and protection against the adverse conditions, while in return microbes offer several interactive relationships ranging from symbiosis to parasitism (Rosenblueth and Martínez-Romero 2006; Wu et al. 2009; Reichling 2010). Sometimes, plants also exhibit defence-related responses towards the microbial pathogens by producing microbial-inhibitory phytochemicals or inducing self-defence in response to the chemicals secreted by microbes (Reichling 2010; Radulović et al. 2013). This association influences the plant's diversity, metabolism, morphology, productivity, physiology, defence system and tolerance against adversities (Lindow and Brandl 2003; Rosenblueth and Martínez-Romero 2006; van der Heijden et al. 2008; Wu et al. 2009; Saharan and Nehra 2011). Similarly, the microbial population residing in the plant is also affected in terms of morphology, diversity and community composition (Lindow and Brandl 2003; Montesinos 2003; Bever et al. 2012). In symbiotic association, the microbes and their host plant both get benefits from each other. The symbiotic interaction confers several benefits to the host plant such as plant growth promotion, protection from the phytopathogens, improved nutrient availability and uptake, plant fitness and improved tolerance to the abiotic stress (Lindow and Brandl 2003; Rosenblueth and Martínez-Romero 2006; Saharan and Nehra 2011). In some plant-microbial interactions, microbes colonize plant tissue without causing any benefits or deleterious effects to the host, and the host plant also accepts such flora as a component of the innate plant system. Such plant-microbial interactions are regarded as a neutral relationship. The microbial pathogens invade the plant tissues, consume nutrients and release toxins that affect plant health. Such interactions can affect plant growth, development, nutrient dynamics, defence system, etc.

2 Plant Niches and Plant-Microbial Communications

The plant-microbial interactions are much more diversified in terms of their physiological and pathological functions. The plant system offers several ecological niches for the colonization of microbes and produces diversified bioactive phytochemicals in response to the microbial interactions (Narasimhan et al. 2003; Reichling 2010; Garcia-Brugger et al. 2006; Radulović et al. 2013). The microbial system also releases several active metabolites for communication with host plants to maintain a wide range of interactive relationships (Shulaev et al. 2008; Ryan et al. 2008; Braeken et al. 2008; Saharan and Nehra 2011). The plant comprises three major niches such as the rhizosphere, phyllosphere and endosphere for the entry of atmospheric and soil microbes. The interactive relationship between plants and microbes is mostly governed by the chemical signals, and they play a major role in the plant's growth, development and fitness (Bais et al. 2004; Shulaev et al. 2008; Braeken et al. 2008; Mandal and Dey 2008).

3 Rhizospheric Root Microbial Communication

The soil microbes generally enter in the plant system via rhizospheric roots. The root surface and exudates are nutrient-plenteous niches and attract the soil microflora (symbionts, pathogens). The thin layer of rhizosphere around plant roots is densely populated with a variety of microbes such as bacteria, fungi, actinomycetes and algae. Among these microbes, bacterial population is found to be the highest in rhizosphere. This microbial diversity around the rhizosphere may influence the plant's physiology as a result of competitive colonization by rhizospheric flora in the plant roots for nutrients and habitat (Morgan et al. 2005). The symbiotic microbes offer protection to the plants against phytopathogen by competing for food and shelter. The interaction of plant microbes requires recognition of each other. Plants and microbes both produce diversified signalling molecules for recognition and communication (Bais et al. 2004; Badri et al. 2008a; Braeken et al. 2008; Mandal and Dey 2008). Root exudates comprise of various signalling molecules such as carbohydrates, proteins, phenolics, flavonoids and isoflavonoids (Narasimhan et al. 2003; Bais et al. 2004). The secretion of these molecules in the root exudates plausibly controlled by the expression of transporter proteins in the root system and the variation in the composition may depend upon the types of microbial interaction with roots (Sugiyama et al. 2006; Loyola-Vargas et al. 2007; Badri et al. 2008a, b). The root exudate of Arabidopsis is composed of sugars, amino acids, organic acids, flavonols, lignins, coumarins, aurones, glucosinolates, anthocyanins, carotenes and indole compounds (De-la-Peña et al. 2008; Mandal and Dey 2008). Liquid chromatography coupled with electrospray mass spectrometry (LC-ESI-MS) enabled the identification of more than 149 metabolites including 125 secondary metabolites in the Arabidopsis root exudates (De-la-Peña et al. 2008). The plant sugars and amino acids have a major role in energy metabolism and polymer biosynthesis which is essential for the growth and development of the plant. The sugars, amino acids and organic acids are chemoattractants which guide the motility of microbes (Smeekens and Rook 1997; Welbaum et al. 2004). Some exudates are also composed of antimicrobial compounds which provide protection to the plant system against phytopathogens (reviewed by Bais et al. 2004). This suggests that plants selectively allow certain microbes with the capability to detoxify antagonizing chemical agents to colonize in the plant. The composition and concentration of such chemoattractants and antimicrobials secreted by the plant system are controlled by genetic and environmental factors (Bais et al. 2004). For instance, the endophytic bacteria exhibited a fivefold increment in chemotactic response induced by rice exudates than the non-plant-growth-promoting bacteria present in the rice rhizosphere (Bacilio-Jiménez et al. 2003). De Weert et al. (2002) have reported that root colonization of *Pseudomonas fluorescens* in tomato root is dependent upon the induction of flagellar chemotaxis by root exudates. The variation in the degree of chemotactic response induced by chemoattractants (sugars, organic acids and amino acids) among the *Azospirillum* strains has been observed by Reinhold and group (1985). This suggests that microbial competence is greatly dependent upon their capability either to take benefit of the specific plant environment or to adjust themselves against the altering conditions. The protection of a delicate unprotected plant from the attack of phytopathogens is mediated by the secretion of phytoalexins, defence proteins and certain phenolic compounds by plant roots in response to the pathogens (Garcia-Brugger et al. 2006; Mandal et al. 2010). This response is known as systemic acquired resistance. Certain phenolic acids and their derivatives, such as cinnamic acids, ferulic acid, hydroxy benzoic acids, syringic acid, salicylic acid, p-coumaric acid, 4-hydroxy aldehyde, tannic acid, vanillic acid, vanillin, vanillyl alcohol and glycosides, produced by the plant have a role in the induction of symbiotic rhizospheric plant-microbial interactions (reviewed by Mandal et al. 2010). Thus, the root exudate is a library of a complex mixture of biologically active molecules exerting several benefits to the plant and also an ideal source of novel biochemicals.

4 Phyllosphere and Microbe Interaction

The rhizospheric and atmospheric microbes move to the above-ground parts of the plant and generally colonize in the outer layer of the plants. These microbes are known as epiphytes (Lindow and Brandl 2003). They generally remained associated to the plant surface. In the phyllosphere, leaves are more exploited for studying the diversity and interactive relationship with microbes than buds and flowers (Beattie and Lindow 1995; Jacques et al. 1995; Hirano and Upper 2000; Andrews and Harris 2000). The phyllosphere is considered to be a very large habitat for microbes (~6.4×10^8 km^2) (Morris et al. 2002). Based on the surface area, the phyllosphere bacterial population of 10^{26} cell could be estimated for the tropical plants (Morris et al. 2002). The microbial population density of phyllosphere is large enough to establish beneficial, detrimental or neutral interactions with phyllosphere. The phyllosphere allows the colonization of diversified microbial genera including bacteria, yeast, fungi, algae, protozoa and nematodes with the highest population of bacteria (~10^8 cells/g of aerial tissue) (Beattie and Lindow 1995; Jacques et al. 1995; Hirano and Upper 2000). The filamentous fungi are ephemeral and mostly colonize in the phyllosphere in the form of spores, while the rapidly sporulating fungi and yeast are actively colonizing the phyllosphere (Andrews and Harris 2000). The microbial population on the plant surface varies in their population size, shape within the plant of the same species over a short time period and change in environmental conditions (Hirano and Upper 1989; Ercolani 1991; Legard et al. 1994). The nutritional and physical conditions of the phyllosphere also affect its microbial ecology (Wilson and Lindow 1994). The peculiar difference in the above- and underground atmosphere suggests the significant difference among the phyllospheric and rhizospheric bacterial population. The failing attempt to colonize two of the most common rhizosphere-specific microbes, i.e. *Rhizobium* and *Azospirillum*, in the plant leaf tissues provides an evidence of environmental-dependent differential specificity in microbial colonization of plant niches (Fokkema and Schippers 1986). Many physicochemical factors restrict the colonization of microbes in the phyllosphere. To inhabit the phyllospheric tissues, microbes capable of modifying the microhabitat and utilizing the nutrients must be selected by the phyllosphere to colonize in it (Lindow and Brandl 2003). Microbes can improve the wettability by producing biosurfactants which facilitate penetration of microbes in the phyllosphere (Lindow and Brandl 2003). For instance, a biosurfactant tolassin produced by *Pseudomonas tolaasii* had facilitated the motility of bacteria on the phyllosphere (Hutchison and Johnstone 1993). A biosurfactant syringomycin production by a common pathogenic as well as nonpathogenic epiphyte, *Pseudomonas syringae*, on the phyllosphere can affect the ion transport by triggering the formation of ion channels which release metabolites from the cell and cause cell lysis (Hutchison et al. 1995). The nonpathogenic strain also induces the release of a low concentration of syringic acid metabolites (Hutchison et al. 1995). Some of the bacterial epiphytes produce plant growth regulator indole-3-acetic acid (IAA) and their derivatives. IAA production by an epiphyte, *Pantoea agglomerans*, has been demonstrated for their role in the maintenance of fitness of the plant during drought condition (Brandl and Lindow 1998). At low concentration, IAA releases the plant cell wall saccharides (Fry 1989). This phenomenon can be correlated with the plant's fitness by IAA-mediated nutrient availability upon the release of plant cell wall saccharides (Fry 1989). The phyllospheric microbial community secretes cell wall exopolysaccharides and forms a slimy and sticky layer. This layer provides protection to the phyllosphere and its bacterial community against desiccation and reactive oxygen species (Kiraly et al. 1997). The sticky matrix of microbial epiphytes may increase the nutrient concentration in order to provide nutrients to the microbial epiphytes to survive on a nutrient-deficient phyllospheric environment (Costerton et al. 1995). The pathogenic bacteria are known to alter the environment of the host plant to facilitate interaction to modulate the metabolism of the host plant for its own benefit. The interaction of *P. syringae* with its host plant has been studied with great care to understand the role of various molecular determinants in their interactions (Hutchison and Johnstone 1993; Hutchison et al. 1995; Costerton et al. 1995). The hypersensitive response and pathogenicity regulated by *hrp* genes in *P. syringae* have been investigated in detail. A complete set of type III secretion protein pathways and related proteins encoded by *hrp* gene cluster is critical for the growth and fitness of *P. syringae* in the phyllosphere (Hirano et al. 1997, 1999; He 1998). The studies also revealed that certain secretary metabolites induce *P. syringae* interaction with the host plant and modulate the host plant system in its favour (Hutchison and Johnstone 1993; Hutchison et al. 1995). Thus, this type III secretion pathway may also have a role in the colonization of the host plant phyllosphere by nonpathological strains.

Certain epiphytic bacteria interact with each other for shelter and nutrients. Such bacterial interactions may provide benefits to host plant by providing protection against phyllosphere pathogen and frost injury to the plant surface. The population epiphytic bacteria always remain affected due to the presence of antagonist on the phyllosphere (Lindow 1985). Certain phyllospheres such as flowers of deciduous trees, leaves of susceptible herbaceous plants and tropical tree species are generally susceptible to frost injury (Lindow 1985). The epiphytic bacteria with ice nucleation activity (Ice^+ bacteria) have a role in frost injury (Lindow 1987). The *P. syringae* is an Ice^+ bacterium which avoids damaging ice formation (Lindow 1987). The population of these bacteria on epiphytes has a positive effect on the ice nucleation temperature essential for the ice nucleation activity (Lindow 1995). Thus, the increase in population leads to an increase in ice nucleation temperature which makes the plant vulnerable to frost injury (Lindow 1987). The population size of such bacteria on the susceptible young phyllospheric region is generally very low and it increases over time. Thus, the protection of the phyllosphere from frost injury can be achieved by removing the Ice^+ bacteria from the phyllosphere (Lindow 1987, 1995). The colonization of bacterial antagonists of Ice^+ bacteria on the phyllosphere can be the potential strategy to control Ice^+ bacterial population (Lindow 1995). The competitive colonization of Ice^- bacteria on the phyllosphere also provides an effective way of biological control (Lindow 1985, 1987, 1995). The use of lyophilized preparation *P. fluorescens* A506 for foliar spray application in order to manage population of Ice^+ bacteria and protect agricultural crop from frost injury has been commercialized (BlightBan A506; Nufarm Americas, Inc., Sugar Land, TX). The precolonization or competitive colonization of antagonistic bacteria on the disease-prone phyllosphere provides an effective strategy of biocontrol of the phyllosphere pathogen. The bacterial blight disease of apple and pear caused by *Erwinia amylovora* is the most devastating bacterial disease to the plant phyllosphere (Mercier and Lindow 2001; Pusey 2002). This pathogen establishes itself on the phyllosphere before the onset of infection, and prevents the colonization of this pathogen. The prior colonization of bacterial antagonist strains *P. fluorescens* A506 and *P. agglomerans*, on the phyllosphere, has been proven to be an effective way to suppress the colonization of *Erwinia amylovora* which resulted in the drastic reduction in disease symptoms (Lindow et al. 1996; Pusey 2002). The microbial population of the phyllosphere always undergoes changes upon the change in environment and also takes up a wide variety of plasmids leading to the increase in the rate of mixing of genes in the bacterial population (Lilley and Bailey 1997; Bailey et al. 2002). This indicates that the phyllosphere, specifically the leaf surface, is the ideal place for the horizontal transfer of genetic information and may provide a significant base for raising the diversified microbial ecology.

5 Endosphere and Microbial Communication

Some of the rhizospheric or epiphytic microbes colonize the internal tissues (endosphere) without causing any adverse effect to the host plants (Bacon and White 2000). These microbes are endophytes. The endophytes are selected naturally to colonize the endosphere. Some endophytes are seedborne, while others come through horizontal transfer. The endophytes are generally regarded as plant symbionts which can offer a variety of benefits to the plants (Miller et al. 1998; reviewed by Strobel et al. 2004; Compant et al. 2005; Rosenblueth and Martínez-Romero 2006; Ryan et al. 2008). The endosphere offers the habitat for bacteria, fungi and algae. The fungal diversity of endophytes in herbs, shrubs and trees has been studied in great detail for its wide variety of biological applications (reviewed by Strobel et al. 2004; Compant et al. 2005; Rosenblueth and Martínez-Romero 2006; Ryan et al. 2008). The mycorrhizal fungi are ubiquitously present in most of the plants (Redecker et al. 2000). The fungi spread through the extraradical mycelia in the soil and increase the availability of nutrients to the plant in nutrient-deprived land zones. The mycorrhizal fungal community also solubilizes inorganic phosphorus, nitrogen and other essential nutrients and translocates to the plants from the soil (Finlay 2008). These fungi also impart other benefits such as increased water uptake and improved soil fertility and resistance towards pathogens, drought and herbivores. Moreover, it also helps the plant in carbon cycling to facilitate the carbon supply to the soil aggregates and other microbial communities (Finlay 2008). Like mycorrhizal fungi, other endophytic fungi also have shown to promote plant growth and protect the host plant from herbivores and pathogens by producing chemically novel bioactive metabolites (Finlay 2008; Smith and Read 2008; Kawaguchi and Minamisawa 2010). Some fungal endophytes have the ability to synthesize biologically important plant metabolites. The taxol-producing endophytic fungi such as *Taxus brevifolia* and *Taxus chinensis* are important examples that demonstrate the ability of microbes to synthesize biologically important phytochemicals (Wani et al. 1971; Guo et al. 2006). Taxol is an important phytochemical isolated from the yew tree (Wani et al. 1971). It is most effectively used in the treatment of tumours. Various endophytic proteobacteria, firmicutes, bacteroidetes and actinomycetes have been isolated from the agriculturally important crops (reviewed by Rosenblueth and Martínez-Romero 2006; Wu et al. 2009; Ryan et al. 2008). The species diversity of the microbial flora and their ability to produce beneficial metabolites are dependent on the types of plant niche, host developmental stage, surrounding environment, climate, etc. (Guo et al. 2006; Rosenblueth and Martínez-Romero 2006; Ryan et al. 2008). The endophytic populations in the tomato plants promoted plant growth, while the rhizospheric microbial community failed to do so (Pillay and Nowak 1997). The plant-growth-promoting bacteria of the genus *Azospirillum* exhibit their beneficial effects in the rhizosphere, and the colonization in the internal cortical tissue is uncommon (Somers et al. 2004). The endophytic microbes are known to produce phytohormones, antimicrobial agents and siderophore; induce systemic resistance; and improve nutrient availability and uptake (Sturz et al. 1997; Pillay and Nowak 1997; Reiter et al. 2003; Somers et al. 2004; and reviewed by Rosenblueth and Martínez-Romero 2006; Ryan et al. 2008). The nitrogen-fixing bacterial endophytes contribute very less in total endophytic populations. The endophytic bacteria isolated from the sweet potato growing in N-limited soil were found to fix atmospheric nitrogen (Reiter et al. 2003). The presence of nitrogenase *(nifH)* genes further confirmed their ability of nitrogen fixation. The legume nodules also harbour endophytic bacteria. The *Rhizobium rhizogenes* and *R. leguminosarum* pv. *trifolii* have been isolated from the nodule of the red clover (Sturz et al. 1997).

6 Microbial Quorum Sensing and Plant-Microbial Interaction

Bacteria produce chemical signals to communicate with each other. This process is known as quorum sensing (QS). The bacterial cells recognize the signals sent by other bacteria which help the bacterial community to proceed for the particular function (Fuqua et al. 2001). The quorum-sensing signals are very much important in plant-bacterial interactions. In recent years, large numbers of signalling compounds have been identified (reviewed by Braeken et al. 2008). The gram-positive bacteria generally secrete peptide-based signals, while gram-negative bacteria release *N*-acyl homoserine lactone (AHL)-based quorum-sensing signals (Fuqua et al. 2001; Waters and Bassler 2005). The synthesis of AHL autoinducer signals depends upon the LuxI-like autoinducer synthase. The autoinducer signals diffuse freely through the bacterial membrane and bind with cognate LuxR-like proteins to form a LuxR-HSL autoinducer complex to trigger transcription of target genes (Fuqua et al. 2001). The AHL-based QS systems in plant symbionts as well as pathogens influence colonization, swarming motility, biofilm formation, plasmid transfer, stress tolerance and synthesis of antimicrobials, extracellular enzymes, exopolysaccharides, biosurfactants, etc. (reviewed by Braeken et al. 2008; Waters and Bassler 2005). For instance, 3-oxo-C_8-HSL produced by *Agrobacterium tumefaciens* has a role in Ti plasmid transfer (Piper et al. 1993). The *P. aureofaciens* strain produces C_6-HSL which induces rhizosphere colonization and protease and phenazine production (Wood et al. 1997; Chancey et al. 1999; Zhang and Pierson 2001). The 3-oxo-C_{12}-HSL produced by *P. putida* IsoF is essential for the development of biofilm structure (Arevalo-Ferro et al. 2005). Some QS signals such as C_8-HSL and 3-oxo-C_6-HSL produced by *P. corrugate CFBP*5454 trigger hypersensitivity response in tobacco and tomato pith necrosis, respectively (Licciardello et al. 2007). The QS signals are important in nitrogen fixation, nodulation initiation, growth and biocontrol (reviewed by Braeken et al. 2008; Waters and Bassler 2005).

7 Potential Applications of Symbiotic Interaction

A variety of beneficial microbes have been identified and characterized for their beneficiary effects from the rhizosphere, phyllosphere and endosphere of the several plant species. The symbiotic interaction with microbes leads to confer positive effects on the plant's health and productivity. The microbes are known to produce a wide variety of bioactive compounds that have potency to promote plant growth, improve nutrient availability, protect against pathogens, protect from the environmental pollutants by detoxifying them, etc. Such properties of symbiotic microbes have great importance for their commercial applications in agriculture, environmental cleaning and pharmaceutical industry.

8 Plant Growth Promotion

The bacteria- as well as fungi-colonized rhizosphere, phyllosphere and endosphere are found to promote plant growth and development. The plant produces five major groups of hormones such as auxins, gibberellins, ethylene, cytokinins and abscisic acid which have a role in the regulation of plant growth. IAA is a phytohormone which is known as native auxin. IAA has a role in various plant developmental processes such as organogenesis, cell expansion, division, differentiation and gene regulation (Ryu and Patten 2008). The bacteria associated to the rhizosphere, phyllosphere and endosphere region of plants synthesize IAA and its variants (indole-3-pytuvic acid, indole-3-butyric acid and indole lactic acid) (Ryu and Patten 2008). The production of IAA variants is an important feature of the plant-microbial interaction which leads to several effects ranging from phytostimulation to diseased condition (Khalid et al. 2004; Narula et al. 2006). IAA produced by bacteria can promote plant growth at low concentration and also act as a signalling molecule to trigger physiological responses including colonization and defence response (Spaepen et al. 2007). The high concentration of IAA inhibits growth (Spaepen et al. 2007). The response to IAA production varies from plant to plant. At low concentration, it also releases saccharides from the plant cell wall and increases availability of nutrients to the plant (Fry 1989). IAA- and indole acetamide-producing bacteria induce growth and yield in wheat crop (Khalid et al. 2004). IAA production by rhizospheric bacteria positively affects the root and shoot weight in wheat plants (Narula et al. 2006). The bacteria from *Rhizobium*, *Microbacterium*, *Sphingomonas* and *Mycobacterium* genera isolated from the roots of the epiphytic orchid *Dendrobium moschatum* have been identified as the most active IAA producers (reviewed by Saharan and Nehra 2011). Rhizobium strains associated to root nodules of *Sesbania sesban* (L.) Merr and *Vigna mungo* (L.) Hepper have also been identified as IAA producers (reviewed by Wu et al. 2009; Saharan and Nehra 2011). The rhizobia also have been used as biostimulants and biofertilizer for wheat production due to their ability to produce IAA, uptake minerals (nitrogen, phosphorus and potassium) from the soil and transport to the plant and increase the length of root and shoot (reviewed by Wu et al. 2009; Saharan and Nehra 2011). IAA production has been reported from the phyllospheric as well as rhizospheric microbes which provide total synergistic effect on the plants (Lindow and Brandl 2003; Saharan and Nehra 2011). The isolates *Bacillus*, *Pseudomonas* and *Azotobacter* from the chickpea were also found to produce IAA. In all *Rhizobium* species identified so far, 85.7 % rhizobia have been reported for their ability to produce IAA (Joseph et al. 2007). *Pseudomonas fluorescens* B16, a plant-growth-promoting rhizobacterium, synthesizes plant growth promotion factor and pyrroloquinoline quinone (Choi et al. 2008).

9 Nutrient Availability and Uptake

Plants need micronutrients for growth and development. The micronutrients can work as cofactors in various enzymatic processes. Iron, phosphorus and nitrogen are essential growth elements for all living systems. Though iron is considered as one of the most abundant metals in the earth, the bioavailability of iron is limited in certain environments such as the soil and plants due to the low solubility of Fe^{3+} ion. The limitation of bioavailable iron in the soil as well as the phyllosphere leads to increased competition for iron availability. In these iron-limited conditions, certain plant-associated microbes synthesize low molecular weight iron chelators known as siderophores (Whipps 2001). Microbes secrete siderophores to scavenge iron from the mineral phases and to make the soluble Fe^{3+} complexes which can be transported to the cells. Within the microbes, bacteria generally produce highly diverse types of siderophores. The gram-negative bacteria belonging to genera *Pseudomonas*, *Enterobacter*, *Burkholderia*, *Rhizobia*, *Yersinia*, *Azotobacter* and *Escherichia* are known to produce siderophores (reviewed by Saharan and Nehra 2011). Based on the types of ligand used for ferric ion chelation, siderophores are classified into three major groups: catecholates, hydroxamates and carboxylates. The rhizobia produce hydroxamate and catecholate types of siderophores. The hydroxamate siderophores such as ferrioxamine B and pseudobactin are produced by rhizospheric bacteria (Sridevi and Mallaiah 2008). The gram-positive bacteria *Bacillus subtilis* and *B. anthracis* produce bacillibactin. The enterobactin, azotobactin, pyoverdine, yersiniabactin and ornibactin are the siderophores produced by *E. coli*, *Azotobacter vinelandii*, *P. aeruginosa*, *Yersinia pestis* and *Burkholderia cepacia*, respectively (reviewed by Saharan and Nehra 2011). Certain fungal strains such as *Ustilago sphaerogena* and *Fusarium roseum* also produce siderophores identified as ferrichrome and fusarinine C, respectively. The actinomyces including *Streptomyces pilosus* and *S. coelicolor* produce desferrioxamine types of siderophores (reviewed by Saharan and Nehra 2011). Various experiments have been carried out in recent years to study the effect of siderophore-producing microbes on plant growth. The endophytic *E. coli* strains isolated from the sugarcane and rye grass have the ability to produce maximum siderophores with improvement in the growth of the plants (Gangwar and Kaur 2009). The seed inoculation of *P. fluorescens* and *P. putida* strains resulted in increase in the plant growth and yield of various crops (Kloepper et al. 1980). The siderophore production ability of the plant-associated microbes is an important trait for plant growth promotion by improving iron availability to the plant. These siderophore-producing microbes may be used as potential plant-growth-promoting agents in agriculture.

Phosphate is another micronutrient essential for growth and development. Same as iron, phosphorus is also present in huge quantity but unavailable to plants due to its insolubility. Some microbes have the capability to solubilize phosphorous and make it available to plants.

The phosphate-solubilizing microbes have a role in the plant growth promotion by improving the nutrient uptake of plants. The bacteria that correspond to genera *Bacillus*, *Rhizobium* and *Pseudomonas* have been reported for their phosphate-solubilizing property. Fungi belonging to *Aspergillus* and *Penicillium* genera are also known as phosphate solubilization microbes (Saharan and Nehra 2011). The microbes solubilize phosphate by exudating organic acids and release phosphate in the solution. The combined use of *Bacillus* strains (M3 and OSU-142) resulted in increase in the yield, growth and nutrition level of raspberry plants (Orhan et al. 2006). In the total population of bacteria isolated from the root-free soil, rhizosphere and rhizoplane of *Prosopis juliflora*, the number of phosphate-solubilizing bacteria is high (Rivas et al. 2006). Arbuscular mycorrhizal (AM) fungi are also known for its phosphate-solubilizing capacity. The application of phosphate-solubilizing microbes in the soil may increase the availability of phosphorous to plants. The utilization of soluble phosphorous enhances vegetative growth and fruit quality in plants. The plant-associated microbes or mixed microbial population exhibiting plant-growth-promoting properties including phosphate solubilization, IAA and siderophore production may have potential in the development of efficient biofertilizer to improve crop yield in agriculture.

10 Nitrogen Fixation

The interaction of nitrogen-fixing bacteria with plant roots is one of the most extensively studied symbiotic associations. Plants of the *Fabaceae* family have been known for root symbiosis with *Rhizobium* spp. or *Bradyrhizobium* spp. (Wu et al. 2009; Kawaguchi and Minamisawa 2010). The bacteria penetrate the host plant via the root hair. The root exudates comprising of various flavonoid and isoflavonoid molecules induce expression of *nod* (nodulation) genes by bacteria in rhizobia (Mandal et al. 2010). The root cells form bacterial cells containing root nodule upon induction by bacterial cells released in the cytoplasm. The bacteria utilize nutrient from the nodules, and in return, they fix atmospheric N_2 to the NH_4^+. NH_4^+ further converts into amides and are transported via vascular tissues to the plant (Kawaguchi and Minamisawa 2010; Saharan and Nehra 2011). Along with *Rhizobium* spp. or *Bradyrhizobium* spp., other bacterial strains such as *Ralstonia*, *Burkholderia* and *Methylobacterium* also have been reported for their ability to fix nitrogen in the tropical *Fabaceae* plants (Kawaguchi and Minamisawa 2010). The use of nitrogen-fixing bacteria as biofertilizer and bioenhancer in nitrogen-poor soil may reduce the use of costly chemical fertilizers. The use of biofertilizer also reduces the accumulation of unwanted residues of chemical fertilizer which affects soil fertility.

11 Biological Control

Plant-associated symbiotic microbes protect plant from the pathogens by several means. The microbial symbionts prevent the entry of pathogens by competing for shelter and food. This competitive colonization of microbial symbionts provides protection to the host plants. Certain microbes are known to produce antibiotics which have growth-inhibitory activity against the pathogens. The antibiotic production ability also exerts protection to the plant from deleterious effects of plant pathogens. The plant-associated bacterial species of genera *Bacillus*, *Pseudomonas*, *Serratia* and *Streptomyces* are known for their ability to produce fungal cell wall lytic enzymes, antifungals and antibacterial compounds (Lindow 1985; Lindow et al. 1996; Whipps 2001; Mercier and Lindow 2001; Pusey 2002; reviewed by Ryan et al. 2008). The banyan endophytic strains belonging to *B. subtilis* and *B. amyloliquefaciens* exhibited broad-spectrum antifungal activity against phytopathogenic fungi including *Aspergillus niger*, *A. parasiticus*, *A. flavus*, *F. oxysporum*, *Alternaria burnsii*, *Sclerotia rolfsii*, *Chrysosporium indicum* and *Lasiodiplodia theobromae* (Pathak et al. 2012). The banyan endophytic *B. subtilis* K1 strain has been reported for the production of surfactin, iturin and fengycin types of lipopeptides with antifungal, biosurfactant, antimicrobial and antiviral activity (Pathak et al. 2012). The *Pseudomonas stutzeri* exhibited inhibitory activity against *Fusarium solani* by producing chitinase and laminarinase (Mauch et al. 1988). The fungal antagonists, *Paenibacillus polymyxa* strains, isolated from wheat, lodge pine, green beans and canola have been reported for the production of antifungal compounds, fusaricidin A, fusaricidin B, fusaricidin C and fusaricidin D (Li et al. 2007). Oocydin A produced by the endophyte, *Serratia marcescens*, isolated from *Rhyncholacis penicillata* exhibited antifungal activity (Strobel et al. 2004). The *P. viridiflava* from grass produces antimicrobial compounds, ecomycins B and C (Miller et al. 1998). These antibiotic-producing strains may have a potential role as biocontrol agents in controlling plant disease caused by pathogens. The preinoculation of biocontrol agents can reduce the damage caused by the bacterial, fungal and viral plant pathogens. Various plant-associated microbes exhibiting broad-spectrum microbial antagonistic activity have been studied for their potential use as biocontrol agents in small scale as well as in the field level. The indigenous *Pseudomonas* strains isolated from rice rhizosphere exhibited an antimicrobial-compound-dependent suppression of bacterial leaf blight and sheath blight diseases caused by *Xanthomonas oryzae* and *Rhizoctonia solani*. *B. luciferensis* exerted a protective effect against Phytophthora blight in paper by increasing root colonization along with protease production and enhancing antimicrobial activities (Rangarajan et al. 2001). The antifungal volatiles from the *P. aeruginosa* Sha8 inhibited the growth of *F. oxysporum* and *Helminthosporium* sp. (Hassanein et al. 2009). Along with the production of antifungal compounds, plant-associated microbes confer protection to the plant by eliciting the plant defence system. This strategy is known as induced systemic resistance (ISR). In the ISR, the inducing bacterium does not exert any apparent damages to the host plant. Another defence response is systemic acquired resistance which is relatively similar to ISR response. In SAR, primary infection of pathogens activates their defence mechanism. The plant-associated microbes produce specific molecules known as elicitors which trigger the plant defence responses. The chitosan from the fungal cell wall, lipopolysaccharides from gram-negative bacteria, elicitins from oomycetes of *Phytophthora cryptogea* and flg22 from bacterial flagellin are examples of microbial elicitors responsible for the induction of systemic resistance in the plant (reviewed by Garcia-Brugger et al. 2006).

12 Improvement in Phytoremediation by Plant-Associated Microbes

The plant has a natural tendency to absorb soil as well as atmospheric pollutants and remediate these environmental pollutants. The plants raise the endophytes harbouring degrading genes when they are grown in contaminated soil. The plant-associated microbes belonging to genera including *Pseudomonas*, *Burkholderia*, *Methylobacter* and *Herbaspirillum* have the ability to degrade a wide range of pollutants including methane, trinitro toluene, chlorinated benzoic acids, nitro aromatics, benzene, toluene, ethylbenzene, xylene, tetrachlorophenol and polychlorinated biphenyl (reviewed by Ryan et al. 2008). The methylobacterium endophyte of the hybrid poplar tree had the capacity to degrade nitro-aromatic compounds including 2,4,6,-trinitro toluene (Van Aken et al. 2004). The inoculation of *Pseudomonas* endophyte along with organochloride herbicide 2,4-dichloro phenoxyacetic acid (2,4-D) in pea plants showed no accumulation of 2,4-D (Germaine et al. 2006). Another experiment of the inoculation of 2,4-D without preinoculation of 2,4-D degrading strain leads to significant detection of herbicide and the signs of phytotoxicity such as reduction in biomass and leaf abscission. This experiment demonstrated the effectiveness of the *Pseudomonas* strain in improving phytoremediation of 2,4-D (Germaine et al. 2006). The phyllosphere and endosphere are the important niches for horizontal gene transfer through plasmid due to the aggregation of diversified microbes carrying degradative genes. The natural transfer of degradative plasmid, pTOM-Bu61, in the endophytes suggests the importance of plant-associated microbes in increasing the diversity of microbes carrying the degradative genes which have importance in phytoremediation of environmental pollutants (Taghavi et al. 2005).

13 Concluding Remarks

In the universe, microbes seem to have coevolved with the plant system and established close association to perform biological functions. The plant system is an important niche to study microbial ecology and their variety of relationships with the plant. The symbiotic relationship between the plant and microbes confers a wide range of benefits to plant as well as microbial communities. In such interaction, the plant provides a platform for the microbes to perform diversified biological activity and produce bioactive metabolites which can have a positive impact on the plant's health. The beneficial plant-associated microbes can serve for their wide array of applications in agriculture, environmental clean-up and pharmaceutical field. These plant-associated microbes with the ability to produce pathogen-inhibitory activity and plant-growth-promoting potential can be used as potential biocontrol agents and biofertilizers in agriculture industries. The capability of these microbes to synthesize a bioactive metabolite with a diverse range of bioactivity such as antimicrobial, antifungal, antitumour and antiviral can be exploited in the development of a potent therapeutic agent in pharmaceutical industries. The capacity of biotransformation and degradation of toxic environmental pollutant leads to exploitation of plant-associated microbes in microbe-mediated phytoremediation to clean up the environment from the pollutants. Though the plant-associated microbes have been studied extensively, their interactive effects and functions have not been comprehensively understood. In detail an understanding of this plant-microbe interactive relationship may open the door to develop efficient biofertilizers, biocontrol agents, novel therapeutics, environmental clean-up agents and efficient nutraceuticals and to make disease-resistant plants.

References

Andrews JH, Harris RF (2000) The ecology and biogeography microorganisms on plant surfaces. Annu Rev Phytopathol 38:145–180

Arevalo-Ferro C, Reil G, Gorg A, Eberl L, Riedel K (2005) Biofilm formation of *Pseudomonas putida* IsoF: the role of quorum sensing as assessed by proteomics. Syst Appl Microbiol 28:87–114

Bacilio-Jiménez M, Aguilar-Flores S, Ventura-Zapata E, Pérez-Campos E, Bouquelet S, Zenteno E (2003) Chemical characterization of root exudates from rice (Oryza sativa) and their effects on the chemotactic response of endophytic bacteria. Plant Soil 249:271–277

Bacon CW, White JF (2000) Microbial endophytes. Marcel Dekker, New York

Badri DV, Loyola-Vargas VM, Broeckling CD, De-la-Peña C, Jasinski M, Santelia D, Martinoia E, Sumner LW, Banta LM, Stermitz F, Vivanco JM (2008a) Altered profile of secondary metabolites in the root exudates of Arabidopsis ATP-binding cassette transporter mutants. Plant Physiol 146:762–771

Badri DV, Loyola-Vargas VM, Du J, Stermitz FR, Broeckling CD, Iglesias-Andreu L, Vivanco JM (2008b) Transcriptome analysis of Arabidopsis roots treated with signalling compounds: a focus on signal transduction, metabolic regulation and secretion. New Phytol 179:209–223

Bailey MJ, Rainey PB, Zhang XX, Lilley AK (2002) Population dynamics, gene transfer and gene expression in plasmids: the role of the horizontal gene pool in local adaptation at the plant surface. In: Lindow SE, Hecht-Poinar EI, Elliot VJ (eds) Phyllosphere microbiology. APS Press, St. Paul, pp 173–192

Bais HP, Park SW, Weir TL, Callaway RM, Vivanco JM (2004) How plants communicate using the underground information superhighway. Trends Plant Sci 9:26–32

Beattie GA, Lindow SE (1995) The secret life of foliar bacterial pathogens on leaves. Annu Rev Phytopathol 33:145–172

Bever JD, Platt TG, Morton ER (2012) Microbial population and community dynamics on plant roots and their feedbacks on plant communities. Annu Rev Microbiol 66:265–283

Braeken K, Daniels R, Ndayizeye M, Vanderleyden J, Michiels J (2008) Chapter 11 Quorum sensing in bacteria-plant interactions. In: Nautiyal CS, Dion P (eds) Molecular mechanisms of plant 265 and microbe coexistence, vol 15, Soil biology. Springer, Berlin/Heidelberg

Brandl MT, Lindow SE (1998) Contribution of indole-3-acetic acid production to the epiphytic fitness of *Erwinia herbicola*. Appl Environ Microbiol 64:3256–3263

Chancey ST, Wood DW, Pierson LS III (1999) Two-component transcriptional regulation of *N* acyl-homoserine lactone production in *Pseudomonas aureofaciens*. Appl Environ Microbiol 65:2294–2299

Choi O, Kim J, Kim JG, Jeong Y, Moon JS, Park CS, Hwang I (2008) Pyrroloquinoline quinone is a plant growth promotion factor produced by Pseudomonas fluorescens B16. Plant Physiol 146(2):657–668

Compant S, Duffy B, Nowak J, Cl C, Barka EA (2005) Use of plant growth-promoting bacteria for biocontrol of plant diseases: principles, mechanisms of action, and future prospects. Appl Environ Microbiol 71:4951–4959

Costerton JW, Lewandowski Z, Caldwell DE, Korber DR, Lappin-Scott HM (1995) Microbial biofilms. Annu Rev Microbiol 49:711–745

de Weert S, Vermeiren H, Mulders IH, Kuiper I, Hendrickx N, Bloemberg GV, Vanderleyden J, De Mot R, Lugtenberg BJ (2002) Flagella-driven chemotaxis towards exudates components is an important trait for tomato root colonization by *Pseudomonas fluorescens*. Mol Plant Microbe Interact 15:1173–1180

De-la-Peña C, Lei Z, Watson BS, Sumner LW, Vivanco JM (2008) Root-microbe communication through protein secretion. J Biol Chem 283:25247–25255

Ercolani GL (1991) Distribution of epiphytic bacteria on olive leaves and the influence of leaf age and sampling time. Microb Ecol 21:35–48

Finlay RD (2008) Ecological aspects of mycorrhizal symbiosis: with special emphasis on the functional diversity of interactions involving the extraradical mycelium. J Exp Bot 59:1115–1126

Fokkema NJ, Schippers B (1986) Phyllosphere vs rhizosphere as environments for saprophytic colonization. In: Fokkema NJ, Van den Heuvel J (eds) Microbiology of the phyllosphere. Cambridge University Press, London, pp 137–159

Fry SC (1989) Cellulases, hemicelluloses and auxin-stimulated growth: a possible relationship. Physiol Plant 75:532–536

Fuqua C, Parsek MR, Greenberg EP (2001) Regulation of gene expression by cell-to-cell communication: acyl-homoserine lactone quorum sensing. Annu Rev Genet 35:439–468

Gangwar M, Kaur G (2009) Isolation and characterization of endophytic bacteria from endorhizosphere of sugarcane and ryegrass. Internet J Microbiol 7. doi:10.5580/181d

Germaine K, Liu X, Cabellos G, Hogan J, Ryan D, Dowling DN (2006) Bacterial endophyte-enhanced phyto-remediation of the organochlorine herbicide 2,4-dichlorophenoxyacetic acid. FEMS Microbiol Ecol 57:302–310

Guo BH, Wang YC, Zhou XW, Hu K, Tan F, Miao ZQ, Tang KX (2006) An endophytic taxol-producing fungus BT2 isolated from *Taxus chinensis* var. *mairei*. Afr J Biotechnol 5:875–877

Garcia-Brugger A, Lamotte O, Vandelle E, Bourque S, Lecourieux D, Poinssot B, Wendehenne D, Pugin A (2006) Early signaling events induced by elicitors of plant defenses. Mol Plant Microbe Interact 19 (7):711–24, Review

Hassanein WA, Awny NM, El-Mougith AA, Salah El-Dien SH (2009) The antagonistic activities of some metabolites produced by *Pseudomonas aeruginosa* Sha8. J Appl Sci Res 5:404–414

He SY (1998) Type III protein secretion systems in plant and animal pathogenic bacteria. Annu Rev Phytopathol 36:363–392

Hirano SS, Upper CD (1989) Diel variation in population size and ice nucleation activity of *Pseudomonas syringae* on snap bean leaflets. Appl Environ Microbiol 55:623–630

Hirano SS, Upper CD (2000) Bacteria in the leaf ecosystem with emphasis on *Pseudomonas syringae*—a pathogen, ice nucleus, and epiphyte. Microbiol Mol Biol Rev 64:624–653

Hirano SS, Ostertag EM, Savage SA, Baker LS, Willis DK, Upper DC (1997) Contribution of the regulatory gene *lemA* to field fitness of *Pseudomonas syringae* pv. syringae. Appl Environ Microbiol 63:4304–4312

Hirano SS, Charkowski AO, Collmer A, Willis DK, Upper CD (1999) Role of the Hrp type III protein secretion system in growth of *Pseudomonas syringae* pv. syringae B728a on host plants in the field. Proc Natl Acad Sci U S A 96:9851–9856

Hutchison ML, Johnstone K (1993) Evidence for the involvement the surface active properties of the extracellular toxin tolaasin in the manifestation of brown blotch disease symptoms by *Pseudomonas tolaasii* on *Agaricus bisporus*. Physiol Mol Plant Pathol 42:373–384

Hutchison ML, Tester MA, Gross DC (1995) Role of biosurfactant and ion channel-forming activities of syringomycin in transmembrane ion flux: a model for the mechanism of action in the plant-pathogen interaction. Mol Plant Microbe Interact 8:610–620

Jacques MA, Kinkel LL, Morris CE (1995) Population sizes, immigration, and growth of epiphytic bacteria on leaves of different ages and positions of field-grown endive (*Cichorium endivia* var. *latifolia*). Appl Environ Microbiol 61:899–906

Joseph B, Patra RR, Lawrence R (2007) Characterization of plant growth promoting Rhizobacteria associated with chickpea (Cicer arietinum L). Int J Plant Prod 1:141–152

Kawaguchi M, Minamisawa K (2010) Plant-microbe communications for symbiosis. Plant Cell Physiol 51:1377–1380

Khalid A, Arshad M, Zahir ZA (2004) Screening plant growth-promoting rhizobacteria for improving growth and yield of wheat. J Appl Microbiol 96:473–480

Kiraly Z, El-Zahaby HM, Klement Z (1997) Role of extracellular polysaccharide (EPS) slime in plant pathogenic bacteria in protecting cells to reactive oxygen species. J Phytopathol 145:59–68

Kloepper JW, Leong J, Teintze M, Schroth MN (1980) Enhanced plant growth by siderophores produced by plant growth promoting rhizobacteria. Nature 286:885–886

Legard DE, McQuilken MP, Whipps JM, Fenlon JS, Fermor TR, Thompson IP, Bailey MJ, Lynch JM (1994) Studies of seasonal changes in the microbial populations on the phyllosphere of spring wheat as a prelude to the release of genetically modified microorganisms. Agric Ecosyst Environ 50:87–101

Li J, Beatty PK, Shah S, Jensen SE (2007) Use of PCR-targeted mutagenesis to disrupt production of Fusaricidin-type antifungal antibiotics in Paenibacillus polymyxa. Appl Environ Microbiol 73:3480–3489

Licciardello G, Bertani I, Steindler L, Bella P, Venturi V, Catara V (2007) *Pseudomonas corrugate* contains a conserved *N*-acyl homoserine lactone quorum sensing system; its role in tomato pathogenicity and tobacco hypersensitivity response. FEMS Microbiol Ecol 61:222–234

Lilley AK, Bailey MJ (1997) The acquisition of indigenous plasmids by a genetically marked pseudomonad population colonizing the sugar beet phytosphere is related to local environment conditions. Appl Environ Microbiol 63:1577–1583

Lindow SE (1985) Integrated control and role of antibiosis in biological control of fire blight and frost injury. In: Windels C, Lindow SE (eds) Biological control on the phylloplane. APS Press, St. Paul, pp 83–115

Lindow SE (1987) Competitive exclusion of epiphytic bacteria by Ice *Pseudomonas syringae* mutants. Appl Environ Microbiol 53:2520–2527

Lindow SE (1995) Control of epiphytic ice nucleation-active bacteria for management of plant frost injury. In: Lee RE, Warren GJ, Gusta LV (eds) Biological ice nucleation and its applications. APS Press, St. Paul, pp 239–256

Lindow SE, Brandl MT (2003) Microbiology of the phyllosphere. Appl Environ Microbiol 69:1875–1883

Lindow SE, McGourty G, Elkins R (1996) Interactions of antibiotics with *Pseudomonas fluorescens* strain A506 in the control of fire blight and frost injury to pear. Phytopathology 86:841–848

Loyola-Vargas VM, Broeckling CD, Badri D, Vivanco JM (2007) Effect of transporters on the secretion of phytochemicals by the roots of *Arabidopsis thaliana*. Planta 225:301–310

Mandal SM, Dey S (2008) LC-MALDI-TOF MS-based rapid identification of phenolic acids. J Biomol Tech 19(2):116–121

Mandal SM, Chakraborty D, Dey S (2010) Phenolic acids act as signalling molecules in plant -microbe symbioses. Plant Signal Behav 5:359–368

Mauch F, Mauch-Mani B, Boller T (1988) Antifungal hydrolases in pea tissue. II. Inhibition of fungal growth by combinations of chitinase and/3-1,3-glucanase. Plant Physiol 88:936–942

Mercier J, Lindow SE (2001) Field performance of antagonistic bacteria identified in a novel laboratory assay for biological control of fire blight of pear. Biol Control 22:66–71

Miller CM, Miller RV, Garton-Kenny D, Redgrave B, Sears J, Condron MM, Teplow DB, Strobel GA (1998) Ecomycins, unique antimycotics from Pseudomonas viridiflava. J Appl Microbiol 84:937–944

Montesinos E (2003) Plant-associated microorganisms: a view from the scope of microbiology. Int Microbiol 6:221–223

Morgan JA, Bending GD, White PJ (2005) Biological costs and benefits to plant-microbe interactions in the rhizosphere. J Exp Bot 56:1729–1739

Narasimhan K, Basheer C, Bajic VB, Swarup S (2003) Enhancement of plant- microbe interactions using a rhizosphere metabolomics-driven approach and its application in the removal of polychlorinated biphenyls. Plant Physiol 32:146–153

Narula N, Deubel A, Gans W, Behl RK, Merbach W (2006) Paranodules and colonization of wheat roots by phytohormone producing bacteria in soil. Plant Soil Environ 52:119–129

Orhan E, Esitken A, Ercisli S, Turan M, Sahin F (2006) Effects of plant growth promoting rhizobacteria (PGPR) on yield, growth and nutrient contents in organically growing raspberry. Sci Hortic 111:38–43

Pathak KV, Keharia H, Gupta K, Thakur SS, Balaram P (2012) Lipopeptides from the banyan endophyte, *Bacillus subtilis* K1: mass spectrometric characterization of a library of fengycins. J Am Soc Mass Spectrom 23:1716–1728

Pillay VK, Nowak J (1997) Inoculum density, temperature, and genotype effects on in vitro growth promotion and epiphytic and endophytic colonization of tomato (*Lycopersicon esculentum* L.) seedlings inoculated with a pseudomonad bacterium. Can J Microbiol 43:354–361

Piper KR, von Bodman SB, Farrand SK (1993) Conjugation factor of *Agrobacterium tumefaciens* regulates Ti plasmid transfer by autoinduction. Nature 362:448–450

Pusey PL (2002) Biological control agents for fire blight of apple compared under conditions limiting natural dispersal. Plant Dis 86:639–644

Radulović NS, Blagojević PD, Stojanović-Radić ZZ, Stojanović NM (2013) Antimicrobial plant metabolites: structural diversity and mechanism of action. Curr Med Chem 20:932–952

Redecker D, Morton JB, Bruns TD (2000) Ancestral lineages of arbuscular mycorrhizal fungi (Glomales). Mol Phylogenet Evol 14(2):276–284

Rangarajan S, Loganathan P, Saleena LM, Nair S (2001) Diversity of pseudomonads isolated from three different plant rhizospheres. J Appl Microbiol 91:742–749

Reichling J (2010) Plant-microbe interactions and secondary metabolites with antibacterial, antifungal and antiviral properties. In: Wink M (ed) Annual plant reviews volume 39: functions and biotechnology of plant secondary metabolites, 2nd edn. Wiley-Blackwell, Oxford

Reinhold B, Hurek T, Fendrik I (1985) Strain-specific chemotaxis of *Azospirillum* spp. J Bacteriol 162:190–195

Reiter B, Bürgmann H, Burg K, Sessitsch A (2003) Endophytic *nifH* gene diversity in African sweet potato. Can J Microbiol 49:549–555

Rivas R, Peix A, Mateos PF, Trujillo ME, Martinez-Molina E, Velazqueze E (2006) Biodiversity of populations of phosphate solubilizing rhizobia that nodulates chickpea in different Spanish soils. Plant Soil 287:23–33

Rosenblueth M, Martínez-Romero E (2006) Bacterial endophytes and their interactions with hosts. Mol Plant Microbe Interact 19:827–837

Ryan RP, Germaine K, Franks A, Ryan DJ, Dowling DN (2008) Bacterial endophytes: recent developments and applications. FEMS Microbiol Lett 278:1–9

Ryu R, Patten CL (2008) Aromatic amino acid-dependent expression of indole-3-pyruvate decarboxylase is regulated by 4 TyrR in Enterobacter cloacae UW5. Am Soc Microbiol 190:1–35

Saharan BS, Nehra V (2011) Plant growth promoting rhizobacteria: a critical review. Life Sci Med Res 2011:1–30

Shulaev V, Cortes D, Miller G, Mittler R (2008) Metabolomics for plant stress response. Physiol Plant 132:199–208

Smeekens S, Rook F (1997) Sugar sensing and sugar-mediated signal transduction in plants. Plant Physiol 115:7–13

Smith SE, Read DJ (2008) Mycorrhizal symbiosis, 3rd edn. Academic, Amsterdam/London

Somers E, Vanderleyden J, Srinivasan M (2004) Rhizosphere bacterial signaling: a love parade beneath our feet. Crit Rev Microbiol 30:205–240

Spaepen S, Vanderleyden J, Remans R (2007) Indole-3-acetic acid in microbial and microorganism-plant signalling. FEMS Microbiol Rev 31:425–448

Sridevi M, Mallaiah KV (2008) Production of hydroxamate-type of siderophore by *Rhizobium* strains from *Sesbania sesban* (L). Int J Soil Sci 3:28–34

Strobel G, Daisy B, Castillo U, Harper J (2004) Natural products from endophytic microorganisms. J Nat Prod 67:257–268

Sturz AV, Christie BR, Matheson BG, Nowak J (1997) Biodiversity of endophytic bacteria which colonize red clover nodules, roots, stems and foliage and their influence on host growth. Biol Fertil Soils 25:13–19

Sugiyama A, Shitan N, Sato S, Nakamura Y, Tabata S, Yazaki K (2006) Genome-wide analysis of ATP-binding cassette (ABC) proteins in a model legume plant, *Lotus japonicus*: comparison with Arabidopsis ABC protein family. DNA Res 13:205–228

Taghavi S, Barac T, Greenberg B, Borremans B, Vangronsveld J, van der Lelie D (2005) Horizontal gene transfer to endogenous endophytic bacteria from poplar improved phyto-remediation of toluene. Appl Environ Microbiol 71:8500–8505

Van Aken B, Peres C, Doty S, Yoon J, Schnoor J (2004) Methylobacterium populi sp. nov., a novel aerobic, pink-pigmented, facultatively methylotrophic, methane-utilising bacterium isolated from poplar trees (*Populus deltoides* x *nigra* DN34). Evol Microbiol 54:1191–1196

van der Heijden MG, Bardgett RD, van Straalen NM (2008) The unseen majority: soil microbes as drivers of plant diversity and productivity in terrestrial ecosystems. Ecol Lett 11:296–310

Wani MC, Taylor HL, Wall ME, Coggon P, Mcphail AT (1971) Plant antitumor agents. VI. Isolation and structure of taxol, a novel antileukemic and antitumor agent from *Taxus brevifolia*. J Am Chem Soc 93:2325–2327

Waters CM, Bassler BL (2005) Quorum sensing: cell-to-cell communication in bacteria. Annu Rev Cell Dev Biol 21:319–346

Welbaum G, Sturz AV, Dong Z, Nowak J (2004) Fertilizing soil microorganisms to improve productivity of agroecosystems. Crit Rev Plant Sci 23:175–193

Whipps JM (2001) Microbial interactions and biocontrol in the rhizosphere. J Exp Bot 52:487–511

Wilson M, Lindow SE (1994) Coexistence among epiphytic bacterial populations mediated through nutritional resource partitioning. Appl Environ Microbiol 60:4468–4477

Wood DW, Gong F, Daykin MM, Williams P, Pierson LS III (1997) *N*-Acyl-homoserine lactone mediated regulation of phenazine gene expression by *Pseudomonas aureofaciens* 30–84 in the wheat rhizosphere. J Bacteriol 179:7663–7670

Wu CH, Bernard SM, Andersen GL, Chen W (2009) Developing microbe-plant interactions for applications in plant-growth promotion and disease control, production of useful compounds, remediation and carbon sequestration. Microb Biotechnol 2:428–440

Zhang Z, Pierson LS III (2001) A second quorum-sensing system regulates cell surface properties but not phenazine antibiotic production in *Pseudomonas aureofaciens*. Appl Environ Microbiol 67:4305–4315

Machine Learning with Special Emphasis on Support Vector Machines (SVMs) in Systems Biology: A Plant Perspective

Tiratha Raj Singh

Abstract

Systems biology has been progressing with integrative genomics and tools such as bioinformatics. Recent developments in high-throughput techniques have led to the accumulation of deluge of biological data. To address specific biological questions and to generate biologically meaningful information from this deluge of data, there was a need to integrate components and system levels at biological point of view. Combined strategies from systems biology and computational biology lead to computational systems biology. Logical applications from machine learning have lots of applications with state-of-the-art techniques to deal with this data. Machine-learning applications in biology gave enhancements to the overall aspects of biological problems and their fast and accurate solutions. This chapter addresses the implications and applications of machine-learning techniques with special emphasis on support vector machines, on plants and associated research areas.

Keywords

Machine learning • SVM • Bioinformatics • Systems biology

Abbreviations

ANN	Artificial neural network
ATF3	Activating transcription factor 3
IFN γ	Interferon gamma
LOO	Leave-one-out
MCMV	Murine cytomegalovirus
miRNA	microRNA
SNPs	Single nucleotide polymorphisms
SVMs	Support Vector Machines
TRN	Transcriptional regulatory network

T.R. Singh (✉)
Department of Biotechnology and Bioinformatics, Jaypee University of Information Technology (JUIT), Waknaghat, Solan, HP, India
e-mail: tiratharaj@gmail.com

1 Background

Recent revolutions in high-throughput data accumulations offer scientists to study biological entities at system level. Shift of paradigm from component-level to system-level understanding of biological processes leads to understanding

K.K. P.B. et al. (eds.), *Agricultural Bioinformatics*,
DOI 10.1007/978-81-322-1880-7_16, © Springer India 2014

complex biological processes involved in the management and manipulation of biological entities. Systems biology therefore provides an opportunity to analyze complex systems by combining prior knowledge with data (Huang 2004; Bruggeman and Westerhoff 2007).

While bioinformatics and systems biology have induced new significant developments associated with biological data, where most of the targets of interest are biological sequences, their two- and three-dimensional structures, or their interaction networks, machine learning naturally appears as one of the main drivers of progress in this context (Huang and Wikswo 2006). Machine learning is also progressing with the advancements of learning with structured data, graph inference, semi-supervised learning, system identification, and novel combinations of optimization and learning algorithms. There is a growing attention in the area of logic-based machine learning which includes availability of large-scale biological data such as interaction and structural data. This serves as a catalyst for supervised learning through training and testing phases.

Machine-learning techniques are ideal for interpreting unannotated deluge amount of genomic DNA and for elucidating various kind of functionalities for the biological data (Baldi et al. 2000). These approaches have some advantages over other contemporary methods in having a built-in robustness when presented with uncorrelated data features. These machine-learning techniques are excellent example for discarding and compacting redundant sequence and other biological information (Baldi and Brunak 2001). Machine-learning techniques are able to cope with nonlinearities and to find more complex correlations in sequence spaces that are not functionally segregated into continuous domains. On the other hand, these techniques can fit fine in the scenario of structural, functional, and evolutionary domains also.

There are obvious evidences of cutting edge research advancements in systems biology for efficient and effective development of computational methodologies for modeling, simulation, and analysis of complex biological processes. Modeling within systems biology is a key application area for machine learning in general. Modeling could range from small component level such as a compound to a very wide biological system such as ecosystem. It involves gene regulatory networks, protein-protein interaction networks, metabolic networks, signal transduction networks, and many other kinds of biological networks. Some important and crucial developments in this field have been noticed and found worthy for the bright future of this rapidly emerging area.

The development of multiscale computational modeling and simulation techniques and their applications in the areas such as aging and drug design has shown promising results and outcomes. Three-dimensional multiscale brain tumor model developed by Zhang et al. (2007) is an endeavor in this direction. A multiscale computational framework to investigate morphogenesis mechanisms in *Xenopus laevis* has been presented (Robertson et al. 2007). There have been extensive studies on aging in *Caenorhabditis elegans*, humans, mice, fruit flies, and other organisms. Several genes have been discovered that extend organisms' lifespan. Increase in longevity possibly by genetic mutations has been reported in various aging studies (Longo and Finch 2003; Kenyon 2005; Longo et al. 2008).

Computational approaches could provide elucidation to the key issue that is how to design lifesaving and cost-effective drugs so that the diseases such as cancer, AIDS, and many other bacterial and viral illnesses could be cured and prevented. Systems-based computational techniques will be useful in designing and development of effective therapeutic drugs, and it is also the topical vision of pharmaceutical companies. Recently, there is an introduction of a methodology for the prediction of pathway responses to combinatorial drug perturbations or drug combinations. The method uses multiple input-output model (Nelander et al. 2008).

In another study, the role of activating transcription factor 3 (ATF3) was investigated for murine cytomegalovirus (MCMV) infection. The study demonstrated negative regulation of

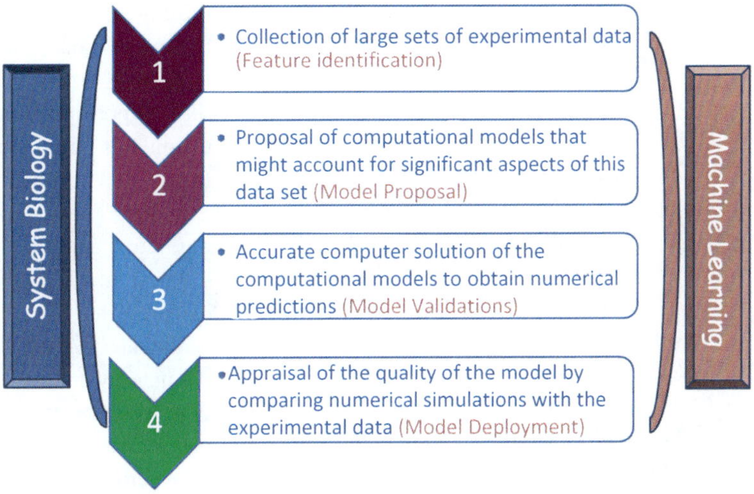

Fig. 1 Schematic representation of general computational approach for the analysis and annotation of biological systems data. Systems biology and machine learning represent two sides of the data and analysis, respectively

interferon gamma (IFN-γ) expression caused by ATF3 in natural killer cells and the mice that had zero ATF3 exhibited high resistance to MCMV infection. Thus, transcription-factor-based methods can also play an important role for devising effective therapeutic and preventive intervention strategies for diseases (Rosenberger et al. 2008). These advancements in all aspects of life could provide learned and insightful predictions (Fig. 1) for future enhancement and positive refinement of living systems. The purpose of this chapter is to elucidate association between machine learning and systems biology and to correlate it with plant-specific research.

1.1 Applications of Machine Learning in Systems Biology

Machine learning aims to extract useful information from an enormous mass of data. This information should be biologically meaningful when this mass has biological nature. Machine-learning methods are computationally rigorous and benefit greatly from progress in computer technology in terms of speed, accuracy, and compatibility. It is interesting to see that bioinformatics and machine-learning methods have incorporated well to have a momentous impact in biology, biotechnology, and medicine. Machine-learning methods have successfully been implemented for solving bioinformatics problems ranging from gene prediction, transcription factor binding site, promoters, nucleosome positioning signals, RNA secondary structure prediction, protein structures, protein family classification, microarray data analysis, single nucleotide polymorphisms (SNPs), etc. (Brown et al. 2000; Donnes and Elofsson 2002; Ward et al. 2003; Morel et al. 2004; Garg et al. 2005; Matukumalli et al. 2006; Yang et al. 2008; Singh et al. 2011; Wu et al. 2011; Gupta and Singh 2013).

Specific applications of machine-learning techniques have been developed for the analysis of biological data at interaction levels which has been generated through systems biology approaches. These approaches have been implemented in various aspects of biological networks ranging from gene regulatory or transcriptional networks (De Jong 2002; Tong et al. 2004), protein-protein interaction networks (Ng et al. 2003), signal transduction networks (Klipp et al. 2002), metabolic networks (Fiehn 2002; Olivier and Snoep 2004), etc. Besides that, several other kinds of biological networks have been incorporated to perform machine-learning-based studies such as phylogenetic networks (Singh 2011) and ecological networks (Laska and Wootton 1998).

2 Support Vector Machines (SVMs)

Support vector machines (SVMs) are an excellent machine-learning technique. SVMs have several attractive features as these are supervised learning methods that can influence labeled data. SVMs are able to minimize the structural risk through a unique hyperplane with maximum margin to separate data from two classes. This feature provides SVMs best generalization ability for blind data set (Yang 2004).

SVM classification of a sample with a vector x of predictors is based on

$$f(x) = \text{sign}\left[\sum_i y_i \alpha_i k(x_i, x) + b\right]$$

where the kernel function k measures the similarity of its two vector arguments. For a linear SVM, the inner product kernel function $f(x)$ is used. If $f(x)$ is positive, then the sample is predicted to be in class +1, otherwise class −1. The summation is over the set of "support vectors" that defines the boundary between the classes. Support vector x_i is associated with a class label y_i that is either +1 or −1. The $\{\alpha_i\}$ and b coefficients are determined by "learning" the data. An SVM attempts to minimize the generalization error for the independent data rather than minimizing the mean square error for the training set; therefore, it is an approximate implementation of the structural risk minimization induction principle. For two-group classification, the SVM separates the classes with a surface that maximizes the margin between them (Joachims 1999).

Because the optimal parameters for each classification task are not known from the beginning, it is necessary to test different parameters in order to find the optimal ones. This is best done by a systematic sampling of the parameter space by grid search. Here, several combinations of two parameters, c and γ, were tested, given the start, stop, and step size for each of the parameters. In general, the RBF kernel is a reasonable first choice which nonlinearly maps samples into a higher dimensional space, unlike the linear kernel, and can handle the case when the relation between class labels and attributes is nonlinear. Four variables are defined and used for this purpose: true-positives (TP), the number of candidates predicted as such; true-negatives (TN), the number of noncandidates predicted as such; false-positives (FP), the number of predicted candidates that actually are noncandidates; and false-negatives (FN), the number of predicted noncandidates that actually are candidates. A perfect correlation of real and predicted values will result in Matthew's correlation coefficient (MCC) value of 1; random predictions would result in lesser values, close to 0; whereas negative correlation would result in values closer to −1. Hence, the value of MCC is another parameter that defines the quality of the model generated and used for training the SVM algorithm. Sensitivity defines the rate of prediction of true-positives from the set of prediction that are truly accepted and falsely rejected, whereas specificity determines the rate of true-negatives from the set of predictions that are truly rejected and falsely accepted. For evaluations, following standard parameters should be used:

Sensitivity = TP/TP + FN
Specificity = TN/TN + FP
Accuracy = TP + TN/TP + FP + TN + FN

Mathew's Correlation Coefficient (MCC)

$$= \frac{(TP*TN) - (FN*FP)}{\sqrt{(TP+FN)*(TN+FP)*(TP+FP)*(TN+FN)}}$$

SVM-based algorithms are particularly appealing for binary type of prediction because of the ability of SVMs to build efficient prediction models when the dimensionality of the data is high and the number of observations is limited. Machine-learning methods need a sufficient amount of data for training. Generation of five different sets randomly from the original set of both categories, for the purpose of training and testing the SVM algorithm for any given set of parameters, is called fivefold cross-validation. Each set contains 20 % test and 80 % train data. Another approach applied is tenfold cross-validations where each set contains 10 % test and 90 % train data. Another approach that is

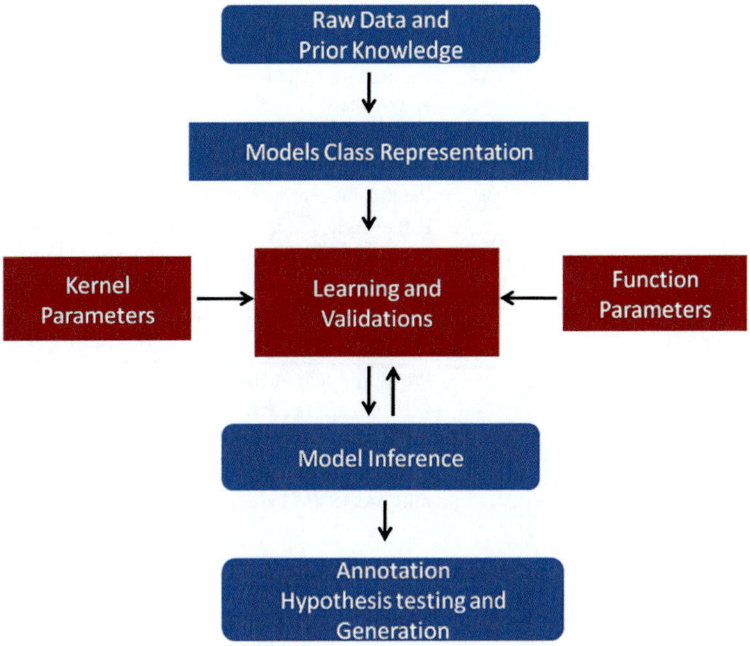

Fig. 2 Pipeline of general machine-learning approach for the analysis and annotation of biological data through various classification schemes and parameter estimations

also important is leave-one-out (LOO) where each set contains 1 % test and 99 % train data, i.e., on each iteration, we have to cross-validate for 1:99 ratio for test and train data, respectively. After validations, suitable parameters have to be decided to finally apply on prediction computations. Concise scheme of inference for general machine-learning methodology with an emphasis on SVM parameters is shown in Fig. 2.

2.1 Support Vector Machines in Systems Biology

The support vector machines (SVMs) have successfully used for the classification of MHC binder prediction, protein secondary structure prediction, SNPs (single nucleotide polymorphisms), microarray data, microRNA (miRNA) precursors, and targets, and many other applications (Brown et al. 2000; Donnes and Elofsson 2002; Ward et al. 2003; Garg et al. 2005; Yang et al. 2008, and references therein). Integrating data from multiple global assays and curated databases is essential to understand the spatiotemporal interactions within cells. There are plenty of studies which are based on SVMs usage for analyzing interaction and association data. There are implementations where framework has been developed for transcriptional regulatory network (TRN) for functional annotations (Zhang et al. 2008).

2.2 Machine Learning and SVM Implementations Toward Their Application in Plants

Machine learning and SVMs have been successfully implemented for plant-associated biological aspects. Their implementations were applied through systems approach in myriad perspectives. SVM-based classification method has been developed where main basis in plant classification is plant's outward characteristics. The result of experiment can be made by applying the SVM algorithm and withdrawing the characteristic data of the plant (Yin-Xiao and Min 2007). A machine-learning-based method for fast and accurate detection and classification of plant diseases was proposed in Tang and Baojun (2009). Otsu segmentation, K-means clustering, and back propagation feed forward neural network were applied for clustering and

Table 1 State-of-the-art methods, tools, and resources for plant-specific research where machine-learning techniques have been applied to infer system-level understanding of plant systems

Methods/tools/resources	Applied for (data types and method)	Web address
Plant-microbe interaction	Microbial interaction (literature based)	http://www.scoop.it/t/plant-microbe-interaction
PathoPlant	Plant-pathogen interactions	http://www.pathoplant.de
PHI-base	Pathogen-host interactions	http://www.phi-base.org/
Pseudomonas-plant interaction	Plant-pathogen interactions	http://www.pseudomonas-syringae.org/pst_home.html
AtPID	Protein-protein interactions	http://www.megabionet.org/atpid/webfile/
FPPI database	Protein-protein interaction	http://csb.shu.edu.cn/fppi
SIGnal plant interactome database	Protein-protein interaction	http://signal.salk.edu/interactome.html
miRPara	miRNAs (SVM based)	http://159.226.126.177/mirpara/
MatureBayes	miRNAs (Bayesian statistics)	http://mirna.imbb.forth.gr/MatureBayes.html
MaturePred	miRNAs (SVM based)	http://nclab.hit.edu.cn/maturepred/

classification of diseases that affect plant leaves (Cui et al. 2009).

There are evidences of application of image-processing techniques and SVM for detecting rice diseases early and accurately (Cui et al. 2009; Tang and Baojun 2009). Tang and Baojun (2009) have sliced rice disease spots and then extracted their shape and texture features and used their characteristic values of classification. The SVM method was employed to classify rice bacterial leaf blight, rice sheath blight, and rice blast. Overall accuracy of 97.2 % was achieved (Tang and Baojun 2009), which indicated effective usage of SVMs in a specific plant disease early detection. Kaundal et al. (2006) proposed an SVM-based prediction model for predicting weather-based plant diseases. The performance of conventional multiple regression, artificial neural network (ANN), and SVM was compared. Authors have concluded that SVM-based regression approach better describes the relationship between the environmental conditions and disease level which could be useful for disease management (Kaundal et al. 2006).

Another interesting method using both color and texture feature to recognize plant leaf image was proposed by Man et al. (2008). Authors first preprocessed the images and then obtained color feature and texture feature plant images. Finally, SVM classifier was trained and used for plant images recognition. With a good accuracy of recognition, experimental results show that plant images could be recognized using both color and texture features (Man et al. 2008).

Initially, miRNA target transcripts were identified through experiments. Experimental methods are expensive and time consuming. Therefore, computational prediction for the miRNA targets could be faster and could save lots of time, money, and energy. Best suitable candidates can then be verified experimentally. This approach has been implemented, and several miRNA target prediction methods and tools were developed in the last decade. Few examples of such tools are MiRTif, miPred, mirTarget, MatureBayes, and many more (Table 1) (Kim et al. 2006; Jiang et al. 2007; Yang et al. 2008; Gkirtzou et al. 2010).

Some interesting tools have been developed which are based on hybrid data sets such as MaturePred (Xuan et al. 2011) and miRPara (Wu et al. 2011). MaturePred can accurately predict plant miRNAs and achieve higher prediction accuracy compared with the existing methods. Further, it has been trained on a prediction model with animal data to predict animal miRNAs. miRPara predicts most probable mature miRNA coding regions from genome-scale sequences in a species-specific manner. Here, sequences from miRBase were classified into animal, plant, and overall categories, and SVM was used to train three models based on an initial set of 77 parameters related to the physical properties of the pre-miRNA and its miRNAs. miRPara achieves an accuracy of up to 80 % against experimentally verified mature miRNAs, making it one of the most accurate methods available (Wu et al. 2011).

3 Recent Advancements and Applications of Machine Learning Along with Systems Biology in Plants

Several advancements in systems biology have been made and applied on plant-specific research which includes identification and analysis of transcription factor binding site (TFBS), miRNAs, elucidation of metabolic and signal transduction pathways for medicinal plants, etc. (Fiehn 2002; Han and Gross 2003, and references therein). In a very recent study, Wang et al. (2012) proposed a novel SVM-based detector, named MiR-PD, to identify pre-miRNAs in plants. The classifier is constructed based on twelve features of pre-miRNAs, inclusive of five global features and seven substructure features. MiR-PD achieves 96.43 % fivefold cross-validation accuracy while trained on 790 plant pre-miRNAs and 7,900 pseudo pre-miRNAs. MiR-PD reports an accuracy of 99.71 % with 77.55 % sensitivity and 99.87 % specificity, suggesting a feasible genome-wide application of this miRNA detector so as to identify species-specific novel miRNAs in plants without relying on phylogenetic conservation (Wang et al. 2012).

Identifying genes with essential roles in resisting environmental stress rates high in agronomic importance. Liang et al. (2011) applied an SVM-RFE (Support Vector Machine-Recursive Feature Elimination) feature selection method for the prediction of drought resistance and water susceptibility genes in *Arabidopsis thaliana*. Authors have used 22 sets of *Arabidopsis thaliana* gene expression data from GEO to predict the key genes involved in water tolerance. To address small sample sizes, they developed a modified approach for SVM-RFE by using bootstrapping and LOO cross-validations (Liang et al. 2011). There are many developments in systems biology through machine-learning techniques which generated a wealth of methods, tools, and resources (Table 1) for plant-specific research domain as discussed earlier. Various kinds of interaction resources were developed during the last decade ranging from plant-pathogen interactions (Bülow et al. 2004; Winnenburg et al. 2006), plant repeat database (Shu and Robin 2004), plant-specific protein-protein interaction database (Li et al. 2011), fungal-pathogen interaction database (Zhao et al. 2009), and many more coming up.

4 Conclusion

As was avowed at the beginning of this chapter, my aim was to discuss a direct association between machine learning and systems biology. Systems biology deciphers annotation at system level rather than at component level while machine learning aids component-level feature extraction into it. This combination provides a wealth of annotation methods being developed during the last decade which helped us in biological understanding of myriad of data types for plants. Despite substantial progress in machine-learning implementations in systems biology and their further applications in plant science, many open questions regarding nature's design of biological systems still remained unsolved. Understanding plant systems at single molecule level through various qualitative and quantitative approaches would definitely generate cooperation between experimentalists and computational researchers.

References

Baldi P, Brunak S (2001) Bioinformatics: the machine learning approach. MIT Press, Cambridge

Baldi P, Brunak S, Chauvin Y, Andersen CAF, Nielsen H (2000) Assessing the accuracy of prediction algorithms for classification: an overview. Bioinformatics 16:412–424

Brown MPS, Grundy WN, Lion D, Cristianini N, Sugnet CW, Furey TS, Ares M Jr, Haussler D (2000) Knowledge-based analysis of microarray gene expression data by using support vector machines. Proc Natl Acad Sci U S A 97:262–297

Bruggeman FJ, Westerhoff HV (2007) The nature of systems biology. Trends Microbiol 15:45–50
Bülow L, Schindler M, Choi C, Hehl R (2004) PathoPlant: a database on plant-pathogen interactions. In Silico Biol 4:0044
Cui D, Zhang O, Li M, Zhao Y, Hartman GL (2009) Detection of soybean rust using a multispectral image sensor. Sens Instrum Food Qual 3:49–56
De Jong H (2002) Modeling and simulation of genetic regulatory systems: a literature review. J Comput Biol 9:67–103
Donnes P, Elofsson A (2002) Prediction of MHC class I binding peptides, using SVMHC. BMC Bioinformatics 3:25
Fiehn O (2002) Metabolomics-the link between genotypes and phenotypes. Plant Mol Biol 48:155–171
Garg A, Bhasin M, Raghava GPS (2005) Support vector machine-based method for subcellular localization of human proteins using amino acid compositions, their order, and similarity search. J Biol Chem 280:14427–14432
Gkirtzou K, Tsamardinos L, Tsakalides P, Poirazi P (2010) Mature Bayes: a probabilistic algorithm for identifying the mature miRNA within novel precursors. PLoS One 5:e11843
Gupta A, Singh TR (2013) SHIFT: server for hidden stops analysis in frame-shifted translation. BMC Res Notes 6:68
Han X, Gross RW (2003) Global analyses of cellular lipidomes directly from crude extracts of biological samples by ESI mass spectrometry: a bridge to lipidomics. J Lipid Res 4:1071–1079
Huang S (2004) Back to the biology in systems biology: what can we learn from biomolecular networks? Brief Funct Genomic Proteomic 2:279–297
Huang S, Wikswo J (2006) Dimensions of systems biology. Rev Physiol Biochem Pharmacol 157:81–104
Jiang P, Wu H, Wang W, Ma W, Sun X et al (2007) MiPred: classification of real and pseudo microRNA precursors using random forest prediction model with combined features. Nucleic Acids Res 35:339–344
Joachims T (1999) Making large-scale SVM learning practical. In: Scholkopf B, Burges C, Smole A (eds) Advances in kernel methods – support vector learning. MIT Press, Cambridge, pp 169–184
Kaundal R, Kapoor AS, Raghava GPS (2006) Machine learning techniques in disease forecasting: a case study on rice blast prediction. BMC Bioinformatics 7:485
Kenyon C (2005) The plasticity of aging: insight from long lived mutant. Cell 120:449–460
Kim SK, Nam JW, Rhee JK, Lee WJ, Zhang BT (2006) miTarget: microRNA target gene prediction using a support vector machine. BMC Bioinformatics 7:411
Klipp E, Heinrich R et al (2002) Prediction of temporal gene expression. Metabolic optimization by re-distribution of enzyme activities. Eur J Biochem 269:5604–5613
Laska MS, Wootton JT (1998) Theoretical concepts and empirical approaches to measuring interaction strength. Ecology 79:461–476
Li P, Zang W, Li Y, Xu F, Wang J, Shi T (2011) AtPID: the overall hierarchical functional protein interaction network interface and analytic platform for Arabidopsis. Nucl Acids Res 39(suppl 1):D1130–D1133
Liang Y, Zhang F, Wang J, Joshi T, Wang Y et al (2011) Prediction of drought-resistant genes in Arabidopsis thaliana using SVM-RFE. PLoS One 6:e21750
Longo VD, Finch CE (2003) Evolutionary medicine: from dwarf model systems to healthy centenarians. Science 299:1342–1346
Longo VD, Leiber MR, Vijg J (2008) Turning antiaging genes against cancer. Mol Cell Biol 9:903–910
Man Q-K, Zheng C-H, Wang X-F, Lin F-Y (2008) Recognition of plant leaves using support vector machine. Commun Comput Inf Sci 15:192–199
Matukumalli LK, Grefenstette JJ, Hyten DL, Choi I-Y, Cregan PB, Tassell CPV (2006) Application of machine learning in SNP discovery. BMC Bioinformatics 7:4
Morel NM, Holland JM, van der Greef J, Marple EW et al (2004) Primer on medical genomics. Part XIV: Introduction to systems biology-a new approach to understanding disease and treatment. Mayo Clin Proc 79:651–658
Nelander S, Wang W, Nilsson B, Pratilas C, She QB, Rossen N, Gennemark P (2008) Models from experiments: combinatorial drug perturbations of cancer cells. Mol Syst Biol 4:216
Ng SK, Zhang Z, Tan SH (2003) Integrative approach for computationally inferring protein domain interactions. Bioinformatics 19:923–929
Olivier BG, Snoep JL (2004) Web-based kinetic modeling using JWS online. Bioinformatics 20:2143–2144
Robertson SH, Smith CK, Langhans AL, McLinden SE, Oberhardt MA, Jakab KR, Dzamba B, DeSimone DW, Papin JA, Peirce SM (2007) Multiscale computational analysis of *Xenopus laevis* morphogenesis reveals key insights of systems level behaviour. BMC Syst Biol 1:46
Rosenberger CM, Clark AE, Treuting PM, Jhonson CD, Aderem A (2008) Atf3 regulates mcmv infection in mice by modulating inf γ expression in natural killer cells. Proc Natl Acad Sci U S A 105:2544–2549
Shu O, Robin BC (2004) The TIGR Plant Repeat Databases: a collective resource for the identification of repetitive sequences in plants. Nucleic Acids Res 32:D360–D363
Singh TR (2011) Phylogenetic networks: concepts, algorithms and applications, book review. Curr Sci 100:1570–1571
Singh TR, Gupta A, Riju A, Mahalaxmi M, Seal A, Arunachalam V (2011) Computational identification and analysis of single nucleotide polymorphisms and insertions/deletions in expressed sequence tag data of *Eucalyptus*. J Genet 90:e34–e38
Tang, YH, Baojun Y (2009) Application of support vector machine for detecting rice diseases using shape and color texture features. In: Proceedings of international conference on engineering computation. IEEE Computer Society, pp 79–83

Tong AH, Lesage G, Bader GD, Ding H, Xu H et al (2004) Global mapping of the yeast genetic interaction network. Science 294:2364–2368

Wang Y, Jin C, Zhou M, Zhou A (2012) An SVM-based approach to discover microRNA precursors in plant genomes. Lect Notes Comput Sci 7104:304–315

Ward JJ, McGuffin LJ, Buxton BF, Jones DT (2003) Secondary structure prediction using support vector machines. Bioinformatics 19:1650–1655

Winnenburg R, Baldwin TK, Urban M, Rawlings C, Köhler J, Hammond-Kosack KE (2006) PHI-base: a new database for pathogen host interactions. Nucleic Acids Res 34:D459–D464

Wu Y, Wei B, Liu H, Li T, Rayner S (2011) MiRPara: a SVM-based software tool for prediction of most probable microRNA coding regions in genome scale sequences. BMC Bioinformatics 12:107

Xuan P, Guo M, Huang Y, Li W, Huang Y (2011) MaturePred: efficient identification of microRNAs within novel plant pre-miRNAs. PLoS One 6:e27422

Yang ZR (2004) Biological applications of support vector machines. Brief Bioinformatics 5:328–338

Yang Y, Wang Y-P, Li K-B (2008) MiRTif: a support vector machine-based microRNA target interaction filter. BMC Bioinformatics 9:S4

Yin-xiao MA, Min YAO (2007) Application of SVM in plant classification. Bull Sci Technol 3:404–407

Zhang L, Athale CA, Deisboeck TS (2007) Development of a three dimensional multiscale agent based tumor model: simulating gene protein interaction profiles, cell phenotypes and multicellular patterns in brain cancer. J Theor Biol 244:96–107

Zhang Y, Xuan J, de los Reyes BG, Clarke R, Ressom HW (2008) Network motif-based identification of transcription factor-target gene relationships by integrating multi-source biological data. BMC Bioinformatics 9:203

Zhao X-M, Zhang X-W, Tang W-H, Chen L (2009) FPPI: *Fusarium graminearum* protein-protein interaction database. J Proteome Res 8:4714–4721

Xanthine Derivatives: A Molecular Modeling Perspective

Renuka Suravajhala, Rajdeep Poddar, Sivaramaiah Nallapeta, and Saif Ullah

Abstract

Xanthines and its derivatives are a group of alkaloids which are well represented in plants. The xanthine analogs play a key role as adenosine receptors and calcium release channels and therefore can be used as behavioral stimulants. Various such stimulants in the form of bronchodilators, diuretics, natriuretics, analgesic adjuvants, and lipolytics are in use. This mini review focuses on the molecular modeling of xanthine derivatives which, in the recent past, has proved to be a promising avenue for structure-based methods of target identification.

Keywords

Xanthines • Adenosine receptors • Apoptosis • Biochemical modulators • Molecular modeling

Abbreviations

GPCR	G-protein-coupled receptors
MD	Molecular dynamics
RESP	Restrained Electrostatic Potential
SAR	Structure-activity relationships
SEAL	Steric and electrostatic alignment
TM	Transmembrane
XD	Xanthine derivatives
XO	Xanthine oxidase

1 Introduction

Xanthines are nitrogenous compounds containing two carbon rings with six and five member rings fused together. Hypoxanthine and xanthine are not included as a part of nucleic acids as they are important intermediates in the synthesis and degradation of purine nucleotides. Xanthines were first synthesized by German chemist Emil Fischer, and the word was later coined in 1899. Xanthines have a same skeleton that of purines which form the building blocks of ribonucleotides (RNA) and deoxyribonucleotides (DNA). Caffeine is an

R. Suravajhala (✉)
Department of Science, Systems and Models, Roskilde University, Univesitetsvej 1, 4000 Roskilde, Denmark
e-mail: renu@ruc.dk

R. Poddar • S. Nallapeta
Bioclues Organization, IKP Knowledge Park, Picket, Secunderabad 500009, AP, India

S. Ullah (✉)
Radiometer Medical Aps, Åkandevej 21, 2700 Brønshøj, Denmark
e-mail: saifullahsaul@gmail.com

K.K. P.B. et al. (eds.), *Agricultural Bioinformatics*,
DOI 10.1007/978-81-322-1880-7_17, © Springer India 2014

important and most commonly used derivative of xanthines, and these are present in most tissues and fluids. Caffeine, theobromine, and theophylline are collectively termed as xanthine derivatives (XD). Several groups of alkaloids including xanthines were commonly used as mild stimulants (Snyder et al. 1981). Caffeine is widely consumed in beverages as many xanthine analogs are shown to have a major impact in medicine (Dally and Fredholm 1998; Dally et al. 1998).

XDs are well known in plants. Topical administration of plant extracts, viz. Cola nuts, cacao beans, caffeine, theobromine, and theophylline has been shown to have several applications in treating wrinkle formation (Mitani et al. 2007). Several derivatives and analogs have played key roles as adenosine receptors (Jacobson and Gao 2006) and calcium release channels in physiological processes (Gerasimenko et al. 2006). Adenosine receptors and behavioral actions of methylated xanthines have been widely referred to as bronchodilators (Pascal et al. 1985). It has been known that Ca^{2+} release from the endoplasmic reticulum (ER) induced by intracellular acids such as bile acid activates inositol triphosphate and ryanodine receptors. Besides induction, many of these derivatives serve as potential therapeutic agents for cure and intervention in Alzheimer's disease (Arendash et al. 2006). Furthermore it was shown that caffeine protects Alzheimer's mice against cognitive impairment and reduces brain β-amyloid production (Johnston and Mrotchie 2006). In addition, these compounds also are known as active analgesics (Sawynok and Yaksh 1993), and many such analgesic adjuvants were described for pharmacological mechanism of action (Akkari et al. 2006). Recent progress in the development of adenosine receptor ligands as anti-inflammatory drugs (Usmani et al. 2005a) has supported the previous studies on behavioral stimulants, diuretics/natriuretics, and lipolytics (Beavo and David 1990). Various effects of XD on lipolysis have shown to be resulted in anxiety, hypertension, certain drug interactions, and withdrawal symptoms (Dally 2007). In addition, methylated XDs act as phosphodiesterase and adenosine antagonist receptors and are rarely found as constituents of nucleic acids. Methylated xanthines have a broad range of biochemical and physiological effects. For example, caffeine is well studied to modulate carcinogenesis at various organ sites of the human, and some authors have shown that the tumors decreased by treatment with precursors, namely, *N*-nitrosomorpholine and urethane (Nomura 1976).

There are four subtypes of the adenosine antagonist receptors, namely, A1, A2A, A2B, and A3, which are activated by extracellular adenosine in response to organ stress or tissue damage. The subtype A2A signals in both the periphery and the CNS, with agonists explored as anti-inflammatory drugs and antagonists explored for neurodegenerative diseases, e.g., Parkinson's disease (PD) (Cristalli et al. 2009). With potent use of xanthines as biochemical modulators and adenosine antagonist receptors, there is paucity of studies with respect to substituted XD. Antagonistic activities of adenosine receptor in the treatment of PD have widely advanced into clinical development. Apart from XD, the role of non-xanthine derivatives (Shah and Hodgson 2010), namely, non-purines, has led to an interest in modeling the XD. Theophylline was modeled and the molecular modeling studies indicated that there is a conformational similarity of the lead structure of xanthine and adenosine derivatives (Mager et al. 1995). The substituents bound to the C8 of xanthine and to the C2 of adenosine derivatives were known to be involved in the discrimination into adenosine A2 antagonists and agonists. Xanthines and their derivatives act on the cell membrane and transport antitumor drugs. The role of XD as biochemical modulators toward anticancer drug has been shown in various cancer lines (Overington et al. 2006). Of late, the XDs studied in cancer cell lines have turned out to be sensitive toward cell lines. The studies on HeLa cells on treatment with high concentration of caffeine and their comparison to low concentrations showed impact on the inhibitory action of cell progression, while the p53-mutated cells showed more sensitivity. A great deal of understanding has begun in this area where cell lines, namely, HeLa cells, V79, melanoma, squamous cell carcinoma, murine cell lines, US9-93, LMS6-93, A549 (p53 transfectants), T24 bladder cell line, have been known to be sensitive to the treatment with high concentration of methylated derivatives of xanthines, mainly caffeine and pentoxifylline (Evgeny et al. 2008). Yet, there is a lacuna in studying these compounds in some cancer cell lines like colorectal, lung, and prostate where no concrete work has been reported. With the increase in concentration of caffeine leading to progression in G phase and/or of inhibition of DNA synthesis, there is a great scope of underlying research on molecular modeling of XD. Such studies would bring efforts in understanding induced apoptosis in the aforementioned cancer cell lines (Lu et al. 1997).

2 Applications of Xanthine Derivatives

2.1 Alzheimer's Disease

A variety of therapeutic interventions are known to be used for treatment of Alzheimer's disease; however, most of them involve cholinergic functions (Bachurin 2003). Nevertheless, A2B adenosine antagonist receptors are studied as possible therapeutics for Alzheimer's disease (Rosi et al. 2003). Also, A2Aand A2B adenosine antagonist receptors and xanthines such as caffeine, ß-amyloid protein are used in the treatment of Alzheimer's disease (Renata et al. 2011)

2.2 Asthma

Xanthines such as caffeine, theophylline, pentoxifylline, lisofylline, and xanthines with alkyl piperazines in 7-position of theophylline and 3-position of theobromine are used as antihistaminics in the potential treatment of chronic bronchitis (Pascal et al. 1985). New antihistaminic theophylline and theobromine derivatives; A1, A2A, and A2B adenosine antagonist receptors; and phosphodiesterase inhibitors are used in the treatment of anti-inflammatory effect (Kramer et al. 1977). Doxofylline, a new methylxanthine derivative, has recently been shown to have similar efficacy to that of theophylline with significantly less side effects in both animal studies and human adults (Sankar et al. 2008).

2.3 Behavioral Targets

Extensive work on the use of caffeine in the treatment as antidepressant, anxiolytic, cognitive enhancement, neuroprotection, and adenosine antagonist receptors and phosphodiesterase inhibitors has been well done in relation to caffeine's behavioral effect (Dally and Fredholm 1998; Tarter et al. 1998).

2.4 Cancer

Caffeine, theophylline, and theobromine are known to inhibit G2 checkpoint. These have been used for repair of DNA damaged and cancer cells that are p53 defective repairing the G1 checkpoint. Thus, caffeine can enhance toxicity of tumor cells to DNA-damaging treatment by blocking the G2 checkpoint. Caffeine and a series of xanthines were known to show G2 checkpoint inhibition (Katsuro and Hiroyuki 2005).

2.5 Diabetes

A2B adenosine antagonist receptors have been suggested in the treatment of type 2 diabetes (Lu et al. 2002). Caffeine and caffeine sodium benzoate have a sunscreen effect enhancing UVB-induced apoptosis and inhibit UVB-induced skin carcinogenesis in SKH-1 mice (Conney et al. 2007). Not only xanthines such as lisofylline and pentoxifylline and phosphoenolpyruvate carboxykinase inhibitors have been well studied in the treatment of diabetes, but also dipeptidyl peptidase IV inhibitors are known to rate-limit the step in hepatic gluconeogenesis (Foley et al. 2003).

2.6 Analgesics

From several decades, caffeine has been used as antianalgesic. A benefit-risk assessment of caffeine as an analgesic adjuvant has been well documented (Zhang 2001).

2.7 Parkinsonism

Caffeine is known to be consumed in the form of tea and is known to be associated with a reduced risk of parkinsonism, and A2A adenosine antagonist receptors and phosphodiesterase inhibitors are studied for the treatment of parkinsonism (Johnston and Mrotchie 2006).

2.8 Renal Effects and Diuretics

Caffeine and theophylline are used in the treatment of edema associated with heart failure (McColl et al. 1956). 1- and 7-substituted methylated xanthines as diuretics have been successful in rat and have been employed in the treatment of renal failures.

2.9 Respiratory Targets

Theobromine and its analogs have been well studied in antitussives treatment (Usmani et al. 2005a, b), while caffeine and theophylline are widely used in the treatment of apnea which occurs normally in premature infants with minimal side effects (Schmidt et al. 2006) alongside chronic obstructive pulmonary disease and cystic fibrosis (Sang et al. 2010).

3 Xanthine Derivatives Have Been Known to Be Modeled

The XDs and their interactions with others have been studied extensively by means of molecular modeling. Xanthines, such as caffeine and theophylline in the recent past, have provided an impetus for use as antagonists. An attempt to design selective adenine receptors or XD through structural biology approach has been made (please see Figs. 1 and 2). Several computational studies examined the binding affinity and modes of caffeine and other xanthine antagonists using homology modeling and de novo design. Specifically, the A3 antagonist family has a peculiarity wherein it lacks common chemical or structural characteristics (Jiang et al. 1996). Hence, common electronic and steric features have been studied in combination of ab initio based methods such as quantum mechanical calculations and steric and electrostatic alignment (SEAL) analysis. In the previous studies, the dihydropyridine

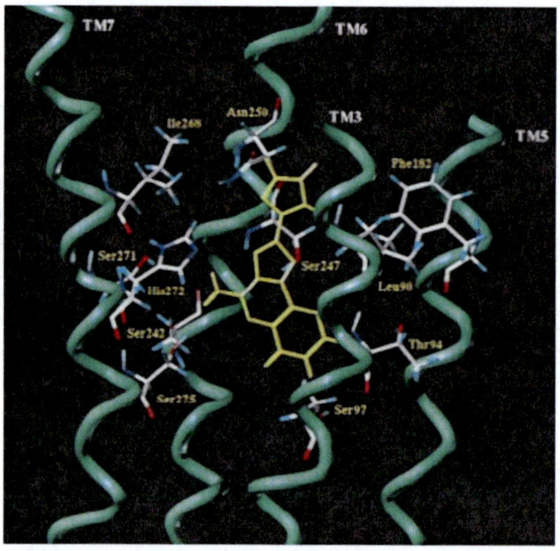

Fig. 1 Side view of the triazoloquinazoline antagonist complex with side chain residues such as Leu90 (*TM3*), Phe182 (*TM5*), Ser242 (*TM6*), Ser247 (*TM6*), Asn250 (*TM6*), Ser271 (*TM7*), His272 (*TM7*), and Ser275 (*TM7*) in proximity (e5 Å) had been docked

analog was prepared which is known to be selective in binding to human A3 receptors, as predicted from recently reported structure-activity relationships (SAR).

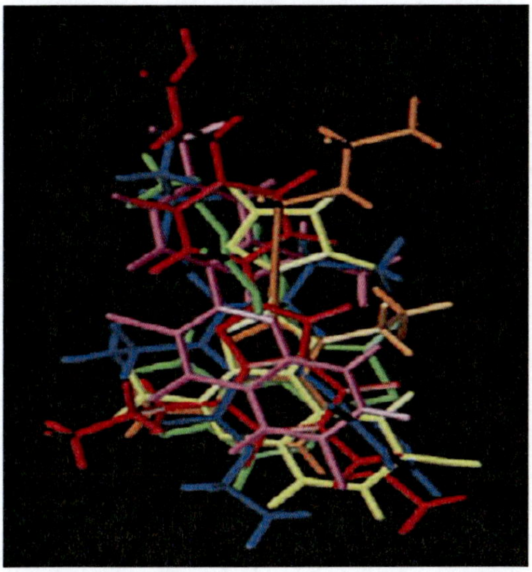

Fig. 2 Superposition of docked A3 antagonist model: N9-alkyl adenosine derivatives (*blue*), (3-(4-methoxyphenyl)-5-amino-7-oxothiazolo [3, 2] pyrimidine (*green*), triazoloquinazoline antagonist (*yellow*), and 6-phenylpyridine derivative (*magenta*) using molecular modeling by Moro et al. (1998) and Li et al. (1998)

Another example is G-protein (heterotrimeric guanine nucleotide-binding protein)-coupled receptors (GPCR) which are known to play a very important role with respect to its pharmaceutical properties. The GPCRs are known to target approximately one third of present-day drugs. Adenosine receptors are a class of GPCRs that respond to adenosine in the central nervous system, and various classes of these receptors play regulatory activities in various tissues. Although high-resolution structural data is considered to be an important step in understanding GPCRs, the lack of structures has certainly not made ligand discovery inadequate. For the past few decades, ligand-based medicinal chemistry approaches are being used to identify thousands of ligands. Caffeine is known to act as a nonselective antagonist on the A1 and A2A receptors (Kolb et al. 2012). It has been found that elucidating the mechanism of action of caffeine has given important insights into designing new xanthine-based compounds for controlling the adenosine A2A receptor.

Structure-based molecular modeling studies performed between flavonoids and xanthine oxidase suggested apigenin was the most potent inhibitor and showed favorable interaction in the reactive site. Molecular modeling of apigenin

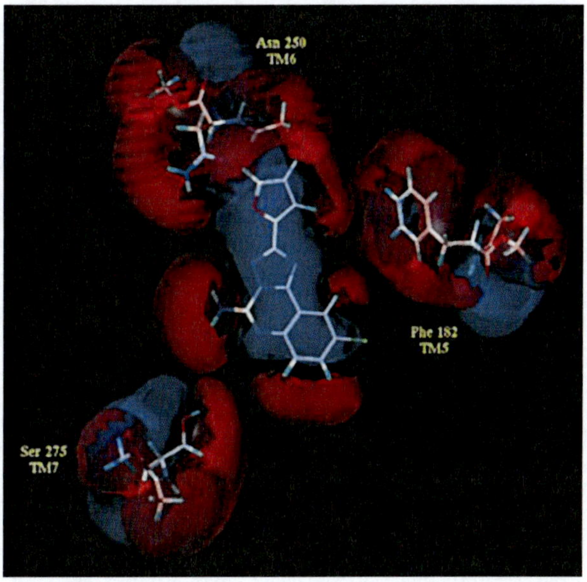

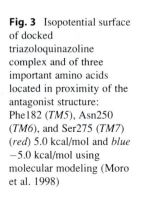

Fig. 3 Isopotential surface of docked triazoloquinazoline complex and of three important amino acids located in proximity of the antagonist structure: Phe182 (*TM5*), Asn250 (*TM6*), and Ser275 (*TM7*) (*red*) 5.0 kcal/mol and *blue* −5.0 kcal/mol using molecular modeling (Moro et al. 1998)

revealed that hydroxyl moiety at C7 and C5 and carbonyl group at C4 contribute toward hydrogen bonds and electrostatic interactions between inhibitors and the active site (Van Rhee et al. 1996). At the same time, 3-substituted hydroxyl benzopyranone ring exhibited weaker inhibitory effect, which is explained by the destabilization of polar hydroxyl stretching into the hydrophobic region of active site and resulting in lower binding affinity. This forms the basis on which it can

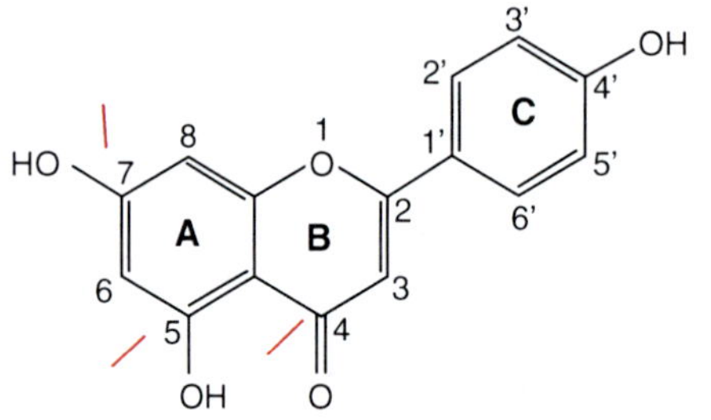

Fig. 4 Structure of apigenin (CID: 5280443). Molecular modeling of apigenin revealed that hydroxyl moiety at C7 and C5 and carbonyl group at C4 (*marked in red color*) contribute toward hydrogen bonds and electrostatic interactions between inhibitors and the active site

be concluded that flavonoid interaction with xanthine oxidase may develop new potential drugs for xanthine oxidase blockade (Lin et al. 2002). Another phytochemical constituent that needs much attention toward pharmacological activities is flavonoids· There is a need to study stereochemistry of binding of flavonoids on xanthine oxidase much extensively. Recently, structural models of ligands in xanthine oxidase were studied by Muthuswamy Umamaheswari et al. They conclude that flavonoids specially butein, fisetin, isorhamnetin, rhamnetin, robinetin, and herbacetin have excellent binding interactions with xanthine oxidase (Umamaheswari et al. 2011).

In the recent past molecular dynamics (MD) simulations are known to be performed on docked poses of caffeine, while adding hydrogen atoms. Tools like Amber packages (Case et al. 2012) were used extensively to expedite these simulations. It has been observed that caffeine has circa. 115,930 atoms, including 26,162 water molecules, 225 POPC lipid molecules, and 17 Cl# anions which were further used for determining and calculating the Restrained Electrostatic

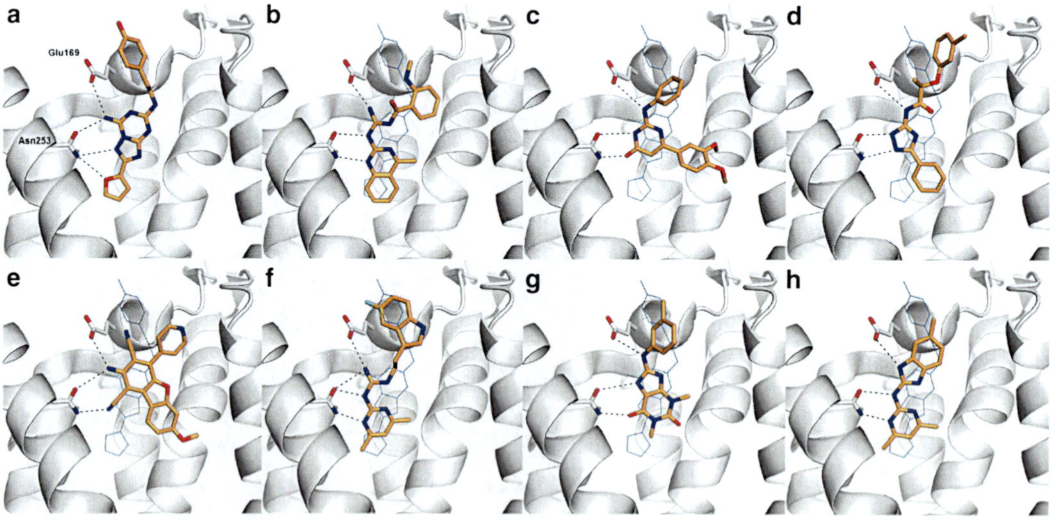

Fig. 5 (**a**) The type A2A adenosine receptor ligand binding site is shown in *white ribbons* with the side chains of Glu169 and Asn253 in sticks. In (**b**–**h**) the crystallographic ligand is shown using *blue lines* and the docking poses for the ligands are depicted with *red* are oxygen atoms and *orange* are carbon atoms where the *black dotted lines* indicate hydrogen bonds (Carlsson et al. 2010)

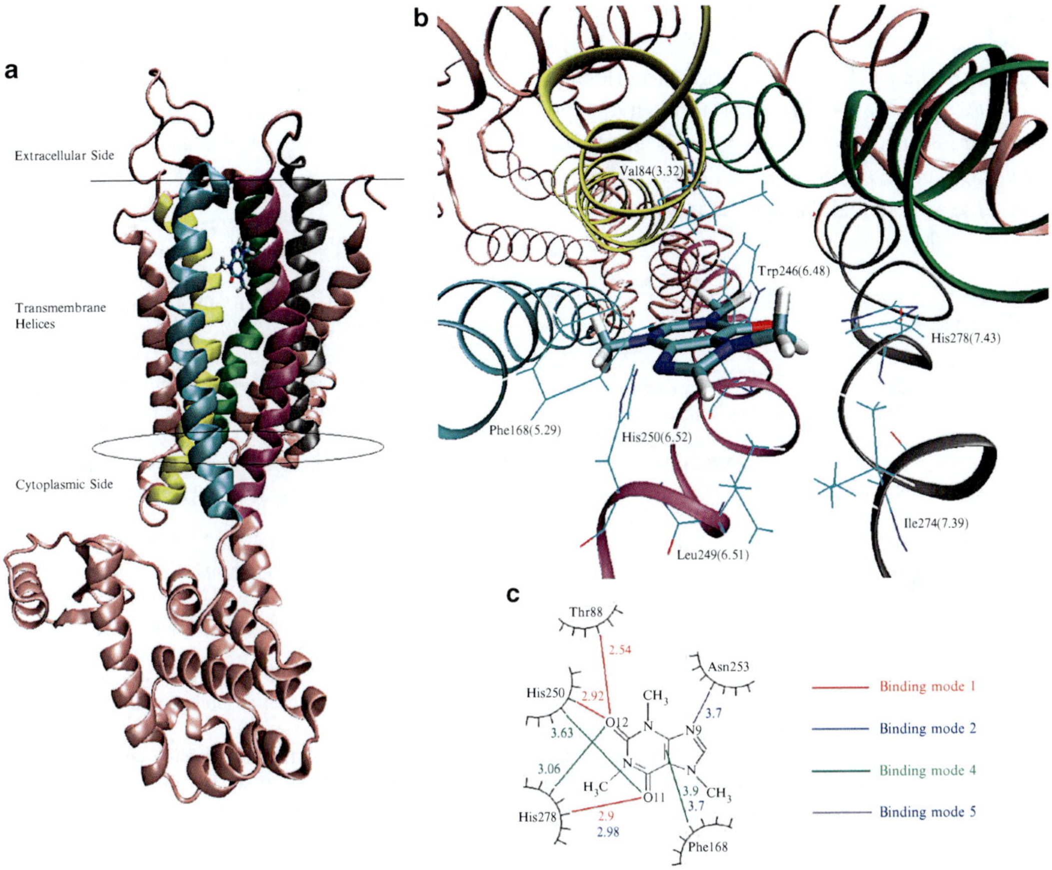

Fig. 6 (**a**) Caffeine binding cavity. The five transmembrane (*TM*) helices that define the caffeine binding cavity are illustrated with different colors: TM II, *green*; TM III, *yellow*; TM V, *cyan*; TM VI, *magenta*; TM VII, *gray*; and everything else is in *pink*. (**b**) Caffeine binding cavity, extracellular views. (**c**) The five transmembrane (*TM*) helices that define the caffeine binding cavity are illustrated with different colors: TM II, *green*; TM III, *yellow*; TM V, *cyan*; TM VI, *magenta*; and TM VII, *gray*. Some important pocket residues are labeled and shown in lines. Everything else is in *pink*. The position number in the *parentheses* tells which TM helix the residue is on and the residue's relative position to the most conserved residue on the helix. For example, Pro248 is the most conserved residue on TM helix VI; its position number is defined as 6.50. Therefore, the position number of Leu249 is 6.51

Potential (RESP) fitting with a series of energy minimizations. Figures 1, 2, 3, 4, 5, and 6 sum up how some XDs have been modeled.

4 Conclusions

In conclusion, XD serves as potential therapeutic agents for cure and intervention of various diseases like diabetes, obesity, hypertension, inflammation, and cancer. Molecular docking studies of XD continue to play a significant role in the quest for the development of potential novel ligands, as it is known to identify variety of receptors. However, the predictions of computer-generated model for the binding site must be tested with further efforts in ligand synthesis and modification of the receptor structure through site-directed mutagenesis. This would eventually be helpful to target them for therapeutic approaches. Combined efforts in computational and molecular biology will lead to the validation and optimization of various other XDs further enabling the rational design of new chemical entities. Purine plays an important role in ureide synthesis through catabolism in plants. With xanthine oxidase (XO) responsible for catalyzing the catabolism of purines to uric acid, the latter can play a key role for nitrogen metabolism, for instance, in legumes. There is a hope that new XD can be synthesized and modeled which can serve as potential therapeutic targets.

References

Akkari R, Burbiel JC, Joachim C, Hockemeyer JM, Christa E (2006) Recent progress in the development of adenosine receptor ligands as anti inflammatory drugs. Curr Top Med Chem 6:1375–1399

Arendash GW, Schleif W, Rezai-Zadeh K, Jackson EK, Zacharia LC, Cracchiolo JR, Shippy D, Tan J (2006) Caffeine protects Alzheimer's mice against cognitive impairment and reduces brain beta-amyloid production. Neuroscience. doi:10.1016/j.neuroscience.2006.07.021

Bachurin SO (2003) Medicinal chemistry approaches for the treatment and prevention of Alzheimer's disease. Med Res Rev 23:48–88. doi:10.1002/med.10026

Beavo JA, David HR (1990) Primary sequence of cyclic nucleotide phosphodiesterase isozymes and the design of selective inhibitors. Trends Pharmacol Sci 11:50–155. doi:10.1016/0165-6147(90)90066-H

Carlsson J, Yoo L, Gao Z-G, Irwin JJ, Shoichet BK, Jacobson KA (2010) Structure-based discovery of A2A adenosine receptor ligands. J Med Chem 53(9):3748–3755

Case DA, Darden TA, Cheatham TE III, Simmerling CL, Wang J, Duke RE, Luo R, Walker RC, Zhang W, Merz KM, Roberts B, Hayik S, Roitberg A, Seabra G, Swails J, Goetz AW, Kolossváry I, Wong KF, Paesani F, Vanicek J, Wolf RM, Liu J, Wu X, Brozell SR, Steinbrecher T, Gohlke H, Cai Q, Ye X, Wang J, Hsieh M-J, Cui G, Roe DR, Mathews DH, Seetin MG, Salomon-Ferrer R, Sagui C, Babin V, Luchko T, Gusarov S, Kovalenko A, Kollman PA (2012) AMBER 12. University of California, San Francisco

Conney AH, Zhou S, Lee MJ, Xie JG, Yang CS, Lou YR, Lu Y (2007) Stimulatory effect of oral administration of tea, coffee or caffeine on UVB-induced apoptosis in the epidermis of SKH-1 mice. Toxicol Appl Pharmacol 224(3):209–213. doi:10.1016/j.taap.2006.11.001

Cristalli G, Muller CE, Volpini R (2009) Recent developments in adenosine A2A receptor ligands. Handb Exp Pharmacol 59–98. doi:10.1007/978-3-540-89615-9_3

Dally JW (2007) Caffeine analogs: biomedical impact. Cell Mol Life Sci 64:16

Dally JW, Fredholm BB (1998) Caffeine – a typical, drug of dependence. Drug Alcohol Depend 51:199–206

Dally JW et al (1998) Pharmacology of caffeine. In: Handbook of substance abuse: neurobehavioral pharmacology. Plenum Press, New York. doi:10.1007/978-1-4757-2913-9_5

Evgeny AR, Ki WL, Nam JK, Haoyu Y, Masaaki N, Ken-Ichi M, Allan HC, Ann MB, Zigang D (2008) Inhibitory effects of caffeine analogues on neoplastic transformation: structure–activity relationship. Carcinogenesis 29 (6):1228–1234

Foley LH, Wang P, Dunten P, Ramsey G, Gubler ML, Wertheimer SJ (2003) Modified 3-alkyl-1,8-dibenzylxanthines as GTP-competitive inhibitors of phosphoenolpyruvate carboxykinase. Bioorg Med Chem Lett 13:3607–3610. doi:10.1016/S0960-894X(03)00722-4

Gerasimenko JV, Sherwood M, Tepikin AV, Petersen OH, Gerasimenko OV (2006) NAADP, cADPR and IP3 all release Ca2+ from the endoplasmic reticulum and an acidic store in the secretary granule area. J Cell Sci 119:226–238

Jacobson KA, Gao ZG (2006) Adenosine receptors as therapeutic targets. Nat Rev Drug Discov 5:247–264. doi:10.1038/nrd1983

Jiang JL, van Rhee AM, Melman N, Ji XD, Jacobson KA (1996) 6-Phenyl-1,4-dihydropyridine derivatives as potent and selective A3 adenosine receptor antagonists. J Med Chem 39:4667–4675. doi:10.1002/chin.199708139

Johnston TH, Mrotchie JM (2006) Drugs in development for Parkinson's disease: an update. Curr Opin Invest Drugs 7:25–32

Katsuro T, Hiroyuki T (2005) Caffeine enhancement of the effect of anticancer agents on human sarcoma cells. Cancer Sci 80(1):83–88

Kolb P, Phan K, Gao ZG, Marko AC, Sali A, Jacobson KA (2012) Limits of ligand selectivity from docking to models: *In silico* screening for A1 adenosine receptor antagonists. PLoS One 7(11):e49910. doi:10.1371/journal.Pone.0049910

Kramer GL, Garst JE, Mitchel SS, Wells JN (1977) Selective inhibition of cyclic nucleotide phosphodiesterases by analogs of 1- methyl-3-isobutyl xanthine, 1977. Biochemistry 16(15):3316–3321

Li AH, Moro S, Melman N, Ji XD, Jacobson KA (1998) Structure activity relationships and molecular modeling of 3,5-diacyl-2,4-dialkyl pyridine derivatives as selective A3 adenosine receptor antagonists. J Med Chem 41(17):3186–3201

Lin CM, Chen CS, Chen CT, Liang YC, Lin JK (2002) Molecular modeling of flavonoids that inhibits xanthine oxidase. Biochem Biophys Res Commun 294 (1):167–172. doi:10.1016/S0006-291X(02)00442-4

Lu YP, Lou YR, Xie JG, Yen P, Huang MT, Conney AH (1997) Inhibitory effect of black tea on the growth of established skin tumors in mice: effects on tumor size, apoptosis, mitosis and bromodeoxyuridine incorporation into DNA. Carcinogenesis 8:2163–2169. doi:10.1093/carcin/18.11.2163

Lu YP, Lou Y-R, Xie J-G, Peng Q-Y, Liao J, Yang CS, Huang M-T, Conney AH (2002) Topical applications of caffeine or (−)-epigallocatechin gallate (EGCG) inhibit carcinogenesis and selectively increase apoptosis in UVB-induced skin tumors in mice. Proc Natl Acad Sci U S A 99(19):12455–12460

Mager PP, Reinhardt R, Richter M, Walther H, Rockel B (1995) Molecular simulation of 8-styrylxanthines. Drug Des Discov 13(2):89–107

McColl JD, Parker JM, Ferguson JKW (1956) Evaluation of some 1- and 7-substitutedmethylated xanthines as diuretics in the rat. J Pharmacol Exp Ther 118:162–167

Mitani H, Ryu A, Suzuki T, Yamashita M, Arakane K, Koide C (2007) Topical application of plant extracts containing xanthine derivatives can prevent UV-induced wrinkle formation in hairless mice. Photodermatol Photoimmunol Photomed 23(2–3):86–94. doi:10.1111/j.1600-0781.2007.00283.x

Moro S, Li A-H, Jacobson KA (1998) Molecular modeling studies of human A3 adenosine antagonists: structural homology and receptor docking. J Chem Inf Comput Sci 38(6):1239–1248

Nomura T (1976) Diminution of tumorigenesis initiated by 4-nitroquinoline-l-oxide by post-treatment with caffeine in mice. Nature 260:547. doi:10.1038/260547a0

Overington JP, Al-Lazikani B, Hopkins AL (2006) How many drug targets are there? Nat Rev Drug Discov 5:993–995. doi:10.1038/nrd2199

Pascal JC, Beranger S, Pinhas H, Poizot A, Désiles JP (1985) New antihistaminic theophylline or theobromine derivatives. J Med Chem 28(5):647. doi:10.1021/jm50001a019

Renata VA, Eliane MSO, Márcio FDM, Grace SP, Tasso MS (2011) Chronic coffee and caffeine ingestion effects on the cognitive function and antioxidant system of rat brains. Pharmacol Biochem Behav 99(4):659–664

Rosi S, McGann K, Hauss WB, Wonk GL (2003) The influence of brain inflammation upon neuronal adenosine A2B receptors, 2003. J Neurochem 86:220–227. doi:10.1046/j.1471-4159.2003.01825.x

Sang SK, Kyung SH, Bo MK, Yeon KL, Jinpyo H, Hye YS, Antoine GA, Dong HW, Daniel JB, Eun MH, Seung HY, Chun KC, Sung HP, Sun HP, Eun JR, Sung JL, Jae-Yong P, Stephen FTC, Justin L (2010) Caffeine-mediated inhibition of calcium release channel inositol 1,4,5-trisphosphate receptor subtype 3 blocks glioblastoma invasion and extends survival. Tumor Stem Cell Biol 70:1173–1183

Sankar J, Lodha R, Kabra SK (2008) Doxofylline: the next generation methyl xanthine. Indian J Pediatr 75(3):251–254

Sawynok J, Yaksh TL (1993) Caffeine as an analgesic adjuvant: a review of pharmacology and mechanisms of action. Pharmacol Rev 45:43

Schmidt B, Roberts RS, Davis P, Doyle LW, Barrington KJ, Ohlsson A, Solimano A, Tin W (2006) Caffeine therapy for apnea of prematurity. N Engl J Med 354:2112–2121. doi:10.1056/NEJMoa054065

Shah U, Hodgson R (2010) Recent progress in the discovery of adenosine A(2A) receptor antagonists for the treatment of Parkinson's disease. Curr Opin Drug Discov Devel 13(4):466–480

Snyder SH, Katims JJ, Annau Z, Bruns RF, Daly JW (1981) Adenosine receptors and behavioral actions of methyl xanthines. Proc Natl Acad Sci U S A 78 (5):3260–3264

Tarter RE, Ammerman RT, Ott PJ (eds) (1998) Plenum, New York, pp 53–68

Umamaheswari M, Madeswaran A, Asokkumar K, Sivashanmugam T, Subhadradevi V, Jagannath P (2011) Discovery of potential xanthine oxidase inhibitors using *in silico* docking studies. Scholars Res Libr Der Pharma Chemica 3(5):240–247

Usmani OS, Belvisi MG, Patel HJ, Crispino N, Birrell MA, Korbonits M, Korbonits D, Barnes PJ (2005a) Theobromine inhibits sensory nerve activation and cough. FASEB J 19:231

Usmani OS, Belvisi MG, Hema JP, Natascia C, Mark AB, Márta K, Dezso K, Peter JB (2005b) Theobromine inhibits sensory nerve activation and cough. FASEB J 19:231–233. doi:10.1096/fj.04-1990fje

Van Rhee AM, Jiang JL, Melman N, Olah ME, Stiles GL, Jacobson KA (1996) Interaction of 1,4-dihydropyridine and pyridine derivatives with adenosine receptors-selectivity for A3 receptors. J Med Chem 39:2980–2989

Zhang WY (2001) A benefit-risk assessment of caffeine as an analgetic adjuvant. Drug Saf 24:1127–1142. doi:10.2165/00002018-200124150-00004